ADVANCES IN MATERIALS RESEARCH

Series Editor-in-Chief: Y. Kawazoe

Series Editors: M. Hasegawa   A. Inoue   N. Kobayashi   T. Sakurai   L. Wille

The series Advances in Materials Research reports in a systematic and comprehensive way on the latest progress in basic materials sciences. It contains both theoretically and experimentally oriented texts written by leading experts in the field. Advances in Materials Research is a continuation of the series Research Institute of Tohoku University (RITU).

1   **Mesoscopic Dynamics of Fracture**
    Computational Materials Design
    Editors: H. Kitagawa, T. Aihara, Jr.,
    and Y. Kawazoe

2   **Advances
    in Scanning Probe Microscopy**
    Editors: T. Sakurai and Y. Watanabe

3   **Amorphous
    and Nanocrystalline Materials**
    Preparation, Properties,
    and Applications
    Editors: A. Inoue and K. Hashimoto

4   **Materials Science
    in Static High Magnetic Fields**
    Editors: K. Watanabe and M. Motokawa

5   **Structure and Properties
    of Aperiodic Materials**
    Editors: Y. Kawazoe and Y. Waseda

6   **Fiber Crystal Growth from the Melt**
    Editors: T. Fukuda, P. Rudolph,
    and S. Uda

7   **Advanced Materials Characterization
    for Corrosion Products
    Formed on the Steel Surface**
    Editors: Y. Waseda and S. Suzuki

8   **Shaped Crystals**
    Growth
    by Micro-Pulling-Down Technique
    Editors: T. Fukuda and V.I. Chani

9   **Nano- and Micromaterials**
    Editors: K. Ohno, M. Tanaka, J. Takeda,
    and Y. Kawazoe

10 **Frontiers in Materials Research**
    Editors: Y. Fujikawa, K. Nakajima,
    and T. Sakurai

11 **High-Temperature Measurements
    of Materials**
    Editors: H. Fukuyama and Y. Waseda

12 **Oxide and Nitride Semiconductors**
    Processing, Properties,
    and Applications
    Editors: T. Yao and S.-K. Hong

Takafumi Yao
Soon-Ku Hong
(Eds.)

Oxide and Nitride Semiconductors

Processing, Properties, and Applications

With 370 Figures

Editors

Professor Takafumi Yao
Tohoku University, Center for Interdisciplinary Research
Aramaki Aza-Aoba, Aoba-ku, Sendai 980-8578, Japan
E-mail: tyao@cir.tohoku.ac.jp

Professor Dr. Soon-Ku Hong
Chungnam National University, School of Nanoscience and Technology
Gung-dong 220, Youseong-gu, 305-764 Daejeon, Republic of Korea
E-mail: soonku@cnu.ac.kr

Series Editor-in-Chief:

Professor Yoshiyuki Kawazoe
Institute for Materials Research, Tohoku University
2-1-1 Katahira, Aoba-ku, Sendai 980-8577, Japan

Series Editors:

Professor Masayuki Hasegawa
Professor Akihisa Inoue
Professor Norio Kobayashi
Professor Toshio Sakurai
Institute for Materials Research, Tohoku University
2-1-1 Katahira, Aoba-ku, Sendai 980-8577, Japan

Professor Luc Wille
Department of Physics, Florida Atlantic University
777 Glades Road, Boca Raton, FL 33431, USA

Advances in Materials Research   ISSN 1435-1889

ISBN 978-3-540-88846-8   e-ISBN 978-3-540-88847-5

Library of Congress Control Number: 2008938478

© Springer Berlin Heidelberg 2009

This work is subject to copyright. All rights are reserved, whether the whole or part of the material is concerned, specifically the rights of translation, reprinting, reuse of illustrations, recitation, broadcasting, reproduction on microfilm or in any other way, and storage in data banks. Duplication of this publication or parts thereof is permitted only under the provisions of the German Copyright Law of September 9, 1965, in its current version, and permission for use must always be obtained from Springer-Verlag. Violations are liable to prosecution under the German Copyright Law.

The use of general descriptive names, registered names, trademarks, etc. in this publication does not imply, even in the absence of a specific statement, that such names are exempt from the relevant protective laws and regulations and therefore free for general use.

Typesetting: Data prepared by SPI Kolam using a Springer $\LaTeX$ macro package
Cover concept: eStudio Calmar Steinen
Cover production: SPI Kolam

SPIN: 12205196   57/3180/SPI
Printed on acid-free paper

9 8 7 6 5 4 3 2 1

springer.com

Preface

Semiconducting oxides and nitrides are becoming the most important subjects in materials science. In particular, zinc oxide (ZnO), gallium nitride (GaN), and related compounds form a novel class of semiconductors which possess unique properties in terms of crystallography, crystal growth, optical properties, electrical properties, magnetic properties, and so forth. These unique properties make these materials quite important in optoelectronics and electronics.

Although for more than three decades oxide and nitride semiconductors have been known to possess unique properties, it is only recently that these materials have been exploited to fabricate novel electronic and optical devices, which have never been possible with other semiconductors. It should be mentioned that revolutionary breakthroughs in materials science have been made before the remarkable development of such devices. In particular, recent breakthrough and advance in epitaxy, bulk growth, and synthesis of nanostructures coupled with exploration and investigation on structural, optical, and electrical properties, enabled us to achieve novel display, general lighting, optical storage, high-speed, -temperature and -power electronics, bio and environmental sensors, and energy generating and saving devices.

The unique structure of this book is that each chapter addresses both oxides and nitrides, which, we believe, will help readers gain comprehensive and comparative information on oxide and nitride semiconductors. This book consists of ten chapters, addressing the basic properties of materials, bulk growth, film growth, polarity issues, nonpolar films, structural defects, optical properties, electrical properties, light emitting diodes, and nanostructures. Thus the book covers processing, properties, and applications of materials based on ZnO, GaN, and related compounds.

We hope that this book will be the new, essential, and easy-to-access book for readers who have interest in and need to get detailed knowledge of processing, properties, and applications of materials based on ZnO, GaN, and related compounds.

The editors express their sincere appreciation to Dr. Claus Ascheron and Ms. Adelheid Duhm of Springer Verlag for their great help and excellent cooperation during the preparation of this book. Finally, the editors would like to express their deep gratitude to Dr. Yoshiyuki Kawazoe, editor of this series, for his encouraging them to prepare this book.

Japan and Korea *Takafumi Yao*
December 2008 *Soon-Ku Hong*

Contents

Preface V

1 Basic Properties of ZnO, GaN, and Related Materials
T. Hanada 1
1.1 Introduction 1
1.2 Crystal Structure 1
    1.2.1 Crystal Structure of Related Materials 1
    1.2.2 Hexagonal Lattice Vectors and Planes 4
1.3 Elastic Strain 6
    1.3.1 Elastic Constants and Elastic Energy Density 6
    1.3.2 Strain in Hexagonal- and Cubic-Structure Films 8
1.4 Electronic Structure 10
    1.4.1 Bandgap and Band-Edge Electronic Structure 10
    1.4.2 Bir-Pikus Hamiltonian 12
    1.4.3 Valence Band-Edge and Bandgap of Strained ZnO 15
    1.4.4 Exciton Binding Energy 17
References 18

2 Solvothermal Growth of ZnO and GaN
D. Ehrentraut, F. Orito, Y. Mikawa, and T. Fukuda 21
2.1 Introduction 21
2.2 Hydrothermal Growth 23
    2.2.1 Hydrothermal Growth Technology for ZnO 23
    2.2.2 Growth Mechanism 28
    2.2.3 Liquid Phase Epitaxy and Growth of Microcrystals .. 30
    2.2.4 Properties of Hydrothermal ZnO 32
    2.2.5 Future Developments 46
2.3 Ammonothermal Growth 47
    2.3.1 A Brief Review of the Growth of GaN Bulk Crystals . 47
    2.3.2 Ammonothermal Growth Technology for GaN 48

VIII   Contents

| | 2.3.3 | Properties of Ammonothermal GaN .... 54 |
| | 2.3.4 | Prospects for Ammonothermal GaN .... 60 |
| References .... 62 |

3 Growth of ZnO and GaN Films
J. Chang, S.-K. Hong, K. Matsumoto, H. Tokunaga, A. Tachibana,
S.W. Lee, and M.-W. Cho 67
3.1   MBE and PLD of ZnO and Related Materials 67
      J. Chang and S.-K. Hong 67
      3.1.1   Introduction 67
      3.1.2   General Features of the PAMBE 68
      3.1.3   PAMBE of ZnO Thin Films on Various Substrates 71
      3.1.4   Other Growth Factors in PAMBE of ZnO
              and their Properties 84
      3.1.5   Bandgap Engineering of ZnO-Based Alloys Grown
              by PAMBE 101
      3.1.6   General Features of the PLD 112
      3.1.7   Droplet Formation in the PLD Process 113
      3.1.8   Growth of ZnO Thin Films on Various
              Substrates by PLD 114
      3.1.9   Bandgap Engineering of ZnO-based Alloys
              Grown by PLD 124
3.2   MOCVD Growth of GaN and Related Materials 130
      K. Matsumoto, H. Tokunaga, and A. Tachibana 130
      3.2.1   Introduction 130
      3.2.2   Heteroepitaxy of GaN and Alloys: Materials Property
              and Growth Condition 130
      3.2.3   Reaction Mechanism: Experiments and Theory 140
      3.2.4   Growth Equipment 146
      3.2.5   Atmospheric Pressure Reactor for Nitride 149
3.3   HVPE of GaN and Related Materials 156
      S.W. Lee and M.-W. Cho 156
      3.3.1   Introduction 156
      3.3.2   Thermodynamic Considerations of HVPE of GaN 156
      3.3.3   Growth System 160
      3.3.4   Thick GaN Film Growth 163
      3.3.5   Fabrication of Free-Standing GaN Substrate 166
      3.3.6   Other III Nitrides Grown by HVPE 174
References 176

4 Control of Polarity and Application to Devices
J.S. Park and S.-K. Hong 185
4.1   Introduction 185
4.2   Polarity 186
4.3   Spontaneous and Piezoelectric Polarization 188

4.4		Determination of Polarity 191	
	4.4.1	Determination Based on Diffraction.................. 192	
	4.4.2	Determination Based on Spectroscopy 195	
	4.4.3	Determination Based on Microscopy 198	
	4.4.4	Determination Based on Differences in Etching and Growth Rates 201	
4.5		Control of Polarity 205	
	4.5.1	Polarity Control of GaN Films 205	
	4.5.2	Polarity Control of ZnO Films 207	
	4.5.3	Polarity of AlN and InN Films 210	
4.6		Effects of Polarity on Material Properties................... 211	
4.7		Device Applications of Polarization Induced Properties......... 213	
	4.7.1	Electronic Devices 214	
	4.7.2	Nonlinear Optical Devices 216	
References 218			

5 Growth of Nonpolar GaN and ZnO Films
S.-K. Hong and H.-J. Lee 225

5.1	Introduction 225	
5.2	Polar Surface, Nonpolar Surface, and Heterostructures 226	
5.3	Growth of Nonpolar GaN Films 229	
	5.3.1 M-plane GaN Films 229	
	5.3.2 A-plane GaN Films................................ 233	
	5.3.3 Semipolar GaN Films 238	
5.4	Lateral Epitaxial Overgrowth of Nonpolar GaN Films........... 239	
	5.4.1 LEO of A-plane GaN 239	
	5.4.2 LEO of M-plane GaN 245	
5.5	Growth of Nonpolar ZnO Films 247	
	5.5.1 A-plane ZnO Films 247	
	5.5.2 M-plane ZnO Films................................ 254	
References 258		

6 Structural Defects in GaN and ZnO
S.-K. Hong and H.K. Cho.. 261

6.1	Introduction 261	
6.2	Dislocation and Stacking Faults 262	
	6.2.1 Dislocations 262	
	6.2.2 Misfit and Threading Dislocations 264	
	6.2.3 Dislocation in Wurtzite Structure 265	
	6.2.4 Stacking Fault in Wurtzite Structure.................. 266	
6.3	TEM of Defects in GaN and ZnO Films 267	
	6.3.1 Defect Contrast in TEM 267	
	6.3.2 Analysis of Threading Dislocation by Cross-sectional TEM 269	
	6.3.3 Analysis of Threading Dislocation by Plan-View TEM ... 274	

	6.3.4	Misfit Dislocation 277
	6.3.5	Nanopipe 281
	6.3.6	Stacking Fault 281
	6.3.7	Inversion Domain Boundary 285
6.4	Dislocation Reduction of Epitaxial Films by Process 286
	6.4.1	Defects in Epitaxial Lateral Overgrowth (ELOG)........ 286
	6.4.2	Defects in PENDEO Epitaxy (PE) 293
	6.4.3	Defects in Facet-Controlled Epitaxial Lateral Overgrowth (FACELO) 297
	6.4.4	Defects in Other Overgrowth Techniques 299
	6.4.5	Other Growth and Process Techniques for Defect Reduction 302

References 306

7 Optical Properties of GaN and ZnO

J.-H. Song 311

7.1	Introduction 311
	7.1.1 Basics 313
	7.1.2 Valence Band Structure 313
7.2	Emission Properties of GaN 315
	7.2.1 Band Edge Emissions 315
	7.2.2 Defect-Related Emissions............................ 320
	7.2.3 Deep Level Emissions 323
	7.2.4 Emission Properties of InGaN/GaN Quantum Wells 324
7.3	Emission Properties of ZnO................................. 331
	7.3.1 Band Edge Emissions 332
	7.3.2 Defect-Related Emissions............................ 334
	7.3.3 Deep-Level Emissions 338
	7.3.4 Emission Properties of MgZnO/ZnO Quantum Wells 338
7.4	Emission Properties of Nonpolar GaN and ZnO 339
	7.4.1 Polarized Emission 340
	7.4.2 Strain Effects 340
7.5	Raman Scattering Properties of GaN and ZnO 344
	7.5.1 Strain Effects 344
	7.5.2 Carrier Concentration............................... 346

References 349

8 Electrical Properties of GaN and ZnO

D.-C. Oh 355

8.1	Introduction 355
8.2	Ohmic Contacts to GaN and ZnO 356
	8.2.1 Principle of Ohmic Contact 356
	8.2.2 Ohmic Contacts to GaN 358
	8.2.3 Ohmic Contacts to ZnO............................. 363
8.3	Schottky Contacts to GaN and ZnO 366
	8.3.1 Principle of Schottky Contact 366

	8.3.2	Schottky Contacts to GaN 369
	8.3.3	Schottky Contacts to ZnO 373
8.4	Electrical Properties 382	
	8.4.1	Electron Transport Mechanism 382
	8.4.2	Electrical Properties of GaN 383
	8.4.3	Electrical Properties of ZnO 391
8.5	Deep Levels of GaN and ZnO 397	
	8.5.1	Characterization of Deep Levels 397
	8.5.2	Deep Levels in GaN 399
	8.5.3	Deep Levels in ZnO 405
References	 411	

9 GaN and ZnO Light Emitters
J.-S. Ha 415
9.1 Introduction 415
9.2 Light-Emitting Diodes Basic 416
9.3 Light-Emitting Diodes Based on GaN 419
    9.3.1 Issues for High Internal Quantum Efficiency 419
    9.3.2 Issues for High External Quantum Efficiency 429
    9.3.3 Packaging 437
    9.3.4 Vertical Light-Emitting Diode 440
9.4 Light-Emitting Diodes Based on ZnO 444
    9.4.1 Hybrid LED 444
    9.4.2 ZnO LEDs with Homo p–n Junction 449
References 454

10 ZnO and GaN Nanostructures and their Applications
S.H. Lee 459
10.1 Introduction 459
10.2 Control of ZnO and GaN Nanostructures 460
    10.2.1 Synthetic Methods 460
    10.2.2 Processing and Assembly 465
10.3 ZnO and GaN Nanostructures for Photonic
    Device Applications 469
    10.3.1 Optical Cavity and Lasing 470
    10.3.2 Nanostructure-Based LED 479
10.4 ZnO and GaN Nanostructures for Electronic and Sensing
    Device Applications 485
    10.4.1 Field Effect Transistors (FET) 485
    10.4.2 Light Sensor 492
    10.4.3 Gas and Solution Sensor 496
    10.4.4 Biosensor 498
References 503

Index 507

List of Contributors

Jiho Chang
Major of Nano-Semiconductor
Korea Maritime University
1 Dongsam-dong
Yeongdo-ku Pusan 606-791, Korea
jiho_chang@hhu.ac.kr

Hyung Koun Cho
School of Advanced Materials
Science & Engineering
Sungkyunkwan University
300 Cheoncheon-dong
Jangan-ku Suwon 440-746
Korea
chohk@skku.edu

Meoung-Whan Cho
Center for Interdisciplinary Research
Tohoku University
6–3 Aramaki Aoba-ku
Sendai 980–8578
Japan
mwcho@cir.tohoku.ac.jp

Dirk Ehrentraut
Institute of Multidisciplinary
Research for Advanced Materials
Tohoku University
2-1-1 Katahira Aoba-ku
Sendai 980-8577
Japan
dirk@imr.tohoku.ac.jp

Tsuguo Fukuda
Institute of Multidisciplinary
Research for Advanced Materials
Tohoku University
2-1-1 Katahira Aoba-ku
Sendai 980-8577
Japan
t-fukuda@tagen.tohoku.ac.jp

Jun-Seok Ha
Center for Interdisciplinary Research
Tohoku University
6–3 Aramaki Aoba-ku
Sendai 980-8578 Japan
jsha@cir.tohoku.ac.jp

Takashi Hanada
Institute for Materials Research
Tohoku University
2-1-1 Katahira, Aobaku
Sendai 980-8577 Japan
thanada@imr.tohoku.ac.jp

Soon-Ku Hong
Department of Materials Science
and Engineering
Chungnam National University
220 Gung-dong
Youseong-ku Daejeon 305-764
Korea
soonku@cnu.ac.kr

Hyo-Jong Lee
Center for Interdisciplinary Research
Tohoku University
6–3 Aramaki Aoba-ku
Sendai 980–8578
Japan
ocma97@hotmail.com

Sang Hyun Lee
Center for Interdisciplinary Research
Tohoku University
6–3 Aramaki Aoba-ku
Sendai 980–8578
Japan
shlee7579@gmail.com

Seog Woo Lee
Center for Interdisciplinary Research
Tohoku University
6–3 Aramaki Aoba-ku
Sendai 980–8578
Japan
seogwoo@cir.tohoku.ac.jp

Koh Matsumoto
TN EMC Ltd., 2–2008 Wada
Tama-shi, Tokyo 206–0001, Japan
Kou.Matsumoto@tnemc.tn-sanso.co.jp

Yutaka Mikawa
Fukuda X'tal Laboratory, c/o ICR
6-6-3 Minami-Yoshinari, Aoba-ku
Sendai 989-3204, Japan
mikawa@fxtal.co.jp

Dong-Cheol Oh
Department of Defense Science
and Technology, Hoseo University
165 Sechul-ri Baebang-myun
Asan 336-795, Korea
ohdongcheol@hoseo.edu

Fumio Orito
Mitsubishi Chemical Corporation
Innovation Center, 4-14-1 Shiba
Minato-ku, Tokyo 108-0014
Japan
orito.fumio@mw.m-kagaku.co.jp

Jin Sub Park
Center for Interdisciplinary Research
Tohoku University
6–3 Aramaki Aoba-ku
Sendai 980–8578
Japan
jspark@imr.tohoku.ac.jp

Jung-Hoon Song
Department of Physics
Kongju National University
Kongju, Chungnam 314-701
Korea
jh-song@kongju.ac.kr

Akitomo Tachibana
Department of Micro Engineering
Kyoto University, Sakyo-ku
Kyoto 606–8501
Japan
akitomo@scl.kyoto-u.ac.jp

Hiroki Tokunaga
Tsukuba Laboratories
Taiyo Nippon Sanso Corporation
10 Oh-kubo
Tsukuba 300–2611
Japan
Hiroki.Tokunaga@tn-sanso.co.jp

1

Basic Properties of ZnO, GaN, and Related Materials

T. Hanada

Abstract. Structural, elastic, and electronic properties of the group-III nitride and the group-II oxide semiconductors are introduced here with basic material parameters. These materials generally have uniaxial anisotropy due to the wurtzite-type crystal structure. The basic formulae on the elastic properties and the electronic structures characterized by the uniaxial anisotropy are presented in this chapter.

1.1 Introduction

The group-III nitride and the group-II oxide semiconductors have direct bandgaps, which cover the ultraviolet to infrared energy range. Particularly, the wide bandgap of AlN, GaN, and ZnO is favorable for short-wavelength light emitting devices and high-power devices. Figure 1.1 shows the relation between bandgap energy and wurtzite lattice constant a or $1/\sqrt{2}$ of zincblende lattice constant a_{ZB} for AlN, GaN, InN, and their alloys [1].

Figure 1.2 shows bandgap energy and lattice constants of wurtzite $Be_xZn_{1-x}O$ [2], $Mg_xZn_{1-x}O$ [3, 4], and $Zn_{1-y}Cd_yO$ [4–6] alloy films. ZnO also attracts attention owing to the large exciton binding energy, which is expected to be favorable for high efficiency light emitting devices up to high temperatures. A set of basic data and related formulae on the structural, elastic, and electronic properties of the group-III nitride and the group-II oxide semiconductors will be introduced in this chapter.

1.2 Crystal Structure

1.2.1 Crystal Structure of Related Materials

Crystal structure of ZnO, AlN, GaN, and InN is usually the hexagonal wurtzite (WZ) type, which has nearly the same tetrahedral nearest-neighbor atomic coordination as cubic zincblende (ZB) type structure. As shown in

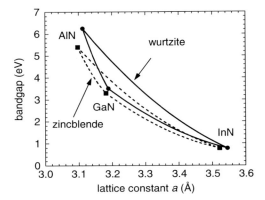

Fig. 1.1. Relation between bandgap energy and lattice constant a for wurtzite or $a_{ZB}/\sqrt{2}$ for zincblende AlN, GaN, InN, and their alloys [1]

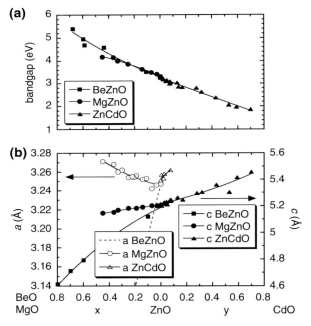

Fig. 1.2. Bandgap energy (a) and lattice constants a and c (b) of wurtzite $Be_xZn_{1-x}O$ [2], $Mg_xZn_{1-x}O$ [3,4], and $Zn_{1-y}Cd_yO$ [4–6] alloy films. *Dashed line* is a line between a of bulk BeO [7] and a of the ZnO film. *Solid lines* are quadratic fittings of measured points (*symbols*)

Figs. 1.3a,b, the WZ structure has AaBbAaBbAaBb... stacking sequence along the [0001] axis, while the ZB structure has AaBbCcAaBbCc... stacking sequence along the [111] axis, where A (a), B (b), and C (c) denote three kinds

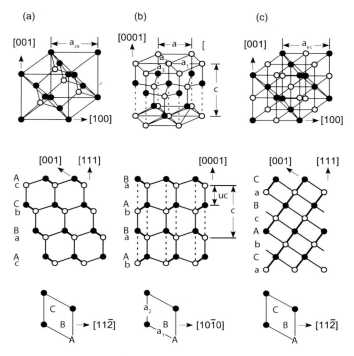

Fig. 1.3. Atomic structure of zincblende (**a**) wurtzite (**b**) and rocksalt (**c**) structures. *Closed circle, open circle*, and *thick solid line* represent cation, anion, and projection of two bonds, respectively

of cation (anion) position in the triangular lattice on the (0001) and (111) planes. If the difference between the WZ and ZB structures is only the stacking sequence, WZ lattice constants a and c have relation as $c/a = \sqrt{8/3} = 1.633$ and internal parameter $u = 3/8 = 0.375$, where uc corresponds to the length of the bonds parallel to [0001].

In the WZ structure, however, there are pairs of cation and anion atoms connected by dashed lines along the [0001] direction in Fig. 1.3b and attracted to each other by electrostatic force. It is considered that these electrostatic interactions make WZ–ZnO, AlN, GaN, and InN stabler than ZB–ZnO, AlN, GaN, and InN because ionicity of these compounds is large among the III–V and II–VI compound semiconductors. In the WZ structure, therefore, the length of the dashed lines in Fig. 1.3b tends to be shorter than the ideal one. In fact, c/a is smaller than ideal 1.633 in most of the WZ type materials as shown in Table 1.1. Furthermore, it is easier to shorten the interlayer distances between A–b and B–a than to shorten those between A–a and B–b because the former can be done mostly with angle deformation of the bond pairs. As a result, u is usually larger than ideal 0.375. These structural deformations

Table 1.1. Lattice constants a, c (Å) and internal parameter u of wurtzite structure, lattice constant a_{ZB} (Å) of zincblende structure, and lattice constant a_{RS} (Å) of rocksalt structure. Al_2O_3 is corundum type

	a	c	u	c/a	a_{ZB}	a_{RS}
BeO	2.698 [7]	4.377 [7]	0.378 [7]	1.622	3.81 [8]	3.648 [8]
MgO	3.43 [9]	4.11 [9]	0.5 [9]	1.198		4.216 [10]
ZnO	3.250 [11]	5.204 [11]	0.382 [11]	1.601	4.60 [12]	4.271 [11]
CdO	3.66 [13]	5.86 [13]	0.35 [13]	1.601		4.77 [13]
AlN	3.112 [1]	4.982 [1]	0.380 [14]	1.601	4.38 [1]	
GaN	3.189 [1]	5.185 [1]	0.376 [14]	1.626	4.50 [1]	
InN	3.545 [1]	5.703 [1]	0.377 [14]	1.609	4.98 [1]	
Al_2O_3	4.758 [10]	12.99 [10]				

induce spontaneous polarization; relative displacement of cation to [000$\bar{1}$] and anion to [0001] from the ideal structure.

Alloys of ZnO with BeO, MgO, and CdO are used to control bandgap of the ZnO system as shown in Fig. 1.2. MgO and CdO are ionic crystals with sixfold coordinated cubic rocksalt (RS) type structure, whose triangular lattice structure in the (111) plane is similar to the hexagonal (0001) plane as shown in Fig. 1.3c. Therefore, WZ $Mg_xZn_{1-x}O$ [3,4] and $Zn_{1-x}Cd_xO$ [4–6] have not been obtained when x approaches unity. It was reported that WZ MgO whose u is fixed at 3/8 is unstable and deforms to be quasi-stable at $u = 0.5$, where a of MgO larger than that of ZnO as shown in Table 1.1 [9]. The increase of a and decrease of c/a with Mg content x of $Mg_xZn_{1-x}O$ shown in Fig. 1.2b is consistent with this prediction. It is considered that u increases with x in the case of $Mg_xZn_{1-x}O$. If $u = 0.5$, MgO deforms to be a fivefold coordinated hexagonal structure, where the bonds shown by the dashed lines in Fig. 1.3b have nearly the same length with other bonds [9]. Bond length and volume per atom of the materials and structures in Table 1.1 are summarized in Table 1.2.

Thermal expansion coefficient of related materials is shown in Table 1.3. Temperature-dependent thermal-expansion coefficients of BeO, ZnO, AlN, GaN, and Al_2O_3 are reported in [15, 16].

1.2.2 Hexagonal Lattice Vectors and Planes

Though three-dimensional space can be indexed by three indices, lattice points of hexagonal lattice are indexed with four indices as $u\ v\ s\ w$, where $u+v+s=0$, to recognize the equivalent directions at a glance. Using three basic lattice vectors $\mathbf{a}_1$, $\mathbf{a}_2$, $\mathbf{a}_3$ on the (0001) plane shown in Fig. 1.4a and basic lattice vector $\mathbf{c}$ perpendicular to them, lattice point indexed by $u\ v\ s\ w$ is located at $u\ \mathbf{a}_1 + v\ \mathbf{a}_2 + s\ \mathbf{a}_3 + w\ \mathbf{c} = (2u+v)\ \mathbf{a}_1 + (2v+u)\ \mathbf{a}_2 + w\ \mathbf{c}$ since $\mathbf{a}_3 = -\mathbf{a}_1 - \mathbf{a}_2$ and $s = -u-v$. For example, lattice point $\mathbf{a}_1$ is represented by 2/3–1/3–1/3 0. Two lattice vectors indexed by $u\ v\ s\ w$ and $u'v's'0$ are orthogonal when $u\ u' + v\ v' + s\ s' = 0$.

1 Basic Properties of ZnO, GaN, and Related Materials   5

Table 1.2. Length of the bond parallel to the c-axis b_c (Å), other bond length b_a (Å) and volume per atom v (Å^3) of wurtzite structure, bond length b_{ZB} (Å) and volume per atom v_{ZB} (Å^3) of zincblende structure, bond length b_{RS} (Å) and volume per atom v_{RS} (Å^3) of rocksalt structure evaluated from the parameters in Table 1.1

	b_a	b_c	v	b_{ZB}	v_{ZB}	b_{RS}	v_{RS}
BeO	1.647	1.655	6.90	1.650	6.91	1.824	6.07
MgO	1.980	2.055	10.47			2.108	9.37
ZnO	1.974	1.988	11.90	1.992	12.17	2.136	9.74
CdO	2.289	2.051	17.00			2.385	13.57
AlN	1.894	1.893	10.45	1.897	10.50		
GaN	1.950	1.950	11.42	1.949	11.39		
InN	2.164	2.150	15.52	2.156	15.44		

Table 1.3. Thermal expansion coefficients normal to the c-axis $\alpha^\perp$ ($10^{-6}\mathrm{K}^{-1}$) and parallel to the c-axis $\alpha^\parallel$ ($10^{-6}\mathrm{K}^{-1}$) at temperature $T(\mathrm{K})$ of wurtzite type materials except rocksalt type MgO and corundum type Al$_2$O$_3$

	T	$\alpha^\perp$	$\alpha^\parallel$
BeO [15]	300	5.99	5.35
BeO [15]	700	8.62	7.79
MgO [10]		10.5	
ZnO [15]	300	4.31	2.49
ZnO [15]	700	7.47	4.26
AlN [15]	300	4.35	3.48
AlN [15]	700	4.95	4.22
GaN [15]	300	3.43	3.34
GaN [15]	700	4.91	4.05
InN [10]		4	3
Al$_2$O$_3$[16]	294	4.3	3.9
Al$_2$O$_3$[16]	703	9.2	9.3

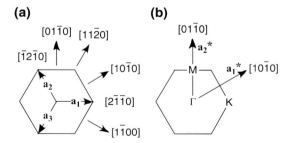

Fig. 1.4. Fundamental lattice vector (**a**) and reciprocal lattice vector (**b**) of hexagonal (0001) plane

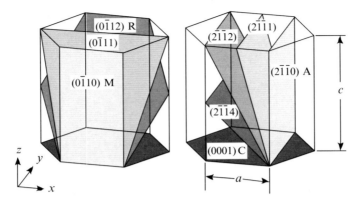

Fig. 1.5. Indices of typical planes in hexagonal structure

Reciprocal lattice vectors $\mathbf{a}_1^*$, $\mathbf{a}_2^*$ and $\mathbf{c}^*$ have the following relations with the real space vectors $\mathbf{a}_1$, $\mathbf{a}_2$ and $\mathbf{c}$: $\mathbf{a}_1 \cdot \mathbf{a}_1^* = \mathbf{a}_2 \cdot \mathbf{a}_2^* = \mathbf{c} \cdot \mathbf{c}^* = 2\pi$; $\mathbf{a}_1^*$, $\mathbf{a}_2^*$, $\mathbf{c}^*$ are, respectively, perpendicular to $\mathbf{a}_2$ and $\mathbf{c}$, $\mathbf{a}_1$ and $\mathbf{c}$, and $\mathbf{a}_1$ and $\mathbf{a}_2$. Therefore, the length of $\mathbf{a}_1^*$, $\mathbf{a}_2^*$ and $\mathbf{c}^*$ are $4\pi/(\sqrt{3}\,a)$, $4\pi/(\sqrt{3}\,a)$, and $2\pi/c$, respectively. Reciprocal lattice point

$$\mathbf{g}_{hkl} = h\mathbf{a}_1^* + k\mathbf{a}_2^* + l\mathbf{c}^* \tag{1.1}$$

is indexed by $h\ k\ i\ l$, where $i = -h - k$. When the indices are defined like this, identical directions on the (0001) plane in the real and reciprocal spaces have the same indices as shown in Fig. 1.4. Real-space plane normal to the reciprocal vector $\mathbf{g}_{hkl}$ is also indexed by $h\ k\ i\ l$. Some real space planes are shown in Fig. 1.5. Generally, the ($h\ k\ i\ l$) plane is not normal to the [$h\ k\ i\ l$] direction in real space unless $l = 0$ or $h = k = 0$.

1.3 Elastic Strain

1.3.1 Elastic Constants and Elastic Energy Density

Nonzero and independent elastic stiffness constants of hexagonal crystals are $c_{11}, c_{12}, c_{13}, c_{33}$, and c_{44} and $c_{66} = (c_{11} - c_{12})/2$. The stress components $\tau_{\alpha\beta}$ are linear function of strain components $\varepsilon_{\alpha\beta}$,

$$\begin{bmatrix} \tau_{xx} \\ \tau_{yy} \\ \tau_{zz} \\ \tau_{yz} \\ \tau_{zx} \\ \tau_{xy} \end{bmatrix} = \begin{bmatrix} c_{11} & c_{12} & c_{13} & 0 & 0 & 0 \\ c_{12} & c_{11} & c_{13} & 0 & 0 & 0 \\ c_{13} & c_{13} & c_{33} & 0 & 0 & 0 \\ 0 & 0 & 0 & c_{44} & 0 & 0 \\ 0 & 0 & 0 & 0 & c_{44} & 0 \\ 0 & 0 & 0 & 0 & 0 & c_{66} \end{bmatrix} \begin{bmatrix} \varepsilon_{xx} \\ \varepsilon_{yy} \\ \varepsilon_{zz} \\ \varepsilon_{yz} \\ \varepsilon_{zx} \\ \varepsilon_{xy} \end{bmatrix}, \tag{1.2}$$

1 Basic Properties of ZnO, GaN, and Related Materials

Table 1.4. Elastic stiffness coefficients (GPa) of the related materials

	Structure	c_{11}	c_{12}	c_{13}	c_{33}	c_{44}
BeO [17]	WZ	454	85	77	488	155
ZnO [18]	WZ	209.7	121.1	105.1	210.9	42.5
AlN [19]	WZ	396	137	108	373	116
GaN [20]	WZ	390	145	106	398	105
InN [19]	WZ	223	115	92	224	48
MgO [21]	RS	286	87			148
ZnO [12]	ZB	193	139			96
AlN [19]	ZB	304	160			193
GaN [19]	ZB	293	159			155
InN [19]	ZB	187	125			86

where x, y, and z denotes orthogonal $[2\bar{1}\bar{1}0]$, $[01\bar{1}0]$, and $[0001]$ directions and subscripts 1, 2, 3, 4, 5, and 6 denotes xx, yy, zz, yz, zx, and xy, respectively. Elastic energy per unit volume is

$$F = \frac{c_{11}}{2}(\varepsilon_{xx}^2 + \varepsilon_{yy}^2) + \frac{c_{33}}{2}\varepsilon_{zz}^2 + c_{12}\varepsilon_{xx}\varepsilon_{yy} + c_{13}(\varepsilon_{xx} + \varepsilon_{yy})\varepsilon_{zz}$$
$$+ 2c_{44}(\varepsilon_{yz}^2 + \varepsilon_{zx}^2) + 2c_{66}\varepsilon_{xy}^2. \tag{1.3}$$

The elastic stiffness constants of related materials are shown in Table 1.4.

When hydrostatic pressure $\tau_{xx} = \tau_{yy} = \tau_{zz}$ is applied,

$$\varepsilon_{xx} = \varepsilon_{yy},$$
$$\varepsilon_{zz} = \frac{c_{11} + c_{12} - 2c_{13}}{c_{33} - c_{13}}\varepsilon_{xx} \tag{1.4}$$

is obtained from (1.2). Shear strains are zero under zero shear stress. When stress is applied in one or two directions, stress in the free direction is zero. For example, a normal strain of hexagonal (0001) film under isotropic in-plane stress obeys relation of

$$\varepsilon_{xx} = \varepsilon_{yy},$$
$$\varepsilon_{zz} = -\frac{2c_{13}}{c_{33}}\varepsilon_{xx} \tag{1.5}$$

because $\tau_{zz} = 0$ and $\tau_{xx} = \tau_{yy}$.

In the WZ structure, spontaneous polarization parallel to c-axis P_{SP} is induced as mentioned in Sect. 1.2.1. External stress also modifies c/a and/or u and induces piezo polarization. Nonzero and independent components of piezoelectric coefficients of the WZ-type materials are e_{31}, e_{33}, and e_{15}. The x, y, and z components of the piezo polarization is

$$P_x = e_{15}\varepsilon_{zx},$$
$$P_y = e_{15}\varepsilon_{yz},$$
$$P_z = e_{31}(\varepsilon_{xx} + \varepsilon_{yy}) + e_{33}\varepsilon_{zz}. \tag{1.6}$$

Table 1.5. Spontaneous polarization $P_{\rm SP}(\rm C/m^2)$ and piezoelectric modulus $e_{ij}(\rm C/m^2)$ of wurtzite type materials

	$P_{\rm SP}$	e_{33}	e_{31}	e_{15}
BeO	−0.045 [14]	0.02 [14]	−0.02 [14]	
ZnO	−0.057 [14]	0.89 [14]	−0.51 [14]	−0.45 [22]
AlN	−0.081 [14]	1.46 [14]	−0.60 [14]	−0.48 [23]
GaN	−0.029 [14]	0.73 [14]	−0.49 [14]	−0.3 [24]
InN	−0.032 [14]	0.97 [14]	−0.57 [14]	

P_x and P_y are induced by shear strains, which break the symmetry of the WZ structure. When strain $\varepsilon_{\alpha\beta}$ in (1.6) are replaced with stress $\tau_{\alpha\beta}$, e_{ij} are replaced with d_{ij} that have the following relations to e_{ij} as

$$e_{31} = d_{31}(c_{11} + c_{12}) + d_{33}c_{13},$$
$$e_{33} = 2d_{31}c_{13} + d_{33}c_{33},$$
$$e_{15} = d_{15}c_{44}. \tag{1.7}$$

$P_{\rm SP}$ and e_{ij} of related materials are shown in Table 1.5.

In the case of the cubic crystal, a number of independent elastic stiffness constants are reduced to three as $c_{33} = c_{11}$, $c_{13} = c_{12}$, and $c_{66} = c_{44}$ in (1.2) and (1.3) because the uniaxial anisotropy about c-axis of the hexagonal crystal vanishes. In this case x, y and z denote orthogonal [100], [010] and [001] directions, respectively. The elastic stiffness constants of ZB polymorphs are also shown in Table 1.4. There is only one nonzero and independent component of piezoelectric coefficients for the ZB-type materials. Piezo polarization of the ZB-type materials is

$$P_x = e_{14}\varepsilon_{yz},$$
$$P_y = e_{14}\varepsilon_{zx},$$
$$P_z = e_{14}\varepsilon_{xy}, \tag{1.8}$$

where

$$e_{14} = d_{14}c_{44}. \tag{1.9}$$

1.3.2 Strain in Hexagonal- and Cubic-Structure Films

When a hexagonal crystalline film grows in $(01\bar{1}n)$ plane, the surface is normal to $\mathbf{p}_\zeta = \mathbf{g}_{01n}/(2\pi) = (\mathbf{a}_2^* + n\mathbf{c}^*)/(2\pi)$. Using the rectangle coordinate in which x, y, and z are parallel to $[2\bar{1}\bar{1}0]$, $[01\bar{1}0]$ and $[0001]$, $\mathbf{p}_\zeta = (0, 1/b, n/c)$ where $b = \sqrt{3}\,a/2$. Furthermore $\mathbf{p}_\xi = (1, 0, 0)$ and $\mathbf{p}_\eta = (0, n/c, -1/b)$ can be selected as two orthogonal vectors on the $(01\bar{1}n)$ plane. Then fundamental unit vectors $\mathbf{e}_x = (1, 0, 0)$, $\mathbf{e}_y = (0, 1, 0)$, and $\mathbf{e}_z = (0, 0, 1)$ are expressed as

1 Basic Properties of ZnO, GaN, and Related Materials

$$\mathbf{e}_x = \mathbf{p}_\xi,$$
$$\mathbf{e}_y = \frac{bc}{n^2b^2+c^2}(nb\mathbf{p}_\eta + c\mathbf{p}_\zeta),$$
$$\mathbf{e}_z = \frac{bc}{n^2b^2+c^2}(-c\mathbf{p}_\eta + nb\mathbf{p}_\zeta). \quad (1.10)$$

When the in-plane strain along $\mathbf{p}_\xi$ and $\mathbf{p}_\eta$ are, respectively, ε_ξ and ε_η and surface-normal strain parallel to $\mathbf{p}_\zeta$ is ε_ζ, $\mathbf{p}_\xi$, $\mathbf{p}_\eta$, and $\mathbf{p}_\zeta$ are deformed to $(1+\varepsilon_\xi)\mathbf{p}_\xi$, $(1+\varepsilon_\eta)\mathbf{p}_\eta$, and $(1+\varepsilon_\zeta)\mathbf{p}_\zeta$. How $\mathbf{e}_x$, $\mathbf{e}_y$, and $\mathbf{e}_z$ are deformed with these strains is known by replacing $\mathbf{p}_\alpha$ with $(1+\varepsilon_\alpha)\mathbf{p}_\alpha$ in (1.10) where $\alpha = \xi$, η, and ζ. In this way, strains in the xyz- rectangle coordinate system are represented with strains ε_ξ, ε_η, and ε_ζ as

$$\varepsilon_{xx} = \varepsilon_\xi,$$
$$\varepsilon_{yy} = \frac{n^2b^2\varepsilon_\eta + c^2\varepsilon_\zeta}{n^2b^2+c^2},$$
$$\varepsilon_{zz} = \frac{c^2\varepsilon_\eta + n^2b^2\varepsilon_\zeta}{n^2b^2+c^2},$$
$$\varepsilon_{yz} = \frac{nbc(\varepsilon_\zeta - \varepsilon_\eta)}{n^2b^2+c^2},$$
$$\varepsilon_{zx} = 0,$$
$$\varepsilon_{xy} = 0, \quad (1.11)$$

and strain energy density of (1.3) is represented with ε_ξ, ε_η, and ε_ζ. Surface-normal strain ε_ζ that minimizes F is given by solving $\partial F/\partial \varepsilon_\zeta = 0$ as

$$\varepsilon_\zeta = -\frac{(n^2b^2+c^2)(n^2b^2c_{13}+c^2c_{12})\varepsilon_\xi + \{n^2b^2c^2(c_{11}+c_{33}-4c_{44})+(n^4b^4+c^4)c_{13}\}\varepsilon_\eta}{c^4c_{11}+n^4b^4c_{33}+2n^2b^2c^2(c_{13}+2c_{44})}, \quad (1.12)$$

where $b = \sqrt{3}a/2$.

When a hexagonal crystalline film grows in $(2\bar{1}\bar{1}n)$ plane, the surface is normal to $\mathbf{p}_\zeta = (2\mathbf{a}_1^* - \mathbf{a}_2^* + n\mathbf{c}^*)/(2\pi) = (1/b, 0, n/c)$, where $b = a/2$. In this case, $\mathbf{p}_\xi = (n/c, 0, -1/b)$ and $\mathbf{p}_\eta = (0, 1, 0)$ can be selected as two orthogonal vectors on the $(2\bar{1}\bar{1}n)$ plane. Therefore, surface-normal strain ε_ζ is also given by (1.12), where $b = a/2$, ε_ξ and ε_η are in-plane strain along $\mathbf{p}_\eta$ and $\mathbf{p}_\xi$, respectively because x and y, $\mathbf{p}_\xi$ and $\mathbf{p}_\eta$ are respectively exchanged with each other in comparison with the case of the $(01\bar{1}n)$ plane.

When a cubic crystalline film grows in $(mn1)$ plane, surface is normal to $\mathbf{p}_\zeta = (m, n, 1)$, where $\mathbf{p}_\xi = (n^2+1, -mn, -m)$ and $\mathbf{p}_\eta = (0, 1, -n)$ can be selected as two orthogonal vectors on the $(mn1)$ plane. In this case x, y, and z components are parallel to [100], [010], and [001] directions, respectively. Surface-normal strain ε_ζ is given by solving $\partial F/\partial \varepsilon_\zeta = 0$ as

$$\varepsilon_\zeta = -2\varepsilon_\| \frac{(m^4+n^4+1)c_{12}+2(m^2n^2+m^2+n^2)(c_{11}+c_{12}-2c_{44})}{(m^4+n^4+1)c_{11}+2(m^2n^2+m^2+n^2)(c_{12}+2c_{44})}, \quad (1.13)$$

where $\varepsilon_\|$ is in-plane strain, which is assumed to be isotropic.

1.4 Electronic Structure

1.4.1 Bandgap and Band-Edge Electronic Structure

The first Brillouin zone of a hexagonal crystal is a hexagonal prism of height $2\pi/c$. The boundary of the zone at $k_z = 0$ (k_z is a component of the wave vector **k** parallel to c-axis) is shown in Fig. 1.4b, where symbols for symmetric points are also shown. The point at $k_z = \pi/c$ just above the Γ, K, and M points are denoted as A, H, and L, respectively.

BeO, ZnO, AlN, GaN, and InN are direct-bandgap semiconductors, whose band edge is located at the Γ point. Temperature dependence of the bandgap energy E_g can be expressed by the Varshni formula

$$E_g(T) = E_g(0) - \frac{\alpha T^2}{T+\beta}. \quad (1.14)$$

Bandgap energy, Varshni parameters α and β are listed in Table 1.6. Bandgap energy of alloy $A_{1-x}B_xC$ is approximately represented by a quadratic function of x as

$$E_g(x) = (1-x)E_g^{AC} + xE_g^{BC} - bx(1-x), \quad (1.15)$$

where b is bowing parameter. Figure 1.1 is drawn using bandgap energies and bowing parameters of III-nitrides shown in Tables 1.6 and 1.7.

Energy dispersion around the bottom of conduction band (mainly consists of anti-bonding state of cation s electrons) can be represented as

$$E_c(\mathbf{k}) = E_{c0} + \frac{\hbar^2}{2m_0 m_e^\perp}(k_x^2+k_y^2) + \frac{\hbar^2}{2m_0 m_e^\|}k_z^2 + a_2(\varepsilon_{xx}+\varepsilon_{yy}) + a_1\varepsilon_{zz}, \quad (1.16)$$

Table 1.6. Bandgap energy E_g (eV) at temperature T (K) and Varshni parameters α (meV/K) and β (K)

	Structure	T	E_g	α	β
BeO	WZ	77	10.63 [25]		
ZnO	WZ	4.2	3.437 [26]		
AlN	WZ	0	6.25 [1]	1.799 [1]	1462 [1]
GaN	WZ	0	3.510 [1]	0.909 [1]	830 [1]
InN	WZ	0	0.78 [1]	0.245 [1]	624 [1]
AlN	ZB	0	5.4 [1]	0.593 [1]	600 [1]
GaN	ZB	0	3.299 [1]	0.593 [1]	600 [1]
InN	ZB	0	0.78 [1]	0.245 [1]	624 [1]

Table 1.7. Bowing parameter (eV) recommended for both wurtzite and zincblende type III-nitrides [1]

	b
AlGaN	0.7
GaInN	1.4
AlInN	2.5

Table 1.8. Electron and hole effective mass (in the unit of m_0) of wurtzite type materials

	$m_e^{\|\|}$	$m_e^{\perp}$	$m_A^{\|\|}$	$m_B^{\|\|}$	$m_C^{\|\|}$	$m_A^{\perp}$	$m_B^{\perp}$	$m_C^{\perp}$
BeO [27]	0.58	0.74	12.01			2.44		
ZnO [28]	0.23	0.21	2.74	3.03	0.27	0.54	0.55	1.12
AlN [29]	0.33	0.25	0.25	3.68	3.68	3.68	6.33	0.25
GaN [29]	0.20	0.18	1.10	1.10	0.15	1.65	0.15	1.10
InN [30]	0.11	0.10	1.67	1.67	0.10	1.61	0.11	1.67

Table 1.9. Deformation potential (eV) of wurtzite type materials

	a_1	a_2	D_1	D_2	D_3	D_4	D_5	D_6
ZnO [31]			3.9	4.13	1.15	-1.22	-1.53	-2.88
AlN [1]	-3.4	-11.8	-17.1	7.9	8.8	-3.9	-3.4	-3.4
GaN [1]	-4.9	-11.3	-3.7	4.5	8.2	-4.1	-4.0	-5.5
InN [1]	-3.5	-3.5	-3.7	4.5	8.2	-4.1	-4.0	-5.5

where E_{c0} is band edge energy of the conduction band, $\hbar$ is Planck constant divided by 2π, m_0 is rest mass of electron in vacuum, $m_e^{\perp}$ and $m_e^{\|\|}$ are effective mass of the conduction electron, a_1 and a_2 are deformation potential of the conduction band, k_x, k_y, and k_z are x, y, and z component of wave vector $\mathbf{k}$, superscript $\|\|$ and $\perp$ denotes parallel and perpendicular to c-axis, respectively. The effective mass and the deformation potential are shown in Tables 1.8 and 1.9.

Figure 1.6 illustrates energy-level splittings for WZ and ZB semiconductors at the top of the valence band, which mainly consists of anion p electrons. The spin–orbit coupling proportional to $l \cdot s = (j^2 - l^2 - s^2)/2$ splits the energy level according to j, which is angular momentum composed of orbital (l) and spin (s) angular momentums. In WZ structure, crystal-field splitting appears owing to the structural anisotropy between parallel to and normal to the c-axis. The l_z and j_z in Fig. 1.6 are c-axis component of l and j, respectively. The crystal-field splitting Δ_{cr} and the spin–orbit splitting Δ_{so} are shown in Table 1.10. As for AlN, the crystal-field split-off band, which is lowest in Fig. 6, moves to the top because Δ_{cr} of AlN is negative.

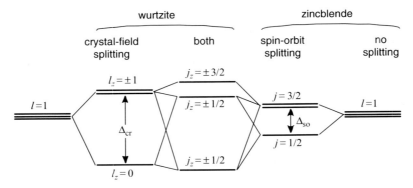

Fig. 1.6. Schematic energy diagram of valence-band top for wurtzite and zincblende semiconductors

Table 1.10. Crystal-field splitting Δ_{cr} and spin–orbit splitting Δ_{so} (meV) of wurtzite type materials

	Δ_{cr}	Δ_{so}
ZnO [32]	43	16
AlN [29]	−58.5	20.4
GaN [29]	72.9	15.5
InN [1]	40	5

1.4.2 Bir-Pikus Hamiltonian

Energy dispersion around valence band maximum at Γ point can be obtained by calculating eigen values of the following Bir-Pikus Hamiltonian matrix $\mathbf{H}_{\mathrm{BP}}(\mathbf{k})$ for WZ semiconductors as a function of $\mathbf{k}$ [29,33].

$$\mathbf{H}_{\mathrm{BP}}(\mathbf{k}) = \begin{bmatrix} F & 0 & -H^* & 0 & K^* & 0 \\ 0 & G & \Delta & -H^* & 0 & K^* \\ -H & \Delta & \lambda & 0 & I^* & 0 \\ 0 & -H & 0 & \lambda & \Delta & I^* \\ K & 0 & I & \Delta & G & 0 \\ 0 & K & 0 & I & 0 & F \end{bmatrix}, \quad (1.17)$$

Where

$$F = \Delta_1 + \Delta_2 + \lambda + \theta, \tag{1.18}$$

$$G = \Delta_1 - \Delta_2 + \lambda + \theta, \tag{1.19}$$

$$\lambda = A_1 k_z^2 + A_2(k_x^2 + k_y^2) + \lambda_\varepsilon, \tag{1.20}$$

$$\lambda_\varepsilon = D_1 \varepsilon_{zz} + D_2(\varepsilon_{xx} + \varepsilon_{yy}), \tag{1.21}$$

$$\theta = A_3 k_z^2 + A_4(k_x^2 + k_y^2) + \theta_\varepsilon, \tag{1.22}$$

Table 1.11. Valence band effective-mass parameters (in the unit of $\hbar^2/(2m_0) = 3.81\,\mathrm{eV\mathring{A}^2}$) of wurtzite type materials

	A_1	A_2	A_3	A_4	A_5	A_6
ZnO [28]	−3.78	−0.44	3.45	−1.63	1.68	−2.23
AlN [29]	−3.95	−0.27	3.68	−1.84	−1.95	−2.91
GaN [29]	−6.56	−0.91	5.65	−2.83	−3.13	−4.86
InN [30]	−9.28	−0.60	8.68	−4.34	−4.32	−6.08

$$\theta_\varepsilon = D_3 \varepsilon_{zz} + D_4(\varepsilon_{xx} + \varepsilon_{yy}), \tag{1.23}$$

$$\Delta = \sqrt{2}\Delta_3, \tag{1.24}$$

$$H = (iA_6 k_z - A_7)(k_x + ik_y) + iD_6(\varepsilon_{zx} + i\varepsilon_{yz}), \tag{1.25}$$

$$I = (iA_6 k_z + A_7)(k_x + ik_y) + iD_6(\varepsilon_{zx} + i\varepsilon_{yz}), \tag{1.26}$$

$$K = A_5(k_x + ik_y)^2 + D_5(\varepsilon_{xx} - \varepsilon_{yy} + 2i\varepsilon_{xy}), \tag{1.27}$$

$\Delta_1 = \Delta_{cr}$, $\Delta_2 = \Delta_3 = \Delta_{so}/3$, A_i are Luttinger-like parameters related to effective mass of holes, D_i are deformation potentials, and superscript * denotes complex conjugate. The strain dependent terms are given by replacing A_i and $k_\alpha k_\beta$ with D_i and $\varepsilon_{\alpha\beta}$, where α, $\beta = x$, y, z. λ_ε is common to all the diagonal elements and corresponds to the shift of the whole valence band, relative to the conduction band minimum under strain. θ_ε corresponds to variation in Δ_1 under strain. The parameters for the related WZ semiconductors are shown in Tables 1.9, 1.10, and 1.11. Sign of D_i is opposite to that of [31] because D_i in [31] is defined as a deformation potential for the bandgap. The six basis functions are as follows:

$$u_1 = \sqrt{\tfrac{1}{2}}|(X+iY),\uparrow\rangle, u_2 = \sqrt{\tfrac{1}{2}}|(X+iY),\downarrow\rangle, \tag{1.28}$$

$$u_3 = |Z,\uparrow\rangle, u_4 = |Z,\downarrow\rangle, \tag{1.29}$$

$$u_5 = \sqrt{\tfrac{1}{2}}|(X-iY),\uparrow\rangle, u_6 = \sqrt{\tfrac{1}{2}}|(X-iY),\downarrow\rangle, \tag{1.30}$$

where $|\uparrow\rangle$ and $|\downarrow\rangle$ are up and down spin functions, respectively, l_z of u_1 and u_2 is 1, u_3 and u_4 is 0, and u_5 and u_6 is −1. Energy and wave function at any $\mathbf{k}$ and strain can be numerically calculated. Fortunately, in some limited cases, $H_{BP}(\mathbf{k})$ can be diagonalized analytically.

First, if $k_x = k_y = 0$, $\varepsilon_{xx} = \varepsilon_{yy}$, and shear strains are zero, then H, I, and K become zero. Symmetry of the WZ structure is kept under this kind of strain, which replaces Δ_1 with $\Delta_1 + \theta_\varepsilon$. Then $H_{BP}(\mathbf{k})$ consists of 1^2, 2^2, 2^2, and 1^2 block matrices, whose basis functions are u_1, u_2 and u_3, u_4 and u_5, and u_6. Eigen energies are

$$E_1^{\parallel} = F = \Delta_1 + \theta_\varepsilon + \Delta_2 + (A_1 + A_3)k_z^2 + \lambda_\varepsilon, \tag{1.31}$$

$$E_\pm^{\parallel} = \frac{G + \lambda \pm \sqrt{(G-\lambda)^2 + 4\Delta^2}}{2}$$

$$= \frac{\Delta_1 + \theta_\varepsilon - \Delta_2 + (2A_1 + A_3)k_z^2 \pm \sqrt{(\Delta_1 + \theta_\varepsilon - \Delta_2 + A_3 k_z^2)^2 + 8\Delta_3^2}}{2} + \lambda_\varepsilon. \tag{1.32}$$

Eigen function of $E_1^{\parallel}$ is u_1 or u_6 ($j_z = 3/2$) having Γ_9 symmetry and those of $E_\pm^{\parallel}$ are a mixture of u_2 and u_3 or a mixture of u_4 and u_5 ($j_z = 1/2$) having Γ_7 symmetry. The three bands at the valence-band top of the WZ semiconductors (Fig. 1.6) are called A, B, and C bands from upper band to lower band. Generally $\Delta_{\mathrm{cr}} > \Delta_{\mathrm{so}} > 0$ is satisfied. When $\Delta_{\mathrm{cr}} + \theta_\varepsilon > \Delta_{\mathrm{so}} > 0$, $E_1^{\parallel}$, $E_+^{\parallel}$, and $E_-^{\parallel}$ with Γ_9, Γ_7 (mainly u_2 or u_5), and Γ_7 (mainly u_3 or u_4) symmetry correspond to the A, B, and C bands, respectively. In the case of AlN, Δ_{cr} is negative. When $-(\Delta_{\mathrm{cr}} + \theta_\varepsilon) > \Delta_{\mathrm{so}} > 0$, $E_+^{\parallel}$, $E_1^{\parallel}$, and $E_-^{\parallel}$ with Γ_7 (mainly u_3 or u_4), Γ_9, and Γ_7 (mainly u_2 or u_5) symmetry correspond to the A, B, and C bands, respectively. In the case of ZnO, it had been thought that Δ_2 and Δ_3 are negative. When $\Delta_{\mathrm{cr}} + \theta_\varepsilon > -\Delta_{\mathrm{so}} > 0$, $E_+^{\parallel}$, $E_1^{\parallel}$, and $E_-^{\parallel}$ with Γ_7 (mainly u_2 or u_5), Γ_9, and Γ_7 (mainly u_3 or u_4) symmetry correspond to the A, B, and C bands, respectively. It was supported by recent optical measurements, however, that Δ_2 and Δ_3 of ZnO are positive [32]. Anyway since $|\Delta_{\mathrm{so}}|$ is small, interband optical transitions between the conduction band and the $E_+^{\parallel}$ ($E_-^{\parallel}$ in the case of AlN) band as well as the $E_1^{\parallel}$ band are strong for the light polarized perpendicular to the c-axis and that of the $E_-^{\parallel}$ ($E_+^{\parallel}$ in the case of AlN) band is strong for the light polarized parallel to the c-axis.

Effective mass of the hole in k_z direction is obtained from the coefficient of k_z^2 term as

$$m_1^{\parallel} = -\frac{1}{A_1 + A_3}, \tag{1.33}$$

$$m_+^{\parallel} = -\frac{1}{A_1 + A_3}, \tag{1.34}$$

$$m_-^{\parallel} = -\frac{1}{A_1}, \tag{1.35}$$

for the $E_1^{\parallel}$, $E_+^{\parallel}$ and $E_-^{\parallel}$ band, respectively, if Δ_3 is disregarded when a comparatively wide range of k_z is considered above room temperature. In the case of AlN, $m_+^{\parallel}$ and $m_-^{\parallel}$ must be replaced with each other.

Next if $k_z = 0$, $A_7 = 0$, and shear strains are zero, then H and I become zero. In order to simplify further, $\Delta_3 = 0$ is assumed since the crystal field Δ_1 mainly characterizes the valence band-edge structures of the WZ crystals. Then, $H_{\mathrm{BP}}(\mathbf{k})$ is decomposed into 2^2, 2^2, 1^2, and 1^2 blocks, whose basic functions are u_1 and u_5, u_2 and u_6, u_3, and u_4 [29]. Eigen energies are

$$E^{\perp}_{\pm} = \frac{F + G \pm \sqrt{(F-G)^2 + 4|K|^2}}{2} = \Delta_1 + \theta_\varepsilon + (A_2 + A_4)(k_x^2 + k_y^2)$$

$$\pm \sqrt{\Delta_2^2 + A_5^2(k_x^2 + k_y^2)^2} + \lambda_\varepsilon, \quad (1.36)$$

$$E_3^{\perp} = \lambda = A_2(k_x^2 + k_y^2) + \lambda_\varepsilon. \quad (1.37)$$

The effective mass of the hole in k_x and k_y direction is obtained as

$$m^{\perp}_{\pm} = -\frac{1}{A_2 + A_4 \pm |A_5|}, \quad (1.38)$$

$$m_3^{\perp} = -\frac{1}{A_2}, \quad (1.39)$$

for the $E_{\pm}^{\perp}$ and $E_3^{\perp}$ band, respectively. Generally (in the case of AlN), $E_+^{\perp}$, $E_-^{\perp}$, and $E_3^{\perp}$ ($E_3^{\perp}$, $E_+^{\perp}$, and $E_-^{\perp}$) correspond to the A, B, and C bands, respectively.

1.4.3 Valence Band-Edge and Bandgap of Strained ZnO

Valence band dispersion curves of ZnO at various isotropic in-plane strain $\varepsilon_{xx} = \varepsilon_{yy}$, where ε_{zz} is given by (1.5), are shown in Fig. 1.7 using ZnO parameters in Tables 1.9, 1.10, and 1.11.

Solid lines are calculated numerically, while dashed lines are calculated for $\Delta_3 = 0$ using (1.36) and (1.37) along k_x axis and (1.31) and (1.32) along k_z axis. Here $\Delta_3 = 0$ is assumed even in (1.31) and (1.32) to let the dashed lines be continuous. Of course (1.31) and (1.32) give identical lines with the solid lines without such assumption. Around $\varepsilon_{xx} = 1.2\%$ spin–orbit interaction becomes relatively important and the error due to the omission of Δ_3 is not small near the Γ point because the substantial crystal-field splitting $\Delta_1 + \theta_\varepsilon$, which includes the effect of strain, approaches to be zero. Valence band structure of the strain free AlN is similar to that of ZnO at $\varepsilon_{xx} = 2\%$ shown in Fig. 1.7 because $\Delta_1 + \theta_\varepsilon$ is negative in this case.

ZnO bandgaps between the bottom of the conduction band and the top of the A, B, and C bands are shown in Fig. 1.8 as a function of isotropic strain in c-plane. Here, fundamental bandgap at zero strain is set to be $E_{g0} = 3.4368$ eV [26]. The solid line shows the case when $\tau_{zz} = 0$, $\tau_{xx} = \tau_{yy}$, and $\varepsilon_{zz} = -0.997\,\varepsilon_{xx}$ from (1.5). Fundamental bandgap E_g is expected to vary as $E_g = E_{g0} - 0.787\varepsilon_{xx}$ eV when (1.31) and the ZnO deformation potentials in Table 1.9 are used. However, the slope is about half compared with the one measured for ZnO (0001) thin films [34]. The deformation potentials in Table 1.9 are measured for bulk ZnO under uniaxial stress [31]. The dashed lines show the case under hydrostatic pressure, where $\varepsilon_{zz} = 1.14\,\varepsilon_{xx}$ from

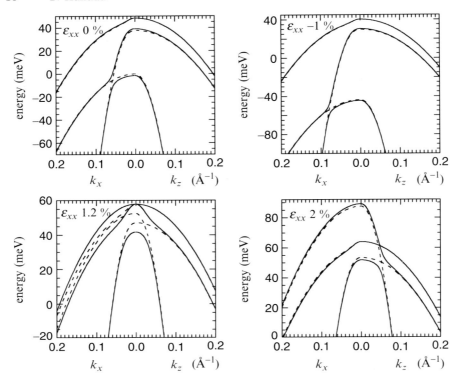

Fig. 1.7. ZnO band structures near the valence-band top at various isotropic in-plane strain $\varepsilon_{xx} = \varepsilon_{yy}$

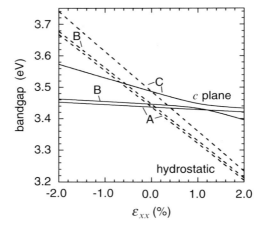

Fig. 1.8. Bandgap energy between conduction band minimum and A, B, and C valence band maximum of ZnO as a function of in-plane strain under hydrostatic (*dashed line*) and isotropic in-plane (*solid line*) stress

(1.4). Fundamental bandgap $E_\mathrm{g} = E_\mathrm{g0} - 11.57\,\varepsilon_{xx}$ eV decreases with increasing interatomic distance. When this strain coefficient is converted using (1.2) and (1.4), hydrostatic pressure coefficient of $25.69\,\mathrm{meV\,GPa^{-1}}$, which is in fair agreement with measured values $23.5 \sim 29.7\,\mathrm{meV\,GPa^{-1}}$ [35], is obtained.

1.4.4 Exciton Binding Energy

When electron-hole pair of exciton is treated like electron and proton of hydrogen atom, binding energy and Bohr radius of 1S exciton are $E_\mathrm{ex0} = E_\mathrm{H}\mu/\varepsilon(0)^2$, and $a_\mathrm{ex0} = a_\mathrm{B}\varepsilon(0)/\mu$, respectively, where $E_\mathrm{H} = 13.6$ eV and $a_\mathrm{B} = 0.529$ Å are binding energy and Bohr radius of 1s electron of hydrogen atom, respectively. Reduced mass μ and static dielectric constant $\varepsilon(0)$ are given as [36]

$$\frac{1}{\mu} = \frac{1}{m_e} + \frac{1}{m_h}, \tag{1.40}$$

$$\frac{1}{m_e} = \frac{2}{3m_e^\perp} + \frac{\varepsilon^\perp(0)}{3\varepsilon^{||}(0)m_e^{||}}, \tag{1.41}$$

$$\frac{1}{m_h} = \frac{2}{3m_h^\perp} + \frac{\varepsilon^\perp(0)}{3\varepsilon^{||}(0)m_h^{||}}, \tag{1.42}$$

$$\varepsilon(0) = \sqrt{\varepsilon^{||}(0)\varepsilon^\perp(0)}. \tag{1.43}$$

The electron and hole effective mass and the dielectric constant of related WZ materials are listed in Tables 1.8 and 1.12. The data in Table 1.12 fulfill the Lyddane–Sachs–Teller relation $\omega_\mathrm{LO}^2/\omega_\mathrm{TO}^2 = \varepsilon(0)/\varepsilon(\infty)$ except 15% discrepancy for InN, where $\varepsilon(\infty)$ is the dielectric constant at high frequency, and ω_LO and ω_TO are longitudinal and transverse optical-phonon frequency, respectively. E_ex0 and a_ex0 calculated for A exciton, where $m_h = m_A$, using these values are shown in Table 1.13. In the cases of ZnO and AlN, E_ex0 estimated with the static Coulomb potential is smaller than the observed exciton binding energy E_ex. When electrons and holes move in a polar crystal,

Table 1.12. Dielectric constant and optical phonon energy (meV) of wurtzite type materials except rocksalt type MgO

	$\varepsilon^{||}(0)$	$\varepsilon^\perp(0)$	$\varepsilon^{||}(\infty)$	$\varepsilon^\perp(\infty)$	$\hbar\omega_\mathrm{LO}$	$\hbar\omega_\mathrm{TO}$
BeO	7.65 [37]	6.94 [37]	2.99 [37]	2.95 [37]	135.2 [37]	87.2 [37]
MgO	9.8 [21]		2.95 [21]		92.2 [21]	49.4 [21]
ZnO	8.49 [38]	7.40 [38]	3.72 [38]	3.68 [38]	72.8 [39]	51.0 [39]
AlN	8.5 [10]	8.5 [10]	4.76 [10]	4.76 [10]	112.6 [40]	81.1 [40]
GaN	9.5 [41]	10.4 [41]	5.35 [10]	5.35 [10]	90.6 [40]	68.3 [40]
InN	15 [10]	15 [10]	8.4 [10]	8.4 [10]	86.0 [40]	59.3 [40]

Table 1.13. Exciton binding energy: E_{ex} (experiment); E_{ex0} (hydrogen like); and E_{ex1} (Pollmann–Büttner–Kane) (meV), Bohr radius: a_{ex0} (hydrogen like) and a_{ex1} (Pollmann–Büttner–Kane) (Å), electron–polaron radius r_e (Å), and hole–polaron radius r_h (Å)

	E_{ex}	E_{ex0}	a_{ex0}	r_e	r_h	E_{ex1}	a_{ex1}
BeO	175 [25]	149	6.6	6.3	2.9	389	3.0
ZnO	60 [26]	37.5	24.2	15.2	8.4	61.0	14.5
AlN	57 [42]	36.3	23.4	11.1	7.2	45.1	18.0
GaN	23.44 [43]	22.0	32.9	15.2	5.6	25.3	27.7
InN		5.9	81.8	20.7	5.2	6.1	78.0

they interact with nearby ions and distort the lattice. Therefore, it is better to describe the electron and hole as polaron in II-oxides and III-nitrides. Pollmann and Büttner investigated the interaction of Wannier excitons with LO-phonon field. When the exciton Bohr radius is sufficiently larger than the polaron radii, effective electron-hole interaction potential approximates [44]

$$V_{PB}(r) = -\frac{e^2}{4\pi\varepsilon(0)r} - \frac{e^2}{4\pi\varepsilon^*(m_h - m_e)r}\left\{m_h \exp\left(-\frac{r}{r_h}\right) - m_e \exp\left(-\frac{r}{r_e}\right)\right\},$$
(1.44)

where r is the distance between the electron and hole, $r_h = \hbar/(2m_h\hbar\omega_{LO})^{1/2}$ is the hole–polaron radius, and $r_e = \hbar/(2m_e\hbar\omega_{LO})^{1/2}$ is the electron–polaron radius, and $1/\varepsilon^* = 1/\varepsilon(\infty) - 1/\varepsilon(0)$. Equation (1.44) indicates that effective dielectric constant approaches $\varepsilon(\infty)$ as the electron–hole distance r decreases to be smaller than the polaron radii. Later, Kane added the correction for the polaron mass and provided tabulated functions to estimate the excitonic–polaron binding energy (E_{ex1}) by variational method [45]. Trial wave function is hydrogen 1s type $\exp(-r/a)/(\pi a^3)^{1/2}$ and minimized – E_{ex1} is obtained at the variational parameter $a = a_{ex1}$. The binding energy E_{ex1} and the Bohr radius a_{ex1} corrected by this method with parameters listed in Tables 1.8 and 1.12 are shown in Table 1.13. The correction by the polaron effect is large for II-oxide owing to the small a_{ex1}/r_e and a_{ex1}/r_h ratios.

References

1. I. Vurgaftman, J.R. Meyer, J. Appl. Phys. **94**, 3675 (2003)
2. Y.R. Ryu et al., Appl. Phys. Lett. **88**, 052103 (2006)
3. A. Ohtomo et al., Appl. Phys. Lett. **72**, 2466 (1998)
4. T. Makino et al., Appl. Phys. Lett. **78**, 1237 (2001)
5. A. Nakamura et al., Jpn. J. Appl. Phys. **43**, L1452 (2004)
6. J. Ishihara et al., Appl. Phys. Lett. **89**, 091914 (2006)
7. R.M. Hazen, L.W. Finger, J. Appl. Phys. **59**, 3728 (1986)

8. C.J. Park et al., Phys. Rev. B **59**, 13501 (1999)
9. S. Limpijumnong, W.R.L. Lambrecht, Phys. Rev. B **63**, 104103 (2001)
10. H. Morkoç et al., J. Appl. Phys. **76**, 1363 (1994)
11. H. Karzel et al., Phys. Rev. B **53**, 11425 (1996)
12. Ü. Özgür et al., J. Appl. Phys. **98**, 041301 (2005)
13. R.J. Guerrero-Moreno, N. Takeuchi, Phys. Rev. B **66**, 205205 (2002)
14. F. Bernardini, V. Fiorentini, D. Vanderbilt, Phys. Rev. B **56**, R10024 (1997)
15. H. Iwanaga, A. Kunishige, S. Takeuchi, J. Mater. Sci. **35**, 2451 (2000)
16. M. Leszczynski, J. Appl. Phys. **76**, 4909 (1994)
17. A. Bosak et al., Phys. Rev. B **77**, 224303 (2008)
18. T.B. Bateman, J. Appl. Phys. **33**, 3309 (1962)
19. A.F. Wright, J. Appl. Phys. **82**, 2833 (1997)
20. A. Polian, M. Grimsditch, I. Grzegory, J. Appl. Phys. **79**, 3343 (1996)
21. C. Kittel, *Introduction to Solid State Physics*, 4th edn. (Wiley, New York, 1971)
22. C. Carlotti et al., Appl. Phys. Lett. **51**, 1889 (1987)
23. K. Tsubouchi, N. Mikoshiba, IEEE Trans. Sonics Ultrason. **SU-32**, 634 (1985)
24. G.D. O'Clock, M.T. Duffy, Appl. Phys. Lett. **23**, 55 (1973)
25. D.M. Roessler, W.C. Walker, E. Loh, J. Phys. Chem. Solids **30**, 157 (1969)
26. S.F. Chichibu et al., J. Appl. Phys. **93**, 756 (2003)
27. Y.N. Xu, W.Y. Ching, Phys. Rev. B **48**, 4335 (1993)
28. W.R.L. Lambrecht et al., Phys. Rev. B **65**, 075207 (2002)
29. M. Suzuki, T. Uenoyama, A. Yanase, Phys. Rev. B **52**, 8132 (1995)
30. Y.C. Yeo, T.C. Chong, M.F. Li, J. Appl. Phys. **83**, 1429 (1998)
31. J. Wrzesinski, D. Fröhlich, Phys. Rev. B **56**, 13087 (1997)
32. D.C. Reynolds et al., Phys. Rev. B **60**, 2340 (1999)
33. G.L. Bir, G.E. Pikus, *Symmetry and Strain-Induced Effects in Semiconductors* (Wiley, New York, 1974)
34. Th. Gruber et al., J. Appl. Phys. **96**, 289 (2004)
35. S.J. Chen et al., J. Appl. Phys. **99**, 066102 (2006)
36. A.V. Rodina et al., Phys. Rev. B **64**, 115204 (2001)
37. E. Loh, Phys. Rev. **166**, 673 (1968)
38. H. Yoshikawa, S. Adachi, Jpn. J. Appl. Phys. **36**, 6237 (1997)
39. T.C. Damen, S.P.S. Porto, B. Tell, Phys. Rev. **142**, 570 (1966)
40. K. Kim, W.R.L. Lambrecht, B. Segall, Phys. Rev. B **53**, 16310 (1996); **56**, 7018 (1997)
41. A.S. Barker Jr., M. Ilegems, Phys. Rev. B **7**, 743 (1973)
42. Y. Yamada et al., Appl. Phys. Lett. **92**, 131912 (2008)
43. K. Torii et al., Phys. Rev. B **60**, 4723 (1999)
44. J. Pollmann, H. Büttner, Phys. Rev. B **16**, 4480 (1977)
45. E.O. Kane, Phys. Rev. B **18**, 6849 (1978)

2

Solvothermal Growth of ZnO and GaN

D. Ehrentraut, F. Orito, Y. Mikawa, and T. Fukuda

Abstract. The solvothermal growth of ZnO and GaN is presented. The hydrothermal method yields large ZnO crystals of excellent crystallinity. The physical properties of the crystals are discussed. Liquid phase epitaxy of ZnO is employed as a tool for fast screening. The ammonothermal method is used to fabricate GaN bulk crystals. General aspects are discussed and the use of acidic mineralizer is focused on. This method is a recent development to challenge the problem of fabrication of lattice-matched substrate crystals for future group-III nitride device technology.

2.1 Introduction

Lattice- and thermally matched substrates are beneficial in the design and fabrication of optimized device structures providing high efficiency and lifetime [1]. The hexagonal phase of the wide bandgap materials, zinc oxide (ZnO, $E_g = 3.3$ eV) and gallium nitride (GaN, $E_g = 3.4$ eV), are of substantial importance for numerous applications in electronics and optoelectronics (see relevant chapters in this book); however, fabrication of free-standing wafers of high crystalline quality has only recently been reported [2–4]. Whereas ZnO wafers of the technologically important minimum size of 2 in. (50 mm) are already being fabricated and 4 in. (100 mm) size has been demonstrated [5], much more effort is required in the case of GaN. Till date, template substrates of SiC or Al_2O_3 are employed to grow a thick GaN film on them using hydride vapor phase epitaxy (HVPE). These films do contain a large number of crystal defects such as grain boundaries, threading dislocations, stacking fault, etc. and the substrate has to be removed in subsequent steps to realize an electrically conductive back contact. Free-standing GaN wafers of 2 in. (50 mm) size and with low defect concentrations are in vital demand by industry.

Among the promising techniques used to grow bulk crystals of ZnO and GaN, the growth from a solution under near- or supercritical conditions has turned out to be highly attractive. This so-called solvothermal growth technique basically employs a polar solvent which dissolves and successively forms metastable products with the solute and later, under slightly different temperature conditions, recrystallizes the desired phase of hexagonal ZnO and GaN on a seed of similar crystal structure.

Mineralizers are essential additives used to increase the solubility of the solute in the solvent. Since a closed system is utilized, i.e., exchange of matter with the ambient is impossible, often the solvent attains a supercritical state, which also improves the solubility of the nutrient.

The solvothermal crystal growth technology, as compared to vapor phase growth technologies, is characterized by the following merits: (1) The operation is near thermodynamic equilibrium for a given temperature–pressure range. This leads to the generation of an extremely low supersaturation in the solution, and consequently a high crystallinity can be expected. (2) The temperature gradient at the growing interface is practically zero, which results in stable growth conditions. (3) The process is scalable: large quantities of crystals can be controlled over long process times, thereby enabling high throughput. Multiple-seed (around 2,000 pieces) growth processes are typically applied for the low-temperature phase of quartz α-SiO$_2$. (4) There is no need for expensive vacuum technology. (5) Environmentally benign conditions for production and capability for complete recycling of the used solution can be achieved. (6) This technology has been in use for a long time (>60 years) to mass-produce crystals such as α-SiO$_2$.

Like with most of the commonly used technologies, there are certain limitations such as the incorporation of constituents or reaction products from the solution on a mesoscopic scale that may occur if the temperature inside the growth vessel is not controlled precisely. The homogeneous incorporation of dopants over long growth cycles may be difficult to manage since the reaction rates of the host crystal to be grown and the dopant under similar conditions are usually different. This effect will result in graded crystals. It should be stated here that the vicinity to thermal equilibrium does not allow the growth of thermodynamically unstable solid solutions. This in turn limits the fabrication of lattice-matched alloyed substrates, which is a general feature for growth techniques from solution. However, the main advantage of solvothermal growth technologies lies in the fabrication of crystals of superior crystallinity for use as substrates in a following step of device fabrication, i.e., the main purpose is to provide lattice-matched, low-strain substrates of large size at reasonable cost. Quite naturally, a high-throughput technique such as hydrothermal (ammonothermal) growth employing a solution holds great promise to satisfy the market demand.

We will describe in detail the hydrothermal and ammonothermal growth techniques in the following sections. Since the hydrothermal growth of large ZnO crystals up to 4 in. (100 mm) in diameter across the (0001) plane has already been demonstrated [5], the understanding of the incorporation of impurities is a very important issue now. Doping issues in ZnO can easily be investigated in ZnO microcrystals and thin films produced by low-temperature hydrothermal growth and liquid phase epitaxy (LPE) [6–9], respectively, currently used for that purpose, since the bulk conditions of ZnO are already present.

This chapter comprehensively reports on the growth of hydrothermal ZnO and ammonothermal GaN single crystals.

2.2 Hydrothermal Growth

Historically, the hydrothermal growth of zinc oxide (ZnO) has been derived from the successful hydrothermal growth of α-SiO$_2$. Both technologies have in common many features such as the use of the polar solvent water (H$_2$O, dielectric constant $\varepsilon = 80$ at $25°$C), operation under near-supercritical conditions, and the growth of species which are both oxides with a 4-fold coordination of nearest neighboring atoms Zn–O and Si–O. The crystal symmetry involves a polar axis along the [100] direction with consequences on the growth rate.

The solubility-enhancing mineralizers are of basic nature involving LiOH, NaOH, and KOH in case of ZnO, and K$_2$CO$_3$, Na$_2$CO$_3$, LiOH, KOH, etc. in the case of α-SiO$_2$. Consequently, the impurities, among others, are related to the mineralizer used.

Because of the similarities in the technology of hydrothermal α-SiO$_2$ and ZnO, it is certainly no surprise that established quartz manufacturers are currently among the leading producers of ZnO.

Research on the hydrothermal growth of ZnO is carried out mainly by using Morey and Tuttle type autoclaves [10–13], whereas for large-scale production, large-capacity autoclaves are used for quartz and ZnO (Figs. 2.1 and 2.2). These are basically modified Bridgman-type autoclaves [10].

2.2.1 Hydrothermal Growth Technology for ZnO

The relatively simple equipment for the hydrothermal growth of α-SiO$_2$ has to be modified for the hydrothermal growth of ZnO. The critical part is the inner surface of the autoclave, which is also the main source of metal impurities in the crystal. In case of α-SiO$_2$, the first growth cycle serves to form an inert and dense layer of NaFe silicates such as acamite (NaFeSi$_2$O$_6$), emeleusite (Li$_2$Na$_4$Fe$_2$Si$_{12}$O$_{30}$), etc. The second growth cycle then gives high-quality crystals of α-SiO$_2$.

In the case of ZnO, an inner liner made of a noble metal such as silver (Ag) or platinum (Pt), or a titanium (Ti) alloy is employed [2, 11, 14–16]. This also means that a new autoclave can be used for production from its first employment onwards. The inner liner is welded and the space between the welded inner liner and the autoclave wall is filled with water to generate a counter pressure. Alternatively, the inner liner may be tightly attached to the inner wall of the autoclave. The latter route is also used for the ammonothermal

Fig. 2.1. Large autoclave for the hydrothermal growth of α-SiO$_2$: (**a**) during the installation and (**b**) after a growth cycle

growth of GaN from acidic mineralizers owing to technological reason that the solvent ammonia requires low temperatures to keep it in the liquid state (see Sect. 2.3).

There are some critical parameters in the technology of the hydrothermal growth of ZnO that influence the crystal quality [11]:

(1) A high concentration of alkaline mineralizers is required to achieve proper solubility of ZnO [17] in the supercritical water (SCW) while keeping the dynamic viscosity of the solution sufficiently low.
(2) Proper temperature difference ΔT between the dissolution and growth zone, which has to be such that $\Delta T \leq 20\,\mathrm{K}$ and control of ΔT is within $\pm 3\,\mathrm{K}$, is essential to suppress the tendency toward spontaneous nucleation and flawed growth in ZnO. This process is certainly rate limiting.
(3) Wall nucleation, spontaneous nucleation, and flawed growth can be strongly reduced by a multistep warm-up process. First, the long autoclaves are slowly heated over a period of 24 h. Second, further heating is maintained keeping the conditions of $\Delta T \leq 25\,\mathrm{K}$ at 150°C and $\Delta T \leq 20\,\mathrm{K}$ above 250°C. This warm-up process certainly differs from manufacturer to manufacturer as different mineralizers are used.
(4) Use of a lithium ion (Li$^+$)-containing mineralizer prevents the formation of growth hillocks in ZnO, and flawless crystals can therefore be grown.

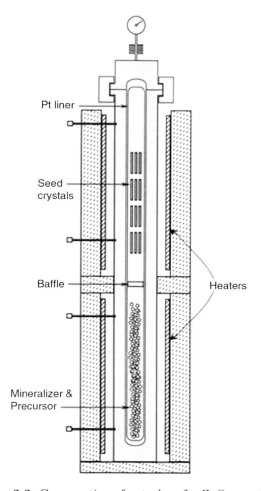

Fig. 2.2. Cross-section of autoclave for ZnO growth

(5) ZnO seeds often possess a damaged surface as result of mechanical machining. Good-quality ZnO is grown when a surface depth of ≥ 50–$70\,\mu$m is etched away from [0001] oriented ZnO seeds. The values for nonpolar ZnO seeds are of the same order.

(6) Particle size of the precursor is important: very small particle size results in low growth rates and flawed growth. The best results were obtained when the precursor size was larger than U.S. Standard Sieve Size No. 10. The use of hydrothermally grown lumps of about 6 mm in size did not yield an improvement in either the growth rate or the quality. These experimental findings on the optimum precursor size were recently confirmed by Chen et al. [18] using numerical simulation to analyze the ammonothermal

growth of GaN (see Sect. 2.3). The flow of solution in the precursor stock and the temperature distribution were found to be optimum for a precursor particle size of 3 mm. In this case, significant convective effects were seen in the precursor stock, and the flow was highly three dimensional. However, reducing the particle size to 0.6 mm yielded a very weak flow inside the precursor stock, and the temperature distribution was controlled by thermal conduction. Consequently, the dissolution of the precursor and the mixing of the precursor-enriched solvent with the solvent above the precursor stock were rather poor [18].

On the other hand, the increase of the particle size to centimeter dimensions makes the surface–volume ratio significantly smaller and the surface available for dissolution by the solvent drastically decreases. Concurrently, the temperature distribution is controlled by the solution flow.

Basic to the growth from autoclaves is a constant mass flow circulating from the feedstock to the seed crystal, which is established by proper heater arrangements around the autoclave. The inner part contains a baffle, which separates the feedstock (saturation zone; bottom of the autoclave) from the seed crystals (growth zone; upper fraction in the autoclave). The parameters for the baffle are the surface of the bore holes (openings typically around 5–15% [3, 16]) and their geometrical arrangement. If necessary, even several baffles can be used to get better control of the mass flow. Unfortunately, the details on the inner construction of autoclaves that produce large ZnO crystals cannot be disclosed at present, as they are the key to successful production of high-quality ZnO crystals and therefore form the intellectual property of the ZnO-producing companies. The same is true for the growth process as well.

The fill of an autoclave – the amount of liquid solution which is introduced into the autoclave – typically ranges from 70 to 90 vol%. A high fill generates a higher internal pressure but provides a greater amount of solvent to dissolve more mineralizer and consequently more ZnO species. We worked with 65–80 vol% in case of our small autoclaves with 16 mm inner diameter and 200 mm inner length to achieve appreciable growth rates.

In our work, we used a Pt inner crucible throughout [2]. The volume between the autoclave and the Pt inner container was filled with a suitable amount of distilled water for pressure balancing to prevent the Pt inner container from serious deformation. The Pt inner container is filled with the already prepared homogeneous solution of water, mineralizer, and ZnO feedstock. The seed holder and the baffle are inside the Pt inner container and are both made of Pt. There is, however, the disadvantage that any inert surface introduced into the autoclave gives rise to parasitic nucleation. One should therefore decrease the size of the seed holder as much as possible to reduce the surface available for parasitic nucleation while ensuring the mechanical stability. What is typically observed for small autoclaves is the negative effect of the seed holder and the crystal on the mass flow. Sometimes crystals are not well shaped, i.e., some facets are not fully developed owing to effective shading and localized turbulence. This problem may be reduced by reducing the number of seed crystals; consequently, the seed holder would become smaller as well.

The above-mentioned trend gets less severe with larger autoclaves as shown in Figs. 2.1 and 2.2. The growth of 2 in. (50 mm) and 3 in. (75 mm) size crystals requires increasing the inner diameter to 200 (length 3 m) and 300 mm, respectively. Figure 2.3 illustrates the result from one growth run. Almost 100 specimens about 2 in. (50 mm) in size have been grown on mainly (0001) and (10$\bar{1}$0) oriented seeds. Excellent facets have been formed, and the average crystal thickness was about 1 cm for crystals grown on (0001) oriented seeds.

A ZnO crystal 3 in. (75 mm) in diameter and a wafer processed from it are shown in Figs. 2.4a and b, respectively. The crystal is about 1 cm in thickness. Recently, several reports on 3 in. (75 mm) ZnO single crystals grown by the hydrothermal method have been published, and autoclaves of up to 500 L

Fig. 2.3. Two-inch (50 mm) ZnO crystals grown during one growth cycle. Reprinted with permission from [3]. Copyright (2006), Elsevier

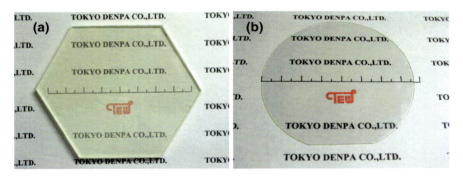

Fig. 2.4. A 3-in. (75 mm) (0001) ZnO crystal and wafer produced by Tokyo Denpa. Reprinted with permission from [3]. Copyright (2006), Elsevier

volume have been employed to grow 100–200 crystals per growth cycle [16,19]. Tokyo Denpa (TEW) seems to has already succeeded in the growth of 4 in. (100 mm) ZnO [5]. These developments are encouraging and large ZnO wafers are very likely to soon enter the market.

2.2.2 Growth Mechanism

The growth mechanism of ZnO on a ZnO seed crystal under hydrothermal conditions has been intensively studied by Laudise et al. [11], Khodakovsky and Elkin [20], and Demianets and Kostomarov [21], to name but a few, and was recently summarized by Ehrentraut [3]. The main zinc species in the solution are $ZnOOH^-$, $Zn(OH)_4^{2-}$, and ZnO_2^{2-}. Their concentrations depend on the OH^- concentration (typically given as the pH value) and temperature of the solution. Most of the OH^- concentration derives from the basic mineralizer, but the autoprotolysis of water also contributes according to:

$$2H_2O \longleftrightarrow H_3O^+ + OH^-. \tag{2.1}$$

The high concentration of $ZnOOH^-$, $Zn(OH)_4^{2-}$, and ZnO_2^{2-} in the solution builds up the driving force to foster the growth process, i.e., the supersaturation necessary to trigger nucleation on the ZnO seed crystal. The following reactions lead to the growth of ZnO:

$$Zn(OH)_4^{2-} \longrightarrow Zn(OH)_2 + 2OH^- \tag{2.2}$$

$$Zn(OH)_2 \longrightarrow 2H^+ + ZnO_2^{2-}. \tag{2.3}$$

Now, the constant of reaction K is considered, which is composed of the ratio of K_p and K_r, the constant of reaction of the products and the reactants, respectively:

$$K = K_p/K_r \tag{2.4}$$

$$K = [H^+]^2[ZnO_2^{2-}]/[Zn(OH)_2] \text{ and} \tag{2.5}$$

$$[ZnO_2^{2-}]/[Zn(OH)_2] = K_r[OH^-]^2/K_w^2. \tag{2.6}$$

where K_w is the ionic product of water. The increase of OH^- in the solution yields an increase of ZnO_2^{2-} [21]. The concentration of the other main species, $ZnOOH^-$ and $Zn(OH)_4^{2-}$, therefore reduces.

The growth on the (0001) and $(000\bar{1})$ polar faces, i.e., Zn- and O-terminated faces of the Zn^+ (surface) and O^{2+} (surface), respectively, possibly involves:

$$\text{on } (0001) \text{ face}: Zn^+(\text{surface}) + ZnO_2^{2-} \longrightarrow 2ZnO(\text{crystal}), \tag{2.7}$$

$$\text{on } (000\bar{1}) \text{ face}: O^{2-}(\text{surface}) + ZnO_2^{2-} + 2H_2O \longrightarrow 4OH^- + ZnO(\text{crystal}). \tag{2.8}$$

Zn^+(surface) and O^{2+}(surface) are the species that are provided by the crystal surface, i.e., the seed crystal itself. The source for ZnO_2^{2-} and H_2O is the hydrothermal solution itself.

Reactions (1)–(8) can now be summarized by saying that during a ZnO growth experiment under hydrothermal conditions one might have the following transport of the Zn-containing species from the precursor to the crystal:

$$ZnO(\text{precursor}) + 4OH^- \longrightarrow Zn(OH)_4^{2-} + O^{2-} \longrightarrow Zn(OH)_2 + 2OH^-$$
$$\longrightarrow ZnO_2^{-2} + 2H^+ \longrightarrow ZnO(\text{crystal}) + 2e^-. \tag{2.9}$$

The growth rate is about 2 times faster on the (0001) face than on the (000$\bar{1}$) face. This fact has been explained by the ratio of the formed ZnO units, ZnO (crystal), as obtained from (7) and (8) [21]. This is close to the reported 2–3 factor [2, 22–25]. Similarly, the growth speed is about 2 times faster on the (0001) face than on the (10$\bar{1}$1) face [23].

The presence of Li^+ in the growth solution significantly slows down the growth rate in the $\langle 0001 \rangle$ direction and slightly increases it for the $\langle 1100 \rangle$ direction [11, 21]. This might be related to a decreased positive surface charge that lowers the probability of incorporating Zn-containing negatively charged species. A high growth rate of 2.053 and 1.097 mm per day for the (0001) and (000$\bar{1}$) face, respectively, was reported by Demianets and Kostomarov [21]. In their case, KOH was solely used as the mineralizer, which is, however, known to reduce the crystal quality [26]. However, already Laudise et al. had reported that Li^+ was needed to reduce the number of crystal defects [11]. This was also confirmed by our work. We found that the ratio of 3M KOH to 1M LiOH yields the best crystalline and highly transparent crystals [3, 6, 27].

To finish the discussion on the effect of mineralizers, it can be summarized by saying that they are indispensable in establishing an effective solubility of ZnO. On the other hand, they are unwanted for the process yield in ZnO crystal growth owing to the probability of their incorporation into the crystal. This is intrinsic to the hydrothermal growth of ZnO and can therefore only be compromised. It is important to state that the hydrothermal growth of ZnO crystal is targeted to supply high-quality ZnO wafers, i.e., highly single crystalline with a very low concentration of structural defects.

The growth rates for high-quality ZnO crystals are in the range of 0.2–0.3 mm per day. Dem'yanets and Lyutin [16] have reported improved ZnO crystals, i.e., low dislocation density growth according to a layer-by-layer growth mechanism rather than a dislocation-induced one.

The growth rate is rather slow compared to that in the pressurized melt growth of ZnO, which is upto 10 mm h^{-1} [28, 29] or by that of the seeded chemical vapor transport (SCVT) growth, which is <70 μm h^{-1} [20]. In addition to the high crystallinity, the high throughput in large autoclaves clearly indicates the favorable commercial potential of hydrothermal technology. It can therefore be assumed that the use of large autoclaves for ZnO, such as those used for the growth of quartz (Fig. 2.1), would produce sufficient ZnO wafers to cover a variety of applications.

2.2.3 Liquid Phase Epitaxy and Growth of Microcrystals

The growth of ZnO bulk crystals by the hydrothermal growth technique is very time consuming. In order to investigate the effect of dopants on the properties of single crystalline ZnO, LPE for homoepitaxial ZnO films and low-temperature hydrothermal technique for ZnO microcrystals have been employed as fast screening tools.

LPE of ZnO has been developed very recently [7,8]. A low-melting solvent such as LiCl, $T_m = 605°C$, or the eutectic of NaCl–KCl–CsCl, $T_m = 480°C$, has been found suitable to grow ZnO films of over $10\,\mu m$ thickness at growth rates around 0.1–$0.5\,\mu m\,h^{-1}$. The growth rate very much depends on the doping ion [3, 30].

The LPE growth process comprises a ZnO substrate which is attached to the lift and rotation unit via a substrate holder (Fig. 2.5). The substrate is immersed into the liquid and rotated in order to achieve a steady flow supplying saturated solution to the growing crystal face. The growth will be terminated by pulling the substrate out of the solution. After the substrate has cooled down to room temperature (RT), cleaning can easily be effected.

The major advantages of the LPE technique are the following:

(1) The LPE equipment is very simple, as shown in Fig. 2.5. Air at atmospheric pressure is employed.

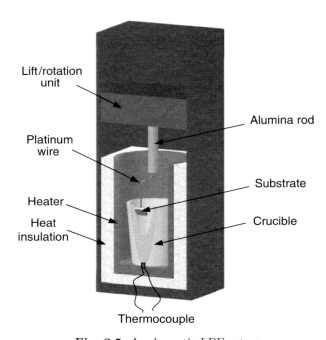

Fig. 2.5. A schematic LPE setup

(2) The entire process provides good control over the solubility of ZnO in the solvent in that the excess ZnO would be deposited at the bottom of the crucible as microcrystals. This also ensures that the ZnO substrate is not etched prior to the growth of the film; thus a very flat interface between the substrate and film can be formed.
(3) The water-soluble solvent ensures that the grown film can be easily cleaned by using distilled water.

The quality of homoepitaxial ZnO films has been discussed in a number of publications [3,6,7,30–33]. X-ray rocking curve (XRC) measurements have revealed low values for the full width at half-maximum (FWHM) of around 25–30 arcsec, and XRC line mappings of In- and Ge-doped ZnO films grown on 1×1 cm^2 size substrates have revealed high uniformity over the entire substrate [6,7]. Figure 2.6 shows the SEM image taken from an intentionally undoped homoepitaxial ZnO film. Parallel steps spaced around 1–5 µm are clearly seen. This result has been confirmed by atomic force microscopy (AFM), which reveals monatomic steps [7].

Doping and the formation of solid solutions with Mg, Cd, Ga, In, Ge, P, Sb, Bi, Cu, etc. have been investigated in a number of publications [6,7,30,33,34], and Fig. 2.7 reflects just a few of them. As can be seen, the nature of the dopand has a tremendous effect on the morphology of the film. Particular facets are preferred, which give rise to the formation of the various morphologies. It has been found that doping of ZnO with In strongly reduces the growth in the c direction, but the m and a directions still show reasonable growth rates.

Fig. 2.6. Intentinally undoped ZnO film grown from $c_{\text{ZnO}} = 13$ mmol displaying large steps. Reprinted with permission from [3]. Copyright (2006), Elsevier

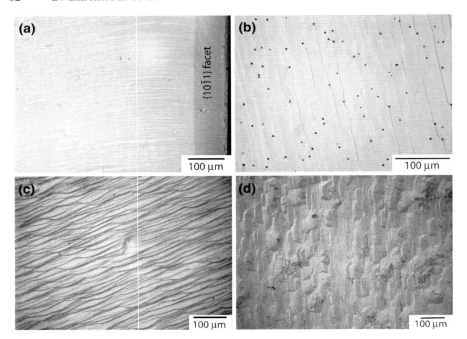

Fig. 2.7. NDIM micrographs from LPE homoepitaxial films: (**a**) MgZnO, (**b**) Cd-doped ZnO, (**c**) Ga-doped ZnO, and (**d**) Ge-doped ZnO. Reprinted with permission from [3]. Copyright (2006), Elsevier

The growth of ZnO microcrystals is typically performed under subcritical to near-supercritical hydrothermal conditions in autoclaves at temperatures of 90–300°C [6, 35, 36]. Upon self-nucleation, well-shaped crystals in the range of 100 nm to several micrometers along the c axis are typically formed (Fig. 2.8). Here we note that the growth conditions are strikingly different from those for the LPE films, i.e., low alkali-metal content, and otherwise relatively close to the hydrothermal growth of ZnO bulk crystals.

We have also used the above-described method for LPE to yield ZnO microcrystals. Those specimens always contain considerable amounts of alkali-metal ions in the range 10^{17}–10^{20} cm^{-3} depending on the nature of the growth solution. Figure 2.8 shows ZnO microcrystals grown from the LiCl flux and under mild hydrothermal conditions.

2.2.4 Properties of Hydrothermal ZnO

Structural Properties

We will now discuss the properties of the hydrothermally grown ZnO crystals, and also compare them with LPE ZnO films and microcrystals. The quality of the (0001) substrates was investigated by XRC measurements

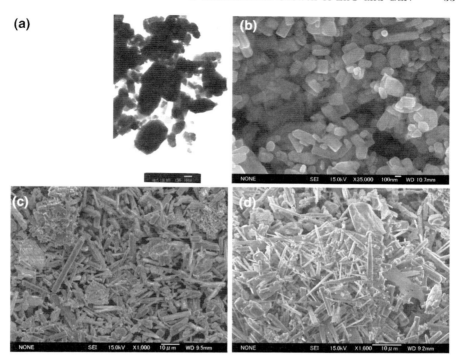

Fig. 2.8. ZnO microcrystals: (**a**) TEM of an undoped ZnO and SEM images from (**b**) Mo-doped ZnO grown under mild hydrothermal conditions, (**c**) Li, In codoped ZnO and (**d**) Li, Sb codoped ZnO prepared from a LiCl flux

(Rigaku RINT-2000 diffractometer, Cu Kα, four-crystal Ge (220) channel monochromator, beam divergence 12 arcsec, scan speed 0.01 min^{-1}, step width 10^{-4} degrees) using the 002 reflection. The FWHM ranges between 19 and 30 arcsec after chemical–mechanical polishing (CMP), which is better than the values reported for ZnO wafers machined from pressure-melt-grown ZnO (49 arcsec [28, 29]) and comparable to ZnO fabricated by SCVT (FWHM around 30 arcsec [37]). The FWHMs of the 002 reflection reported by other groups are 43 and 37 arcsec for polished (0001) and (000$\bar{1}$) surfaces of hydrothermal ZnO, respectively [25], and 57 arcsec for slightly N-doped hydrothermal ZnO [38]. By X-ray reciprocal space mapping using the 0002 symmetric reflection, a highly symmetric single peak with FWHM from the ω scan of 15 arcsec was measured [7]. This was confirmed by other groups [39], who reported about 12–15 arcsec by triple-axis $\omega - 2\theta$ scans.

X-ray topography (Cu Kα, 40 kV, 10 mA, detected by a film IX80; Berg–Barrett geometry) was studied on an epi-ready (0001) ZnO wafer 2 in. (50 mm) in diameter and 500 μm in thickness. The XRC FWHM from the 0002 reflection of the sample was 19 arcsec, as reported above. About 2,500 scans were assembled to yield the contrast-enhanced image of Fig. 2.9. The 114 reflection

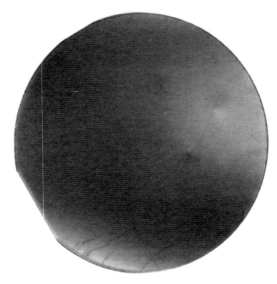

Fig. 2.9. Transmission X-ray topograph of a 2 in. (50 mm), epi-ready (0001) ZnO wafer produced by Tokyo Denpa. Reprinted with permission from [3]. Copyright (2006), Elsevier

Fig. 2.10. X-ray topograph of an m-plane ZnO wafer from GoodWill, Russia. The $12\bar{3}0$ reflection has been employed

at $2\theta = 98.6°$ and $\omega = 49.3°$ was employed. The wafer appears very homogeneous over the entire area. Slight contrast effects are seen, which are presumably be due to slightly different lattice parameters caused by fluctuations in the impurity concentration or stoichiometry. It has been reported that dislocations are the principal imperfections in hydrothermal ZnO crystals [24]. They lie in the basal plane and would run parallel to or 20° inclined to the growth direction $\langle 10\bar{1}0 \rangle$. The point defects at the interface of the seed-grown crystal are supposed to initiate the dislocations.

Figure 2.10 shows the X-ray topography (W Lα, 50 kV, 30 mA, detected by a film IX50; Berg–Barrett geometry) was measured on a CMP-polished m-plane ZnO wafer of around 500 µm thickness. The slight contrast effect

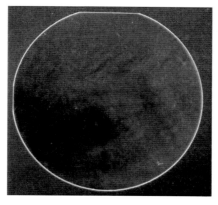

Fig. 2.11. Observation under crossed polarizer reveals little random strain in a 2 in. (50 mm) (0001) ZnO wafer. Reprinted with permission from [3]. Copyright (2006), Elsevier

shows that there are only a few defects propagating in the crystal. This was later supported by photoluminescence (PL) measurements (Fig. 2.15).

Figure 2.11 shows the observation of the epi-ready (0001) 2 in. (50 mm) ZnO wafer under crossed polarizers. Only a small random strain was observed, which supports the result from the X-ray topography that the homogeneity of the wafer is very high.

Measurements employing contact-mode AFM under ambient air conditions have revealed a root mean square (RMS) roughness of 0.285 and 0.155 nm for the (0001) and (000$\bar{1}$) face, respectively, for ZnO wafers that had been treated by CMP. The roughness was further reduced to about 0.12 nm if proper thermal annealing conditions were applied [7].

The etch pit density (EPD) was determined for the (0001) and (000$\bar{1}$) polar faces of several wafers. A concentrated aqueous solution of H_3PO_4 was applied for 5 min at 25°C. The etching behaviors are strikingly different as displayed by the shape of the etch figures. While those on the (0001) face clearly exhibit a 6-fold axis with pyramidal facets, those on the (000$\bar{1}$) face are less facetted. This behavior can be attributed to the polar character of the $\langle 0001 \rangle$ axis. The EPD was about 300 cm^{-2} after CMP and is further lowered to less than 80 cm^{-2} by annealing [27]. This demonstrates the currently best hydrothermal ZnO wafers commercially available.

The etch rate of the annealed wafers was relatively low, which is related to the improved crystallinity and reduced dislocation density. An aqueous solution of 0.7% HCl was applied for 5 min at 60°C.

Etching with 1% HCl yielded an etch rate of 10 μm min^{-1} according to another report [40], and a mixture of 1 mL conc. H_3PO_4, 1 mL conc. acetic acid, and 10 mL H_2O etched at 1.5 μm min^{-1}. The use of 30% HNO_3 aqueous solution produces hexagonal pyramids on the (0001) face of mechanically polished ZnO [41].

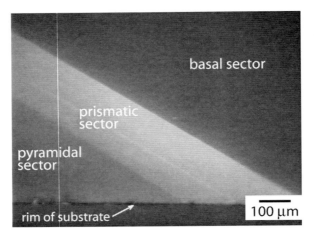

Fig. 2.12. The basal, prismatic, and pyramidal growth sectors as revealed by fluorescence ($\lambda_{\text{exc}} = 365$ nm) microscopy. Reprinted with permission from [3]. Copyright (2006), Elsevier

Earlier reports on hydrothermal ZnO [42] disclosed an EPD of about 10^2 and 10^3 cm^{-2} on (0001) and (000$\bar{1}$) faces, respectively, using an aqueous solution of H$_3$PO$_4$. Mainly dislocations were said to be responsible for the higher EPD on the (000$\bar{1}$) face.

The occurrence of growth sectors in the hydrothermal ZnO crystal is caused by the incorporation of impurities [11, 22, 24, 25, 43]. Nonradiative recombination centers are therefore found. Growth sectors are typically formed as a result of the differences in growth rate and mechanism for faces with different crystallographic orientations. In Fig. 2.12, the basal, prismatic, and pyramidal growth sectors in a (0001) wafer are shown as revealed by fluorescence microscopy ($\lambda_{\text{exc}} = 365$ nm). The intensities decrease from the prismatic to the pyramidal to the basal sector, which also was observed in the broad emission band at <2.8 eV in spectra derived by cathodoluminescence (CL) by Mass et al. [43]. Samples cut from the crystal volume directly above the (0001) seed did not show growth sectors other than that found in the low-impurity basal sector. XRC measurements on 2 in. (50 mm) wafers using the 0002 reflection strongly support these findings. The FWHM is around 18–22 and ≤50 arcsec for the part from only the basal sector and for both prismatic and pyramidal sectors, respectively.

Impurities

Secondary ion mass spectrometry (SIMS) has been employed to study the impurity distribution in the depth of a wafer that had been subjected to CMP and annealed. In our measurements, the primary beam species was Cs$^+$ (5 kV, 350 nA) and the sputtering speed around 120–150 nm min^{-1}. We found that the impurity levels remained constant with increasing scan depth, and the following concentrations were revealed: Li, $2 \times 10^{16}\,\text{cm}^{-3}$; Na, $8 \times 10^{15}\,\text{cm}^{-3}$; K, $3 \times 10^{15}\,\text{cm}^{-3}$; Mg, $10^{16}\,\text{cm}^{-3}$; Al, $4 \times 10^{15}\,\text{cm}^{-3}$; Si, $7 \times 10^{17}\,\text{cm}^{-3}$; Fe, $8 \times 10^{15}\,\text{cm}^{-3}$; and Cd, $10^{17}\,\text{cm}^{-3}$.

By inductively coupled plasma mass spectrometry (ICP-MS), impurity concentrations of Fe, Al, Li, and K in a 2 in. (50 mm) ZnO wafer from TEW were investigated in refs. [2, 27]. Specimens grown and cut from the $(000\bar{1})$ and (0001) faces of the seed crystal showed different impurity levels. Also, at increasing distance from the $(000\bar{1})$ face of the seed, less impurity in the grown crystal was seen, particularly for the case of Li, whereas the concentration of K remained nearly unchanged for both faces. Fe and Al both show higher concentrations in wafers grown on the $(000\bar{1})$ face of the seed crystal <11 and <1 ppm for Fe and <8 and <0.5 ppm for Al for the $(000\bar{1})$ and (0001) face, respectively. Mass et al. [43] reported about 1–10 ppm of Li and K and not above 1–2 ppm for Na. The other impurities were Al, Fe, Si, and C from 1 to 10 ppm in the crystal.

We did compare the results obtained from our hydrothermally grown ZnO crystals to those of a ZnO crystal grown by SCVT using H_2 and N_2 as carrier gases [3]. The SCVT-grown ZnO incorporates 0.03–0.1 ppm K and 0.2–1 ppm Na, whereas Fe and Al were found at concentrations below the detection limit of 1 and 7 ppm, respectively, as recorded by atomic absorption spectroscopy (AAS). The use of Cl_2 and C as transport agents in SCVT incorporates about 0.053% Cl and 0.05% C when using graphitized ampoules [37]. This compares with pressure-melt-grown ZnO, which contains 4 ppm Fe, 8 ppm Pb, and 2.5 ppm Cd [28, 29].

Positron annihilation spectroscopy on hydrothermal ZnO from Tokyo Denpa has revealed about $10^{16}\,\text{cm}^{-3}$ zinc vacancies (V_{Zn}) as neutral defect complexes and about $10^{17}\,\text{cm}^{-3}$ oxygen vacancies (V_O) as neutral oxygen [44]. Clear effects of V_{Zn} and V_O and Li-related defects can be seen in the low-temperature PL spectrum obtained from excitation with a 325 nm laser ($P_{out} = 1.6\,\text{mW}$). Figure 2.13 shows the PL from a bulk sample with the surface prepared by CMP. A broad band peaking at around 2.3 eV appears in the visible region, and the bands at 2.53, 2.35, and 2.17 eV have been assigned to V_{Zn} and V_{Zn} and Li-related defects, respectively [45].

Defects related to hydrogen in the hydrothermal, as-grown ZnO with a carrier concentration about $10^{17}\,\text{cm}^{-3}$ at room temperature have been studied by infrared absorption spectroscopy [46]. A number of bands are located in the vicinity of the characteristic O–H stretching local vibrational modes, and the band at $3{,}577.3\,\text{cm}^{-1}$ was related to a defect containing one O–H bond primarily aligned with the c axis of the crystal.

The same absorption band at $3{,}577.3\,\text{cm}^{-1}$ at 12 K, and shifting to $3{,}547\,\text{cm}^{-1}$ at 300 K, was characterized by electron paramagnetic resonance (EPR) in a hydrothermal ZnO sample [47]. This band, however, was assigned to an OH^- ion located on an oxygen site adjacent to a Li^+ ion on the zinc site. The concentration of the neutral complex Li–OH^- was estimated to be

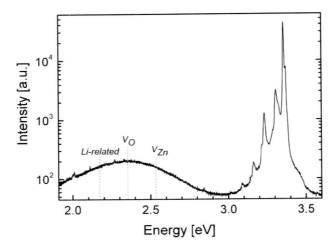

Fig. 2.13. Photoluminescence from a bulk sample with the surface prepared by chemical–mechanical polishing

$7.6 \times 10^{17}\,\mathrm{cm^{-3}}$. Charge compensation by singly ionized acceptors was concluded to play a major role for the hydrogen in ZnO. A band at $3.546\,\mathrm{cm^{-1}}$ at 300 K [17] has already been suspected to be caused by hydrogen located in a bond-centered position between oxygen and zinc. This band might contain some unresolved components from OH^- involving an acceptor on the zinc site. Furthermore, detection of Co^{2+} at $6{,}005\,\mathrm{cm^{-1}}$ and Ni^{2+} at $4{,}240\,\mathrm{cm^{-1}}$ have been reported. Evidence of Fe^{3+} and Mn^{2+} was also found in the same hydrothermal ZnO sample [47]. Photoinduced EPR measurements revealed a signal due to neutral Li acceptors. The concentration of photoinduced neutral Li acceptors was approximately $1.3 \times 10^{15}\,\mathrm{cm^{-3}}$.

Figure 2.14 demonstrates that the concentration of Al can be reduced by a factor of 2–3 just by proper annealing [3]. Similar finding relates to Li, the concentration of which can be reduced by about one order of magnitude. However, in both cases we see an increase of the concentration in the region near the surface.

Doping

The in situ doping of hydrothermal ZnO has turned out to be quite tricky and uncontrollable. Attempts have been made with bi- and trivalent metal ions and also N [3, 26, 38, 48]. A maximum of 5.5 mol% Mg can be introduced in Zn sites. Generally, impurities affect the modification of the crystal habit through shifting growth rates for stable facets. Indium doped in concentrations of $10^{20}\,\mathrm{cm^{-3}}$ almost completely suppresses the growth in the $\langle 0001 \rangle$ directions but greatly improves the growth toward $\langle 10\bar{1}0 \rangle$. Nitrogen was incorporated up to $10^{18}\,\mathrm{cm^{-3}}$, but only a weak donor-acceptor pair (DAP) signal was observed.

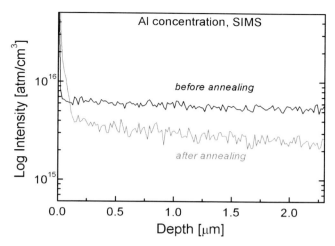

Fig. 2.14. SIMS depth scans in the near-surface region demonstrating the effect of annealing on the Al concentration. Reprinted with permission from [3]. Copyright (2006), Elsevier

Recently, Demianets et al. [26] reported on the modification of the crystal habit upon incorporation of some bi- and trivalent metal ions such as Ni^{2+}, Cd^{2+}, Mn^{2+}, Co^{2+}, Fe^{2+}, Fe^{3+}, and In^{3+}. The ions Ni^{2+}, Cd^{2+}, Mn^{2+}, Co^{2+}, and In^{3+} substitute for the Zn sites in the crystal lattice, and Ni^{2+}, Mn^{2+}, Fe^{2+}, and Fe^{3+} were said to cause some crystal coloration. In the case of In^{3+}, the formation of a point defect of the type In_{Zn}^{+} was found to be a shallow donor. This effect was recently used in combination with Li doping to achieve superfast luminescence decay by donor–acceptor pair recombination [30]. Owing to the relatively low growth temperature under hydrothermal conditions, formation of $In_{Zn}^{+} Li'_{Zn}$-type associates does not happen.

Optical Properties

Optical transmittance of some hydrothermal (0001) ZnO wafers at room temperature from different manufacturers (polished sample of 0.5 mm thickness, no visible scratches on surface) has been measured. It was found that they do not really differ from each other; however, a slightly higher absorption in the range 390 nm $\leq \lambda \leq$ 500 nm was found for the sample from a Russian manufacturer [3]. This sample appeared a little more yellowish than the specimen from Tokyo Denpa. The latter sample shows a transmittance of 80% at $\lambda = 410$ nm and 87% at $\lambda = 700$ nm.

Indium doping lowers the transmittance to about 60% as revealed on a thin, 1% In-doped hydrothermal ZnO crystal (platelet of 3 mm diameter and 0.2 mm thickness with surfaces as grown) [3]. The as-grown (0001) surfaces were used, and loss due to surface scattering might marginally distort the result.

The refractive indices for the ordinary (n_o) and the extraordinary (n_e) beam were measured for the mentioned different samples of hydrothermal ZnO. The laser beam was coupled into the sample by a rutile prism. In contrast to the optical transmittance, identical values were obtained for each sample, which can be fitted by a second-order exponential decay. The value of n_e decreases from 2.23 to 1.95 with λ increasing from 400 to 1,550 nm.

The PL in Fig. 2.13 was obtained by excitation using a continuous-wave He–Cd laser ($\lambda_{exc} = 325$ nm, $P_{out} = 1.6$ mW) at 4 K. The signal was detected by a charge-coupled device (CCD) camera (Princeton Instruments Inc.) after dispersion with a 30-cm triple grating monochromator. The broad emission band from 1.7 to 2.8 eV peaks around 2.3 eV. The nature of this broad emission involves donor–acceptor pair recombination due to Li [49,50]. As already noted previously, V_{Zn} as neutral defect complexes and V_O as neutral oxygen are typically observed in hydrothermal ZnO [44].

The PL emission in the region near the band edge from the Zn- and O-polar faces is shown in Fig. 2.15(a). We clearly find the pronounced peak at around

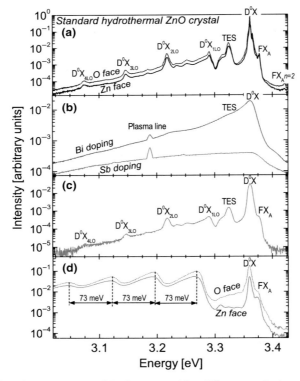

Fig. 2.15. Photoluminescence of ZnO prepared by different techniques: (**a**) standard hydrothermal (0001) ZnO, (**b**) Bi- and Sb-doped ZnO microcrystals prepared under mild hydrothermal conditions, (**c**) Li, Sb codoped microcrystals prepared from the LiCl flux, and (**d**) a Li, Sb codoped homoepitaxial ZnO film grown by LPE

3.361 eV and its four phonon replicas of each 73 meV, which are due to excitons bound to neutral donor (D^0X), their two-electron satellite transition (TES) at 3.33 eV, and the emission from free A-exciton recombination (FE_A with $n=1$ and 2) at 3.375 and 3.39 eV, and B-exciton recombination (FE_B) around 3.42 eV. This points to a reasonably good optical quality. The strong emission at 3.33 eV almost disappears in the LPE-grown films, which was also attributed to the defects [33].

The result from PL from hydrothermal ZnO over the entire energy range of 2–3.5 eV is quite consistent with samples grown by SCVT [51]. Interestingly, the broad, yellow band peaking around 2.3 eV also appears in the SCVT sample although the growth conditions are very different. This might lead to the assumption that the levels of active impurities in our hydrothermal ZnO are comparable to those of SCVT ZnO. Hence, this suggests that the hydrothermal growth technique is an economical way to yield a large quantity of high-quality, large ZnO crystals.

In Fig. 2.15 is also shown the PL emission from low-temperature hydrothermally grown Sb- and Bi-doped ZnO microcrystals, Fig. 2.15(b); Sb, Li codoped ZnO microcrystals, Fig. 2.15(c); and Sb, Li codoped ZnO homoepitaxial film, Fig. 2.15(d). The doping with Sb and Bi under mild hydrothermal conditions does not yield high-quality material [6]. The D^0X peak is weakly developed but somewhat higher in intensity for the Bi-doped sample in comparison to the Sb-doped sample in Fig. 2.15(b).

Now, if Li is added as dopant and the process temperature is raised to 640°C, we see, in Fig. 2.15(c), a much better structure of the emission pattern quite similar to that of the intentionally undoped hydrothermal ZnO bulk crystal in Fig. 2.15(a). Again, this changes by using a lattice-matched substrate as shown in Fig. 2.15(d). We find similar patterns with FX_A and D^0X, but see a new strong emission at around 3.28 eV and replicas each -73 meV. Its nature, however, is not yet clear. The quality seems better for the film grown on the Zn-polar face of the substrate; the emission intensity, however, is slightly higher for the film grown on the O-polar face. Further investigation is currently focused on.

In Fig. 2.16 is presented the development of the PL emission from a hydrothermal a plane, i.e., nonpolar ZnO wafer, over increasing temperature from 10 to 200 K [52]. The D^0X shifts to lower energies and the emission from the free exciton strongly reduces in intensity. This sample generally shows high optical quality.

Time-Resolved Photoluminescence

With excitation by a femtosecond laser pulse (Ti-sapphire laser, 150 fs pulse at $\lambda_{exc} = 260$ nm; Clark-MXR CPA2001), the luminescence data were recorded using a polychromator and a two-dimensional CCD sensor at the output of a streak camera (Hamamatsu C2830) to allow for simultaneous evaluation of

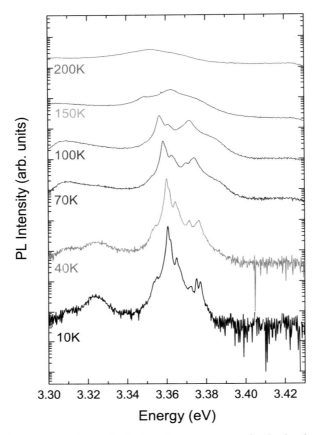

Fig. 2.16. Temperature-dependent photoluminescence of a hydrothermal a-plane wafer

time-resolved spectra and luminescence decays [30]. Room-temperature time-resolved spectroscopy (RTTRS) from homoepitaxial ZnO films doped with In was used to measure and compare the spectra with that from an undoped ZnO film and from a purely In-doped hydrothermal ZnO sample. The time window selected was 0–100 ps. A similar spectral pattern is observed for all samples at short wavelengths below 380 nm. Considerable emission intensity in the In-doped sample is observed until 450 nm. At a time window of 300–700 ps, the longer wavelength emission round 400–420 nm is clearly prevalent.

The emission around 400–420 nm can be assigned to the DAP recombination. The LPE films contain a considerable amount of Li, up to $10^{19}\,\mathrm{cm}^{-3}$, derived from the growth solution. The ratio of In to Li is about 10^2 smaller for the LPE-grown films in comparison to the hydrothermal In-doped bulk sample. Considering the value of the bandgap, the binding energy of In^{3+} donor ions of about 50 meV, the value of binding energy of the Li acceptor between

300 and 500 meV, and further effect of Coulombic interaction depending on the donor–acceptor distance, the position of the expected emission band is just round 400–420 nm, in agreement with the observed emission spectrum of the In-doped sample at longer times.

Another support for this assignment can be obtained from the emission characteristics of the In-doped bulk sample, from which practically no emission is observed above 400 nm at any time window (not shown here). The In concentration of both LPE and the bulk sample was very similar, i.e., 1.5×10^{20} and 6×10^{19} cm^{-3}, respectively, but the latter sample contained 2 orders less Li.

In Fig. 2.17 is shown the RTPL decay of 0.1 mol% In-doped ZnO film upon femtosecond laser excitation [30]. The wavelength region 380–390 nm was extracted. The decay follows a two-exponential course, which is found, with varying relative intensities of two involved decay components, at all the LPE films studied. The faster component shows typically 30–60 ps decay time, while that of the slower component ranged from 250 to 800 ps. The intensity of the latter component increased toward longer emission wavelengths.

The two-component decay below 400 nm is interpreted as due to the bound and free excitons in thermal equilibrium. The slower decay component above 400 nm in the In-doped LPE sample is apparently related to the decay kinetics of the DAP recombination and point to the very fast character of this process as well. Moreover, in case of such pair, a radiative recombination process rather a nonexponential decay is typically expected.

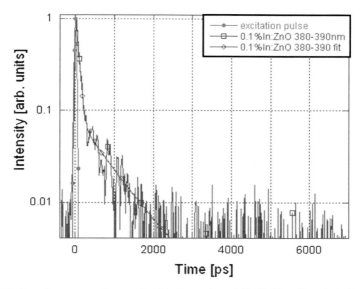

Fig. 2.17. Luminescence decay of a Li, In codoped ZnO film. Reprinted with permission from [28]. Copyright (2006), The Royal Society of Chemistry

The decay process in the In, Li-containing LPE films was ascribed to a tunnel-assisted donor–acceptor recombination process using the model developed for tunneling mechanism in phosphorescence transitions in Zn_2SiO_4 : Mn [53].

Such a fast subnanosecond emission process based on donor–acceptor recombination could be well exploited for scintillator application, as it is sufficiently shifted from the ZnO absorption edge, therefore minimizing the reabsorption losses. The timing characteristics obtained with our LPE-grown films are clearly superior to those obtained in $(n\text{-}C_6H_{13}NH_3)2PbI_4$, a natural thin-film multiple quantum well structure, announced recently as the material of choice to obtain a superfast scintillation time response [54].

Electrical Properties

There is some concern of the impact of surface properties on the reliability of results from measurements of electrical properties using surface contacts. It is known that the ZnO surface does adsorb molecules from the ambient atmosphere, such as CO_2, CO, O_2, H_2, etc. [55]. The formation of a highly conductive layer on the surface of highly resistive ZnO crystals was reported by Markevich et al. [56] and Schmidt et al. [57, 58]. This was explained as due to the adsorbed oxygen atoms on the surface of a ZnO crystal, which leads to the capture of electrons. This results in a negative surface charge and a depletion layer with reduced conductivity. The thickness of this layer on undoped ZnO was $<1\,\mu m$ [56].

The temperature-dependent Hall-effect technique with the van der Pauw geometry was used to examine the carrier concentration (N, Fig. 2.18a), carrier mobility (μ_H, Fig. 2.18b), and electrical resistivity (R, Fig. 2.18c) from a $10 \times 10\,mm^2$ specimen cut from a high-quality Tokyo Denpa ZnO crystal. The surface was polished to an RMS roughness of about 0.2 nm [3]. Ti/Au contacts were made by thermal evaporation. A melt-grown sample was measured for comparison.

The carrier concentration in Fig. 2.18(a) is very much lower in the hydrothermal sample than in the pressure-melt-grown one and decreased with increasing $1/T$ from $N = 2 \times 10^{16}\,cm^{-3}$ at 500 K ($10^3/T = 20$) to $N = 4 \times 10^{13}\,cm^{-3}$ at 100 K. Polyakov et al. [59] obtained similar results of $N = 1.3\text{--}4.6 \times 10^{13}\,cm^{-3}$ and $N = 6.4 \times 10^{11}\,cm^{-3}$ at 300 and 77 K, respectively, for measurements on four samples of Tokyo Denpa's ZnO as purchased. In both cases, the N value is clearly the effect of low impurity, point defects, and dislocation concentrations, in very good agreement with our results from XRD and impurity analysis. The above species are known to produce electrically active centers [60]. A slight hysteresis slope at $10^3/T < 4$ was obtained from the measurement during heating up and cooling down, which is likely the effect related to surface conductivity [56–58]. For a detailed discussion on surface conductivity, see refs. [56–58]. It has been noted in [61] that annealing at 1,150°C would convert a ZnO substrate from highly resistive to n-conductive.

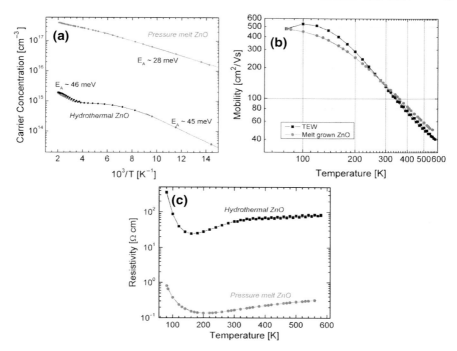

Fig. 2.18. Electrical properties as function of temperature for a hydrothermal (Tokyo Denpa, TEW) and melt-grown ZnO wafer. Reprinted with permission from [3]. Copyright (2006), Elsevier

Shallow donors are expected to dominate the conduction by either a reduction in the concentration of the compensating acceptors or by an increase in the concentration of shallow donors.

For comparison, SCVT-grown ZnO shows a slightly lower carrier concentration at $10^3/T = 20$, $N = 3 \times 10^{14}$ cm^{-3} [60].

The Hall mobility peaks at 100 K, $\mu_H = 530$ cm^2 V^{-1} s^{-1}, and drops down to about 40 cm^2 V^{-1} s^{-1} at 580 K. The higher mobility in comparison to the melt-grown sample, which peaks at 480 cm^2 V^{-1} s^{-1} at 80 K and 430 cm^2 V^{-1} s^{-1} at 100 K, is due to the lower impurity concentration as indicated by the lower carrier concentration (Fig. 2.18a). A previous hydrothermally grown sample [60] had already shown very similar results of μ_H for the measured temperature range of 200–400 K. μ_H decreased from about 300 to 100 cm^2 V^{-1} s^{-1}. The SCVT-grown ZnO shows a higher μ_H upto almost 2,000 cm^2 V^{-1} s^{-1} at 40 K. Compared to GaN, one would find a lower mobility because of the higher effective mass and the larger optical phonon scattering parameter [60].

The electrical resistivity of the hydrothermal ZnO sample in Fig. 2.18c is about 2 orders of magnitude higher than the sample grown from the melt, with a minimum of 20 Ω cm at 60 K and 0.1 Ω cm at 200 K, respectively.

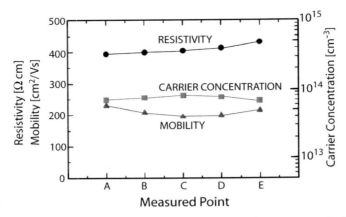

Fig. 2.19. Resistivity, carrier concentration, and mobility across a hydrothermal 2 in. (50 mm) epi-ready ZnO wafer. Reprinted with permission from [3]. Copyright (2006), Elsevier

Hydrothermal ZnO from Russian SPC GoodWill [62] shows an electrical resistivity of 500–1,000 Ω cm, which is higher than that of the TEW material. Other results on TEW ZnO [59] reported a large variation of R between 96 and 5×10^5 Ω cm, which was speculated to come from the Li content in the samples investigated. It is possible that different growth sectors were present in the specimens and therefore Li was incorporated in very different concentrations there.

The uniformity of R, N, and μ_H over a 2 in. (50 mm) wafer from TEW was measured and is shown in Fig. 2.19, which gives the values for $R = 380$ Ω cm $\pm 15\%$, $N = 8 \times 10^{13}$ cm^{-3} $\pm 20\%$, and $\mu_H = 200$ cm^2 V^{-1} s^{-1} $\pm 10\%$, respectively [27].

2.2.5 Future Developments

A key issue is the content of lithium in the ZnO crystals, which must be reduced. Two approaches can be considered: (1) the use of a Li-free mineralizer and (2) postgrowth treatment to reduce the levels of Li, Na, and K in the crystal fraction near the surface [59]. As for the latter, current state of the art is a Li level of around 10^{16} cm^{-3} [30], which needs to be reduced to the order of $\leq 10^{14}$ cm^{-3}.

Other important donor impurities such as Al and Fe, as in the case of Li, need also to be as low as possible in the epitaxy-ready ZnO wafer. This can be achieved by employing high-purity precursors. Also, processes may be considered that passivate them, or by gettering in that volume fraction of the wafer which has no impact on the device performance. The latter technique is applied to silicon, where oxygen and related defects are concentrated in the central part of the body of a wafer.

Generally, the industrial demand for ZnO wafer is mainly driven by further progress in the development of ZnO-based thin-film devices: efficient p-type doping, vacuum ultraviolet (VUV) applications, transparent electronics (TCO), etc. A recent advance shows promises for ZnO as scintillator with ultrafast decay characteristics for use in thin-film devices with enhanced two-dimensional resolution and in extreme UV lithography and imaging applications [30, 63].

2.3 Ammonothermal Growth

2.3.1 A Brief Review of the Growth of GaN Bulk Crystals

The growth of hexagonal GaN (h-GaN) crystals from the following techniques other than the ammonothermal technique has been reported. The melt growth of h-GaN requires an extremely high pressure of >6 GPa and temperature of 2,220°C to crystallize specimen of 100 μm from the stoichiometric melt [64]. At a pressure and temperature of 1.2–2 GPa and 1,400–1,700°C, respectively, very thin (around 100 μm range) GaN platelets of 1 cm^2 in size were grown from molten gallium [65]. Both techniques are difficult to operate from the point of view of industrial mass production owing to the extreme pressure–temperature window.

Sodium-based solutions were successfully employed around 5 MPa at 750°C to synthesize h-GaN platelets and small prismatic crystals through self-nucleation, i.e., without using a h-GaN crystal substrate [66]. However, this technique suffers from evaporation of the solvent due to its high vapor pressure. More recently, the growth of single-crystalline GaN on a 2-in. (50-mm) HVPE-grown GaN substrate was demonstrated by employing LPE and a flux of Ga and Na [67]. The same group is now able to grow 3-mm thick specimens [68]. Low-pressure solution growth (LPSG) uses a gallium-containing solution and NH_3 as the nitrogen source [69, 70]. A multiple wafer concept, i.e., simultaneous growth of three wafers of 2 in. (50 mm) size, can also be worked out [71].

The most prominent growth technique from the vapor phase is HVPE. Mitsubishi Chemical Corp. has demonstrated the growth of bulk crystals upto about 12 mm in thickness along the (0001) direction [72]. This enables processing the largest nonpolar GaN wafers available to date, though this size is still much too small to be employed in industrial application.

Other vapor growth techniques include the vapor growth of free-standing crystals in an NH_3-containing atmosphere [73, 74]. This technique, however, is still in its infancy.

In most of the above techniques parasitic nucleation at the solid–liquid or solid–vapor interface occurs, which often hampers the development of thicker GaN crystals. While the above short list does not claim to be comprehensive, an overview of the techniques that have yielded GaN crystals has been given by Denis et al. [75].

2.3.2 Ammonothermal Growth Technology for GaN

General Aspects

The ammonothermal growth technique is a recent development in comparison to most other techniques for the growth of GaN crystals. On the basis of earlier findings [76], Dwiliński et al. were the first to publish the growth of small h-GaN crystals under ammonothermal conditions using a basic mineralizer [77, 78]. Later on, in the early 2000, more research groups in the US and Japan applied the ammonothermal technique with an acidic mineralizer to synthesize GaN in the crystalline form [79–84]. Dwiliński et al. have shown in 2007 a 1-in. (25-mm) GaN wafer of excellent crystallinity which was sliced from a larger GaN crystal [4].

The principles of the ammonothermal growth do not differ greatly from those of the hydrothermal technique as described in Sect. 2.2. A closed autoclave is used to contain the solution and seed crystals. The differences in the equipment are greatly due to the mineralizers employed (Fig. 2.20). As mentioned earlier, the choice is on either acidic (NH_4X with X = F, Cl, Br, I) or basic (mostly used are $NaNH_2$ and KNH_2) mineralizers. The solubility of GaN in an acidic or basic ammonothermal environment is of the regular or retrograde type, respectively. Consequently, the positions of the seed crystal and the feedstock are opposite. The solubility is treated in a subsequent paragraph.

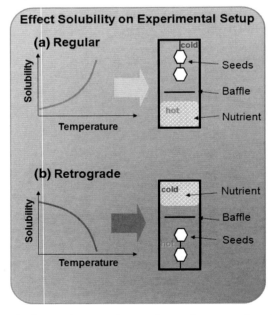

Fig. 2.20. Principal growth setup for regular and retrograde solubility of GaN

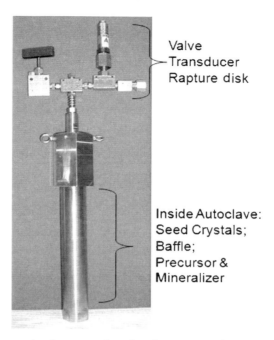

Fig. 2.21. Photograph of an autoclave for the ammonothermal growth of GaN at research scale

The autoclave contains a baffle, seed crystals, the precursor, and the mineralizer (Fig. 2.21). The autoclave is typically made of a Ni-based alloy with no further inner wall in the case of a basic mineralizer. For the acidic mineralizer, it is important to have a Pt inner liner in the high-pressure autoclaves in order to prevent them from corrosion and successive incorporation of impurities from the inner wall. This inner liner must be firmly fitted into the inner wall of the autoclave to avoid any leakage. In our case with the acidic mineralizer, we could profit from the knowledge obtained during the intensive research on the hydrothermal growth of ZnO from Pt-lined autoclaves (see Sect. 2.2.1).

During recent times, much progress in the ammonothermal growth of h-GaN from supercritical (SC) ammonia (NH_3) under an elevated temperature of 500–675°C and a pressure regime of ≤300 MPa has been reported as an alternate route toward true bulk h-GaN crystals [83, 85–87]. Ammonia (under atmospheric pressure melting point $T_m = -77.7°C$, boiling point $T_b = -33.35°C$) takes on an SC state at $T_c = 132°C$ and $p_c = 11.2 MPa$.

Ammonia is not as good a solvent as water for ionic substances. The dielectric constant ε, a measure of the energy of solvatation compared to the lattice energy, is smaller, i.e., 16.5 (at 25°C) for ammonia compared to 80 for water. This forces the formation of low-crystallinity phases, but can be overcome by working at high pressures, as ε increases toward higher density of the solvent [76].

The experimental routine that is applied in case of an acidic mineralizer can be described as follows [87,88]: A temperature ranging from 500 to 550°C is established in the dissolution zone below the baffle to dissolve the precursor, which is GaN grains. The growth zone is called the fraction above the baffle, which is characterized by a lower temperature to force the crystallization of GaN.

We worked with a fill of around 40–70% of NH_3 of >99.999% purity which builds up system pressures of ≤150 MPa. The acidic mineralizer, NH_4X (X = Cl, Br, I), and the precursor were placed at the bottom part of the autoclave. The autoclave is then sealed, evacuated, and successively flooded with nitrogen to remove oxygen. The nitrogen is finally replaced by liquid NH_3 and heating of the autoclave is started.

The mineralizers are of 99.999% phase purity and used in the molar ratio of mineralizer and ammonia in the range $\approx 10^{-2}$. Gallium metal (Ga; 99.9999% purity) had been used initially as the precursor but was gradually replaced by polycrystalline grains of hexagonal GaN. The latter is a byproduct obtained from the synthesis of hexagonal GaN by HVPE. The ratio of the molar concentration of the mineralizer and precursor, calculated as $(1-x)\text{GaN} + x\text{Ga}$, was up to 0.15.

Heating at rates of 30–80 K h^{-1} to the target temperature were followed by a dwell at constant temperature to allow the GaN crystals to grow. The growth period lasts for 28 days. Finally, a controlled cooling down to room temperature (RT) finishes the growth cycle.

The produced GaN crystals are routinely characterized by Nomarski differential interference microscopy (NDIM), SEM (JEOL JSM-7000F), energy-dispersive spectroscopy (EDS; JEOL JSM-7000F, 15 kV, 1 nA, detection range 0–20 keV, accumulation time 60 s), and SIMS (primary beam species Cs$^+$ and O_2^+ at 5 kV and around 200–400 nA) to identify elements and their concentrations in comparison to the standard reference material from Mitsubishi Chemical Corp., which was an HVPE-grown GaN. Powder X-ray diffraction (XRD) (Rigaku RINT-2000, 40 kV and 40 mA) was employed using Cu Kα1 radiation in combination with a Ge (220) channel monochromator (12 arcsec beam divergence, scan speed 0.01° min^{-1}, step width 10^{-4} degrees) to evaluate the crystallinity. The internal standard reference material for XRD was silicon (JCPDS-International Centre for Diffraction Data card no. 27–1402). Steady-state PL was performed using a He–Cd laser (Omnichrome 3056-M-A01: $\lambda = 325$ nm, $P_{\text{out}} = 10$ mW) for excitation and a cryostat (Daikin UV202CL) for temperature control. Spectra were obtained by a photomultiplier using a monochromator (Jobin-Yvon Spex HR320: 1,200 lines mm^{-1} gratings, 320 mm focal length). Time-resolved PL was measured using a He–Cd laser (Kimmon IK5351R-D: $\lambda = 325$ nm, $P_{\text{out}} = 25$ mW) for excitation at RT. Rectangular light pulses (width = 1 ms, repetition rate = 500 Hz, $\tau = 2.5$ ms) were used for the measurements. Spectra were obtained by a CCD camera (Princeton Instruments 576G) using a monochromator (Acton Research Corp., Spectra-pro 150: 300 lines mm^{-1} gratings, reciprocal dispersion = 22 nm mm^{-1}).

Solubility

The solubility of GaN strongly depends on the chemical environment in the ammonothermal solution. Retrograde type solubility was reported for basic mineralizers such as KNH$_2$ [86]. By contrast, an acidic mineralizer such as NH$_4$Cl shows regular type solubility [89]. Figure 2.20 basically depicts the effect of the solubility type on the setup for the crystal growth. The precursor and the seed crystals are placed in the hotter and cooler zone, respectively, for the acidic ammonothermal growth and vice versa for the case of basic ammonothermal growth.

The solubility of GaN was measured in a high-pressure cell of 10 mL volume as described in refs. [89, 90]. The experiments were conducted for 120 h and required a uniform temperature distribution inside the cell. This is particularly critical during heating up and cooling down of the entire system, since even a small gradient might be enough to trigger nucleation of GaN or condensation of the mineralizer thereby deteriorating the result. Similar to the ammonothermal growth of GaN, polycrystalline HVPE GaN crystals have been used as precursor.

We have calculated the energy of formation as equal to 15.9 kcal mol^{-1}, which was derived from the single exponential growth of the concentration of GaN over temperature for the range 250–600°C [89]. This value is independent of the concentration of the mineralizer in NH$_3$ and will provide a tool to calculate the overall energy needed to control the solubility of GaN and therefore the growth of GaN under acidic ammonothermal conditions.

Growth Rate

The dependence of the growth rate on the polarity is connected with the abundance of a particular growth species in front of the growing face of a crystal. Many details of the chemical constitution of the acidic ammonothermal system at high pressure and temperature are still unknown. Wang and Callahan [86] have recently summarized the results on the synthesis of GaN available from the literature. They reported that the most favorable conditions for the growth of GaN under basic ammonothermal conditions with either pure KNH$_2$ or mixed KNH$_2$ and KN$_3$ as mineralizers are at temperatures around 550°C in the hot zone and 25–45 kpsi (172–310 MPa) pressure. The ammonia fill must be >60%.

Pentaaminechlorogallium (III) dichloride [Ga(NH$_3$)$_5$Cl]Cl$_2$ was recently suspected to be an effective precursor for the growth of GaN under acidic ammonothermal conditions. This species has been synthesized at 840 K [91], i.e., it is stable under the temperature of acidic ammonothermal growth of

GaN. The structure is described as cationic $[Ga(NH_3)_5Cl]^{2+}$ octahedra that are surrounded by distorted cubes of Cl^- anions. The latter are weakly bonded and therefore relatively easy to disconnect from the cationic octahedra in a given environment. Also, the free Cl^- anions might easily form NH_4Cl, i.e., a mineralizer molecule again with the NH_4 derived from the autoprotolysis of NH_3, according to:

$$2NH_3 \leftrightarrow NH_4^+ + NH_2^-. \tag{2.10}$$

If we now assume that the double positively charged $[Ga(NH_3)_5Cl]^{2+}$ does exist in the solution of SC NH_3 and NH_4Cl, it might be the reason for the higher growth rate of the $(000\bar{1})$ face in comparison to the (0001) face. Quite in analogy with the hydrothermal ZnO system with basic mineralizers such as LiOH and KOH [3], possible positively charged growth species such as $[Ga(NH_3)_5Cl]^{2+}$ are likely to be attracted by the negatively charged $(000\bar{1})$ face rather than the positively charged (0001) face.

The estimation of the maximum stable growth rate V_{max} at which high crystal quality is still obtained has been attempted. Above V_{max}, entrapment of inclusions and formation of related crystal defects are going to happen. Thus far, nothing is known for GaN produced under ammonothermal conditions, and ZnO also is not well treated in this way. The concept of the maximum stable growth rate is based on Nernst's principle [92], which has been modified later by Carlson [93]:

$$V_{max} = \left(\frac{0.214 D u \sigma^2 C_e^2}{Sc^{1/3} \rho^2 L}\right)^{1/2}, \tag{2.11}$$

where:
- D is the diffusion coefficient,
- u solution flow rate,
- σ the thickness of the diffusion boundary layer in front of the crystal face,
- C_e the equilibrium solute concentration obtained from the solubility,
- ρ the density of the fluid,
- L the length of the crystal surface, and
- Sc the Schmidt number.

Sc gives the ratio of friction and mass transport by diffusion and can be calculated as:

$$Sc = \eta/\rho D \tag{2.12}$$

Using the values for $\eta = 4.6 \times 10^{-5}$ Pa s, $\rho(673\,K) = 0.39\,g\,cm^{-3}$, and $D = 2 \times 10^{-4}\,cm^2 s^{-1}$, calculated after [94], we obtain $Sc = 5.77$. Generally, $Sc = 0.2$–3 for gas mixtures and 100–$1{,}000$ for mixtures of liquids. The calculated value $Sc = 5.77$ suggests that we are dealing with a gas phase rather than a purely liquid phase.

The calculation of V_{max} is based on the bulk transport of the solute and therefore only realistic for macroscopic crystal lengths. As can be seen from equation (11), a higher supersaturation would yield a higher V_{max}. This fact has been proven experimentally [95]. As shown above, increasing the content of mineralizer in NH_3 from almost zero to >1 mol% gives rise to an improved growth rate from nearly zero to >50 µm per day.

In reality, very high flow rates are not realistic in closed systems without forced convection upon stirring, etc. Consequently, the determining parameter in the calculation is the maximum flow rate that can be established in a sustainable way. The effect of the other parameters of the equation is basically through their temperature dependences, and is therefore relatively small.

Typical cases of flux growth show V_{max} = 72–180 µm h^{-1} (1,700–4,320 µm per day) [96]. In the production of large size hydrothermal quartz and ZnO crystals, a growth rate of 500–600 and 200–300 µm per day, respectively, is typically observed for the (0001) face of high-quality crystals [3].

Now, with the parameters for the case of ammonothermal growth of GaN, making the assumptions that $D = 2 \times 10^{-4}$ cm^{2}s^{-1}, $u = 10$ cm s^{-1} as obtained from simulation [97], $\sigma = 0.1$ cm, C_e (773 K) = 1 (i.e., 1 mol GaN/1 mol NH_4Cl), $Sc = 5.77$, ρ (673 K) = 0.39 g cm^{-3}, and $L = 1$ cm, the value of $V_{max} = 15.48$ µm h^{-1} (or 371.5 µm per day) is obtained for a crystal of 1 cm length. Accordingly, a GaN seed of 2 in. (50 mm) diameter could be grown with $V_{max} = 164.6$ µm per day.

The uncertainties are due the uncertainty in D (ranging $10^{-4} - 6 \times 10^{-4}$ cm^2 s^{-1}) and the approximation of σ, and Sc was calculated on the calculated D. D and σ, however, roughly compensate so that Sc and also C_e play a central role in the calculation of V_{max}.

The main increase of V_{max} will come from an improved flow rate u, which in turn reduces σ, which then allows an increase in C_e. This can be achieved by a suitable temperature gradient ΔT between the dissolution and growth zones as shown for the case of supercritical water [98]. Increasing the ΔT from 15 to 25 K almost doubled the main flow rate. We have found that a higher flow rate can be easily established in a larger autoclave such as the one shown in Fig. 2.22.

The growth rate has a high impact on the economy of a crystal growth process and strongly determines the price of wafers cut from a crystal. It is general knowledge that the growth of a single crystal from a solution is a much slower process than, say, growth from the melt and the reasons for that have been discussed extensively in the literature. If, however, multiple crystals are grown during a growth cycle, then the growth from a solution gets much more efficient in terms of total crystal yield.

The growth rate as determined from experiments is a function of the concentration of the mineralizer and the absolute temperature in the autoclave. A growth rate exceeding 50 µm per day for long-term growth cycles requires ≥1 mol% NH_4Cl per 1M NH_3, and higher growth rates close to 100 µm per day have already been achieved in very recent experiments. As already mentioned, a growth rate around 100 µm per day would be well suited for future mass production of GaN by the ammonothermal route [86].

Fig. 2.22. Photograph of an autoclave inside the heater assembly for the ammonothermal growth of large GaN crystals at research scale

It is of interest to note that the growth rates of GaN obtained for the acidic and basic ammonothermal system are very similar, i.e., the different chemistries behind do not prefer one over the other method.

2.3.3 Properties of Ammonothermal GaN

The first crack-free ammonothermal GaN crystal that we grew around the year 2005 is shown in Fig. 2.23. The seed crystal was a $1\,\mathrm{cm}^2$ (0001) HVPE-grown GaN with a mechanically untouched surface in its as-grown state. The quality of the surface of the seed crystal has a tremendous impact on the nucleation and successive growth of the GaN [87, 88]. A mechanically untouched surface would certainly provide the best surface condition with no damage. However, proper CMP or purely chemically etching, if one can get very smooth surfaces without steps, has shown to deliver a satisfactory quality [87]. A similar observation has been made in the growth on m-plane seed crystals as we will discuss later.

The grown GaN crystal in Fig. 2.23 after growth was completely covered with a film mainly containing the mineralizer. This film could have been removed easily upon treatment with distilled water and 2-propanol.

Fig. 2.23. The first thick GaN crystal grown under acidic ammonothermal conditions. The seed was an HVPE GaN crystal with an as-grown surface

Figure 2.24 provides some images of a c-plane ammonothermal crystal grown on a HVPE GaN seed crystal. Here the result was obtained a from remodeled heater system which allowed better control of the overall thermal conditions of the autoclave. Although the surface of the crystal has not been polished, just as grown, a highly transparent ammonothermal crystal is seen. The thickness was around 15 μm for both faces, Ga- and N-polar. The XRC FWHM from the 002 reflection (Fig. 2.25) is very homogeneous with values around 108 and 339 arcsecs for the Ga-polar and the N-polar face, respectively. Such substrates can certainly be used for successive deposition of simple device structures. Our work on this is currently in progress.

It has been analyzed recently [87] that the observed cobble patterning of the as-grown (0001) surface might be related to screw dislocations, quite in analogy to quartz. It can be seen from Fig. 2.24(c, d) that the density of this hillocks is about 3 times larger for the $(000\bar{1})$ face than for the (0001) face. Hashimoto et al. [99] have recently confirmed by transmission electron microscopy (TEM) that the threading dislocation density is higher for the N-polar face, i.e., about 10^6 and 10^7 cm^{-2} for the Ga- and the N-polar face, respectively.

The cross-sectional SEM in Fig. 2.26 shows the flat interface between the HVPE seed crystal and the ammonothermal GaN crystal. Voids have not been found and the pattern of cracking just continues from the HVPE to the ammonothermal material.

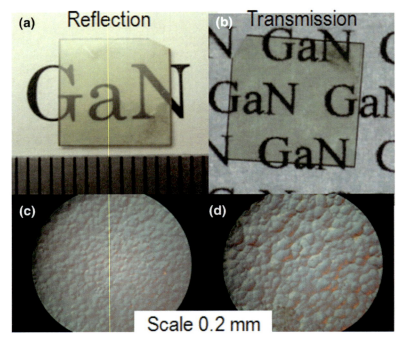

Fig. 2.24. Optical micrographs of high-quality ammonothermal GaN grown on HVPE GaN seed shown in (**a**) reflection, (**b**) transmission, (**c**) NDIM from Ga-polar face, and (**d**) NDIM from N-polar face

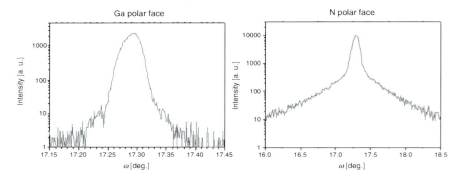

Fig. 2.25. XRC from the Ga- and N-polar face of an ammonothermally grown GaN crystal

Nonpolar substrates have drawn much attention recently since it has been demonstrated that improved internal quantum efficiency can be achieved owing to the lack of an electrostatic field [100]. We have used m-plane HVPE seed crystals that are available in sizes of $1\,\text{cm}^2$ and of excellent quality with the (0002) XRC FWHM of around 100 arcsecs and a dislocation density of

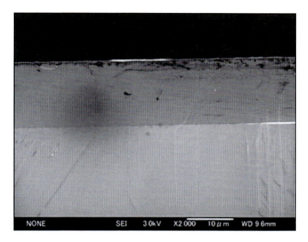

Fig. 2.26. Cross-sectional SEM showing the flat interface region between the HVPE seed and the ammonothermal crystal

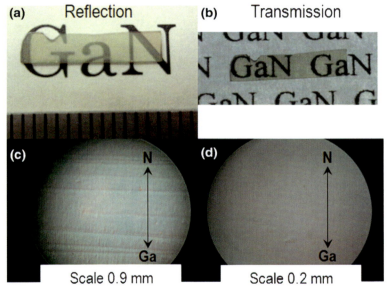

Fig. 2.27. Optical micrographs of high-quality ammonothermal GaN grown on m-plane HVPE GaN seed shown in (**a**) reflection, (**b**) transmission, (**c**), NDIM at higher magnification displaying growth features due to mechanical damaging and (**d**) NDIM revealing a very flat surface

2×10^5 cm^{-2} [72]. Figure 2.27 demonstrates the result from the ammonothermal growth. A very smooth surface morphology is obtained. It must, however, be noted here that the damage from slicing is still sometimes visible under NDIM observation (Fig. 2.27c).

Fig. 2.28. The first 2 in. (50 mm) GaN grown by the acidic ammonothermal route grown from the autoclave shown in **Fig. 2.22**. The seed was an HVPE GaN crystal

We have recently demonstrated the acidic ammonothermal growth on a 1 in. (25 mm) seed crystal [90], and show now in Fig. 2.28 the first result of the acidic ammonothermal growth of GaN on a 2 in. (50 mm) HVPE (0001) GaN seed under similar conditions as above, i.e., temperature and pressure around 500°C and 100 MPa, respectively. The radius of curvature of the seed crystal was around 9 m, and the miscut orientation ±0.15°. The autoclave used for the process measures about $10 \times 100\,\text{cm}^2$ in inner diameter × inner length, presently the largest one in use for the ammonothermal growth of GaN from an acidic mineralizer (Fig. 2.22). Although the crystal quality is still not satisfactory in spite of the few attempts, this result is very promising insofar as we have been able to show the practical feasibility of growing large GaN crystals under moderate process temperature and pressure conditions. This is an important issue for future industrial production. It should be noted here that strain-free (no bending) and large, i.e., $\geq$2 in. (50 mm) GaN seed crystals are not yet available.

Impurities in the crystal are a major concern, and their sources are manifold when autoclave technology is employed. The materials of the autoclave, gas pipes, baffle, and the seed crystal holder as well as the precursor, mineralizer, and ammonia itself contribute to this. In summary of the many experiments we have done thus far, we note that the levels of impurities such as Cr, Fe, and Ni are in the order around 10^{15}–10^{18}, 10^{17}–10^{20}, and 10^{17}–$10^{20}\,\text{cm}^{-3}$, respectively. Platinum from the inner liner contributed the maximum with $10^{16}\,\text{cm}^{-3}$. Silicon (Si) and oxygen (O) are detected at levels around 10^{19} and 10^{18}–$10^{20}\,\text{cm}^{-3}$, respectively. The values for Si and O are somewhat comparable to those form ammonothermal GaN fabricated from basic mineralizers, and were around 10^{19} cm^{-3} for both elements [4], and around 10^{18} for Si and 10^{20} cm^{-3} for O have been reported [4,86]. In all cases, it was found that the N-polar face contains higher amounts of impurities, quite in agreement to ZnO where the O-polar face traps impurities more easily [3].

PL measurements were made to evaluate the optical quality of some of the ammonothermal GaN samples [89,101,102]. A general observation made by PL is that GaN nucleated on the N-polar face seems to be of higher quality than that nucleated on the Ga-polar face. Moreover, we have recently reported on the substantial improvement of the optical quality of acidic ammonothermal GaN and shown that the best ammonothermal GaN samples are comparable to standard HVPE-grown GaN wafers [89].

In Fig. 2.29 is compared the PL spectrum at 10 K obtained from ammonothermal GaN with that from HVPE-grown GaN. Emission from the

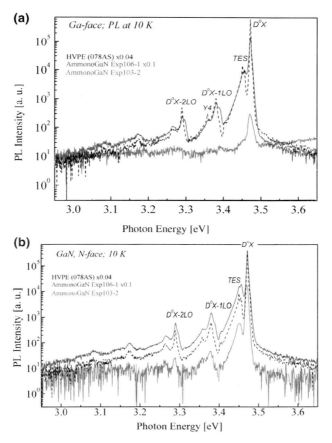

Fig. 2.29. Photoluminescence at 10 K from (**a**) the Ga-polar face and (**b**) the N-polar face in comparison to a high-quality HVPE crystal and a low-quality ammonothermal crystal from an earlier experiment

exciton bound to the neutral donor (D^0X) line of the hexagonal structure at 3.472 eV was observed from the HVPE GaN and at lower intensity also from the ammonothermal GaN. However, an additional peak at 3.357 eV was sometimes found, which was assigned to the Y_4 line [103]. Temperature dependence of this line, however, shows characteristics similar to that of the D^0X emission from *2H*-GaN and to the reported exciton bound to point defects trapped at threading dislocations. It was therefore suggested that the Y_4 line might be related to another crystal structure when comparing the intensity distribution of the (0006) $\omega - 2\theta$ XRD satellite peak and the Y_4 emission. From bandgap energy consideration, the Y_4 line could be linked to the D^0X emission from the *6H* structure of GaN, which is a metastable phase [102].

A deep level luminescence peak at 1.93 eV, red luminescence, was observed from the ammonothermal GaN crystal at 10 K [101]. The characteristics of the temperature dependence of the PL intensity and TRPL suggest that the high oxygen concentration of the ammonothermal GaN may be the origin of this observed red luminescence. This is in good agreement with the high oxygen levels in most of the samples as shown above.

2.3.4 Prospects for Ammonothermal GaN

The year 2007 has proved to be very exciting year for the ammonothermal growth of GaN, and for the first time a 1 in. (25 mm) wafer was shown by Ammono Sp. zo.o. (Poland) [4]. The ammonothermal growth of GaN is now considered to be very competitive to other, more established technologies of growth from the vapor phase. We may compare the ammonothermal technology to the precedent developments in ZnO and SiO_2. Supposing a comparable technological time route for scaling-up the autoclave, GaN seed crystal size, and optimization of the process regime including a better understanding of the rate-determining process, one could expect that commercially available wafers of ammonothermal GaN of 2 in. (50 mm) may soon be available. Here we must remember the development of the hydrothermally grown ZnO, which has already achieved the growth of 3 in. (75 mm) ZnO crystals.

The technological developments in the hydrothermal growth of ZnO were purely based on the achievements from the hydrothermal growth of SiO_2. In contrast, the ammonothermal growth of GaN has much profited from the problems of both SiO_2 and ZnO, such as the design of large autoclaves including choice of the alloy, establishing of the thermal gradient, or a better understanding of the process control. Furthermore, numerical simulation of mass and heat transport under ammonothermal ambient to optimize growth conditions has seen quite some progress recently [18]. We therefore may expect the growth of large quantities of large GaN crystals by the ammonothermal technique in the near future [90].

By comparison of the growth rates of some semiconductor crystals with those of SiO_2 and ZnO, Fig. 2.30, it becomes clear that the melt growth

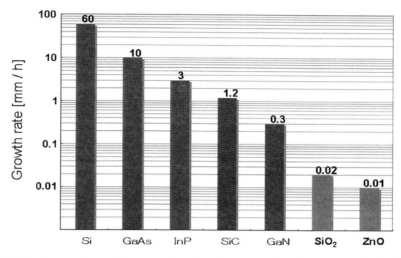

Fig. 2.30. Comparison of the growth rates of some semiconductor crystals including HVPE-grown GaN with those of solution-grown SiO_2 and ZnO

is clearly dominant in terms of growth rates of up to 60 mm h^{-1} for Si. Hydrothermal ZnO, however, grows at about 10 µm h^{-1} only. The target number for GaN would be about half that of ZnO; however, only ≤2 µm h^{-1} is currently achieved. The present low growth rate, roughly 10% of that of quartz, seems to impede the application of the ammonothermal growth technique to produce GaN economically. However, several factors need to be taken into account:

(1) *Scalability* is a main advantage that strongly speaks for the ammonothermal rather than for any of the other above-mentioned technologies to produce the GaN crystal. Analogous to quartz, by using, e.g., 1,000 seeds to grow 1,000 crystals from one autoclave at the same time will drastically bring down the price per GaN wafer. Also, the size of the seed is basically limited only by the size of the inner diameter of the autoclave. We have recently [90] estimated that the price for an ammonothermal GaN wafer is comparable to that of InP, despite the low growth rate and high-pressure equipment.

(2) The *growth rate* of quartz in the early work of Spezia [104] was estimated to a maximum of 1 µm h^{-1} [105]. However, the growth rate of quartz has been improved by a factor of about 20 through a better understanding of the rate-determining parameters along with a better knowledge of the chemistry in the hydrothermal solution. Right now we are facing a similar situation in the ammonothermal growth of GaN.

(3) In contrast to the melt growth of most semiconductors, which has been under research for up to half a century already, the research on am-

monothermal growth of GaN crystals has only commenced less than 15 years ago [106].

The major tasks for the successful ammonothermal growth of GaN have recently been pointed out as follows [90]:
(a) Better knowledge of the solubility of the GaN feedstock and the mineralizers
(b) Availability of large, high-quality seeds
(c) Stable feeding with the GaN feedstock to achieve long-term growth
(d) Scaling-up of the autoclave size for ≥ 2 in. (50 mm) crystals.

Further challenges in GaN wafer manufacturing lie in semi and nonpolar crystal orientation, i.e., $\{10\bar{1}0\}$ and $\{11\bar{2}0\}$. This will allow the fabrication of device structures free of electrostatic fields to yield improved quantum efficiency [100, 107].

The dislocation density of GaN wafers must be decreased to $10^4\,\mathrm{cm}^{-2}$, which is attainable as demonstrated for the case of LPE-grown GaN [108] and, more recently, also for basic ammonothermal GaN [4]. Large GaN seed crystals of high crystalline perfection (no strain, no bending) must also become available.

It was noted recently, in 5–10 years from now a large fraction of the GaN wafers for devices is likely to be grown by the ammonothermal technique [90].

References

1. T. Koyama, S.F. Chichibu, J. Appl. Phys. **95**, 7856 (2004)
2. E. Ohshima, H. Ogino, I. Niikura, K. Maeda, M. Sato, M. Ito, T. Fukuda, J. Cryst. Growth **260**, 166 (2004)
3. D. Ehrentraut, H. Sato, Y. Kagamitani, H. Sato, A. Yoshikawa, T. Fukuda, Prog. Cryst. Growth Charact. Mater. **52**, 280 (2006)
4. R. Dwiliński, R. Doradziński, J. Garczyński, L.P. Sierzputowski, A. Puchalski, K. Yagi, Y. Kanbara, in *Proceedings of 5th International Workshop on Bulk Nitride Semiconductors (IWBNS-V)*, Sept. 24–28, 2007, Itaparica, Bahia, Brazil, Workshop Program and Abstracts
5. Tokyo Denpa Co. Ltd., Tokyo; http://www.tew.co.jp
6. D. Ehrentraut, M. Miyamoto, H. Sato, J. Riegler, K. Byrappa, K. Fujii, K. Inaba, T. Fukuda, T. Adschiri, Simple processing of ZnO from solution: homoepitaxial film and bulk single crystal, Cryst. Growth Des. **8**, 2814 (2008)
7. D. Ehrentraut, H. Sato, M. Miyamoto, T. Fukuda, M. Nikl, K. Maeda, I. Niikura, J. Cryst. Growth **287**, 367 (2006)
8. H. Sato, D. Ehrentraut, T. Fukuda, Jpn. J. Appl. Phys. Part 1 **45**, 190 (2006)
9. D. Andeen, J.H. Kim, F.F. Lange, G.K.L. Goh, S. Tripathy, Adv. Funct. Mater. **16**, 799 (2006)
10. K. Byrappa, in *Hydrothermal Growth of Crystals in Handbook of Crystal Growth*, ed. by D.T.J. Hurle. Bulk Crystal Growth, 2a Basic Techniques (North-Holland, Amsterdam, 1994), pp. 467
11. R.A. Laudise, E.D. Kolb, A.J. Caporaso, J. Am. Ceram. Soc. **47**, 9 (1964)

12. G.W. Morey, P. Niggli, J. Am. Chem. Soc. **35**, 1086 (1913)
13. O.F. Tuttle, Am. J. Sci. **246**, 628 (1948)
14. E.D. Kolb, A.S. Coriell, R.A. Laudise, A.R. Hutson, Mater. Res. Bull. **2**, 1099 (1967)
15. N. Sakagami, K. Shibayama, Jpn. J. Appl. Phys. **20**, 201 (1981)
16. L.N. Dem'yanets, V.I. Lyutin, J. Cryst. Growth **310**, 993 (2008)
17. C.H. Seager, S.M. Myers, J. Appl. Phys. **94**, 2888 (2003)
18. Q.-S. Chen, V. Prasad, W.R. Hy, J. Cryst. Growth **258**, 181 (2003)
19. E.V. Kortunova, P.P. Chvanski, N.G. Nikolaeva, J. Phys. IV France **126**, 39 (2005)
20. I.L. Khodakovsky, A.E. Elkin, Geochemistry **10**, 1490 (1975)
21. L.N. Demianets, D.V. Kostomarov, Ann. Chim. Sci. Mat. **26**, 193 (2001)
22. T. Sekiguchi, S. Miyashita, K. Obara, T. Shishido, N. Sakagami, J. Cryst. Growth **214/215**, 72 (2000)
23. R.A. Laudise, A.A. Ballman, J. Phys. Chem. **64**, 688 (1960)
24. D.F. Croxall, R.C. Ward, C.A. Wallace, R.C. Kell, J. Cryst. Growth **22**, 117 (1974)
25. M. Suscavage, M. Harris, D. Bliss, P. Yip, S.-Q. Wang, D. Schwall, L. Bouthillette, J. Bailey, M. Callahan, D.C. Look, D.C. Reynolds, R.L. Jones, C.W. Litton, MRS Internet J. Nitride Semicond. Res. 4S1, G3.40 (1999)
26. L.N. Demianets, D.V. Kostomarov, I.P. Kuz'mina, S. V. Pushko, Crystallography Rep. Suppl. 1 **47**, S86 (2002)
27. K. Maeda, M. Sato, I. Niikura, T. Fukuda, Semicond. Sci. Technol. **20**, S49 (2005)
28. J. Nause, B. Nemeth, Semicond. Sci. Technol. **20**, S45 (2005)
29. Byrappa, T. Adschiri, Prog. Cryst. Growth Charact. Mater. **53**, 117 (2007)
30. D. Ehrentraut, H. Sato, Y. Kagamitani, A. Yoshikawa, T. Fukuda, J. Pejchal, K. Polak, M. Nikl, H. Odaka, K. Hatanaka, H. Fukumura, J. Mater. Chem. **16**, 3369 (2006)
31. O. Schmidt, P. Kiesel, D. Ehrentraut, T. Fukuda, Noble M. Johnson, Appl. Phys. A Mater. Sci. Process. **88**, 71 (2007)
32. H. Sato, D. Ehrentraut, M. Miyamoto, K.J. Kim, O. Schmidt, P. Kiesel, T. Fukuda, J. Electrochem. Soc. **154**, H142 (2007)
33. I.C. Robin, A. Ribeau, S. Brochen, G. Feuillet, P. Ferret, D. Ehrentraut, T. Fukuda, Appl. Phys. Lett. **92**, 141101 (2008)
34. D. Ehrentraut, A. Yoshikawa, T. Fukuda, J. Optoelectronics Adv. Mater. **9**, 1198 (2007)
35. M. Yoshimura, W. Suchanek, K. Byrappa, MRS Bull. **25**, 17 (2000)
36. L.N. Dem'yanets, L.E. Li, T.G. Uvarova, J. Mater. Sci. **41**, 1439 (2006)
37. J.-M. Ntep, S. Said Hassani, A. Lusson, A. Tromson-Carli, D. Ballutaud, G. Didier, R. Triboulet, J. Cryst. Growth **207**, 30 (1999)
38. B. Wang, M.J. Callahan, L.O. Bouthillette, Chunchuan Xu, M.J. Suscavage, J. Cryst. Growth **287**, 381 (2006)
39. H. Wenisch, V. Kirchner, S.K. Hong, Y.F. Chen, H.J. Ko, T. Yao, J. Cryst. Growth **227–228**, 944 (2001)
40. M.J. Vellekoop, C.C.O. Visser, P.M. Sarro, A. Venema, Sens. Actuators A **23**, 1027 (1990)
41. C.J. Youn, T.S. Jeong, M.S. Han, J. H. Kim, J. Cryst. Growth **261**, 526 (2004)
42. N. Sakagami, M. Yamashita, T. Sekiguchi, S. Miyashita, K. Obara, T. Shishido, J. Cryst. Growth **229**, 98 (2001)

43. J. Mass, M. Avella, J. Jiménez, M. Callahan, E. Grant, K. Rakes, D. Bliss, B. Wang, Mater. Res. Soc. Symp. Proc. **878E**, Y1.7.1 (2005)
44. F. Tuomisto, Proc. SPIE **6474**, 647413 (2007)
45. T. Moe Børseth, B.G. Svensson, A. Yu. Kuznetzov, P. Klason, Q.X. Zhao, M. Willander, Appl. Phys. Lett. **89**, 262112 (2006)
46. E.V. Lavrov, Physica B **340–342**, 195 (2003)
47. L.E. Halliburton, L. Wang, L. Bai, N.Y. Garces, N.C. Giles, M.J. Callahan, B. Wang, J. Appl. Phys. **96**, 7168 (2004)
48. B. Wang, M.J. Callahan, L.O. Bouthillette, Cryst. Growth Des. **6**, 1256 (2006)
49. O.F. Schirmer, D. Zwingel, Solid State Commun. **8**, 1559 (1970)
50. D. Zwingel, J. Lumin. **5**, 385 (1972)
51. N.C. Giles, N.Y. Garces, L. Wang, L.E. Halliburton, Proc. SPIE **5359**, 267 (2004)
52. Measurements by C. Czekalla and H. von Wenckstern, University of Leipzig, Germany (2008)
53. P. Avouris, T.N. Morgan, J. Chem. Phys. **74**, 4347 (1981)
54. K. Shibuya, M. Koshimizu, H. Murakami, Y. Muroya, Y. Katsumura, K. Asai, Jpn. J. Appl. Phys. **43**, L1333 (2004)
55. W. Göpel, Surf. Sci. 1977, 62, 165; P. Esser, W. Göpel, Surf. Sci. **97**, 309 (1980)
56. I.V. Markevich, V.I. Kushnirenko, L.V. Borkovska, B.M. Bulakh, Solid State Commun. **136**, 475 (2005) and references therein
57. O. Schmidt, P. Kiesel, C.G. Van de Walle, N.M. Johnson, J. Nause, G.H. Döhler, Jpn. J. Appl. Phys. **44**, 7271 (2005)
58. O. Schmidt, P. Kiesel, D. Ehrentraut, T. Fukuda, N.M. Johnson, Appl. Phys. A **88**, 71 (2007)
59. A.Y. Polyakov, N.B. Smirnov, A.V. Govorkov, E.A. Kozhukhova, S.J. Pearton, D.P. Norton, A. Osinsky, Amir Dabiran, J. Electron. Mater. **35**, 663 (2006)
60. D.C. Look, Mater. Sci. Eng. B **80**, 383 (2001)
61. S. Graubner, C. Neumann, N. Volbers, B.K. Meyer, J. Bläsing, A. Krost, Appl. Phys. Lett. **90**, 042103 (2007)
62. SPC GoodWill, Russia
63. M. Tanaka, M. Nishikino, H. Yamatani, K. Nagashima, T. Kimura, Y. Furukawa, H. Murakami, S. Saito, N. Sarukura, H. Nishimura, K. Mima, Y. Kagamitani, D. Ehrentraut, T. Fukuda, Appl. Phys. Lett. **91**, 231117 (2007)
64. W. Utsumi, H. Saitoh, H. Kaneko, T. Watanuki, K. Aoki, O. Shimomura, Nat. Mater. **2**, 735 (2003)
65. S. Porowski, J. Cryst. Growth **189/190**, 153 (1998)
66. M. Aoki, H. Yamane, M. Shimada, S. Sarayama, F.J. DiSalvo, Crys. Growth Des. **1**, 119 (2001)
67. F. Kawamura, H. Umeda, M. Morishita, M. Kawahara, M. Yoshimura, Y. Mori, T. Sasaki, Y. Kitaoka, Jpn. J. Appl. Phys. **45**, L1136 (2006)
68. F. Kawamura, M. Tanpo, Y. Kitano, M. Imade, M. Yoshimura, Y. Mori, T. Sasaki, in *Proceedings of 5th International Workshop on Bulk Nitride Semiconductors (IWBNS-V)*, Sept. 24–28, 2007, Itaparica, Bahia, Brazil, Workshop Program and Abstracts
69. R.A. Logan, C.D. Thurmond, J. Electrochem. Soc. **119**, 1727 (1972)
70. E. Meissner, B. Birkmann, S. Hussy, G. Sun, J. Friedrich, G. Mueller, Phys. Stat. Sol. C **2**, 2040 (2005)

71. E. Meissner, S. Hussy, I. Knoke, P. Berwian, J. Friedrich, G. Müller, in *Proceedings of 5th International Workshop on Bulk Nitride Semiconductors (IWBNS-V)*, Sept. 24–28, 2007, Itaparica, Bahia, Brazil, Workshop Program and Abstracts
72. F. Orito, K. Katano, S. Kawabata, Y. Kagamitani, D. Ehrentraut, C. Yokoyama, H. Yamane, T. Fukuda, in *Proceedings of 5th International Workshop on Bulk Nitride Semiconductors (IWBNS-V)*, Sept. 24–28, 2007, Itaparica, Bahia, Brazil, Workshop Program and Abstracts
73. R.S. Qhalid Fareed, S. Tottori, K. Nishino, S. Sakai, J. Cryst. Growth **200**, 348 (1999)
74. D. Siche, H.-J. Rost, K. Boettcher, D. Gogova, R. Fornari, J. Cryst. Growth **310**, 916 (2008)
75. A. Denis, G. Goglio, G. Demazeau, Mater. Sci. Eng. R **50**, 167 (2006)
76. H. Jacobs, D. Schmidt, in *Current Topics in Materials Science*, ed. by E. Kaldis. High-Pressure Ammonolysis in Solid-State Chemistry, vol. 8 (North-Holland, New York, 1982)
77. R. Dwiliński, R. Doradziński, J. Garczyński, L.P. Sierzputowski, M. Palczewska, A. Wysmolek, M. Kamińska, MRS Internet J. Nitride Semiconcutor Res. **3**, 25 (1998)
78. R. Dwiliński, R. Doradziński, J. Garczyński, L.P. Sierzputowski, J.M. Baranowski, M. Kamińska, Mater. Sci. En. B **50**, 46 (1997)
79. A.P. Purdy, Chem. Mater. **11**, 1648 (1999)
80. D.R. Ketchum, J.W. Kolis, J. Cryst. Growth **222**, 431 (2001)
81. G. Demazeau, G. Giglio, A. Denis, A. Largeteau, J. Phys. Cond. Matter **14**, 11085 (2002)
82. A. Yoshikawa, E. Ohshima, T. Fukuda, H. Tsuji, K. Oshima, J. Cryst. Growth **260**, 67 (2004)
83. T. Hashimoto, K. Fujito, M. Saito, J. S. Speck and S. Nakamura, Jpn. J. Appl. Phys. **44**, L1570 (2005)
84. B. Wang, M.J. Callahan, K.D. Rakes, L.O. Bouthillette, S.-Q. Wang, D.F. Bliss, J.W. Kolis, J. Cryst. Growth **287**, 376 (2006)
85. R. Dwiliński, J. M. Baranowski, M. Kamińska, R. Doradziński, J. Garczyński, L. Sierzputowski, Acta Phys. Polonica A **90**, 763 (1997)
86. B. Wang, M.J. Callahan, Cryst. Growth Des. **6**, 1227 (2006)
87. Y. Kagamitani, D. Ehrentraut, A. Yoshikawa, N. Hoshino, T. Fukuda, S. Kawabata, K. Inaba, Jpn. J. Appl. Phys. **45**, 4018 (2006)
88. D. Ehrentraut, N. Hoshino, Y. Kagamitani, A. Yoshikawa, T. Fukuda, H. Itoh, S. Kawabata, J. Mater. Chem. **17**, 886 (2007)
89. D. Ehrentraut, K. Kagamitani, C. Yokoyama, T. Fukuda, J. Cryst. Growth **319**, 891 (2008)
90. T. Fukuda, D. Ehrentraut, J. Cryst. Growth **305**, 304 (2007)
91. H. Yamane, Y. Mikawa, C Yokoyama, Acta Cryst. E **63** i59 (2007)
92. W. Nernst, Z. Physik. Chem. **47**, 52 (1904)
93. A. Carlson, in: *Growth and Perfection of Crystals*, ed. by R.H. Doremus, B.W. Roberst, E. Turnbull (Wiley, New York, 1958), p. 421
94. C.-H. He, Y.-S. Yu, Ind. Eng. Chem. Res. **37**, 3793 (1998)
95. D. Ehrentraut, Y. Kagamitani, in *Proceedings of IWBNS-V*, Itaparica, Salvador, Brazil, September 24–28, 2007, Workshop Program and Abstracts
96. H.J. Scheel, D. Elwell, J. Cryst. Growth **12**, 153 (1972) Also references therein

97. Y. Masuda, private communication
98. K. Nagai, J. Asahara, J. Jpn. Assoc. Cryst. Growth **27**, 68 (2000)
99. T. Hashimoto, F. Wu, J.S. Speck, S. Nakamura, Jpn. J. Appl. Phys. **46**, L525 (2007)
100. P. Waltereit, O. Brandt, A. Trampert, H.T. Grahn, J. Menniger, M. Ramsteiner, M. Reiche, K.H. Ploog, Nature **406**, 865 (2000)
101. K. Fujii, G. Fujimoto, T. Goto, T. Yao, Y. Kagamitani, N. Hoshino, D. Ehrentraut, T. Fukuda, Phys. Stat. Sol. A **204**, 3509 (2007)
102. K. Fujii, G. Fujimoto, T. Goto, T. Yao, Y. Kagamitani, N. Hoshino, D. Ehrentraut, and T. Fukuda, Phys. Stat. Sol. A **204**, 4266 (2007)
103. M.A. Reshchikov, H. Morkoc, J. Appl. Phys. **97**, 061301 (2005)
104. G. Spezia, Atti. Accad. Sci. Torino **35**, 95 (1900)
105. F. Iwasaki, H. Iwasaki, J. Cryst. Growth **237–239**, 820 (2002)
106. R. Dwiliński, A. Wysmolek, J. Baranowski, M. Kamińska, R. Doradziński, J. Garczyński, L.P. Sierzputowski, Acta Phys. Pol. **88**, 833 (1995)
107. A. Chakraborty, T.J. Baker, B.A. Haskell, F. Wu, J.S. Speck, S.P. DenBaars, S. Nakamura, U.K. Mishra, Jpn. J. Appl. Phys. **30**, L945 (2005)
108. Y. Mori, T. Sasaki, Oyo Buturi **75**, 529 (2006)

3
Growth of ZnO and GaN Films

J. Chang, S.-K. Hong, K. Matsumoto, H. Tokunaga, A. Tachibana, S.W. Lee, and M.-W. Cho

Abstract. Zinc oxide (ZnO) and gallium nitride (GaN) are wide bandgap semiconductors applicable to light emitting diodes (LEDs) and laser diodes (LDs) with wavelengths ranging from ultraviolet to blue light. Now ZnO and GaN are key materials for optoelectronic device applications and their applications are being rapidly expanded to lots of other technology including electronics, biotechnology, nanotechnology, and fusion technology among all these. As a fundamental starting point for the development of this new technique, epitaxy of ZnO and GaN films is one of the most important key technology. Hence, development of the growth technique for high quality epitaxial films is highly necessary. Among the various kinds of epitaxy technique for semiconductor films developed so far, physical vapor deposition (PVD)-based epitaxy technique has been revealed to be the appropriate way for the high quality ZnO film and related alloy growths, while chemical vapor deposition (CVD)-based epitaxy technique has been proved to be the best method for the high quality GaN film and related alloy growths.

In this chapter, growth of epitaxial ZnO and GaN films, related alloys, and their key properties are presented. Among the various kinds of epitaxy technique for these materials developed so far, plasma-assisted molecular beam epitaxy (PAMBE) of ZnO, pulsed laser deposition (PLD) of ZnO, metal organic chemical vapor deposition (MOCVD) of GaN, and hydride vapor phase epitaxy (HVPE) of GaN are also presented. Growth features and mechanisms for each epitaxy technique for ZnO, GaN, and related alloys are discussed in detail in addition to the discussions on typical properties of the grown films.

3.1 MBE and PLD of ZnO and Related Materials
J. Chang and S.-K. Hong

3.1.1 Introduction

Single crystalline epitaxial ZnO film growth and its applicability to optoelectronic devices by observing lasing were reported at the end of 1990s. Bagnall et al. reported room temperature (RT) optical pumped lasing from the epitaxial ZnO film grown on (0001) sapphire substrate by using Plasma-Assisted Molecular Beam Epitaxy (PAMBE) in 1997 [1]. On the other hand, Tang et al. observed also RT ultraviolet (UV) laser emission from self-assembled ZnO microcrystallite thin film grown on (0001) sapphire substrate by using Laser Molecular Beam Epitaxy (a kind of Pulsed Laser Deposition(PLD)) in 1998 [2]. In fact, ZnO has received significant attention for being a promising material for excitonic devices after these reports.

Growth of high quality epitaxial ZnO film on highly mismatched (0001) sapphire substrate with a large lattice mismatch of about 18% was achieved by successfully employing the MgO buffer, which effectively reduced the lattice misfit to about 8% [3, 4]. Two-dimensional (2D) layer-by-layer growth was achieved by PAMBE using the MgO buffer [3]. As another approach, growth of high quality epitaxial ZnO film by using a lattice matched (0001) ScAlMgO$_4$ substrate, which has a lattice misfit of 0.09% with ZnO, was reported [5].

Growth of alloys based on ZnO is also a key technology to fabricate optoelectronic devices. Researchers have focused on $Zn_XCd_{1-x}O$ and $Zn_XMg_{1-x}O$ systems since alloys of ZnO with MgO and CdO provide a wide range of bandgap engineering spanning from 2.4 to 8.2 eV that is highly favorable for device applications. In addition, polarity of ZnO films and control of polarity were important issues because the material's properties have been strongly affected by the polarity. The polarity issue was important in the early stage of ZnO researches, since the usually grown epitaxial ZnO films grown by PVD method showed O-polarity (anion polarity), while the high quality GaN epitaxial films grown by CVD showed Ga-polarity (cation polarity).

Various deposition technologies have been employed in the growth of ZnO. Magnetron sputtering [6–8], PLD [9–12], helicon-wave sputtering [13], PAMBE [14–16], CVD [17, 18], charged liquid cluster beam [19], and spray pyrolysis [20] have been extensively applied. Among these lots of growth techniques, PAMBE and PLD are known as the most suitable method for the growth of high-quality, single-crystalline ZnO films. In this section, PAMBE and PLD of ZnO and related materials are described. Important features of growth systems are presented. Characteristics of growth of ZnO on various substrates by using PAMBE and PLD are discussed. Finally, bandgap engineering and heterostructures of ZnO-based ternary compounds are reviewed.

3.1.2 General Features of the PAMBE

During the past decade the early MBE technique has been developed into a variety of different growth techniques: gas source MBE, and its derivative using metal organic compounds; phase-looked epitaxy; atomic layer epitaxy; migration-enhanced epitaxy, and PAMBE. Currently, PAMBE has become the most common approach in the growth of ZnO thin films among the MBE techniques.

PAMBE System for ZnO Film Growth

A schematic of PAMBE system for the growth of the ZnO films is illustrated in Fig. 3.1 [16]. The growth chamber is evacuated by a series of connections of an oil-free turbomolecular pump and a rotary pump. This pumping system produces a background vacuum of around 5×10^{-10} Torr. During PAMBE, zinc is evaporated from a standard Kundsen cell (K-cell) and oxygen is provided by a Radio Frequency (RF) plasma source (13.56 MHz). The important parameters for growth are substrate temperature, O_2 flow rate, plasma power,

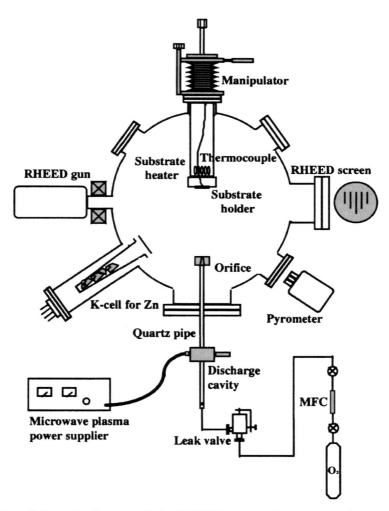

Fig. 3.1. Schematic diagram of the PAMBE system. Reprinted with permission from [16]. Copyright (1998), American Institute of Physics

and Zn flux. Prior to deposition, ultra-high vacuum in the range of low -10^{-10} Torr is generated before oxygen is introduced to create working pressures in the range 8×10^{-6}–9×10^{-5} Torr. Typical growth temperatures are in the range of 300–800°C. Some of the advantages of MBE techniques are the possibility to change the II/VI flux ratio separately and have a precise control of fluxes of the Zn and O sources and in situ monitoring of the growth process by reflection high-energy electron diffraction (RHEED), i.e., PAMBE has some advantages over other growth techniques in terms of versatility and controllability. Early results of *PAMBE* growth of ZnO films and their properties can be found in some review articles [21, 22].

Source Supplying for PAMBE of ZnO

The main difference between a conventional MBE and PAMBE is that some of the sources are plasma source instead of the standard effusion solid source. For ZnO growth, oxygen plasma source was used as an O source to provide a reactive oxygen radical beam, which includes energetic particles as well as atoms and molecules.

As the oxygen molecules could not be efficiently thermal-dissociated on a substrate surface with the temperatures of 300–800°C, the plasma source was used to generate oxygen radicals resulting in a significant enhancement of a growth rate for ZnO growth. Figure 3.2 shows a picture of commercially available plasma sources.

Also special effusion cells with an additional heater around the orifice region are required for supplying Zn as shown in the Fig. 3.3. In oxygen ambient, the metal source intends to have a lower vapor pressure because of the formation of oxide layer on the surface of source materials. The depth of oxidization may change according to both the utilization history and the operation condition of the effusion cell. Furthermore, when using a standard effusion cell, a threshold-like behavior of the beam flux versus the cell temperature sometimes was frequently observed, therefore, a stable beam flux is hardly achieved. The oxidizing problem of metallic Zn source is more severe for the growth conditions with the low Zn beam flux and high oxygen partial pressure. Another serious problem is that the source material might be condensed at the orifice of the crucible, where the temperature is lower than inside. This causes a rapid decrease in the beam flux with operation time. To prevent this problem, a double filament cell or a hot lip cell is efficient in avoiding blocking of the orifice by the condensation of metallic Zn.

The problems mentioned above could be solved by heating the orifice and reducing the outlet area of the effusion cell. With a smaller orifice, or an insert cap with a number of pinholes, oxygen has less possibility to get into the crucible. Furthermore, for a specified flux, the pressure inside the crucible can be much higher, so that the entering O_2 may encounter a metal atom before reaching the solid surface. Less oxidization of the source materials can

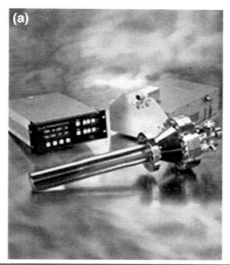

Fig. 3.2. Pictures for commercial plasma sources. (**a**) A plasma source and a matching box (www.veeco.com) and (**b**) A 2.75 in. plasma source. (www.SVTA.com)

be expected when the operation temperature of the Zn source is higher because of the higher vapor pressure and possible dissociation of the oxidized layer on the metallic Zn surface.

3.1.3 PAMBE of ZnO Thin Films on Various Substrates

Various substrates including single crystalline and amorphous such as ZnO, GaN, SiC, Glass, ITO, Quartz, Al_2O_3, CaF_2, Si, GaAs, $ScAlMgO_4$, $Mg\,Al_2O_4$, and $SrTiO_3$ were used for the growth of ZnO thin films. For the PAMBE of ZnO, several substrates have been used. Table 3.1 summarizes lattice misfits of well-known substrates for PAMBE of ZnO thin films. Among these substrates, sapphire is the most widely used for the ZnO epitaxy due to its relatively low cost and stability, although it has the critical problem of large lattice misfit. In this chapter, PAMBE of ZnO thin films on c-sapphire, a-sapphire, GaN template, and ZnO substrates are discussed.

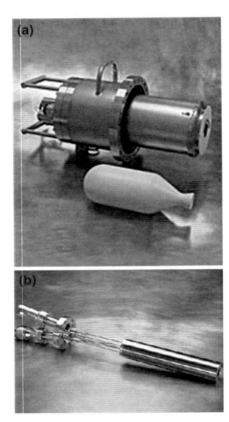

Fig. 3.3. (a) A picture of the Sumo cell (www.veeco.com). PBN crucible for the Sumo cell is also shown. (b) A high temperature cell (www.veeco.com), which also has a specially designed hot-rip structure

Table 3.1. A Lattice misfits between ZnO and several substrates for ZnO heteroepitaxy

Material	Crystal structure	Lattice constant (nm)		Lattice misfit with ZnO(%)
		a	c	
ZnO(0001)	WZ	0.3246	0.5207	–
GaN(0001)	WZ	0.3189	0.5185	1.9
6H–SiC(0001)	WZ	0.3081	1.5118	5.1
MgAl$_2$O$_4$(111)	Cubic	0.8083	–	13.6
Al$_2$O$_3$(0001)	RH	0.4758	1.2990	18

PAMBE ZnO on *C*-plane Sapphire Substrate

PAMBE growth of epitaxial ZnO films on (0001) sapphire substrates was reported by Johnson et al. [23] in 1996. They demonstrated PAMBE growth of

ZnO on sapphire, 6H–SiC, and GaN by using RF oxygen plasma and Zn as the oxygen and the Zn sources, respectively. Their typical growth conditions were: substrate temperatures of 400–500°C, Zn fluxes of 5×10^{-7}–1×10^{-5} Torr, plasma powers of 150–400 W with pressures of 5×10^{-7}–1×10^{-4} Torr, maximum growth rate of 1.16 µm h^{-1}, and film thicknesses of about 6 µm. They focused on reporting the possibility of PAMBE growth of ZnO, thus a systematic comparison of properties of their ZnO films depending on substrate types is not available from their published article, but information on typical properties is available [23]. They observed room temperature (RT) photoluminescence (PL) from the PAMBE ZnO films on GaN/SiC templates and showed a strong band edge emission at 3.298 eV with a Full Width at Half Maximum (FWHM) of 129 meV and a broad deep level emission centered at 2.21 eV. At 4.2 K, they observed a strong near band edge emissions at 3.362 eV, with a FWHM of 8.9 meV. They reported that deep level emission was not observed at temperatures below 100 K.

Vigue et al. [24] reported the defect structure of PAMBE grown ZnO films with thickness of 0.8–1.5 µm. They focused their analysis on the upper part of the film (not interfacial region) and found edge dislocations in major with a Burgers vector of 1/3 <11$\bar{2}$0>. Total dislocation densities in their films were 1×10^{10}–4×10^{10} cm^{-2}. Because of the large lattice misfit between ZnO and sapphire, growth of ZnO epitaxial films on substrates with smaller lattice misfit seems to be highly needed to improve the properties of ZnO films.

In 1997 and 1998 Chen et al. reported ZnO epitaxial growth on (0001) sapphire substrates by using an oxygen microwave plasma source [16]. The epitaxy process exhibited 2D nucleation at the initial growth stage, which was followed by three-dimensional (3D) growth as indicated by the RHEED pattern shown in Fig. 3.4.

An in situ oxygen plasma preexposure of the sapphire substrate was found to be crucial for the initial 2D nucleation. The sapphire substrates were exposed to the oxygen plasma for 30 min at 600°C before the growth. As shown in the Fig. 3.5, the preexposure produces an oxygen terminated surface, which allows the first monolayer of ZnO to adjust itself to the oxygen sublattice of the Al$_2$O$_3$ substrate and thus to reduce the lattice mismatch. The X-Ray Diffraction (XRD) rocking curve (RC) of ZnO showed a narrow (0002) peak (FWHM ∼0.005°) with a broad tail extending from the inhomogeneous interfacial region with the in-plane mosaicity. The PL spectra exhibited dominant bound-exciton emission with a FWHM of 3 meV at low temperature (LT) and a free-exciton emission combined with a very weak deep level emission at RT.

For films grown on (0001) c-plane sapphire, the orientation relationship along the direction normal to the surface was ZnO (0001)//sapphire (0001), but two different types of in-plane rotated domains were observed resulting in the following alignments: ZnO [1$\bar{2}$10]//sapphire [1$\bar{1}$00] and ZnO [10$\bar{1}$0]// sapphire [16, 25]. This corresponding 30°-rotation between the ZnO and the sapphire a axes, is presumably due to the preferential bondings of Zn atoms with sapphire O atoms.

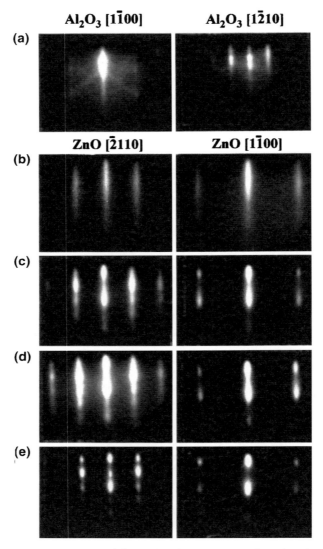

Fig. 3.4. RHEED patterns of (**a**) the Al_2O_3 substrate after the thermal and O_2 plasma treatments and the ZnO epilayer after the deposition of about (**b**) 2, (**c**) 8, (**d**) 40, and (**e**) 300 nm. The growth temperature for this sample is 650°C. Reprinted with permission from [16]. Copyright (1998), American Institute of Physics

PAMBE ZnO on *A*-Plane Sapphire Substrate

In order to eliminate the rotated domains often observed in the ZnO films grown on (0001) sapphire, (11$\bar{2}$0) sapphire substrates have been used for ZnO

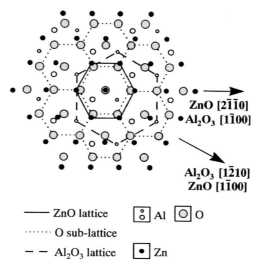

Fig. 3.5. Schematic diagram showing the epitaxial relationship of ZnO (0001) grown on Al$_2$O$_3$ (0001). Reprinted with permission form [16]. Copyright (1998), American Institute of Physics

epitaxy. The lattice parameter $a = 0.3250$ nm of ZnO and parameter $c = 1.299$ nm of sapphire are related almost exactly by a factor of 4, with a lattice mismatch less than 0.08% at RT. In 2000, Fons et al. [26, 27] conducted the investigation of ZnO epitaxy growth on a-plane (11$\bar{2}$0) sapphire, including a detailed analysis of the initial stages of growth [28]. They grew their ZnO films using the MBE with an elemental K-cell of Zn (7N) and oxygen (6N) supplied via a RF radical source. The RF radical source was fitted by using an electrostatic ion trap which was operated at 300 V during the growth [26–28].

Prior to growth, they cleaned the substrate by heating to ∼600°C in ultrahigh vacuum for 15 min. In situ Auger electron spectroscopy (AES) observation on the cleaned surface of substrate showed the presence of only Al and O. RHEED observations of the substrate surface showed streaky patterns with no signs of surface reconstruction. Samples were grown for 2 h with a growth rate of 0.4 μm h^{-1} at the substrate temperatures of 375–500°C. The RHEED patterns along the [1$\bar{2}$10] sapphire changed from a sharp, streaky pattern at the onset of growth, subsequently disappeared, and then gradually transformed into a sharp streaky pattern indicative of (0001) ZnO growth after several minutes of growth. The pattern remained sharp and streaky until the end of growth. No surface reconstruction was observed for either the sapphire substrate or the ZnO epilayer. The RHEED observations revealed that the ZnO <11$\bar{2}$0>//sapphire [0001] direction as agreed with the high-sensitivity pole figure measurements. The concluded the out-of plane and in-plane orientation relationships are ZnO [0001]//sapphire [11$\bar{2}$0] and ZnO

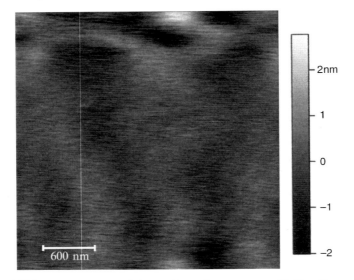

Fig. 3.6. AFM measurements of the surface of an as-grown ZnO epilayer grown on a-sapphire. Reprinted with permission from [26]. Copyright (2000), Elsevier

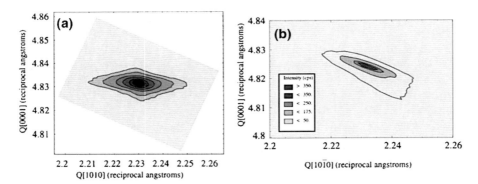

Fig. 3.7. (10$\bar{1}$4) X-ray reciprocal space maps of ZnO epilayers grown on (**a**) (0001) and (**b**) (11$\bar{2}$0) sapphire. Reprinted with permission from [26]. Copyright (2000), Elsevier

[11$\bar{2}$0]//sapphire [0001], respectively, without the rotated domains. Figure 3.6 shows atomic force microscope (AFM) images of the as-grown ZnO film. The images showed that the films are flat with a root mean square (RMS) roughness of approximately 0.4 nm [26], which was expected from the in situ RHEED observation showing the streaky pattern.

Figure 3.7 shows a X-ray reciprocal space map about the ZnO(10$\bar{1}$4) for the ZnO films grown on (0001) and (11$\bar{2}$0) sapphire substrates, respectively.

Typical lateral coherence lengths [29] of the ZnO film grown on (0001) sapphire were of the order of 50 nm, while those of the ZnO film grown on (11$\bar{2}$0) sapphire were of the order of 0.5 µm [26]. This dramatic increase in coherence length was believed to have resulted from the change in the film/substrate relaxation mechanism. In the case of ZnO grown on (0001) sapphire, ZnO islands exhibit a distribution in relative orientation about the (0001) axis, i.e., the twist mosaicity. Reduction of the twist results in the increase in coherence length of the ZnO films grown on (11$\bar{2}$0) sapphire. The twist mosaicity of the ZnO film grown on (11$\bar{2}$0) sapphire was reduced by approximately a factor of two, i.e., from 0.21 to 0.13°, compared with the ZnO films on (0002) sapphire [26].

The c-surface of sapphire is composed of alternate layers of (sixfold symmetric) oxygen and (threefold symmetric) Al atoms, while in the wurtzite structure of ZnO, both O and Zn are sixfold symmetric about the ZnO c-axis. In this idealized case, as shown in Fig. 3.8a, ZnO oxygen atoms would bond to the underlying sapphire Al atoms to form a structure with a lattice mismatch of 31.8% [27]. For simplicity, in Fig. 3.8a only Al atoms of sapphire are shown. The observed in-plane orientation relationship for MBE over the observed temperature range from 200 to 600°C, however, is <10$\bar{1}$0>//<11$\bar{2}$0>. In this configuration, the Zn atoms are ostensibly located above the (sixfold symmetric) sapphire oxygen atoms. This corresponds to a 30° rotation between the

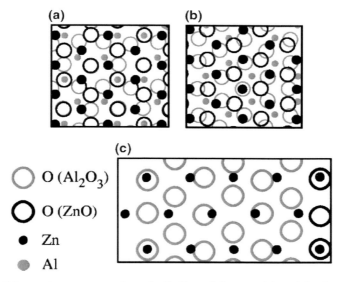

Fig. 3.8. Schematic representation of c-ZnO on (**a**) c sapphire with aligned a axes, (**b**) c-sapphire with a axes rotated by 30°, and (**c**) a sapphire. In the figures, the ZnO a axis is parallel to the short side of the paper. Reprinted with permission from [27]. Copyright (2000), American Institute of Physics

ZnO and sapphire a-axes, as can be seen in Fig. 3.8b note that only the sapphire O atoms are shown, presumably due to the relatively large ionic radius of the O ions (140 pm) versus Al ions (53 pm) resulting in preferential bonding of Zn atoms with sapphire O atoms. This orientation relationship may be metastable in nature. It has been reported that the relative orientation of ZnO on c-sapphire changes from the above 30° a-axis rotation to a parallel configuration as growth temperature is increased for PLD, which suggests the importance of changing surface conditions as a function of the temperature [30]. Figure 3.8c shows a schematic representation of a possible ZnO configuration on the a-surface of sapphire. For simplicity, only the Zn atoms of ZnO and the O atoms of sapphire are drawn. In the drawing, the c-sapphire and the a-ZnO directions are horizontal. The ideal a-surface of the sapphire is twofold symmetric while the ZnO c-surface is sixfold symmetric; the two surfaces are essentially incommensurate with the exception that the ZnO a-axis and the c-sapphire are related almost exactly by a factor of 4 (mismatch less than 0.08% at room temperature). This coincidental match up along the sapphire [0001] direction, the strong tendency of ZnO to grow in the c-orientation, and the incoherence of the interface in directions other than sapphire [0001], leads to the unique configuration in which the ZnO <11$\bar{2}$0>//sapphire <0001> direction. An alternative point of view is that the anisotropy of the a-surface is essential for growing high-quality c-oriented ZnO. Indeed, it has been demonstrated that lowering the symmetry of the c-sapphire surface via the use of vicinal substrates leads to a reduction, but not elimination of rotated domains.

Effects of Substrate Offset Angle on ZnO Films Grown on C-Plane Sapphire

The effects of substrate offset angle on PAMBE growth of ZnO on c-plane sapphire were investigated by Sakurai et al. [31]. They used the sapphire substrates with offset angles of $\theta = 0.04 \sim 2.87°$, tilted from c-plane toward a-axis direction. The substrates were cleaned in vacuum at 700°C for 10 min. The ZnO epitaxy was performed for 2 h at 600°C with the oxygen flow rate of 0.3 sccm at 400 W and with the Zn partial pressure of 8×10^{-7} Torr.

As shown in the Fig. 3.9, the twin-crystal RHEED patterns from both (10$\bar{1}$0) and (11$\bar{2}$0) planes and surface faceting observed from the c-plane-oriented substrates were suppressed by enlarging the offset angles from near 0° to 2.87°, tilted from c-plane toward the a-axis direction. PL spectra at RT were dominated by the near-band edge (NBE) emission and no deep level emissions were observed for all samples. No significant differences in FWHM of NBE peak or in NBE peak intensity were observed between samples with different offset angles. For samples with the smaller substrate offset angles, the RHEED patterns showed the twin crystal patterns from the initial growth stage; the ZnO [11$\bar{2}$0] patterns mixed with the ZnO [10$\bar{1}$0] patterns as shown

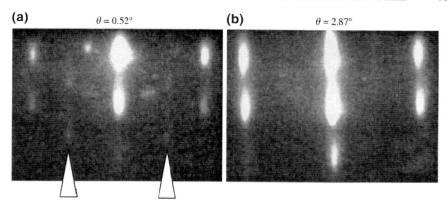

Fig. 3.9. ZnO [10–10] RHEED patterns during growth (after 10 min growth) for different offset angles. ZnO [11$\bar{2}$0] patterns (*shown by triangles*) were seen for samples with smaller offset angles. Reprinted with permission from [31]. Copyright (2000), Elsevier

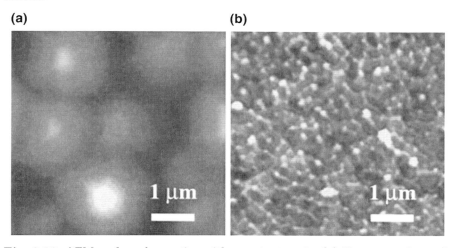

Fig. 3.10. AFM surface observation with a contact mode. (**a**) H-pretreated sample (grain size of 2.5 μm) and (**b**) O-pretreated sample (grain size of 0.2 μm). Reprinted with permission from [32]. Copyright (2003), The Institute of Pure and Applied Physics

in Fig. 3.9a. This indicates that the ZnO films contain different domains that are rotated 30° with each other in the c-plane, thus causing degradation in the film quality.

Pretreatment of Sapphire Substrate

During the MBE process, different surface chemistries of substrates can lead to different growth modes. Figure 3.10 shows different surface morphologies of the ZnO films grown on the H- and the O-pretreated sapphire substrates [32]. The surface chemistry of a-plane sapphire is controlled from O-rich to Al-rich by changing the pregrowth treatment from oxygen plasma to atomic hydrogen. Such a change in surface treatment causes a significant difference in growth mode presumably due to a difference in the surface migration of adatoms: 2D growth is more favorable on an atomic-H-treated surface. Accordingly, ZnO films grown on an atomic-H-treated surface show a smoother surface morphology consisting of larger hexagonal islands with a typical size of 2.5 µm, which should be compared with an island size of 0.2 µm for the ZnO films on an O-plasma-treated surface [32].

Also, a very thin nitrogen-polar AlN layer formed by nitridation acted as a template for the following ZnO growth, resulting in the elimination of the rotated domains, which were often observed in the films grown without nitridation. For the nitridized substrates, the FWHMs of (0002) and ($10\bar{1}2$) ω scans decreased substantially from 912 to 95 arcsec and from 2,870 to 445 arcsec, respectively [33].

PAMBE of ZnO on GaN Template

Ko et al. reported PAMBE growth of ZnO on GaN/Al_2O_3 template prepared by metal organic chemical vapor deposition (MOCVD) [34]. Their ZnO films were grown at 510°C and 700°C with the film thicknesses of 1.7 and 1.1 µm, respectively. The Zn beam flux, measured by a quartz thickness monitor, was 0.1–0.3 nmsec^{-1}, while the oxygen flow rate was kept at 3.5 sccm with a plasma power of 400 W, which resulted in working pressure of 5.5×10^{-5}–1.0×10^{-4} Torr [34]. Prior to the growth, the substrate was heated for 30 min to 1 h below 800°C in the preparation chamber. Then the substrate was exposed to the zinc beam before starting the ZnO growth in the growth chamber. An LT ZnO buffer layer was deposited at 300°C. They found that LT buffer layer growth followed by high-temperature (HT) annealing was effective in growing the high quality ZnO films and observed the RHEED intensity oscillation during overgrowth of ZnO on the HT annealed LT ZnO buffer [42]. They proposed that a LT ZnO buffer layer will improve the epitaxial growth process of ZnO on epi-GaN. A ZnO buffer layer grown at a LT would accommodate a lattice strain through the annealing thereby facilitating the smooth ZnO growth. Another advantage of an LT ZnO buffer layer was reducing the reaction of oxygen atoms with Ga-terminated epi-GaN surface. Figure 3.11 shows changes of RHEED patterns before and after annealing for different thicknesses of ZnO buffer layers grown at 300°C; (a) ∼30 Å, (b) ∼200 Å.

The thickness of an LT buffer layer was estimated using RHEED intensity oscillations. The RHEED pattern of an LT buffer layer with different thickness was composed of diffused streaks and elongated spots, which indicate that the surface of the buffer layer is characterized by small 2D islands compared to the coherent length of the electron beam. This RHEED pattern changes into

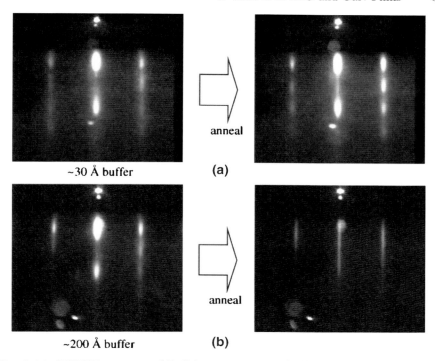

Fig. 3.11. RHEED patterns of ZnO low-temperature buffer layers with thicknesses of (**a**) ∼30 and (**b**) ∼200 Å. The patterns of *left* and *right* sides are before and after annealing of low-temperature buffer layer, respectively. Reprinted with permission from [34]. Copyright (2000), Elsevier

a clear elongated spotty pattern through annealing up to 800°C as shown in Fig. 3.11a, which implies that the surface gets reconstructed to extended island surface with improved lattice ordering. As the buffer layer thickness increases, improvement in surface morphology and crystallinity of the buffer layer was remarkable as shown in Fig. 3.11b. The RHEED pattern of a 200 Å-thick buffer layer changes into sharp streaks by annealing up to 710°C, indicating a smooth and high-quality ZnO surface. Such an improvement in surface morphology has been observed for buffer layers 100–400 Å thick. 2D ZnO growth with RHEED intensity oscillations was observed on such smooth ZnO surfaces as shown in Fig. 3.12.

Figure 3.12 shows RHEED intensity oscillations at the initial stage of ZnO growth on such buffer layers at various substrate temperatures from 300 to 450°C. The incident angle of the electron beam was kept at 1.15°, in which the specula spot was placed between Bragg diffraction spots. The amplitude of the RHEED intensity oscillation at 300°C is large, but its intensity decreases rapidly accompanied by a change in RHEED pattern from streaky to spotty,

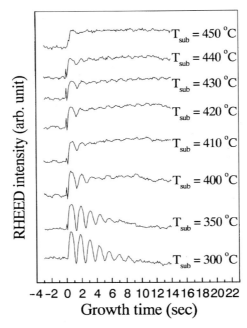

Fig. 3.12. RHEED intensity oscillations versus growth time in growth temperature from 300°C to 450°C. The Zn beam flux was 1.5 Å s^{-1} of deposit rate. Oxygen flowed into 3.5 sccm and RF plasma power was 400 W. Angle of incident electron for RHEED to growing surface was kept at 1.15° in all RHEED intensity measurements. Reprinted with permission from [34]. Copyright (2000), Elsevier

due to the onset of 3D growth. The large amplitude in the oscillation at LT is due to the formation of large density of nucleation sites on the terrace due to short migration length of zinc and oxygen atoms. The amplitude of the oscillation becomes smaller as the substrate temperature increases. The oscillation eventually disappeared at 450°C. The specula beam intensity did not decrease but stayed at a constant value. The RHEED pattern maintained a streaky pattern during ZnO growth at HT. This evolution of the RHEED patterns indicates a growth mode transition from layer-by-layer growth to step-flow growth. The step-flow growth would be preferable for the growth of high-quality ZnO flims.

Vispute et al. studied cathodoluminescence (CL) properties of their 0.5 μm thick ZnO film on GaN/Al$_2$O$_3$ template in 1998 [12]. They observed a free exciton peak at 3.376 eV with a FWHM of 20 meV and a bound exciton peak at 3.36 eV in the CL at 8K. They pointed out the absence of deep level emission on the ZnO film on GaN/Al$_2$O$_3$ template, which was observed in the ZnO film directly grown on sapphire without the GaN layer.

Homoepitaxy

When the effects of the substrate orientation and buffer layers are considered, homoepitaxy of ZnO seems to be the most efficient way to improve the overall properties [35]. High quality ZnO films have been grown on Zn-polar ZnO substrates by PAMBE. On increasing the O/Zn ratio from the stoichiometric to the O-rich flux condition, the growth mode and the surface morphology changed from 3D growth with a rough surface to 2D growth with a smooth surface. The FWHM of $(10\bar{1}0)$ ω-rocking curve was 100 arcsec, and the $n=2$ state of A-exciton was clearly observed in the PL at 4.2 K.

On the other hand, in the early research stage of the homoepitaxy using the available ZnO substrates careful substrate preparation process of the ZnO substrate was inevitably required to grow the ZnO films. In 2001, Wenisch et al. have reported on the comparison of several commercial ZnO substrates and the effects of surface preparation of the ZnO substrates [36]. The commercial 10×10 mm^2-sized n-type ZnO substrates grown by the hydrothermal and seeded physical vapor transport (SPVT) technique were used for their experiments. The first striking difference is the color of the crystals. With a bandgap in the ultraviolet range, pure ZnO substrates should be water transparent. This holds for the SPVT crystals, but not for the hydrothermal ones, because they are pale yellow.

As shown in Fig. 3.13, they tried to characterize the structural quality of those substrates. In the double-axis (0002) HRXRD ω−2θ rocking curves, typical FWHM values are around 120 arcsec for the hydrothermal substrates ("as supplied") and 60 arcsec for the SPVT substrates. For the hydrothermal substrates, a clear shoulder at the low angle side of the main peak about $-0.1° \sim -0.2°$ separated was always observed, which corresponds to the larger lattice constants. Since it can be removed by additional commercial polishing, they concluded that this shoulder stems from a damaged surface layer. In addition, when the (0002) and (0004) reflections are measured, they found the signal more dominant in the case of the (0002) reflection, which is more surface sensitive, being consistent with a surface layer. For the epi-polished SPVT substrates, such a shoulder was never seen, indicating the superiority of the surface preparation. Typical RMS roughnesses for (5×5) μm^2 large were 3–5 nm for the hydrothermal substrates before and 1.5–3 nm after additional polishing. For the SPVT substrates 0.6–1.5 nm RMS roughness values were measured. However, the main difference was the existence of scratches on the (as supplied) hydrothermal substrates, which can be removed by the additional polishing. Here, it should be noted that the results for the substrate preparation were those using the ZnO substrates in early development stage of the bulk substrates. Now, very high quality ZnO substrates are available as described in Chap. 2.

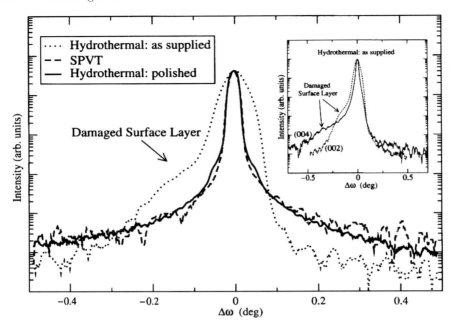

Fig. 3.13. Double-axis HRXRD ω−2θ (0 0 2) rocking curves of different ZnO substrates. The curves are normalized and shifted to facilitate direct comparison. Reprinted with permission from [36]. Copyright (2001), Elsevier

3.1.4 Other Growth Factors in PAMBE of ZnO and their Properties

The buffer layer is another critical growth factor affecting the growth of ZnO films and properties besides the substrates themselves. Additionally II/VI ratio strongly affects the structural and optical properties of the grown films. In this section, buffers and the effects of II/VI ratio were addressed. On the other hand, polarity was a big issue in the early stage of PAMBE of ZnO. This is because the PAMBE grown ZnO films on sapphire substrates always showed the O-polarity, although the polarity control of ZnO films on GaN templates was possible by pretreatments of GaN templates in 2000, until the investigation of polarity control by using the changes in the MgO buffer thickness in 2004. The polarity control of PAMBE ZnO films on sapphire substrates by using the various buffers in addition to the early results on the polarity control of ZnO films on GaN templates are presented. Finally recent results on strains in the (0001) ZnO films on sapphire substrates are addressed.

Buffer Growth for PAMBE of ZnO

The buffer layer is another critical factor affecting the growth of ZnO layers besides substrates. ZnO growth on sapphire without a buffer layer usually changed from the initial 2D growth mode to 3D island growth as confirmed by the appearance of a spotty RHEED pattern as shown in Fig. 3.14 [3].

2D layer-by-layer growth of ZnO films on c-plane sapphire (0001) substrates were achieved by introducing a thin MgO buffer [3] as shown in the RHEED intensity oscillations of Fig. 3.15.

The incorporation of MgO buffer was very effective in improving the surface morphology during the initial stages of growth and leaded to an atomically flat surface. The (3×3) surface reconstruction of ZnO was observed from RHEED pattern as shown in Fig. 3.14f. The rotated domains with a 30° in-plane crystal-orientation misalignment were completely eliminated and the total dislocation density was reduced as expected from the small FWHMs of 13 and 84 arcsec for (0002) and ($10\bar{1}5$) ω-rocking curves, respectively, as shown in Fig. 3.16b and d [3]. The Pendellösung fringes which appeared in Fig. 3.16a indicates a very flat surface and interfaces. The orientation relationship of the ZnO films on the sapphire substrate with the MgO buffer was determined to be Al_2O_3 [$1\bar{1}00$]//MgO [$1\bar{1}0$]//ZnO [$1\bar{2}10$] and Al_2O_3 [$1\bar{2}10$]//MgO [$11\bar{2}$]//ZnO [$1\bar{1}00$].

Effect of II/VI Ratio

In 2002, Ko et al. investigated the effect of II/VI ratio on the properties of the heteroepitaxial ZnO films grown on (0001) c-plane sapphire substrates in detail [37]. ZnO layers were grown under various Zn/O ratios by PAMBE, where the Zn beam flux (F_{Zn}) varied from $0.5 < F_{Zn} < 7.9$ Å s^{-1} with an O_2 flow rate of 2.5 sccm and RF power of 300 W used. The ZnO/MgO double buffer layers were used to accommodate the lattice mismatch between ZnO and c-plane sapphire. The growth was monitored by in situ RHEED and the thickness of the ZnO samples were about 400 nm. They investigated incorporation of Zn and O adaptoms during the growth by investigating the growth rates for the ZnO films grown under the O-rich and the Zn-rich conditions, where the growth rates were determined from the RHEED intensity oscillations [37]. Under the O-rich conditions, the growth rate was limited by the incorporation of Zn adatoms, while it was restricted by the incorporation of O adatoms under the Zn-rich conditions. The adatom incorporation rate F was given by following Eq. [37].

$$F_i = Ap_i \exp(E^i{}_{Ad}/kT). \qquad (3.1)$$

Here, A is the system-dependent scale factor depending upon the vacuum chamber and gauge characteristics, p_i is the ion gauge adatom pressure, $E^i{}_{Ad}$ is an energy for incorporation of i adatom, and T is the substrate temperature. From the plots of the growth rates as a function of reciprocal substrate

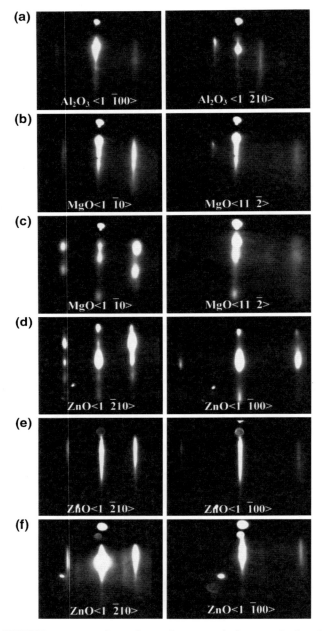

Fig. 3.14. RHEED patterns show the surface morphology evolution during initial growth stages. (**a**) The Al_2O_3 (0001) surface after oxygen plasma treatment. The MgO buffer layer (**b**) before and (**c**) after 2D–3D transition. The low temperature grown ZnO layer on MgO buffer (**d**) before and (**e**) after annealing; (**f**) the ZnO epilayer after a few minutes growth on the buffer layer. Reprinted with permission from [3]. Copyright (2000), American Institute of Physics

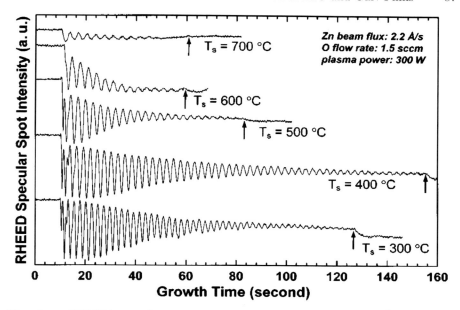

Fig. 3.15. RHEED specular spot intensity oscillations observed in [1$\bar{2}$10] azimuth at various growth temperatures. *Arrows* indicate where the growth is interrupted. Frequency doubling in initial oscillations at low temperatures is due to the presence of single atomic layer RHEED oscillation. Reprinted with permission from [3]. Copyright (2000), American Institute of Physics

temperature, the activation energies of Zn and O adatoms were found to be $E_{\text{Ad}}^{\text{Zn}} = 0.139\,\text{eV}$ and $E_{\text{Ad}}^{\text{O}} = 0.028\,\text{eV}$, thus showing a smaller incorporation barrier for O adatoms than Zn adatoms [37].

The reconstruction patterns of ZnO were observed during growth under various Zn/O ratios during the growth by RHEED. Figure 3.17 shows a phase diagram of ZnO surface (solid rectangles) [37]. In Fig. 3.17, the stoichiometric flux conditions are also depicted in the phase diagram (open rectangles). The transition from (1 × 1) to (3 × 3) reconstruction occurred at a temperature as low as 150°C. With an increase in the temperature, a large Zn beam flux was required for this transition. As the temperature exceeds 700°C, the (3 × 3) reconstruction disappears and (1 × 1) surface structure appears. Little dependence of the Zn beam flux was observed at this transition line. This phase-transition diagram suggests a distinctive surface stoichiometry corresponding to the (3 × 3) reconstruction. It is found the (3 × 3) reconstruction is stable in a wide range of temperature from 200°C to 700°C. It does not change if exposing the surface in either Zn beam flux or O plasma in a wide range of temperature, which implies that a simple O vacancy or Zn adatom model cannot be used to interpret the atomic configuration of the (3 × 3) reconstruction.

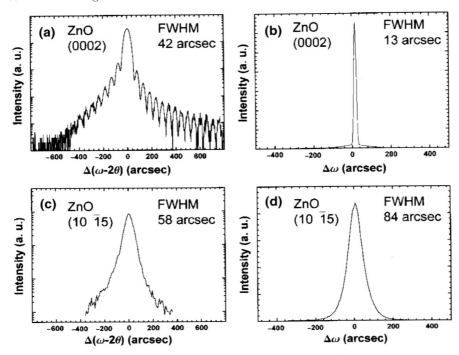

Fig. 3.16. XRD ω–2θ scans of ZnO (**a**) (0002) and (**c**) ($1\bar{1}05$) diffraction peaks. The intensity is plotted in logarithm scale in order to show the Pendellösung fringes clearly. XRD rocking-curves of ZnO (**b**) (0002) and (**d**) ($1\bar{1}05$) diffraction peaks, respectively. Reprinted with permission from [3]. Copyright (2000), American Institute of Physics

A (3×3) reconstruction is also observed for the GaN surface. However, in that case, a Ga adatom-on-adlayer is able to explain the observed (3×3) reconstruction. To investigate whether the surface is caused by O vacancy, they performed annealing of the sample in vacuum. Before annealing, the surface displays a (1×1) RHEED pattern. No evolution to the (3×3) pattern was observed up to the temperature of 1,000°C, which is much higher than the temperature (400–700°C) where the (3×3) reconstruction can be observed. To investigate whether the (3×3) reconstructed surface is due to Zn adatom, they did an additional experiment such that a metal Zn layer was first deposited on the (1×1) ZnO surface at 150°C. The thickness of the deposited Zn was evaluated to be about 3–4 atomic layers. The first Zn layer wetting the ZnO surface, a sharp (1×1) RHEED pattern could be observed at this stage. Then a 2D–3D transition was observed, with a subsequent spotty RHEED pattern. After depositing the metal Zn layer, the sample temperature was gradually increased while the surface was monitored by RHEED. A reevaporation process was observed, where the spotty RHEED changed to (1×1) streaky pattern of a clean ZnO surface. No (3×3) reconstruction pattern was observed at the

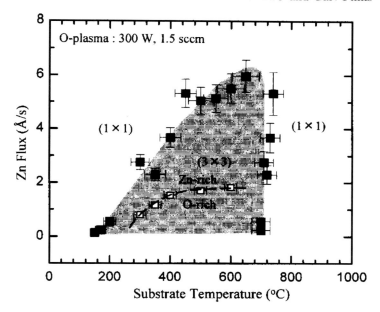

Fig. 3.17. Phase-transition diagram of ZnO surface (*solid rectangles*), obtained by RHEED. The stoichiometric flux conditions are also plotted (*open rectangles*). Reprinted with permission from [37]. Copyright (2002), American Institute of Physics

intermediate stages during reevaporation. From these additional experiments they concluded again that the (3 × 3) surface reconstruction was not originated from either O vacancy or Zn adatom. Therefore, a more complicated configuration model with the reconstructed atoms beyond the single top layer should be considered.

Figure 3.18a shows a (0002) X-ray rocking curve (XRC) and a 2θ/ω XRD profile of the ZnO film grown in a stoichiometric flux condition [37]. In the previous report, the line shape of the (0002) XRC of ZnO epilayers grown on c-plane sapphire substrates without an MgO buffer was separated into two parts: a sharp center peak and a broad tail [16]. The two parts were explained in terms of a two-layer structure including a high-quality ZnO layer on a highly defected interface region. However, the line-shape of the (0002) XRC in Fig. 3.18a is quite improved similar to diffraction of bulk crystals. This improvement of diffraction shape comes from the use of an MgO buffer layer as discussed in the section, "Buffer growth for PAMBE of ZnO". Figure 3.18b shows the FWHM of (0002) symmetric (solid rectangles) and (30–32) skew symmetric (open rectangles) XRCs of ZnO films grown under various Zn/O ratios as a function of the Zn beam flux. The ZnO layers grown under low-Zn beam flux (i.e., highly O-rich condition) showed broader X-ray diffraction. The FWHM values of XRCs decreased on increasing

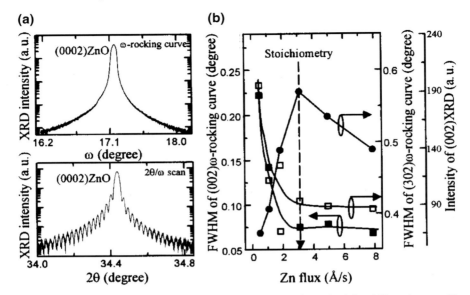

Fig. 3.18. (a) (0002) ω-rocking curve (*upper figure*) and 2θ/ω diffraction profile (*lower figure*) for ZnO layer grown under a stoichiometric flux condition. (b) FWHM values of (0002) symmetric (*solid rectangles*) and (30$\bar{3}$2) skew symmetric (*open rectangles*) X-ray rocking curves and intensity (*solid circles*) of (0002) X-ray diffraction of ZnO layers grown under various Zn/O ratios. Reprinted with permission from [37]. Copyright (2002), American Institute of Physics

the Zn beam flux up to 1.8 Å s^{-1} (near stoichiometry) and almost saturated. The FWHM values from the (30$\bar{3}$2) skew symmetric rocking curve showed the same tendency to those of the (0002) symmetric rocking curves.

They investigated changes of surface morphology depending on the II/VI ratio as shown in Fig. 3.19 [37]. The samples used for AFM investigation were prepared at 700°C with different Zn/O ratios. Hexagonal shaped 2D islands with irregular and rough steps are observed on the ZnO films grown under the O-rich condition. However, as the Zn/O ratio increases from the O rich to the stoichiometric flux condition, the rms values of surface roughness are decreased from rms value of 2.0 to 1.6 nm. Hexagonal islands became connected with each other and the step edge became regular. The step height of the 2D island was equal to the bilayer thickness of Zn–O along the c axis (5.2 Å). The average terrace width was around 800 Å. When moving to Zn-rich conditions, the surface showed hexagonal pits on an atomically flat surface (Fig. 3.19c). The roughness of the ZnO film grown under Zn-rich conditions was measured to be 6 nm for the area of $5 \times 5\,\mu m^2$ for the regions with the pits but 0.8 nm for the area without pits [37]. However, the relations between the source flux ratio and the surface morphology for ZnO films are very different from the GaN films grown by PAMBE. In the cases of GaN grown on GaN template and

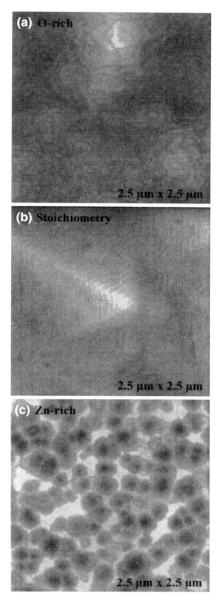

Fig. 3.19. ZnO surface morphologies, observed by AFM, grown under (**a**) O-rich, (**b**) stoichiometric, and (**c**) Zn-rich flux conditions. Hexagonal shape dark spots in (**c**) are pits. Reprinted with permission from [37]. Copyright (2002), American Institute of Physics

c-plane sapphire, smooth surface morphologies were obtained under Ga-rich conditions and pit formation were observed under stoichiometric and N-rich flux conditions. However, as discussed, for ZnO, smooth surfaces were obtained in the stoichiometric flux condition (Fig. 3.19b) and pits were formed under the Zn-rich condition (Fig. 3.19c). The relationship between pit formation and source flux ratio in wurtzite ZnO and GaN seems to be related to polarity of the material surface. ZnO films grown on *c*-plane sapphire in their study were O-polar surface. Pits on O-polar ZnO surface were formed under Zn-rich conditions as mentioned.

Figure 3.20a shows low temperature (10 K) PL spectra measured at the near band edge (NBE) from ZnO films grown under Zn-rich, stoichiometric, and O-rich flux conditions, and Fig. 3.20b show intensities of neutral donor bound exciton ($D°X$) emission and deep level emission at around 2.3 eV from ZnO epilayers grown under various Zn/O ratios. The NBE emission dominates the PL spectrum with weak deep level emission in all ZnO samples. Emission peak at 3.3592 eV from an O-rich grown sample evolved into three peaks at 3.3557, 3.3597, and 3.3648 eV with an increase in Zn/O ratio. These emission lines were assigned to excitons bound to an ionized donor, bound to a neutral donor, and bound to a neutral acceptor. The FWHM value of the ($D°X$) emission of ZnO films grown under O-rich conditions was broader than that of ZnO layers grown under stoichiometric or Zn-rich flux condi-

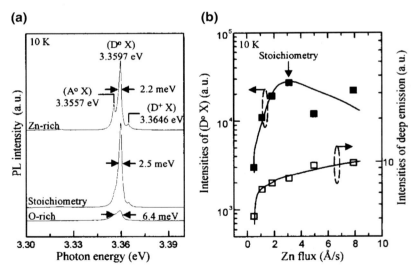

Fig. 3.20. (a) 10 K PL spectra in near band-edge region of ZnO layers grown under Zn-rich, stoichiometric, and O-rich flux conditions. (b) Shows intensities of ($D°X$) emission (*solid rectangles*) and deep level emission (*open rectangles*) from ZnO layers grown under various Zn/O ratios. Reprinted with permission from [37]. Copyright (2002), American Institute of Physics

tions, which indicated larger residual strain in ZnO films grown under O-rich conditions. The intensity of (D°X) emission from the ZnO films was sharply increased with Zn beam flux up to the stoichiometric flux condition, beyond which the intensity of (D°X) emission decreased. The intensity of deep level emission showed an abrupt increase up to the stoichiometric flux condition followed by a gradual increase in intensity. From the point that the FWHM value of XRCs of ZnO layers was reduced with an increase in the Zn/O ratio as shown in Fig. 3.18b, which was caused by a reduction in threading dislocation density, they explained the degradation of optical property in terms of dislocation density. The increase of PL intensity with the Zn beam flux up to a stoichiometric flux condition upon reducing the threading dislocation density suggested that the threading dislocation plays a role in the creation of nonradiative recombination channels.

On the other hand, in 2003, Kato et al. studied the effect of II/VI ratio on the morphology of homoepitaxially grown Zn polar ZnO [38]. Undoped ZnO films were grown on Zn-face ZnO substrates at various O/Zn flux ratios by PAMBE at 450°C. Elemental zinc (7N grade) and oxygen radio frequency (RF) plasma (O_2 gas of 6N grade) were used as the molecular beam sources. The ZnO (0001) substrates were sliced from the bulk ZnO grown by a hydrothermal technique. The ZnO substrate surface was lapped, polished, and then mechano–chemically polished to prepare a mirror surface without any scratches. Prior to the film growth, the substrates were cleaned by heating at 800°C for 30 min under ultrahigh vacuum ($<10^{-9}$ Torr) in the growth chamber. During the film growth, the oxygen flow rate was 3 sccm, the RF power was 250 W, and the chamber pressure was 5×10^{-5} Torr. The O/Zn ratio was varied by adjusting the Zn beam flux (J_{Zn}) from 3.3×10^{13} to 1.9×10^{15} atoms (cm^{-2} s^{-1}) under a constant oxygen plasma condition.

The RHEED and AFM observations revealed that the growth mode and surface morphology strongly depended on the O/Zn ratio. Figure 3.21 shows the RHEED patterns of ZnO films with different O/Zn ratios [38]. The ZnO films grown at $J_{Zn} = 1.6$–2.3×10^{14} atoms (cm^{-2} s^{-1}), namely, high O-rich flux condition, showed a RHEED pattern with streaks, indicating 2D growth (Figs. 3.21a, b). The film grown at $J_{Zn} = 4.6 \times 10^{14}$ atoms (cm^{-2} s^{-1}) showed a pattern with elongated spots superimposed on streaks (Fig. 3.21c). The film grown at $J_{Zn} = 9.5 \times 10^{14}$ atoms (cm^{-2} s^{-1}), namely, near the stoichiometric flux condition, showed a pattern with spots, indicating 3D growth (Fig. 3.21d).

Figure 3.22 shows the AFM images of ZnO surfaces grown at various J_{Zn} values [39]. The ZnO film grown at $J_{Zn} = 1.6 \times 10^{14}$ atoms (cm^{-2} s^{-1}) had a smooth surface with a RMS roughness of 0.68 nm (Fig. 3.22a). As J_{Zn} increased from 2.3×10^{14} to 4.6×10^{14} atoms (cm^{-2} s^{-1}), the RMS increased, as evidenced by some rifts along the a-axis on the otherwise smooth surface of the ZnO film (Fig. 3.22b), and eventually by small islands about 50–200 nm in diameter (Fig. 3.22c). In contrast, the ZnO film grown at $J_{Zn} = 9.5 \times 10^{14}$ atoms (cm^{-2} s^{-1}), namely, near stoichiometric flux condition, showed a textured morphology and a rough surface with an RMS of 22.1 nm (Fig. 3.22d). However, these homoepitaxial ZnO films with Zn polarity did not show the

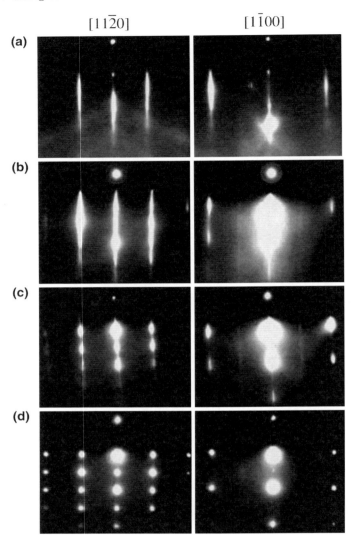

Fig. 3.21. RHEED patterns along [11$\bar{2}$0] and [1$\bar{1}$00] azimuths of ZnO films grown on Zn-face ZnO at Zn beam fluxes (J_{Zn}) of (**a**) 1.6×10^{14}, (**b**) 2.3×10^{14}, (**c**) 4.6×10^{14} and (**d**) 9.5×10^{14} atoms (cm^{-2} s^{-1}). Reprinted with permission from [38]. Copyright (2003), The Institute of Pure and Applied Physics

hexagonal morphology observed in ZnO films with O polarity on sapphire substrates [37].

This difference in morphology between the Zn-polar and O-polar films had been explained in terms of the Zn migration on an O-terminated surface having Zn- or O-polarity as follows. Each O atom on a Zn-polar ZnO surface has three dangling bonds along the c-axis, whereas each O atom on an O-polar

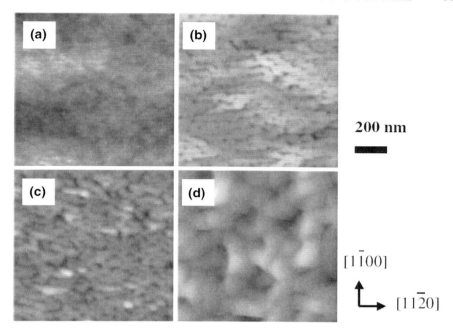

Fig. 3.22. AFM images ($1 \times 1\,\text{mm}^2$) of ZnO surfaces grown on Zn-face ZnO at Zn beam fluxes (J_{Zn}) of (**a**) 1.6×10^{14}, (**b**) 2.3×10^{14}, (**c**) 4.6×10^{14}, and (**d**) 9.5×10^{14} atoms $(\text{cm}^{-2}\,\text{s}^{-1})$. Reprinted with permission from [39]. Copyright (2003), The Institute of Pure and Applied Physics

ZnO surface has a single dangling bond along the c-axis. As Zn migration length on a Zn-face should be smaller than that on an O-face due to the three-bond configuration, the Zn sticking coefficient on O-terminated Zn-face ZnO is higher than that on O-face ZnO. As a result, rather than lateral growth, growth along the c-axis is dominant in Zn-polar ZnO. In contrast, rather than adhesive growth, step-flow growth is dominant in O-polar ZnO. Under O-rich flux conditions with a low Zn beam flux, because a Zn adatom bonded to an O atom on the surface is stable due to the three-bond configuration, the Zn-polar surface is smooth. At a high Zn beam flux, however, because a Zn adatom bonded to an O atom on the surface is unstable, the Zn-polar surface is rough.

Polarity Control by Pregrowth Treatment

The control of the film polarity is an important issue in the growth of ZnO, since it has great influences on the physical properties of grown films. The ZnO films grown by PAMBE on as-polished (0001) sapphire substrates with or without a MgO buffer layer have O polarity [16, 39, 40] mainly due to the growth conditions. The sapphire surface should be O terminated, since sapphire substrates are treated at elevated temperatures (600–700°C) by O plasma prior to the growth. MgO buffer under the oxygen ambient most likely has O-stabilized MgO surface. As demonstrated for atomically flat sapphire (0001) substrates, the polarity of ZnO grown by laser MBE is closely correlated with the in-plane orientation of ZnO, i.e., O polarity for the orientation relationship of ZnO $[10\bar{1}0]//Al_2O_3$ $[11\bar{2}0]$, and Zn polarity for the orientation relationship of ZnO $[11\bar{2}0]//Al_2O_3$ $[11\bar{2}0]$ in the case of Zn polarity [30]. However, it should be noted that almost all of the reported epitaxial ZnO films on sapphire substrates have the orientation relationship of ZnO $[10\bar{1}0]//Al_2O_3$ $[11\bar{2}0]$ except the special case of the results by Ohkubo et al. [30] since this orientation results in the reduced lattice misfit of 18.3% from 31.8%. Therefore, the polarity of ZnO films on sapphire substrates was regarded to have the O polarity before the reliable polarity controlling methods had been reported.

On the other hand, the first reliable heteroepitaxial growth of polarity controlled ZnO films by PAMBE was reported by Hong et al. [41, 42]. They successfully grew Zn- and O-polar ZnO films on Ga-polar GaN templates by engineering the ZnO/GaN heterointerfaces. As confirmed by both coaxial impact collision ion scattering spectroscopy (CAICISS) [41] and convergent beam electron diffraction (CBED) [42], Zn- or O-polar ZnO films were successfully grown with Zn preexposure or oxygen plasma preexposure on the surface of Ga-polar GaN templates prior to the ZnO growth, respectively. The mechanism controlling the polarity could be understood by the different interfaces. The Zn pretreatment provides a well-ordered GaN surface without any interface layer at the ZnO/GaN interface, which results in the atomic sequence of N–Ga–O–Zn and the Zn-polar ZnO. On the other hand, the oxygen plasma pretreatment results in the formation of a monoclinic Ga_2O_3 interfacial layer, which has inversion symmetry, between the ZnO and the GaN. Accordingly, the atomic sequence of N–Ga–O of Ga_2O_3–Ga_2O_3 layer-O of Ga_2O_3–Zn–O and the O-polar ZnO could be grown.

Polarity Control of ZnO Films on C-Plane Sapphire by Using the MgO Buffer

In 2004, Kato et al. reported the polarity-controlled ZnO films on sapphire substrates with an MgO buffer layer. The CBED results showed that the growth of Zn-polar ZnO occurred when the MgO layer was thicker than 3 nm, whereas O-polar ZnO film was grown when the MgO layer was thinner than 2 nm [43]. RHEED observations revealed that the MgO growth was Stranski–Krastanov mode, and that the growth mode transition from two- to three-dimensional occurred when the layer was thicker than 1 nm. The polarity conversion occurs due to the different atomic structure between the thin and thick MgO layer across a thickness of about 2 nm.

Figure 3.23 shows a schematic of the atomic arrangements of ZnO on c-sapphire with two types of MgO buffer layers with a different thickness.

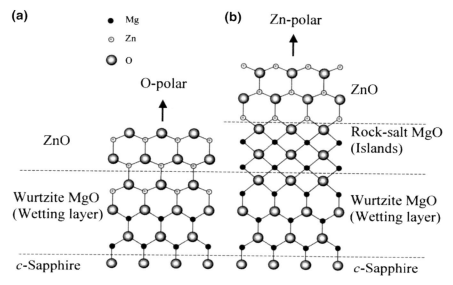

Fig. 3.23. Schematic of atomic arrangement of ZnO on c-plane sapphire with (**a**) a 1 nm-thick MgO buffer layer, and with (**b**) an MgO buffer layer thicker than 3 nm. Reprinted with permission from [43]. Copyright (2004), American Institute of Physics

As the MgO growth occurred under the O-rich flux conditions, growth mainly proceeded at the O-terminated surface. In the initial growth stage up to 1 nm, the wurtzite MgO was grown on O-terminated sapphire as the wetting layer. As the topmost O atoms in the wurtzite MgO have a single dangling bond, each Zn atom in contact with O atoms has three dangling bonds along the c axis, which results in the growth of O-polar ZnO film. On the other hand, because the MgO wetting layer has a compressive strain, the structure changes from wurtzite to rock salt due to relaxation as the layer thickness increases. When the layer thickness exceeds 2–3 nm, the MgO layer changed to rock salt structure. As the topmost O atoms in the rock salt MgO have three-dangling bonds, each Zn atom in contact with O atoms has a single dangling bond along the c axis, which results in the growth of Zn-polar ZnO film. The method of controlling the polarity of ZnO films on sapphire substrates by changing the MgO buffer thickness is effective in controlling the polarity, however, in order to do this we need a careful and accurate control of the MgO layer thickness.

Polarity Control of ZnO Films on C-Plane Sapphire by Using the Cr-Compound Intermediate Layer

Reliable and very easy methods for the selective growth of polarity controlled ZnO films on c-plane sapphire substrates were reported by Park et al. [44]. They employed the Cr-compound intermediate layers to control the crystal

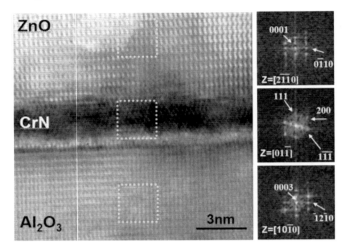

Fig. 3.24. Cross-sectional HRTEM micrograph of the ZnO/CrN/Al$_2$O$_3$. DDPs for the ZnO, CrN, and Al$_2$O$_3$ are shown in the *right*, which are obtained by the FFT of the marked *square* regions of the image. Reprinted with permission from [44]. Copyright (2007), American Institute of Physics

polarity of ZnO films on (0001) Al$_2$O$_3$. The ZnO films grown on rock salt structure CrN/(0001) Al$_2$O$_3$ showed Zn polarity, while those grown on rhombohedral Cr$_2$O$_3$/(0001) Al$_2$O$_3$ showed O polarity.

Figure 3.24 shows the cross-sectional HRTEM micrograph for a ZnO film grown on a CrN/Al$_2$O$_3$ substrate. Digital diffraction patterns (DDPs) obtained by fast Fourier transformation (FFT) of the HRTEM image are also shown in the figure. The DDPs are obtained from the marked square regions on the HRTEM image. The corresponding DDPs from ZnO and Al$_2$O$_3$ reveal typical diffraction patterns of ZnO and Al$_2$O$_3$ for the (11$\bar{2}$0) and (10$\bar{1}$0) zone axes, respectively. Based on the camera constant determined from the DDPs for ZnO and Al$_2$O$_3$, the DDP from the intermediate layer is indexed and determined to be the diffraction pattern for the rock salt structure CrN with the (110) zone axis. The CrN layer is single crystalline with a thickness of 2.5 nm. Here, it should be noted that there was no additional oxide interfacial layer between the ZnO and CrN, which means that the oxidation of the CrN layer was protected by the Zn preexposure before the growth of the ZnO [41]. From the TEM study, they have determined the epitaxial relationship between the ZnO, CrN, and Al$_2$O$_3$ as ZnO (0001)//CrN (111)//Al$_2$O$_3$ (0001) and ZnO (2$\bar{1}\bar{1}$0)//CrN (01$\bar{1}$)//Al$_2$O$_3$ (10$\bar{1}$0).

Figure 3.25 shows the cross-sectional HRTEM micrograph for a ZnO film grown on an oxidized CrN/Al$_2$O$_3$ substrate. Here, two intermediate layers between the ZnO and Al$_2$O$_3$ are clearly seen. The DDPs for the ZnO, Al$_2$O$_3$, and two intermediate layers are shown in the inset of Fig. 3.25. The lower intermediate layer was determined to be CrN based on the DDP

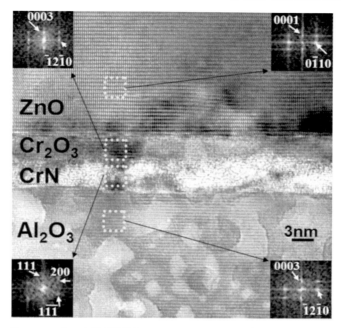

Fig. 3.25. Cross-sectional HRTEM micrographs of $ZnO/Cr_2O_3/CrN/Al_2O_3$. DDPs for the ZnO, Cr_2O_3, CrN, and Al_2O_3 are shown in the insets, which are obtained by the FFT of the marked *square* regions of the image. Reprinted with permission from [44]. Copyright (2007), American Institute of Physics

analysis with the same procedures mentioned in Fig. 3.24. The DDP from the upper intermediate layer was indexed and determined to be a diffraction pattern for rhombohedral structure Cr_2O_3 with the $(01\bar{1}0)$ zone axis. The Cr_2O_3 layer was single crystalline with a thickness of 3.5 nm. Therefore, they have concluded that the top CrN layer was oxidized and changed to a Cr_2O_3 layer. The epitaxial relationship between the layers was determined to be; ZnO $(0001)//Cr_2O_3(0001)//CrN\,(111)//Al_2O_3(0001)$ and ZnO $(2\bar{1}\bar{1}0)//Cr_2O_3(10\bar{1}0)//CrN\,(01\bar{1})//Al_2O_3(10\bar{1}0)$.

The underlying mechanism controlling the polarities of ZnO films using the Cr compound intermediate layer could be explained as follows. The CrN surface is mostly N terminated because the CrN growth is conducted under an N-rich growth condition. In the initial growth of ZnO on CrN, they employed the Zn preexposure to prevent it from oxidation of the CrN. Therefore N–Zn bondings at the interface could be expected. Since the topmost N atoms in rock salt CrN have three dangling bonds, each Zn atom bonded with N atoms has only a single dangling bond along the growth direction and each O atom bonding to the Zn atoms has three dangling bonds. As a result, the ZnO film on the CrN showed the Zn polarity. On the other hand, in the case of the Cr_2O_3, oxygen terminated surface could be expected based on the reported surface phase diagram as a function of oxygen partial pressure and temperature [45]. $\alpha - Cr_2O_3$ has the corundum structure of space group $R\bar{3}c$ with hexagonal close-packed (0001) layers of O atoms and two-thirds of the octahedral interstitial sites filled by Cr atoms [46]. Zn atoms at the interface will occupy octahedral sites bonding to three underlying oxygen atoms in the oxygen layer of Cr_2O_3. Then oxygen atoms bond to the Zn atoms while occupying the tetrahedral sites of the ZnO as similar to the case of the AlN formation on the O terminated (0001) Al_2O_3 by nitridation [47]. Therefore, every oxygen atom has one dangling bond along the c-direction, which results in the O-polar. As a result, the ZnO film on Cr_2O_3 showed the O polarity.

Residual Strain in ZnO Films Grown on *C*-Plane Sapphire Substrate

ZnO films grown on c-plane sapphire substrates always undergo serious residual strain induced by the lattice mismatch of 18.3% [48] and the difference of coefficient of thermal expansion (CTE), which is calculated as $(CTE_{Al_2O_3} - CTE_{ZnO})/CTE_{Al_2O_3} = -34\%$ [49,50]. The strain in films, which is accumulated during film growth, often causes significant deterioration in terms of surface morphology, optical, and structural properties [51,52]. Hence, the residual strain should itself be considered precisely to assess physical properties of the strained film. There have been several reports on the strain evolution and relaxation of ZnO films grown on Al_2O_3 [52–54]. It had been reported that ZnO layers suffered from in-plane compressive stress up to 200 nm thick, as measured by in situ X-ray diffraction (XRD) [52]. The compressive strain started to relax when the ZnO layer exceeded 4.5 nm and monotonically decreased up to 200 nm. Another XRD study of c-ZnO layers grown on a-plane Al_2O_3 revealed that as the thickness of the ZnO film increases from 3 to 600 nm, the c-lattice constant was changed from an elongated value to the bulk one, which indicated that the ZnO films suffered from biaxial compressive stress caused by lattice misfit [53]. Those reports indicate that almost strain free ZnO films can be obtained when the thickness exceeds a few hundred nanometers. However, a controversial report is also available. It has been reported that the peak position of near-band edge emission from ZnO grown on c-Al_2O_3 was redshifted by as large as 50 meV for 40 nm-thick ZnO layers compared to that of bulk ZnO and that it approached the peak position of bulk ZnO as the thickness increased to 500 nm [54]. They interpreted such behaviors in terms of residual tensile strain in the ZnO layers.

Park et al. [55] reported on the lattice relaxation mechanism of ZnO films grown on c-Al_2O_3 substrates by PAMBE. The lattice relaxation of ZnO films with various thicknesses up to 2,000 nm was investigated by using both in situ time-resolved RHEED observations during the initial growth and absolute lattice constant measurements (Bond method) by XRD. The residual strain in the films was explained in terms of lattice misfit relaxation (compression) at the growth temperature and thermal stress (tension) due to the difference

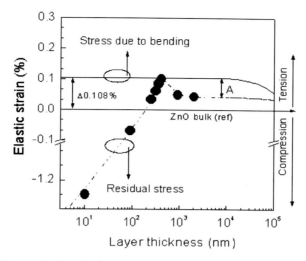

Fig. 3.26. The evolution of elastic strain in the ZnO films as a function of film thickness. *Solid dots* are experimental results. The *dashed line* is estimated from experimental results, and the *solid line* is theoretical calculation of bending stress. The discrepancy, denoted as A, indicates the incomplete relaxation of thermal stress via the formation of microcracks. Reprinted with permission from [55]. Copyright (2007), American Institute of Physics

of growth and measurement temperatures. When the t_{ZnO} was thinner than 5.5 nm, a compressive strained film was obtained during the growth and even at the measurement temperature, since the compressive strain due to lattice mismatch was much larger than the tensile strain due to temperature difference. When the t_{ZnO} was at around 200 nm, the amount of compressive strain was decreased owing to the relaxation of the misfit strain, but the relaxation was not completed yet, and the film still had a compressive strain at the growth temperature. However, the CTE mismatch compensates it during the decrease of temperature and finally, a strain-free film was obtained at the measurement temperature. The relaxation of the lattice misfit strain was estimated to be completed when the t_{ZnO} was as thick as 500 nm. Therefore, one can obtain strain-free ZnO films during growth, but the resultant residual strain at the measurement temperature becomes its maximum value. Also as shown in the Fig. 3.26, in the thick films (>1 μm), the residual tensile strain begins to relax by bending and microcrack formation.

3.1.5 Bandgap Engineering of ZnO-Based Alloys Grown by PAMBE

In this chapter, *bandgap engineering* of ZnO-based alloys is also discussed. Growth and properties of $Zn_xMg_{1-x}O$, $Zn_yCd_{1-y}O$, and $Zn_zBe_{1-z}O$ are presented. Most of works on ZnO-based heterostructure and bandgap engineering were performed on ZnMgO and ZnO/ZnMgO. Recently bandgap engineering and heterostructure employing ZnBeO were reported.

ZnMgo and ZnCdO Alloys

A crucial step in designing modern optoelectronic devices is the realization of bandgap engineering to create barrier layers and quantum wells in device heterostructures. The MBE growth of $Mg_xZn_{1-x}O$ layers has been demonstrated by many researchers allowing modulation of band gap in a wide range. $Mg_xZn_{1-x}O$ alloy has been considered as a suitable material for the barrier layers in ZnO/(Mg, Zn)O superlattice (SL) structures [56] Alloying ZnO with MgO ($Eg = 7.7$ eV) enables widening of band gap of ZnO. According to the phase diagram of the ZnO–MgO binary system, the thermodynamic solid solubility of MgO in ZnO is less than 4 mol %. [57]. In addition, ZnO has a wurtzite structure ($a = 3.24$ Å and $c = 5.20$ Å), while MgO has a cubic structure ($a = 4.24$ Å).

The energy gap $E_g(x)$ of the ternary semiconductor $A_xZn_{1-x}O$ (where A = Mg or Cd) is determined by the following Eq [58]:

$$E_g(x) = (1-x)E_{ZnO} + xE_{AO} - bx(1-x), \tag{3.2}$$

where b is the bowing parameter and E_{AO} and E_{ZnO} are the band-gap energies of compounds AO and ZnO, respectively. The bowing parameter b depends on the difference in electronegativities of the end binaries ZnO and AO. The band gap can be increased (decreased) by incorporating Mg and Cd into ZnO [59–64].

Figure 3.27 shows the a-lattice parameter as a function of room-temperature Eg values in $Cd_yZn_{1-y}O$ and $Mg_xZn_{1-x}O$ alloys [64]. A (Cd, Zn)O/(Mg, Zn)O SL, having a perfect lattice match between layers and a maximum barrier height of 0.09 eV, can be obtained by choosing an appropriate combination of Cd and Mg concentrations, because both a parameters are monotonically increasing functions of alloy composition. This is a major advantage when compared to (In, Ga)N/(Al, Ga)N SLs since in the case of wurtzite structure, if the lattice constant of the well layer differs from that of the barrier layer, strain field exists inside the well layers which causes polarization charge. The same parametric plots for (In, Ga)N and (Al, Ga)N are shown by the dashed curves in Fig. 3.27.

RHEED can be powerful for the process monitoring and control in heterostructure growth. One of the most important advantages of RHEED oscillation investigation is that the growth rate can be precisely determined in situ from oscillation. Thus, the growth rate related information can be determined during growth. One example is that we can do in situ control the Mg incorporation rate when growing MgZnO ternary alloy as demonstrated by Chen at al. [65]. They observed the RHEED oscillation from the growth of MgZnO ternary alloy for the first time as shown in Fig. 3.28.

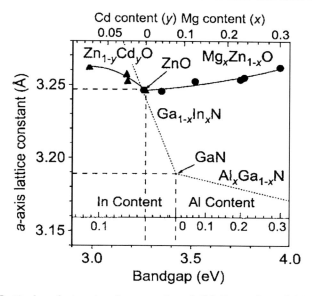

Fig. 3.27. Optical and structural properties of $Cd_yZn_{1-y}O$ and $Mg_xZn_{1-x}O$ alloy films mapped out in a plane of a-axis length and room-temperature band-gap energy. The same curves for (In, Ga)N and (Al, Ga)N alloys are also shown. The alloy compositions are shown on the top axis. Reprinted with permission from [64]. Copyright (2001), American Institute of Physics

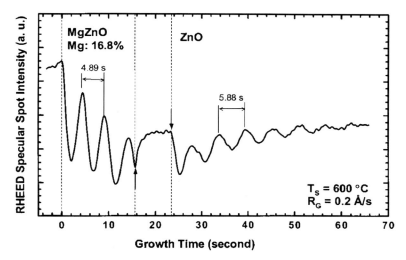

Fig. 3.28. Use the RHEED intensity oscillations to evaluate the Mg incorporation rate in situ. The *upward arrow* indicates stop of MgZnO growth and the *downward arrow* indicates the start of ZnO growth. Reprinted with permission from [65]. Copyright (2000), American Vacuum Society

Since the growth was carried out under oxygen rich conditions, the increase of the growth rate during MgZnO growth came from the incorporation of Mg. For this sample the incorporated Mg composition was determined to be 16.8% based on the RHEED analysis.

On the other hand, El-Shaera et al. have reported on the MBE growth of high quality ZnMgO layers on high-quality ZnO buffer layers grown on (0001) sapphire substrates [66]. The thickness of the ZnO buffer layers was 300 nm, with FWHM of the (0002) XRC as low as 25 arcsec. Single phase $Zn_xMg_{1-x}O$ ($x < 0.22$) thin films having wurtzite structure and c-axis orientation were prepared on high-quality ZnO wafer. Initially, layer-by-layer growth of ZnO was obtained on the ZnO wafers as proven by recording RHEED intensity oscillations.

Figure 3.29a shows the RHEED intensity oscillations obtained at 500°C during ZnO growth on a 300 nm ZnO wafer grown on (0001) Al_2O_3. Also,

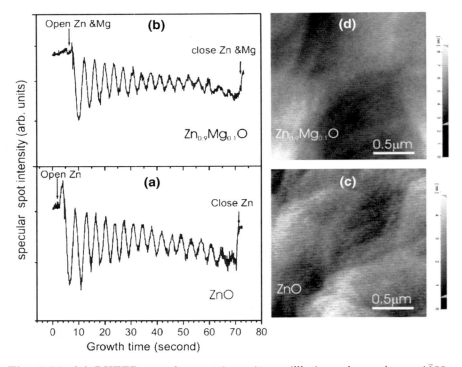

Fig. 3.29. (a) RHEED specular spot intensity oscilllations observed on $<1\bar{1}00>$ azimuth on ZnO, *arrows* indicate where the growth is interrupted. (b) RHEED specular spot intensity oscilllations observed on $<1\bar{1}00>$ azimuth on $Zn_{0.91}Mg_{0.09}O$, *arrows* indicate where the growth is interrupted. (c) AFM image of 300 nm thick ZnO wafer grown on sapphire substrates. (d) AFM image of 200 nm $Zn_{0.91}Mg_{0.09}O$ layers grown on ZnO wafer. The scan area for both images (c) and (d) is $2 \times 2\,\mu m^2$. The rms values of roughness are (c) 0.26 nm for ZnO and (d) 0.7 nm for $Zn_{0.91}Mg_{0.09}O$. Reprinted with permission from [66]. Copyright (2007), Elsevier

layer-by-layer $Zn_xMg_{1-x}O$ ($x < 0.22$) growth was obtained. Figure 3.29b shows the RHEED intensity oscillations detected during $Zn_{0.91}Mg_{0.09}O$ growth at a substrate temperature of 450°C. Figure 3.29c and d show the AFM measurement images of the 300 nm thick ZnO and a 200 nm $Zn_{0.91}Mg_{0.09}O$ surface. The scan area is $2 \times 2\,\mu m^2$. The root mean square (rms) roughness values determined from the scans are 0.26 and 0.7 nm of as grown 2 in. ZnO layers and $Zn_{0.91}Mg_{0.09}O$ layers, respectively. The low temperature PL spectra show a systematic blue-shift of the UV peak position and that the length of the c-axis evaluated from XRD measurements decreases with increasing Mg content as shown in Fig. 3.30.

Another approach to achieve high quality ZnMgO alloy has been reported by Tanaka et al. [67]. They reported on the fabrication of $Mg_xZn_{1-x}O$ quasi-ternary alloys, which consist of wurtzite MgO/ZnO SLs. By changing the thicknesses of ZnO layers and/or of MgO layers of the SL, the bandgap energy was artificially tuned from 3.30 to 4.65 eV. The highest bandgap, consequently realized by the quasi-ternary alloy, was larger than that of the single ZnMgO layer, keeping the wurtzite structure. The bandgap of quasi-ternary alloys was analyzed by the Kronig–Penny model supposing the effective masses of wurtzite MgO as $0.30\,m_0$ and $(1-2)\,m_0$ for electrons and holes, respectively.

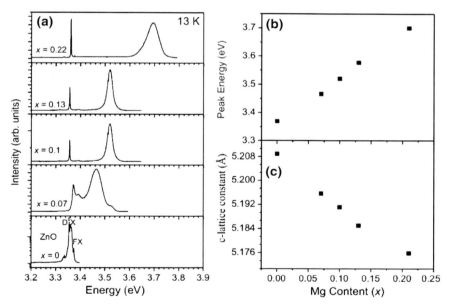

Fig. 3.30. (a) PL spectra of ZnO and $Zn_xMg_{1-x}O$ (x up to 0.2) measured at 13 K excited by the 325.0 nm line of a He–Cd. (b) The band gap evaluated from PL measurements, and (c) c-lattice length constant evaluated from XRD measurements dependence on the Mg content. Reprinted with permission from [66]. Copyright (2007), Elsevier

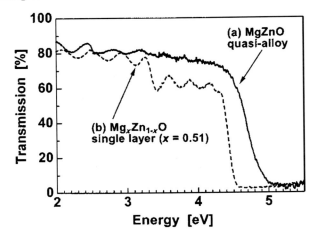

Fig. 3.31. An example of optical transmission spectrum of a MgZnO quasi-ternary alloy with wide band gap, in comparison to that of the MgZnO single layer of wurtzite structure and possessing the highest band-gap energy of 4.45 eV. Reprinted with permission from [67]. Copyright (2005), American Institute of Physics

After starting the growth of MgO on ZnO, no significant change in the RHEED pattern was found until the growth time of 2 min (about 8 nm in thickness). However, as the growth proceeded, the pattern changed to a spotty one, suggesting severe degradation and/or incorporation of the rock salt structure. This implies that the MgO on ZnO takes a wurtzite structure until its thickness exceeds a critical value. The bandgap of the MgZnO quasi-ternary alloys was determined from the transmission spectra of the films. An example of the spectrum, in comparison with that of a single MgZnO layer, whose thickness is about 650 nm, is shown in Fig. 3.31. Calculating the bandgaps from the spectra, it is shown that the arbitral tuning of the band-gap energies from 3.30 to 4.65 eV is achieved by changing the thicknesses of ZnO and MgO layers. The maximum band gap attained was 4.65 eV, which is higher than the previous achievement with a wurtzite $Mg_{0.51}Zn_{0.49}O$ [68].

The XRD θ–2θ spectra of the quasi-ternary alloys and the MgZnO single layer are depicted in Fig. 3.32. For the quasi-ternary alloys with the band gap 3.30–4.65 eV, only the peaks similar to the (0002) peak from the wurtzite MgZnO were observed, suggesting that the MgZnO quasi-ternary alloys keep the wurtzite structure. The peak position slightly shifts to a higher degree with the increase of the average Mg content in the quasi-ternary alloys. It should be noted that the effective masses for wurtzite MgO have been ambiguous so far. Therefore, they treated the effective masses of MgO as fitting parameters and the band gaps of various MgO/ZnO SL structures were analyzed. The theoretical values of the bandgaps well coincided with the experimental values in various structure samples, when they assumed the effective masses of electrons and holes in wurtzite MgO were assumed to be 0.30 m_0 and $(1-2)m_0$, respectively.

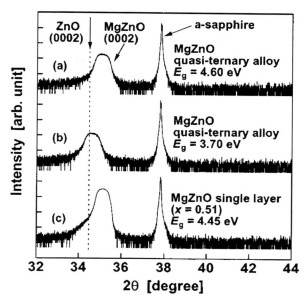

Fig. 3.32. XRD θ − 2θ spectra of MgZnO quasi-ternary alloys, in comparison to that of the MgZnO single layer of wurtzite structure and possessing the highest band-gap energy of 4.45 eV. Reprinted with permission from [68]. Copyright (2003), The Institute of Pure and Applied Physics

The PAMBE growth of ZnMgO opened new possibilities for high electron mobility transistors based upon ZnO-based materials [69]. A two-dimensional electron gas was observed in Zn polar ZnMgO/ZnO (ZnMgO on ZnO) heterostructures grown by PAMBE. The electron mobility of the ZnMgO/ZnO heterostructures dramatically increased with increasing Mg composition and the electron mobility ($\mu \sim 250\,\mathrm{cm^2\,Vs^{-1}}$) at RT reached a value more than twice that of an undoped ZnO layer ($\mu \sim 100\,\mathrm{cm^2\,Vs^{-1}}$). The carrier concentration in turn reached values as high as $\sim 1 \times 10^{13}\,\mathrm{cm^{-2}}$ and remained nearly constant regardless of Mg composition. Strong confinement of electrons at the ZnMgO/ZnO interface was confirmed by C–V measurements with a concentration of over $4 \times 10^{19}\,\mathrm{cm^{-3}}$.

The sample structure is shown in Fig. 3.33. Zn polar ZnO and ZnMgO/ZnO heterostructures were grown by PAMBE. Zn polar ZnO layers on c-plane sapphire were grown by controlling the thickness of the MgO buffer, located between the ZnO layer and the sapphire substrate [70]. The thicknesses of the MgO buffer and the ZnMgO and ZnO layers were 10 nm, 100 nm, and 1 μm, respectively, as shown in Fig. 3.33a. Figure 3.32b shows the relationship between the Mg composition and the Hall mobility at RT for a series of ZnMgO/ZnO heterostructures. The mobility rapidly increased from 100 to 250 cm² Vs⁻¹ for Mg compositions up to 20% and decreased slightly for compositions up

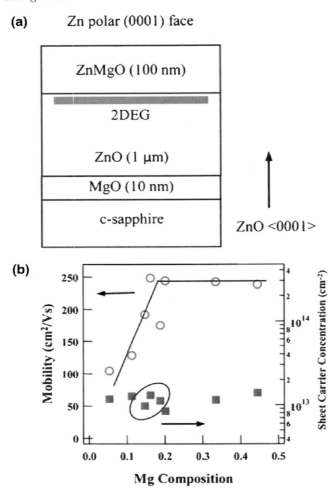

Fig. 3.33. (a) Layer structure for a Zn polar ZnMgO/ZnO heterostructure. The growth direction is (0001) for Zn polarity. (b) Mobility and sheet carrier concentration for Zn polar ZnMgO/ZnO heterostructure as a function of the Mg composition by Hall measurements. Reprinted with permission from [69]. Copyright (2006), American Institute of Physics

to 44%. The lower mobility values observed for low Mg compositions are attributed to an increase in scattering due to penetration of the electrons into the ZnMgO layer due to the small ΔEc.

On the contrary the sheet carrier concentration remained nearly constant for all Mg composition at $1 \times 10^{13}\,\text{cm}^{-2}$ as shown in Fig. 3.33b. The enhancement in mobility and constant sheet carrier composition independent of Mg composition are similar to values reported for AlGaAs/GaAs heterostructures except for the one order of magnitude higher carrier concentration [71]. The higher carrier concentration is due to polarization effects, and the order of magnitude of the carrier concentration was similar to that found in AlGaN/GaN heterostructures for which high sheet carrier concentration is achieved by polarization effects [72, 73]. In the AlGaN/GaN system, however, the carrier concentration strongly depends on the Al composition. The polarization effects for both systems are strong; [74, 75] therefore the results suggest that the dependence of polarization effects on alloy composition is weak for ZnMgO/ZnO, but strong for AlGaN/GaN.

Shibata et al. reported on the photoluminescence characterization of $Zn_{1-x}Mg_xO$ epitaxial thin films grown on ZnO by PAMBE [76]. They reported that high-quality $Zn_{1-x}Mg_xO$ alloys are very brilliant light emitters, even more brilliant than ZnO, particularly in the high-temperature region due to the localization of excitons, because of the compositional fluctuation in $Zn_{1-x}Mg_xO$ alloys.

They studied three samples altogether, which correspond to $x = 0.05, 0.11$, and 0.15. All samples were nominally undoped, and their electrical conduction was n type. PL spectra obtained at temperature $T = 1.4$ K are shown in Fig. 3.34. The sharp emission lines observed in the region of 3.32–3.37 eV were attributed to the band edge emissions from ZnO lower layers, whereas broad emission bands observed in the region of 3.40–3.70 eV were assigned to the

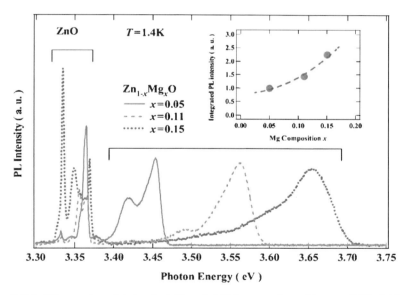

Fig. 3.34. PL spectra obtained at $T = 1.4$ K in samples with $x = 0.05$, 0.11, and 0.15. The integrated signal intensity of the emission bands from $Zn_{1-x}Mg_xO$ layers is estimated and shown in the *inset*. Reprinted with permission from [76]. Copyright (2007), American Institute of Physics

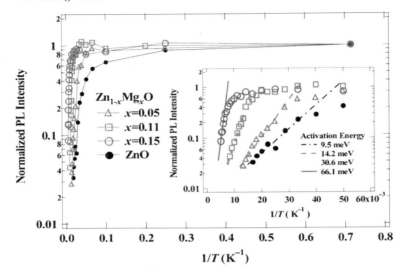

Fig. 3.35. I as a function of T shown in the form of an Arrhenius plot, where the vertical axis is normalized by the value of I at $T = 1.4$ K. The *inset* shows selectively the high-temperature region on an expanded scale. Reprinted with permission from [76]. Copyright (2007), American Institute of Physics

emissions from $Zn_{1-x}Mg_xO$ top layers. They pointed out that the integrated signal intensity of the emission bands from $Zn_{1-x}Mg_xO$ layers increases considerably on increasing the Mg concentration.

The PL spectra obtained from $Zn_{1-x}Mg_xO$ layers have a tail towards the lower energy side, particularly in the sample with $x = 0.15$. These phenomena can be understood in terms of the formation of the tail states in the density of states of excitons in the mixed semiconductors with alloy potential fluctuation [77].

Figure 3.35 shows I as a function of T in the form of an Arrhenius plot. The inset shows the high-temperature region on an expanded scale. The activation energy required for the nonradiative recombination(NR) process in the highest temperature region was estimated by least-squares fitting the results to the equation $\log\{I(T)/I(1.4)\} = (1/T)(\varepsilon/k\text{B}) \log e + C$, where ε and C are the activation energy and a constant, respectively. The activation energy increases in samples of higher magnesium concentration x. (for $x = 0$: 9.5 meV, $x = 0.05$: 14.2 meV, $x = 0.11$: 30.6 meV, $x = 0.15$: 66.1 meV). Therefore, it is very probable that the suppression of the NR process is mainly due to the increase in the activation energy required for the NR process. Hence, it strongly suggests that exciton localization takes place and that the degree of localization increases with increasing x. The essential origin of the localization was inferred to be the spatial fluctuation of the local composition of the Mg in the alloys, which resulted in the spatial fluctuation of the potential energy for the excitons.

ZnBeO Alloys

A new alloy of ZnBeO for ZnO-based LED fabrication was proposed by Ryu et al. in 2006 [78]. They deposited the ZnBeO films on c-plane sapphire substrates by hybrid beam deposition growth method. The energy bandgap of BeZnO can be changed from the ZnO bandgap (3.4 eV) to that of BeO (10.6 eV). As depicted in Fig. 3.36, BeZnO alloys are good candidates for achieving bandgap modulation to values larger than ZnO, and that (Zn, Cd)(Se and/or S, O) quaternary alloys will be good candidates for achieving smaller band gap values [78]. The a-axis lattice constant values for ZnO and BeO are 3.249 and 2.698 Å, respectively [79]. In addition, Ryu et al. investigated the improved lattice matching between ZnO and BeZnO by adding an appropriate amount of Mg into a BeZnO alloy [78]. They characterized their BeZnO films with an UV visible infrared spectrometer to study energy bandgaps and found that the cutoff for transmission was continuously shifted to shorter wavelengths as the Be concentration was increased. The energy bandgap values (Eg) of BeZnO were derived from extrapolating the graph of α^2 vs $(h\nu - E_g)$, where α is the absorption coefficient and ν is the photon frequency. The bandgap of BeZnO was increased as a function of Be concentration. Although the energy band gap for some BeZnO samples

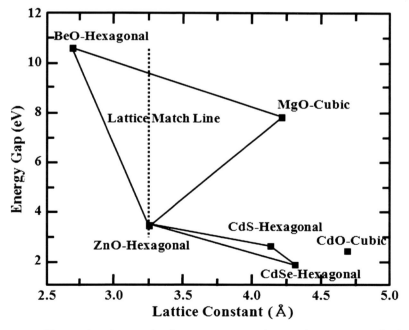

Fig. 3.36. Energy band gaps, lattice constants, and crystal structures of selected II–VI compounds. Reprinted with permission from [78]. Copyright (2006), American Institute of Physics

could not be determined by absorption measurement since they are transparent up to 200 nm wavelength, which is the lower wavelength limit of the spectrometer, the optical transmission measurements demonstrate that the energy bandgap of BeZnO can be varied over the range from 3.3 eV (ZnO) to 10.6 eV (BeO) [78].

3.1.6 General Features of the PLD

PLD is the simplest growth technique especially for the oxide films. In the PLD technique oxidation occurs primarily in the ZnO ablation plume and this alleviates the difficulties encountered with other techniques where oxidation proceeds via surface reactions [80]. In PLD, thin films are obtained by vaporizing a material using high-energy laser pulses. Compared with other techniques, it is generally easier to obtain the stoichiometric films [80–86]. A schematic diagram of the typical PLD system is shown in Fig. 3.37 [87].

The main advantages of the PLD are its ability to create high-energy source particles, permitting high quality film growth at low substrate temperatures, typically ranging from 200 to 800°C, its simple experimental setup, and operation in high ambient gas pressures in the 10^{-5}–10^{-1} Torr range.

Lasers of PLD System for ZnO Growth

ZnO thin film growth by PLD has been reported using a variety of lasers. First reports on ZnO thin film growth by PLD, used Excimer lasers ($\lambda = $ 193–351 nm) [88, 89]. Most of the reports on PLD ZnO thin film are focused on the use of UV excimer lasers (KrF: $\lambda = 248$ nm and ArF: $\lambda = 193$ nm); and Nd:Yttrium Aluminium Garent (YAG) pulsed lasers ($\lambda = 355$ nm) [90]. Also, Okoshi et al. [91] reported on the PLD ZnO by using a Ti:sapphire

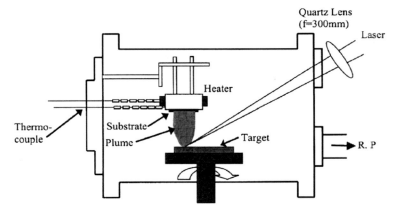

Fig. 3.37. Schematic diagram of a pulsed-laser-deposition system. Reprinted with permission from [87]. Copyright (2001), American Institute of Physics

laser ($\lambda = 790$ nm) and Dinh et al. used visible light from a Cu-vapor laser (emitting at $\lambda = 510$ and 578 nm with a 2:1 intensity ratio) with 50 ns pulse-length operating at $4,400$ Hz [92].

Reported results on laser ablated thin films from wide band gap insulators indicate that the surface morphology depends on the optical properties of the target material [93]. The higher optical absorption coefficient (α) of the target material, the lower is the droplet density on the surface of the deposited films. Since the optical absorption of ZnO increases with decreasing wavelength it has been clearly seen that films deposited with ArF ($\lambda = 193$ nm), KrF ($\lambda = 248$ nm) excimer lasers, and frequency quadrupled ($\lambda = 266$ nm) Nd:YAG laser yield a superior morphology compared with the films deposited using the second harmonic mode ($\lambda = 532$ nm) and the fundamental frequency ($\lambda=1,064$ nm) of Nd:YAG lasers [94]. Therefore, the ArF, KrF excimer lasers, or the frequency quadrupled Nd:YAG lasers are commonly used for PLD of ZnO films.

Target Materials for PLD of ZnO

Since the composition of the deposited film is directly related to the target composition in case of the PLD, a high purity target is very important for the PLD. For ZnO thin film deposition, a different type of target material including sintered ceramic discs prepared from pressed powders [95], single crystals of ZnO [96] and a pure metallic zinc (Zn) target to be used under reactive oxygen atmosphere [97, 98] have been used. Among these targets, sintered ceramic targets are most commonly used. The target size is typically 1–2 in. in diameter, however in special cases, it is possible to prepare and use targets as small as 1×1 cm^2 when the available amount of powder is very small. Special care is needed to reduce the impurity content in targets and the source powders for preparing the target by sintering [81]. It is also required to clean the target surface before using in the PLD system.

3.1.7 Droplet Formation in the PLD Process

Droplet formation during the growth is recognized as a significant problem in films grown by PLD, and their density and size distribution are closely related with the ejection of liquid droplets by the target during irradiations. The liquid droplets lead to the formation of macroscopic particulates on the surface of deposited films [99]. As mentioned before, it is known that deposition with a longer wavelength laser generates more droplets. Droplet formation is a severe problem in the PLD. Therefore, a careful adjustment of growth condition is required. Ablating the highly dense ZnO target with a laser of short wavelength and ensuring a uniform power density across the laser spot tends to reduce the problem of droplet formation [100]. Various methods were developed to reduce the droplet problem. However, the origin of such particulate matter formation during the ablation process is still not well understood. There are a few suggestions based on phase explosions and splashing effects [101,102]. The presence of droplets on the film surface is more or less a direct consequence of the physical nature of the target and the related thermal effects occurring within the target during the laser–matter interaction.

Even with the general problems of the droplet in PLD, actually, ZnO is regarded as a suitable material for PLD. Since the ZnO target melts and sublimes at 2,248 K without a phase transformation a droplet free growth is possible for ZnO film growth. However, experiments indicate that its molten state does exist for a finite duration [103]. Use of a femtosecond laser effectively reduces the droplet density [104] because the laser energy is consumed fast, which minimizes additional thermal effects [91]. The result can be explained in terms of the difference in the kinetic energy of the species ejected from the target [105]. Interaction of a laser pulse with a solid material can be divided into two sub categories, depending on the pulse duration. Laser–solid interaction within nanoseconds or longer laser pulses of sufficient duration couple not only into the electronic but also the vibration wave function of the material, but with femto/picosecond pulses the duration of the excitation is too short to couple directly into the vibrational wave function [106]. Thus different ablation behavior could be induced in solids after changing the pulse duration of the interacting laser.

In order to reduce the droplet density lots of investigations have been carried out on a number of materials using a dual-laser ablation method as well [103,107,108]. Laser irradiation from a CO_2 laser (10.6 μm, 200 ns) and a KrF laser strikes alternately (or partially overlap) on the target, and are temporally adjusted. The laser system and operation are complicated and the experiment demands precise temporal synchronization between the two laser pulses [103]. The infrared laser initially melts the target surface, but does not ablate and the UV laser pulse immediately ablates the molten layer to remove any large-size droplets, and additional heating up in the plume with an extended infrared laser prevents the gaseous specie to form small particles.

3.1.8 Growth of ZnO Thin Films on Various Substrates by PLD

For the growth of ZnO by PLD, various kinds of substrates including amorphous or single crystalline substrates including glass, fused quartz, sapphire, $ScAlMgO_4$, and $SrTiO_3$ etc. were used. However, single crystalline epitaxial growth was generally possible when using the sapphire, $ScAlMgO_4$, and $SrTiO_3$ substrates. Therefore, in this section, single crystalline ZnO films grown on these substrates by PLD have been reviewed and discussed.

ZnO on Sapphire Substrate

Ohtomo et al. have fabricated ZnO thin films on sapphire substrates at temperatures ranging from 350 to 1,000°C by PLD [109]. On increasing the growth temperature, a lateral grain size evaluated by X-ray reciprocal space

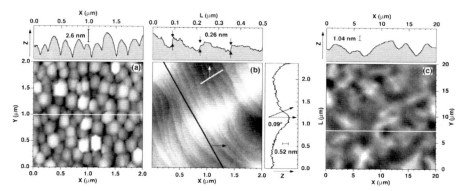

Fig. 3.38. (a) AFM image ($2 \times 2\,\mu m^2$) of films grown at 550°C. (b), (c) AFM images of the same film after annealing at 1,000°C with scales of $2 \times 2\,\mu m^2$ and $20 \times 20\,\mu m^2$, respectively. The cross-sections are shown at the *top* of each image. The marks in the profiles show z-axis scales. As seen in (b) and the right-hand side profile, the terrace plane is slightly *curved*, indicating that the (0 0 0 1) plane is bending. Reprinted with permission from [109]. Copyright (1999), American Institute of Physics

mapping, increased, resulting in improved electron mobility. By annealing the grown films under 1 atm of oxygen at 1,000°C, the films having much larger grain size ($>5\,\mu m^2$) and higher mobility ($\sim 120\,cm^2\,Vs^{-1}$) comparable with those for bulk single crystals were obtained. After annealing at 1,000°C, all the films showed very smooth surface regardless of the growth temperature. Figure 3.38 shows surface morphologies of the as grown film at 500°C and the surfaces after the annealing at 1,000°C. The surface is composed of atomically flat terraces and 0.26 nm high steps (a charge neutral unit of ZnO), as seen in the top panel profile of Fig. 3.38b.

The ZnO films with very high hall mobility were reported by employing the multistep growth process, in which a LT interlayer growth followed by HT annealing and overgrowth were successively repeated several times to improve the crystallinity [110]. The detailed studies suggest that the insertion of a 30 nm thin ZnO relaxation layer deposited at LT was critical to achieve the high electrical quality of ZnO. The AFM images in Fig. 3.39, with a scan area of $333\,\mu m^2$, clearly shows different surface morphologies for the samples with different electron mobilities. Figure 3.39a shows the surface of a three-step-grown ZnO film with a thickness of $0.6\,\mu m$, which showed the low mobility of $26\,cm^2\,Vs^{-1}$ and residual carrier concentration of $1.7 \times 10^{17}\,cm^{-3}$. Clearly resolved hexagonally faceted columnar grains with a height variation of 5–15 nm and a lateral size of 200–300 nm dominate the surface morphology. In contrast, Fig. 3.39b shows an atomically flat surface and a large lateral grain size of 0.5–$1\,\mu m$ for the sample with a thickness of a $1\,\mu m$, which showed the high mobility of $119\,cm^2\,Vs^{-1}$ and residual carrier concentration of $2.8 \times 10^{16}\,cm^{-3}$. These results indicate that the electrical properties of ZnO films grown by PLD have a strong relation with the surface roughness and grain sizes.

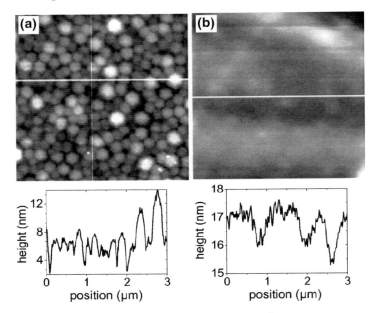

Fig. 3.39. AFM surface images of scan size $333\,\mu m^2$ together with line scans of three-step-grown ZnO films exhibiting different mobility and thickness of $26\,cm^2\,V^{-1}\,s^{-1}$ and $0.6\,\mu m$ (**a**), and $119\,cm^2\,V^{-1}\,s^{-1}$ and $1\,\mu m$ (**b**), respectively. Reprinted with permission from [110]. Copyright (2003), American Institute of Physics

Lin et al. have developed a new method of growing ZnO films developed using a fast-pulsed laser deposition (FPLD) [111]. A diode-pumped solid-state laser (355 nm wavelength, 15 ns pulse width) running at 10 kHz was used to ablate the ZnO target. The energy per laser pulse was fixed at $200\,\mu J$, and the laser beam scanned over the target surface was tightly focused so that the fluence was maintained at approximately $2\,Jcm^{-2}$. The films were grown at a rate of approximately $4\,nm\,min^{-1}$ over a wide range of substrate temperatures from 600 to 900°C. They used RF (operated at 300 W) atomic source as oxygen source and found that the film quality strongly depended on the substrate temperature as well as the oxygen pressure. The optimized film grown at conditions with the substrate temperature of 800°C and with the oxygen pressure of 1×10^{-5} Torr showed the in-plane orientation relationship, $ZnO[10\bar{1}0]//$sapphire $[11\bar{2}0]$, without no other rotational domains. Figure 3.40a shows (0002) θ–2θ scan, where persistent Pendellösung fringes were clearly observed, indicating the very flat surfaces. Figure 3.40b shows the (0002) ω-XRC, where the narrow FWHM of 10 arcsec was observed, implying a very small tilt in the c-planes and perfect ordering along the growth direction [111].

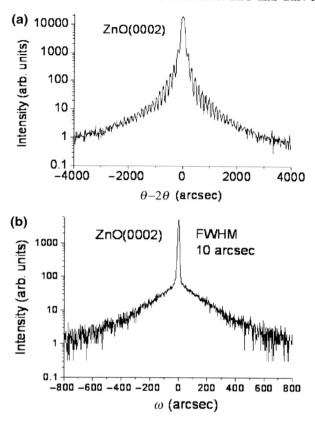

Fig. 3.40. (a) X-ray θ–2θ scan of the ZnO(0002) reflection showing pronounced Pendellösung oscillations. (b) X-ray ω-scan of the ZnO(0002) reflection showing a width of only 10 arcsec. Reprinted with permission from [111]. Copyright (2005), The Institute of Pure and Applied Physics

ZnO on (0001) ScAlMgO$_4$ Substrate

In order to overcome the problem caused by the high lattice misfit when growing the ZnO films on sapphire substrates, we can use the ZnO substrate (homoepitaxy) or the lattice-matched substrate (heteroepitaxy). As the lattice-matched substrate for the ZnO film growth, hexagonal (0001) ScAlMgO$_4$ (SCAM) substrate is proposed [109]. The hexagonal ScAlMgO$_4$ has the lattice constant of $a = 0.3246$ nm and $c = 2.5195$ nm and provides in-plane lattice mismatch as small as 0.09% with ZnO. ScAlMgO$_4$ is considered to be natural superlattice composed of alternating stacking layers of wurtzite (0001)-face (Mg, Al)$_x$ and rock salt (111)-face ScO$_y$ layers, and has a strong cleavage habit along the (0001) plane. The substrate is insulating, and is suitable for characterization of optical and electronic properties of ZnO due to its large band-gap

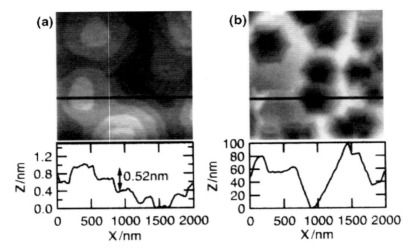

Fig. 3.41. Atomic force microscopy topography and cross-sectional profiles of ZnO films (thickness ~2 μm) grown at 1,000°C on (**a**) cleaved ScAlMgO$_4$(0001) and (**b**) sapphire(0 0 0 1) substrates. Reprinted with permission from [113]. Copyright (2000), Elsevier

of 6.2 eV. Investigation on surface termination of the (0001) SCAM substrate revealed that the surface actually terminates with the ScO$_y$ layer [112]. The ZnO films grown on (0001) SCAM substrate showed the great improvement of surface morphology as shown in Fig. 3.41 [113]. The films grown at 1,000°C on cleaved ScAlMgO$_4$ substrates showed a very flat surface (roughness ~0.2 nm) that consists of atomically flat terraces and is round shaped and has 0.26 nm high steps as shown in Fig. 3.41a. On as-polished ScAlMgO$_4$ substrates, the surface was also very smooth but the 0.26 nm high steps were aligned in parallel with equal spacing representing a miscut of the substrate (~0.2°). The step height corresponds to a half of c-axis length (0.52 nm) of ZnO and the charge neutral unit. On the other hand, the films grown at the same temperature of 1,000°C on sapphire substrates showed a rough surface (roughness ~20 nm) as shown in Fig. 3.41b, where the hexagonal pits (peak to valley ~100 nm) were clearly observed.

The crystal quality of the films was examined by XRC measurements for (0002) and (10$\bar{1}$1) reflection peaks [113]. As shown in Fig. 3.42, crystallinity was also improved by the use of ScAlMgO$_4$ substrates. FWHMs for the (0002) and (10$\bar{1}$1) XRCs of the ZnO film grown on ScAlMgO$_4$ substrate were 39 and 34 arcsec, respectively.

Optical properties of the ZnO films on (0001) SCAM substrates were investigated by LT PL measurements [114], where the sample grown at 600°C was selected because it was the typical deposition condition for obtaining the excitonic lasing samples [114]. The ZnO films grown on ScAlMgO$_4$ substrate gave narrower PL spectrum than that on sapphire substrate as shown in

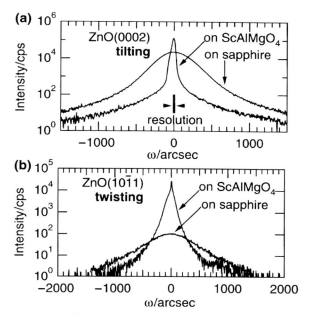

Fig. 3.42. XRD rocking curves for ZnO films grown at $1,000°C$ on $ScAlMgO_4(0001)$ and sapphire(0001) substrates. (a) ZnO(0002) and (b) ZnO (10$\bar{1}$1) rocking curves representing out-of-plane tilting and in-plane twisting, respectively. Reprinted with permission from [113]. Copyright (2000), Elsevier

Fig. 3.43a. The emission peak was observed at 3.364 and 3.361 eV for the films grown on sapphire and $ScAlMgO_4$ substrate, respectively. These peaks correspond to donor bound exciton transitions (D^0X) [115]. As shown in Fig. 3.43b, the absorbance spectrum showed clear splitting into two peaks, which can be attributed to the resonance of A-(3.378 eV) and B-excitons (3.385 eV).

The growth mode and properties of the ZnO films could be further improved by employing the ZnO buffer followed by HT annealing before the HT ZnO growth [116]. AFM images for the ZnO films before and after the annealing are shown in Fig. 3.44a and b, respectively. The surface of as-deposited buffer layer showed subnanoscale corrugation, and the rms roughness was as large as 0.2 nm. The annealed surface showed the 0.26 nm-high steps corresponding to one molecular layer of ZnO and atomically flat terraces (rms roughness of 0.07 nm), as shown in Fig. 3.44b. Most of the trials of $10 \times 10\,\mu m^2$ AFM inspection for the annealed film resulted in a step-free and atomically smooth surface. It was argued that since the cleaved SCAM surface is atomically flat in a macroscopic scale ($>100\,\mu m$), ZnO surface with a step-free and an atomically smooth surface on such a scale was possible.

Optical properties of the ZnO films with the HT annealed ZnO buffer were examined by the PL and optical reflection spectroscopy at 5 K [117].

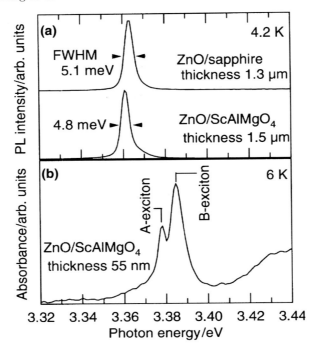

Fig. 3.43. (a) Photoluminescence spectra measured at 4.2 K for ZnO films grown at 600°C on sapphire and ScAlMgO$_4$ substrates and (b) Absorbance spectrum measured at 6 K for the thin (55 nm) film. Reprinted with permission from [114]. Copyright (1998), Elsevier

Figure 3.45(a) shows the PL spectra taken under excitation with a 325 nm line of a He–Cd laser for the following ZnO samples: (from upper to lower) an epilayer directly grown on the SCAM substrate [116], that grows on the buffer layer, and one of the best quality bulk single crystals made by Eagle-Picher Technologies, LLC. [118]. The ZnO film grown on the buffer layer at 800°C showed the dominant luminescence lines at 3.360, 3.365, and 3.372 eV, where the origin of which could be assigned to neutral donor bound excitons (D^{0}X). There were also a free A-exciton (FE$_A$) line at 3.377 eV and a shoulder due to the transverse B-exciton (B$_{\omega T}$) emission at 3.382 eV. These features are very similar to those of bulk single crystal. The ZnO film directly grown on SCAM showed the D^{0}X line at 3.367 eV and the FE$_A$ line at 3.379 eV. Both energy positions were slightly shifted by a compressive strain due to coherent growth. Figure 3.45b shows the reflection spectra. The emission line positions for FE$_A$ agree with the structures observed in the reflection spectra. By inserting the ZnO buffer layer, ground-state anomaly begins to oscillate very sharply and the high-reflectivity range of B-exciton becomes wide. The latter could be understood as a sign of the smaller damping and the thinner

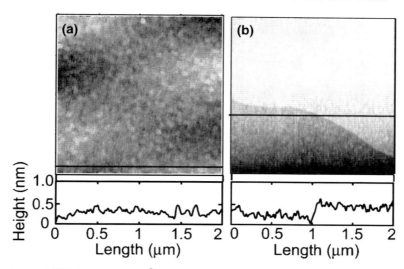

Fig. 3.44. AFM images (4 μm^2) and cross-sectional profiles for as grown (**a**) and annealed (**b**) ZnO buffer layers. Annealed surface shows very wide atomically flat terraces (rms roughness of 0.07 nm) and 0.26 nm-high steps corresponding to the half-unit cell of ZnO, whereas as-grown film has a featureless surface, with rms roughness of 0.2 nm. Reprinted with permission from [116]. Copyright (2003), American Institute of Physics

surface dead layer. The Bohr radius of $n = 1$ exciton is 18 Å and that of $n = 2$ is 72 Å. The clear observation of $n = 2$ resonance in the ZnO film deposited on the buffer layer manifests itself in which the coherent length extends more in ZnO crystal. Since such an excited-state ($n = 2$) excitons have much smaller oscillator strength and are easily exposed to the influence of various scattering events, their clear observation definitely indicated the reduced inhomogeneity or reduced defect.

ZnO on SrTiO$_3$ Substrate

SrTiO$_3$(STO) is the one of the mostly used substrates for the epitaxial growth of functional oxide films. The STO has a cubic perovskite structure (lattice constant, a = 0.3905 nm) different from the sapphire or the ScAlMgO$_4$ substrate. Epitaxial growth of ZnO films by the PLD has been reported by several researchers [119–121] by using the (001), (110), and (111) substrates. Comparison of the epitaxial growth of ZnO films on (001), (011), and (111)-orientated SrTiO$_3$ single-crystal substrates by the PLD was studied by Wei et al. [121]. For the PLD growth, they used a KrF excimer laser of 248 nm wavelength and 30 ns pulse duration. Laser energy density and laser frequency were kept as 2 Jcm^{-2} and 1 Hz, respectively. In order to minimize the effect from miscut angle (0.2°–0.3°) and surface roughness of SrTiO$_3$ substrates, they performed

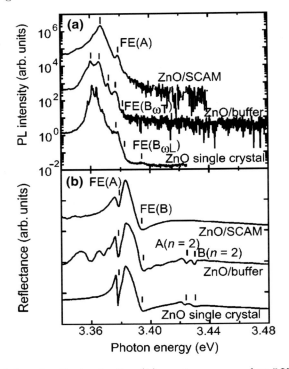

Fig. 3.45. PL (**a**) and optical reflection (**b**) spectra measured at 5 K for a ZnO thin film directly deposited on SCAM substrate, the ZnO thin film deposited on the ZnO buffer layer at 800°C with the laser repetition of 5 Hz. Free exciton emission lines are observed at 3.377 eV (FE$_A$) and 3.382 eV (FE$_{B\omega T}$) for the ZnO thin film grown on the buffer layer. The FE$_A$ (3.379 eV) line for the ZnO thin film grown on SCAM was shifted by the compressive strain induced by the in-plane lattice mismatch. The $n = 2$ exciton structure is observed in the optical reflection spectrum for the ZnO thin film grown on the buffer layer. Reprinted with permission from [117]. Copyright (2003), American Institute of Physics

chemical cleaning of the substrates by using buffered hydrofluoric acid solution (pH = 4.5). They obtained the uniform Ti–O terminated surface by this treatment. Before the heteroepitaxial growth of ZnO films, they grew about ten STO monolayers (homoepitaxial) on the atomically flat substrates.

The growth behavior was in situ RHEED observations as shown in Fig. 3.46 for the ZnO film growth on (001) STO substrate [121].

In the streaky pattern from the (100) azimuth of SrTiO$_3$ (001) substrate, as shown in Fig. 3.46a, the positions of the (0$\bar{1}$), (00), and (01) reflections were labeled, indicating an atomically smooth surface for STO substrate. At the initial growth stage, the diffraction pattern became dark due to the interface

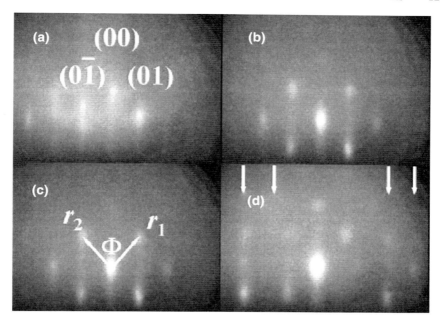

Fig. 3.46. Evolution of RHEED patterns during the ZnO films on SrTiO$_3$ (001); (**a**) SrTiO$_3$ (100), (**b**) 5 min, (**c**) 60 min, and (**d**) rotation of (**c**) by several degrees. Reprinted with permission from [121]. Copyright (2007), American Institute of Physics

transition from the perovskite to the wurtzite structure. As the deposition proceeded, several Bragg-reflection spots were superimposed on the sharp streak diffraction, suggesting that the film was grown in the Stranski–Krastanov growth mode. As shown in Fig. 3.46c, the measured ratio r_2/r_1 of the spacing of diffraction planes and the angle Φ between the corresponding spots is, respectively, about 1 and 80°, if on choosing the central spot to be the origin. Therefore, the azimuth and the orientation of the film were estimated to be [$\bar{1}$11] and (110), respectively. After rotating the sample by several degrees, additional diffraction streaks and spots were observed, labeled by arrows in Fig. 3.46d, indicating that there might be polydomain structure or other phases in the film. From the RHEED observations and XRD measurements, they determined the epitaxial relationships and these were summarized schematically in Fig. 3.47 [121]. In the case of ZnO on SrTiO$_3$ (001), four orthogonal domains coexisted in the ZnO epilayer, i.e., ZnO (110)//SrTiO$_3$ (001) and ZnO ($\bar{1}$11)//SrTiO$_3$ <100>. For (011)- and (111)-orientated substrates, single-domain epitaxy with c axial orientation was observed, in which the in-plane relationship was ZnO [110]//SrTiO$_3$ [110] irrespective of the substrate orientations. The crystalline quality of ZnO on SrTiO$_3$ (111) was the best among the investigated ZnO films.

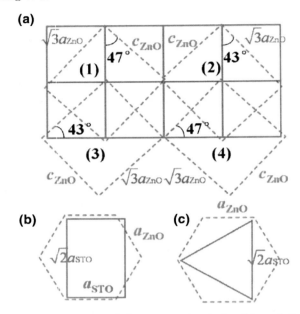

Fig. 3.47. Schematic drawing of the epitaxial relationships of ZnO on SrTiO$_3$: (**a**) ZnO(110)//SrTiO$_3$(001); (**b**) ZnO(001)//SrTiO$_3$(011); and (**c**) ZnO(001)// SrTiO$_3$(111). The *solid* and *dashed lines* are corresponding to the unit cells of SrTiO3 and ZnO, respectively. Reprinted with permission from [121]. Copyright (2007), American Institute of Physics

3.1.9 Bandgap Engineering of ZnO-based Alloys Grown by PLD

In this section, bandgap engineering of ZnO-based alloys is discussed. The growth and properties of $Zn_xMg_{1-x}O$ and $Zn_yCd_{1-y}O$ are presented. Most of works on ZnO-based heterostructure and bandgap engineering were performed on ZnMgO based alloys.

MgZno Alloy

Ohtomo et al. reported the growth and optical properties of $Mg_xZn_{1-x}O$ alloy for the fabrication of heteroepitaxial ultraviolet LEDs based on ZnO [122]. Their $Mg_xZn_{1-x}O$ thin films were grown on sapphire (0001) substrates by the PLD. Predetermined amounts of ZnO (5 N) and MgO (4 N) powders were mixed, calcined, and sintered to prepare the targets with an Mg content ranging from $x = 0$ to $x = 0.18$. The XRD spectra of the targets showed no detectable MgO peaks for $x < 0.10$, whereas clear MgO peaks for $x > 0.13$. The targets were placed at a distance of 4 cm from the substrate and ablated by KrF excimer laser pulses (254 nm, 10 Hz, 20 ns) with a fluence of 1 Jcm^{-2}. The films (300 nm thick) were deposited at a growth temperature of 600°C in 5×10^{-5} Torr of pure oxygen (99.9999%).

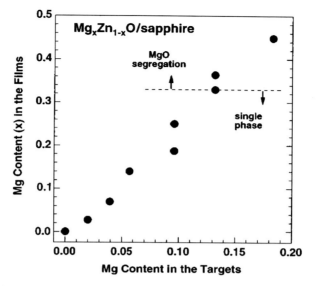

Fig. 3.48. Mg content in the $Mg_xZn_{1-x}O$ epitaxial films as a function of that in the targets. Reprinted with permission from [122]. Copyright (1998), American Institute of Physics

Figure 3.48 shows the Mg content in the films as a function of these in the targets [122]. The Mg content in the films was systematically larger than that of the targets by a factor of 2.5. This difference could be understood as the result of much higher vapor pressures of ZnO and Zn than those of MgO and Mg at the substrate temperature. The Zn related species can easily be desorbed from the growing surface and lead to the condensation of Mg-related species on the growing surface, which result in higher Mg concentration in the films compared with that of the targets.

They could grow single-phase thin films having a wurtzite structure and c-axis orientation with x up to 0.25, as verified by the XRD analysis. When the Mg content was larger than 0.36, small peaks due to an impurity phase [(111) oriented MgO] were observed. At $x = 0.33$, a very weak signal could be detected where MgO (222) peak should appear. However the intensity was much smaller than the ZnO (0002) peak by a factor of 10^{-4}, therefore, they concluded that the solubility limit of the MgO in ZnO for the films as 33 mol%. The thermodynamic solubility limit of MgO in ZnO has been reported to be less than 4 mol%, according to the phase diagram of the ZnO–MgO binary system [123]. Therefore, such $Mg_xZn_{1-x}O$ films with high Mg concentrations can be considered as metastable phases. Figure 3.49 shows transmittance spectra measured at RT by conventional ultraviolet-visible spectrometer. The absorption edge shifted as a function of x when $x < 0.36$, saturating at higher Mg concentration. These results were in good agreement with the appearance of the MgO impurity phase detected by XRD. The bandgap (E_g) of the grown

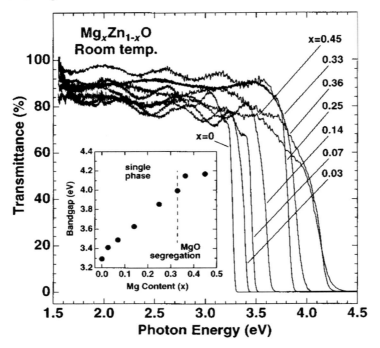

Fig. 3.49. Transmittance spectra of $Mg_xZn_{1-x}O$ films measured at room temperature. The *inset* shows the bandgap (Eg) determined from the spectra assuming an $\alpha^2 \propto (h\nu - Eg)$ dependence, where a and $h\nu$ are the absorption coefficient and the photon energy, respectively. Reprinted with permission from [122]. Copyright (1998), American Institute of Physics

$Mg_xZn_{1-x}O$ films were evaluated by plotting an α^2 vs E_g plot for the spectra to fit the data assuming an $\alpha^2 \propto (h\nu - E_g)$ relationship, where α is the absorption coefficient and $h\nu$ is the photon energy. The determined bandgap was shown as a function of x in the inset of Fig. 3.49. They showed that the bandgap of the $Mg_xZn_{1-x}O$ films, E_g linearly increased up to 4.15 eV for $0 < x < 0.36$. This indicates that the $Mg_xZn_{1-x}O$ alloy is a suitable material for potential barrier layers in ZnO-based optoelectronic devices [122].

On the other hand, the growth and microstructural study of epitaxial $Zn_{1-x}Mg_xO$ composition spread over wide ranges were performed [124, 125]. Here, the $Zn_{1-x}Mg_xO$ epitaxial films with the composition spreads, where the composition across the chips was linearly varied from ZnO to MgO, were prepared by the PLD on (0001) sapphire substrates [125]. They investigated structural variations and solubility limits of the ZnO and MgO phases by a scanning X-ray microdiffractometer, where the measurements were taken across the spread for different compositions. The XRD results are summarized in Fig. 3.50, where the changes of intensities of the detected diffraction peaks with their compositions are plotted. Starting from the ZnO end, the intensity

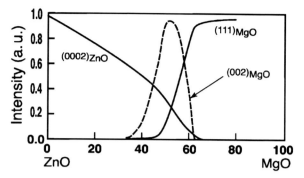

Fig. 3.50. The normalized intensity of X-ray diffraction peaks measured for a compositional spread as a function of composition. The peaks are (0002) of ZnO, and (200) and (111) of MgO. Reprinted with permission from [124]. Copyright (2005), American Institute of Physics

of the (0002) of ZnO decreased with a mole fraction x in the $Zn_{1-x}Mg_xO$ and disappeared at about x = 0.6. The single (0002) peak suggested that the $Zn_{1-x}Mg_xO$ film has (0001) ZnMgO//(0001)Sapphire orientation relationship. Since only the (0002) ZnO peak was detected in a composition range between 0 and 0.33 of MgO, the result suggests a near 33% solubility of MgO in ZnO. The decrease in the intensity of the (0002) ZnO peak supports the substitutional solubility of MgO, since the structure factor value, F_{0002} decreases with substitution of Zn by Mg in the ZnO structure.

The MgO phase was identified by its (111) peak and appeared as a single phase in the composition range 0.6 < x < 1.0 as shown in Fig. 3.50. Since the (111) peak was the only detected peak, the result suggested that the (111) MgZnO//(0001)Sapphire orientation relationship and a 40% solubility of ZnO in MgO. In the composition range of 0.4 < x < 0.6, another strong peak was identified in coexistence with ZnO (0002) and MgO (111). The value of the peak was close to that of the MgO (200), which suggested the presence of MgO in two different orientations, i.e., one with MgO (111) normal to the substrate plane and another with MgO (100) [124].

Heitsch et al. investigated surface roughness and luminescence properties of (0001) $Mg_xZn_{1-x}O$ films (0 < x < 0.19) grown on a-plane sapphire substrates by PLD [126]. The shift of the PL maximum with increasing x over the investigated Mg concentration range is depicted in Fig. 3.51 for samples of ∼250 nm thickness [126]. Figure 3.51a shows a good agreement of the PL results at 300 K with the dependence of the bandgap energy E_g of $Mg_xZn_{1-x}O$ determined by spectroscopic ellipsometry [127]. The redshift of the PL emission with respect to E_g is due to the exciton binding energy and the Stokes shift. The position of the PL emission maximum with increasing x can be fitted by using the following eq,

$$E_{PL}(300K) = (3.31 \pm 0.01)eV + (2.0 \pm 0.1)eVx. \qquad (3.3)$$

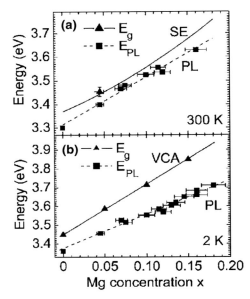

Fig. 3.51. (a) Band gap dependence (*solid line*) determined by ellipsometry and PL peak energy dependence (*filled square*) on the Mg concentration of $Mg_xZn_{1-x}O$ thin films at 300 K. The *dashed line* is the linear fit to the PL data given in (3.2). The *solid triangle* (*filled triangle*) denotes the band gap energy of the $Mg_{0.045}Zn_{0.955}O$ thin film. (b) PL peak energy dependence (*filled square*) on the Mg concentration of $Mg_xZn_{1-x}O$ thin films at 2 K and band gap energies at 0 K calculated with virtual crystal approximation (*filled triangle*). The *dashed line* is the linear fit to the PL data given in (3.3), and the *solid line* is the linear fit to the theoretical band gap values. Reprinted with permission from [126]. Copyright (2007), American Institute of Physics

In Fig. 3.51b, for different x, the PL emission maximum energies at 2 K and the bandgap energies at 0 K, were calculated using the virtual crystal approximation (VCA) [128], which are depicted. A linear fit could be found for the PL data, which can be given by

$$E_{PL}(2K) = (3.38 \pm 0.01)\text{eV} + (1.8 \pm 0.1)\text{eV}x. \quad (3.4)$$

Here, the band gap energies of $Mg_xZn_{1-x}O$ were calculated by VCA neglecting disorder effects, because their influence on the bandgap energy is assumed to be small for the investigated Mg concentration range. Considering the exciton binding energy, the Stokes shift, and the exciton localization in the potential minima at low temperatures, the calculated bandgap energies showed a reasonable agreement with the experimentally determined PL transition energies [126]. Also, the observed increasing difference between the theoretical bandgap values and the PL transition energies with increasing x displays, the expected behavior in Fig. 3.51b was understood as the result of the increasing Stokes shift [129, 130].

CdZnO Alloy

As discussed in the section, "MgZnO alloy", $Mg_xZn_{1-x}O$ alloy films have been considered as a suitable material for barrier layers due to its wider bandgap than that of ZnO. For narrower bandgaps, which are desirable for wavelength tenability and attaining bandgaps corresponding to the visible spectrum, the $Cd_yZn_{1-y}O$ alloy would be a good candidate because of the small direct band gap of CdO (2.3 eV) [131]. Makino et al. demonstrated the single phase $Cd_yZn_{1-y}O$ alloy films grown by PLD on sapphire (0001) and $ScAlMgO_4$ (0001) substrates with Cd content of up to 7% [131]. They investigated that both a- and c-lattice constants increased as the Cd content increases and the ratio of c/a also monotonically increased in contrast to that for the $Mg_xZn_{1-x}O$ alloy. Figure 3.52 shows Cd concentration (y) dependence of the RT absorption spectra in the as-grown films. The spectrum of $Cd_{0.07}Zn_{0.93}O$ encompassed a broad shoulder on the lower-energy side, indicating the formation of a cadmium-rich phase, although the density of this is relatively low. The band-gap energies (E_g) are plotted in the inset of Fig. 3.52. As shown in Fig. 3.52, the bandgap energy E_g decreased as the Cd content increased and could be estimated as $E_g(y) = 3.29 - 4.40y + 5.93y^2$. The band gap was observed to be decreased from 3.28 down to 3.0 eV by introducing 7% of Cd.

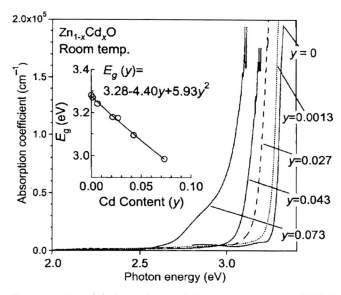

Fig. 3.52. Concentration (y) dependence of absorption spectra of $Cd_yZn_{1-y}O$ epilayers obtained at room temperature. The *curves*, from right to left correspond to those of the samples with $y = 0$, 0.0013, 0.027, 0.043, and 0.073. Reprinted with permission from [131]. Copyright (2001), American Institute of Physics

3.2 MOCVD Growth of GaN and Related Materials

K. Matsumoto, H. Tokunaga, and A. Tachibana

3.2.1 Introduction

High quality GaN growth was first demonstrated by using a two step growth technology in 1986 by H. Amano et al. [132]. They had grown AlN nucleation layer at low temperature, followed by GaN growth at high temperature by MOCVD. This low temperature grown AlN worked as a wetting layer for the succeeding growth of GaN with a proper density of nuclei. Later, S. Nakamura et al. have also grown high quality GaN by using a low temperature grown GaN nucleation layer [133]. After evolution of the growth technology of GaN, other types of nucleation technology have been developed. High temperature grown AlN nucleation layer is also a good template for high quality GaN growth [134]. The central idea of heteroepitaxy of GaN is a multilayered buffer structure, in which each stacked layer has a different role; that is alignment of crystal axis, dislocation filtering and balance or compensation of a large thermal mismatch between the substrate and overlayers.

One of the major difficulties of nitride-semiconductors by MOCVD is a parasitic reaction in vapor phase, which causes a particulate generation, followed by the removal of precursors from the growing region. Especially, It is especially very difficult to grow aluminum containing alloys. Therefore, it is important to understand the mechanism of vapor phase reaction. InGaN growth is also the key technology for the optical devices. Since InGaN alloy has a large miscibility gap, it is difficult to grow high indium containing alloys. It is known that alloy composition of InGaN follows thermodynamic prediction.

In this chapter, the major aspect of MOCVD growth of GaN and related materials: heteroepitaxy of GaN on sapphire and Si substrate, p-type doping of GaN, AlGaN growth and parasitic reaction, InGaN growth are described. Experimental and theoretical aspect of understanding parasitic reaction is addressed in detail. Recent progress of atmospheric pressure MOCVD tool is also described.

3.2.2 Heteroepitaxy of GaN and Alloys: Materials Property and Growth Condition

Because of difficulties in obtaining the high quality of GaN substrates in the early stage of researches on GaN film growth, most of the early studies on epitaxial growth of GaN were performed by heteroepitaxy, where sapphire substrates were generally used. Currently, high quality GaN substrates applicable to the GaN epitaxy and device fabrication are available through the breakthrough in bulk growth for GaN substrate and in thick GaN growth for free standing GaN substrate. However, the GaN substrates are still expensive and growing high quality GaN films on low priced different substrates (heteroepitaxy) are highly needed in fabrication of LEDs and other devices. Therefore, it is a fact that most of the important GaN epitaxy technologies such as defect reduction by employing the buffer or by the lateral over growth, p-type doping, growth of alloys, heterostructures, and bandgap engineering etc., which can be successfully applicable even for GaN homoepitaxy, have been established and developed based on heteroepitaxy.

Heteroepitaxy of GaN

The central idea of heteroepitaxy of GaN is the multilayered buffer structure, in which each stacked layer has a different role; that is alignment of crystal axis, dislocation filtering and balance or compensation of a large thermal mismatch between the substrate and overlayers. Each layer need not be a different material system and the same material grown with different growth conditions can be used. For example, a two-step growth process, which uses a LT AlN or a GaN buffer layer has usually been employed for the epitaxial growth of GaN on a sapphire substrate in MOCVD. The LT grown layer at about 450°C is used for a c-axis orientation at the initial growth stage with an appropriate number of nuclei, followed by 3D growth of GaN at HT of 1,050°C. It is noteworthy that the successive growth layer at HT automatically takes a 3D growth mode by adopting proper growth conditions. Each initial growth nuclei is to be laterally grown and coalesced. During this process, many of threading dislocations (TDs) are bent and disappear. A schematic drawing of this process is illustrated in Fig. 3.53.

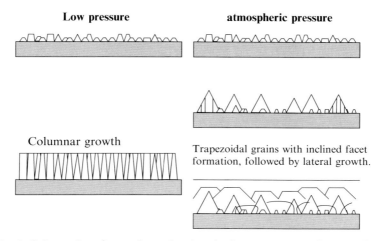

Fig. 3.53. Schematics of growth mechanism for low pressure and atmosphere pressure MOCVD growth of GaN. GaN films with low dislocation density can be grown by lateral growth in case of the atmosphere pressure MOCVD, where the dislocations are bended and hardly to be threaded to the upper region of the GaN film

It is important to optimize the density and the size of the nuclei at the initial growth stage. A retarded coalescence of large grains is useful to eliminate TD. Trapezoidal 3D grains are surrounded by inclined faces, followed by lateral growth, which well resembles epitaxial lateral over growth (ELOG) [135, 136]. These successive processes occur spontaneously in case of atmospheric pressure growth. On the other hand, under low pressure, each LT-grown nucleus tends to grow upward than laterally. Inclined facet formation at the initial growth stage at HT may take an important role in reducing the number of grains and enlargement of their size. The appearance of the facet means that a grain is surrounded by a low growth rate face. This indicates that the growth rate of each grain is slowed down, while there are many precursors migrating on the growing surface. In this situation, the precursors gather on the facets of the larger sized grains and they escape from the smaller sized grains, which eventually results in continuous growth of the larger grains with much higher possibility. Miyake et al. reported a systematic study on selective epitaxy of GaN as a function of various growth pressures and ammonia (NH_3) to trimethyl gallium (TMG) ratio [137]. They showed that the inclined facet was likely to be formed at LT or at a higher growth pressure condition, while vertical facet appears at lower pressure (Fig. 3.54). This finding can well explain the spontaneous ELOG of GaN at atmospheric pressure growth mentioned in Fig. 3.53. By careful optimization of two step growth condition at

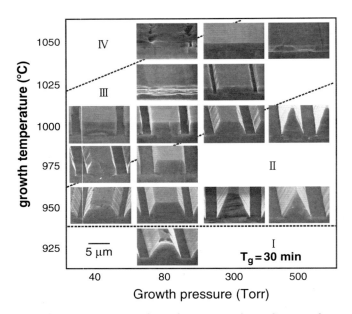

Fig. 3.54. GaN films with various facet formations depending on the growth temperatures and pressures. Reprinted with permission from [137]. Copyright (1999), The Institute of Pure and Applied Physics

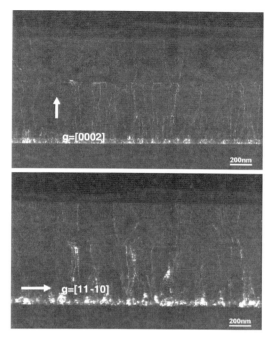

Fig. 3.55. Cross-sectional TEM micrographs of GaN on Si with multistructured buffers. The micrographs were obtained under weak beam conditions with different **g** vectors to determine the types of dislocations. Reprinted with permission from [142]. Copyright (2007), Elsevier

atmospheric pressure MOCVD, Hoshino et al. reported the growth of GaN on c-plane sapphire substrate with the FWHM values of XRC for GaN (0002) and (10$\bar{1}$2) reflections of 198 and 232 arcsec, respectively [138].

A good example of a multibuffer structure can be found in GaN growth on Si reported by Ubukata et al. [139]. Figure 3.55 shows a cross-sectional bright-field TEM image of GaN on Si. The whole structure is as follows; Si substrate/AlN/AlGaN/(AlN/GaN) strained layer superlattice/GaN. The period of the strained layer superlattice (SLS) was measured to be 24 nm (AlN: 4 nm; GaN: 20 nm). The TEM image shows that the (AlN/GaN) SLS has an abrupt and smooth interfaces.

They investigated dislocation structures and observed high densities of both edge-type and screw-type TDs at the interface between the epitaxial layers and the Si substrate. Edge type TDs with a Burgers vector of [10,11] still remained in the GaN top region, and this indicated that SLS has a negligible effect on blocking the edge type TDs while screw type TDs are filtered in the SLS and at the interface of SLS and AlGaN. The major role of SLS was to balance the large tensile strain in GaN arising from the thermal mismatch between GaN film and Si substrate. It is known that GaN on AlN structure has a compressive strain. Thereby, (GaN/AlN) SLS can compensate a tensile strain. The first AlN nucleation layer beneath the SLS was grown at 1,000°C in this case. The initial nucleation layer must cover whole Si surface and be good enough to prevent from GaN alloying with Si at subsequent growth. The detailed mechanism of the dislocation filtering is beyond the focus of this chapter, but, a growth mode at each interface of GaN and AlN looks like to have an important role. Judging from TEM, the initial growth mode of GaN on top of SLS would be a 3D growth. Since high quality GaN can also be grown on a HT grown AlN template, insight into this interface would be helpful to understand the mechanism of reduction of TDs. Sakai et al. have obtained GaN with TD of 5×10^7 cm^{-2} on a highly c-axis oriented AlN on sapphire [140]. In order to avoid a parasitic reaction between trimethyl aluminum (TMA) and NH$_3$, the structure in Fig. 3.55 was grown at 340 Torr (45 kPa). Since the growth mode of strained GaN is a function of growth pressure and temperature, the effect of SLS and the optimum structure would be a function of these growth conditions.

P-Type Doping of GaN

P-type doping of GaN is realized in a certain growth condition. Usually bis-cyclopenta-dienyl-magnesium (Cp$_2$Mg) is used for p-type dopant in MOCVD.

Figure 3.56 shows an example of Mg doping into GaN. Mg concentration in GaN is linearly increased as Cp$_2$Mg flow rate is increased [141]. However,

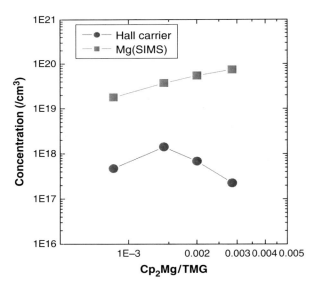

Fig. 3.56. Hole carrier concentration and the atomic Mg concentration Mg-doped GaN films plotted as a function of Cp$_2$Mg/TMG ratio. Reprinted with permission from [141]. Copyright (2004), Elsevier

free hole concentration has a broad maximum at the solid Mg concentration of 4×10^{19} cm^{-3}. At this Mg concentration, the well known defect called inversion domain boundary (IDB) start to appear [142]. The IDB caused by high Mg doping concentration has an inverted pyramid shape, in which Ga face and N face is inverted. The cause of IDB formation is supposed to be a surface segregation of Mg at high doping condition. Excess Mg get together to form the IDB. The IDB tends to nucleate at edge type dislocation and vertically align to form a hollow pipe [143]. This hollow pipe can work as a diffusion pass of Mg into the active region, which result in a device degradation. The surface condensation of Mg results in a long tailing of Mg concentration in the successive grown layer after the Mg doping is finished by switching the valve off [144].

Another important growth parameter for high quality p-type GaN:Mg growth is a growth pressure and growth rate. At the low pressure, it is known that maximum growth rate of the Mg doped GaN film is limited to orders of a few $0.1 \, \mu m \, h^{-1}$ [145]. Therefore, sometimes InGaN MQW active layer is degraded during the growth of p-type contact layer growth at low pressure MOCVD [146]. Figure 3.57 shows the growth pressure dependence of Hall carrier concentration of free hole with a constant Cp$_2$Mg/TMG ratio of 1.5×10^{-3} and a constant growth rate of $3 \, \mu m \, h^{-1}$ [141].

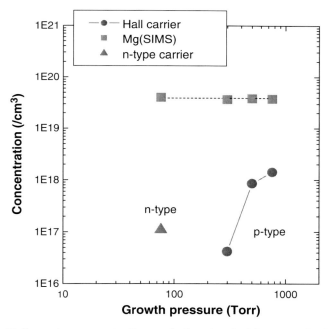

Fig. 3.57. Hall carrier concentration and the atomic Mg concentration for the Mg-doped GaN films as a function of growth pressure. Reprinted with permission from [141]. Copyright (2004), Elsevier

On decreasing the growth pressure, the Hall carrier concentration was decreased, and eventually, the electronic conductivity changed to n-type at 76 Torr (10 kPa) even after postgrowth annealing, while the incorporated Mg concentration was constant under all the growth pressure conditions. From the LT PL measurements, the neutral donor exciton lines for the samples grown at 300 Torr (40 kPa) were several times larger than those for the samples grown at 760 Torr (100 kPa), indicating that the compensation ratio increased with decreasing the growth pressure. It is known that by reducing growth rate and employing high NH_3/TMG ratio, high quality p-type GaN film is obtained at the growth condition with reduced pressure [145].

AlGaN Growth

Al containing alloys are important for device applications. For visible LEDs, AlGaN is used for a carrier stop layer to prevent electrons from migrating into p-type region. For LDs, AlGaN layers with an Al composition of 5–10% are used for cladding layers. For high electron mobility transistor (HEMT), AlGaN/GaN heterostructures with an Al composition of 25–30% are used. However, there is a difficulty in growing AlGaN layer with a high Al concentration because of a parasitic reaction [147]. When TMA and NH_3 are mixed together, they are ready to form particulates with small thermal excitation energy. This parasitic reaction takes place in vapor phase, followed by the removal of precursors from the growth region. Since the detailed mechanism of this reaction is described in the Sect. 3.2.3, only an overall growth characteristic of AlGaN is described here. Figure 3.58 shows the dependence of aluminum concentration and growth rate of the $Al_xGa_{1-x}N$ films on the velocity of gas flow, where the flow is designed as the system for the conventional GaAs growth [148]. TMG and TMA flow rates were 56 and 8.8 µmol min^{-1}, respectively. It is clear that we must increase the total gas flow rate in order to acquire higher aluminum concentration in $Al_xGa_{1-x}N$ films if we employ the atmospheric pressure growth. However, by increasing the flow rate to increase the Al concentration, growth rate of the $Al_xGa_{1-x}N$ films was decreased due to the dilution of the metal organic source as shown in Fig. 3.58. On the other hand, the $Al_xGa_{1-x}N(x = 0.25)$ film with a high aluminum concentration of 25% film without decreasing the growth rate could be grown by employing the low pressure growth with the growth pressure of 76 Torr (10 kPa). Here, it should be noted that these growth characteristic of AlGaN is a strong function of growth reactor. There was a return flow in the reactor used in Fig. 3.58.

Figure 3.59 shows the dependence of the AlGaN growth rate on the growth pressure by using a high flow speed reactor, in which a three-layer laminar flow gas injection was used [141]. The details of the reactor configuration are described later in this chapter. The growth rate of GaN is also shown as a broken line in Fig. 3.59. The total mole flow rate of (TMA + TMG) was kept constant for all the samples, and only the TMA partial pressure was changed between 0 and 0.3. The growth rate was nearly constant as a function

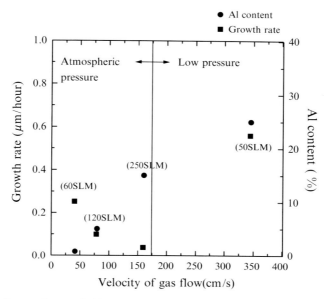

Fig. 3.58. Dependence of Al concentration and growth rate of AlGaN films as a function of the velocity of gas flow. Reprinted with permission from [148]. Copyright (2000), Elsevier

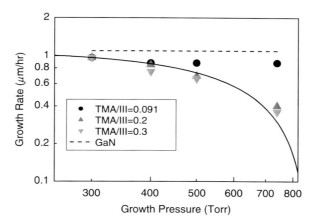

Fig. 3.59. Growth rate of AlGaN films as a function of growth pressure with various TMA/(TMA + TMG) ratio using the reactor system with the three-layer laminar flow gas injection. Reprinted with permission from [141]. Copyright (2004), Elsevier

of growth pressure at an input of TMA/(TMA + TMG) = 0.091, at which aluminum composition incorporated into the solid was 0.08. The growth rate of $0.88\,\mu m\,h^{-1}$ was obtained under this TMA flow condition.

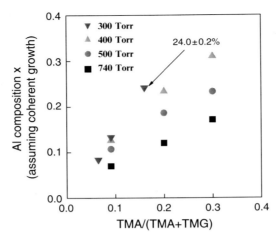

Fig. 3.60. Al composition in the AlGaN films as a function of various TMA/(TMA + TMG) ratio under the various growth pressures. Each data point directly corresponds to those in Fig. 3.59. Reprinted with permission from [141]. Copyright (2004), Elsevier

Such high growth rate obtained for this TMA/(TMA + TMG) rate of 0.091 seems to be a less parasitic reaction. This means that the suppression of parasitic reaction was possible by employing the high flow speed three-layer laminar flow gas injection. However, the growth rate of AlGaN with an input of TMA/(TMA + TMG) of 0.2 and 0.3 showed a steep decrease of growth rate when the growth pressure exceeds about 500 Torr (66 kPa). This is caused by a parasitic reaction of TMA and NH_3. When there is a parasitic reaction, aluminum composition of AlGaN decreases as well as the growth rate. The aluminum composition of the AlGaN films for each growth pressure is shown in Fig. 3.60 [141]. Each data point in Fig. 3.60 directly corresponds to that in Fig. 3.59. In Fig. 3.60, the aluminum composition increased proportionally with the TMA partial pressure at each growth pressure. The aluminum incorporation efficiency increased as the growth pressure decreased, which was caused by selective evaporation of gallium. A similar result was also reported in a vertical high speed rotation disc reactor [149]. At higher temperature, GaN is etched by hydrogen. This can be explained by thermodynamic consideration [150].

InGaN Growth

Thermodynamic consideration is very essential to understand the growth of InGaN. Matsuoka et al. discovered that a full range of InGaN was grown under this condition free from hydrogen [151]. Hydrogen etches InN very efficiently at a growth temperature of multi quantum well (MQW) structure. The solid indium composition is nearly determined by thermodynamic equilibrium. Figure 3.61 shows a solid indium composition of InGaN grown at

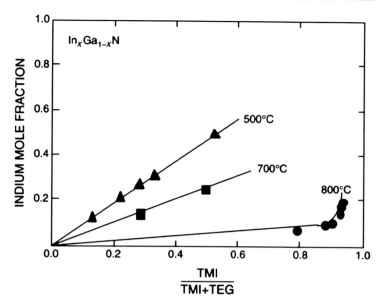

Fig. 3.61. In composition in the InGaN films as a function of input ratio of TMI/(TMI + TEG) grown at different temperature. Reprinted with permission from [151]. Copyright (1992), Springer

various temperatures as a function of input ratio of TMI/(TMI + TEG) in vapor phase, where TMI and TEG stands for trimethyl indium and triethyl gallium, respectively. The experimental results for a thick InGaN can be fitted with a regular solution model for free standing InGaN [150, 151]. However, we must be aware that the solid indium composition in thin films is affected by lattice constraint. By assuming a coherent growth of InGaN on GaN, the InGaN layer is highly strained. This additional strain energy shifts the peak of mixing enthalpy toward higher indium composition region than that of free standing InGaN, which helps incorporating indium in blue green region and inhibits growth of an InGaN with a high indium composition near InN [152].

Another important feature of InGaN growth is the effect of the underlying layer. Akasaka et al. reported that internal quantum efficiency (IQE) of InGaN MQW grown on SiC was improved by insertion of the $In_{0.04}Ga_{0.96}N$ underlying layer [153]. They reported the IQE of as high as 71% for the MQW on the $In_{0.04}Ga_{0.96}N$ underlying layer. Since the indium concentration was only 4%, the enhancement of IQE was not likely to come from a strain compensation. Similar experiment on MQW on sapphire substrate also showed that nonradiative life time was longer for MQW with the $In_{0.02}Ga_{0.98}N$ underlying layer than that without it [154].

Growth characteristic of InGaN MQW on nonpolar m-plane was strongly affected by surface morphology [155]. Yamada et al. reported that electroluminescence (EL) wavelength of InGaN LED fabricated on m-face GaN was shifted towards a short wavelength compared with the LED on c-plane sapphire substrate with a comparable growth condition, while EL wavelength of LED on m-face GaN with an off-cut angle of 5.4° showed the same wavelength to that of LED on c-plane sapphire substrate [155]. An inclined facet by a small angle appeared on the surface for the samples grown using nearly the just cut substrates, while the smooth surface morphology was obtained for the samples grown on off substrates with the off-cut angle of more than 5.4°. These results demonstrate that the surface morphology of GaN and the indium incorporation in MQW are strongly affected by the off-cut angle of the GaN substrates.

3.2.3 Reaction Mechanism: Experiments and Theory

It has been reported that the formation of adducts (Lewis acid–base complexes) between organometallics and NH_3, and the subsequent elimination of methane from such adducts (like as $TMG:AsH_3$), are the most probable reactions for oligomer formation [156, 157]. By the gas-phase decomposition of $TMG/H_2/ND_3$, Mazzarese et al. have shown that the reaction by-product of TMG is solely CH_3D [158]. This demonstrates that the major pathway for methane elimination from TMG involves an adduct process very similar to that of the $TMG:AsH_3$ system [159]. As mentioned in the Sect. 3.2.2, a major obstacle of growing AlGaN alloys arises from a particulate generation in vapor phase. Creighton et al. have reported nano particle generation in MOCVD growth ambient [160]. They visualized by using laser scattering that nano particles condense on the top of the thermal boundary layer. They studied the appearance of the nano particle band as a function of temperature. The $TMAl:NH_3$ system showed nano particle generation at a relatively low temperature of 200°C. They also reported that $TMG:NH_3$ adducts form and decompose reversibly in a hot cell of Fourier transform infrared spectroscopy (FTIR), while $TMAl:NH_3$ adducts irreversibly form a higher order oligomer, which was confirmed by observing the N–Al bond vibration mode. They proposed that nano particles were responsible for removing precursors from the growth ambient. In the following sub sections, experimental and theoretical models of reaction mechanisms are described.

Experimental Characteristics of Parasitic Reaction

Figure 3.62 shows AlN growth rate at atmospheric pressure as a function of growth temperature [161]. AlN growth rate sharply dropped at the growth temperature higher than 500°C. This shows that the parasitic reaction involving TMA requires some excitation energy to initiate the process. Since the upstream gas phase temperature is far below the susceptor temperature, the onset temperature of the parasitic reaction of the TMA–NH_3 system would be less than 500°C.

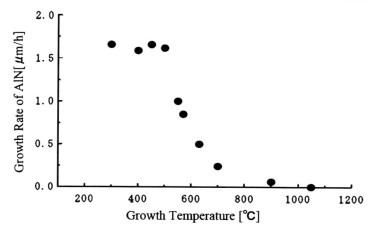

Fig. 3.62. AlN growth rate as a function of growth temperature at atmosphere pressure. AlN growth rate sharply dropped at the growth temperature higher than 500°C. Reprinted with permission from [161]. Copyright (2004), Elsevier

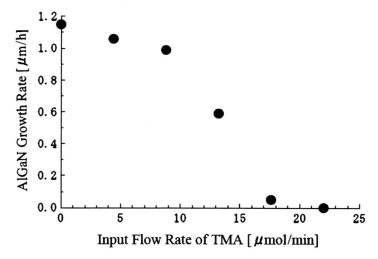

Fig. 3.63. AlGaN growth rate as a function of input flow rate of TMA with a constant flow rate of TMG. Reprinted with permission from [161]. Copyright (2004), Elsevier

Figure 3.63 shows AlGaN growth rate as a function of input flow rate of TMA at a constant TMG flow rate [161]. The AlGaN growth rate was decreased as the input flow rate of TMA was increased. This implies that intermolecular collision has an important role in the enhancement of the parasitic reaction. The above results can be well explained if we assume that nano particles are generated in vapor phase, as is proposed by Creighton et al. [160].

Quantum Chemical Study of Organometals and NH$_3$

Quantum chemical calculation is a powerful tool for understanding a reaction mechanism. In this section, the major pathways are briefly explained [161]. First, the coordination interaction between organometallics and NH$_3$ is as following. The M(CH$_3$)$_3$ molecule (**1**; each one, respectively, denotes **1a**, **1g**, and **1i** for M = Al, Ga, and In, and so forth) forms a very stable complex with NH$_3$, (CH$_3$)$_3$M:NH$_3$ (**2**), due to the M–N coordinate bond through:

$$M(CH_3)_3(\mathbf{1}) + NH_3 \rightarrow (CH_3)M:NH_3(\mathbf{2}). \quad (3.5)$$

The stabilization energy of a complex formation is the largest in the TMA:NH$_3$ system. They are $-23.17\,\text{kcal mol}^{-1}$ for aluminum, $-18.92\,\text{kcal mol}^{-1}$ for Ga, and $-18.22\,\text{kcal mol}^{-1}$ for indium, respectively. They form a coordination bond without energy barrier.

When we consider reactions at the upstream of a horizontal reactor, it would be possible to assume that all the organometallics are converted into adducts of the forms **2a**, **2g**, and **2i**. In this case, bimolecular reaction is important. We shall discuss here the intermolecular reaction between two complexes **2** and **2′** given as

$$(CH_3)_3M:NH_3(\mathbf{2}) + (CH_3)_3M':NH_3(\mathbf{2'}) \rightarrow \text{Transition state 1}[\mathbf{TS1}] \quad (3.6)$$
$$\rightarrow (CH_3)_3MNH_2 - M'(CH_3)_2 - NH_3(\mathbf{3}) + CH_4.$$

In the following, we explain that TMA:NH$_3$ adduct reacts very quickly to form olygomers by thermal excitation, while TMG:NH$_3$ adduct decomposes into TMG and NH$_3$. The optimized geometry of **TS1** as well as the schematic reaction pathway is shown in Fig. 3.64 [161].

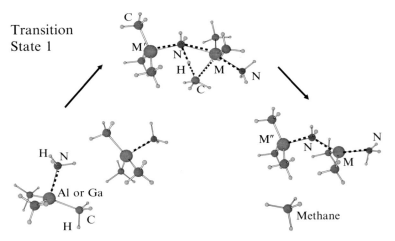

Fig. 3.64. Schematic of bimolecular reaction pathway between adducts showing the transition state of oligomer formation. Reprinted with permission from [161]. Copyright (2004), Elsevier

Table 3.2. Potential energy (in kcal mol^{-1}) of $(CH_3)_3M:NH_3$ system for bimolecular reaction mechanism

$(CH_3)_3M:NH_3 + (CH_3)_3M':NH_3 \rightarrow (CH_3)_3M:NH_2M'(CH_3)_2:NH_3 + CH_4$			
M/M'	Adduct	TS1	Product
Al/Al	−46.34	−10.56	−68.82
Al/Ga	−42.09	−2.90	−61.63
Ga/Al	−42.09	−5.51	−63.80
Ga/Ga	−37.83	2.39	−56.74
Ga/In	−37.14	−3.19	−56.70
In/Ga	−37.14	2.93	−55.67
In/In	−36.45	−2.78	−56.15

The energy of the $(CH_3)_3M$ and NH_3 is set to zero

Table 3.2 summarizes the potential energy of the $M(CH_3)_3:NH_3$ system for the bimolecular mechanism, which is defined as a mechanism of methane elimination from two $M(CH_3)_3$-precursor-derived species, where the relative energies are standardized by setting the energy of the $M(CH_3)_3+M'(CH_3)_3+2NH_3$ system to zero. When the combination of metals M and M' are Al–Al, Ga–Al, and Al–Ga, the TS1 values are −10.56, −5.51, and −2.90 kcal mol^{-1}, respectively. Since each potential energy is lower than zero, **TS1** is favored more than the individual molecular state of TMG or TMA and NH_3. This means that the oligomer formation reaction would favorably proceed to **3**. On the other hand, when the M−M' combination is Ga−Ga, TS1 is 2.39 kcal mol^{-1}. This means that TMG:NH_3 adduct will more favorably decompose to individual molecules than be excited to **TS1**. Figure 3.65 schematically illustrates this mechanism of what happens in the upstream of a horizontal reactor.

Next, we will discuss what would happen in the growth region, where the gas temperature is high enough for every adduct starting to decompose. We will discuss the unimolecular elimination of methane from TMG under excess NH_3 condition. Under excess NH_3 condition, TMG forms a stable complex due to its coordination bond with two ammonia molecules, $H_3N:(CH_3)_3Ga:NH_3$ (**4g**), without a potential energy barrier through:

$$Ga(CH_3)_3(\mathbf{1g}) + NH_3 \rightarrow (CH_3)_3Ga:NH_3(\mathbf{2g})(CH_3)_3Ga:NH_3(\mathbf{2g}) \quad (3.7)$$
$$+NH_3 \rightarrow H_3N:Ga(CH_3)_3:NH_3(\mathbf{4g}).$$

The elimination of a methane molecule can occur in the presence of excess ammonia, by the intramolecular reaction of the complex **4g** or the intermolecular collision between the complex **2g** and an ammonia molecule.

$$H_3N:Ga(CH_3)_3:NH_3(\mathbf{4g}).\text{or}[(CH_3)_3Ga:NH_3(\mathbf{2g}) + NH_3] \quad (3.8)$$
$$\rightarrow \text{transition state 2 } [\mathbf{TS2}] \rightarrow H_3N:Ga(CH_3)_2:NH_2(\mathbf{5g}) + CH_4.$$

As shown in Fig. 3.66, the geometry of **TS2** in the local part where the elimination of a methane molecule occurs is quite similar to that of **TS1**. Here, it

144 J. Chang et al.

TMA: can go beyond TS1 Irreversible
TMG: reversible

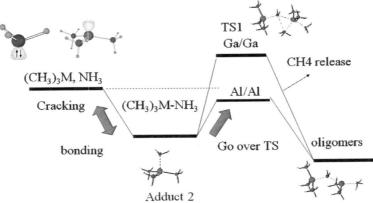

Fig. 3.65. Schematic illustration of the reaction mechanism. If **TS1** is larger than a starting molecular state, adduct 2 would be dissociated by thermal excitation. However, If **TS1** is lower than a starting molecular state, they would go beyond **TS1** to oligomer state. Reprinted with permission from [161]. Copyright (2004), Elsevier

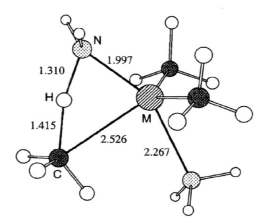

Fig. 3.66. Probable transition state of TMG decomposition pathway. Two NH_3 molecules are bonded under excess NH_3 supply. Reprinted with permission from [161]. Copyright (2004), Elsevier

is noteworthy that Bergmann et al. observed a fragment of the molecule **4g** by Q–MS spectroscopy [162].

In Fig. 3.67, energy diagram of the reaction described in reaction (3.8) is shown. It is found that potential energy barrier is reduced through **TS2** in the presence of excess ammonia for each $M(CH_3)_3 + 2NH_3$ system. In particular,

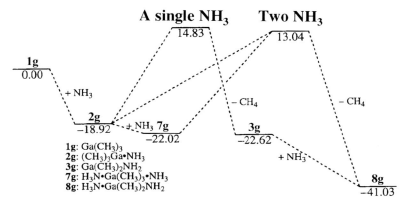

Fig. 3.67. Energy diagram for a different NH$_3$ association number with TMG. Reprinted with permission from [161]. Copyright (2004), Elsevier

Table 3.3. Potential energy (in kcal mol^{-1}) for different NH$_3$ association number and excitation state for M(CH$_3$)$_3$ and NH$_3$

	Ground state (a) (CH$_3$)$_3$M: NH$_3$	Ground state (b) H$_3$N:(CH$_3$)$_3$M: NH$_3$	TS of (a) with one NH$_3$ molecule	TS of (b) with two NH$_3$ molecules	Oligomer H$_3$N:M(CH$_3$)$_2$
Al(CH$_3$)$_3$	−23.17	−28.86	7.37	4.67	−49.86
Ga(CH$_3$)$_3$	−18.92	−22.02	14.83	13.04	−41.03
In(CH$_3$)$_3$	−18.22	−25.83	13.96	7.17	−35.26

TS energy decreases with an excess NH$_3$

the potential energy barrier for TMA is reduced to 4.67 kcal mol^{-1} (Table 3.3). Accordingly, it is considered that the gas phase reaction of TMA proceeds rapidly in the presence of excess ammonia. Since the stabilization energy of the coordination by the second ammonia molecule (reaction (3.7)) is relatively small, the mainstream of reaction (3.8) would be the intermolecular reaction between the complex **2** and an ammonia molecule together with the effective energy transfer due to the collision.

Hirako et al. carried out a computer fluid dynamic simulation of the TMG–NH$_3$ system based on the quantum chemical calculation of each reaction step for a laminar flow horizontal reactor [163]. They considered the detailed temperature distribution of the reactor wall and the wafer susceptor and incorporated the recent results of quantum chemical studies. They showed that the major reaction pathway for the first methane elimination occurs through a reaction path with both single NH$_3$ and two NH$_3$ associations. They also calculated reaction products and their spatial distribution as functions of growth pressure, and showed that the Ga–N molecule concentration near the substrate increases with growth pressure. In contrast to Ga–N, the DMGa–NH$_2$ monomer was another major byproduct in the entire pressure range, and its concentration was almost constant regardless of growth pressure. The $(DMG : NH_2)_2$ concentration was very low and slowly increased with growth pressure. However, it did not show any abrupt jump as a function of growth pressure. Dauelsberg et al. also reported experimental and simulation study of both vertical and horizontal flow reactor as a function of growth pressure [164]. Lobanova et al. also studied the effect of V/III ratio on the growth of AlN [165]. They reported that high V/III ratio resulted in a low growth rate of AlN because of lower energy barrier of TMA cracking under excess NH_3. As we need knowledge about reactor design for better understanding growth mechanism, we will briefly describe the evolution of reactor design in the Sect. 3.2.4.

3.2.4 Growth Equipment

Until early 1980s, the necessity of a perfect laminar flow for MOCVD reactor was established, because abrupt heterointerface control was required for growing alloy compound semiconductors and quantum well devices. To this end, there are two types of reactor designs. One is a vertical flow reactor, which flows carrier gas in perpendicular to the substrate. The other is horizontal flow reactor, which flows carrier gas in parallel to the substrate.

At first, we will describe a vertical flow reactor. A vertical flow reactor is also called a stagnation point flow reactor. Stagnation point is a singular point at the center of the plain, at which the flow in a semi-infinite space impinges vertically (Fig. 3.68). At an ideal vertical flow in a semi-infinite space, materials that transport to the plain is uniform over the entire plain. In this configuration, provided that we can make a uniform vertical flow, we will obtain a uniform film over the substrate. With this reason, stagnation point flow reactor, which utilizes a shower head gas distribution, is the most popular design in the application of Si based LSI process.

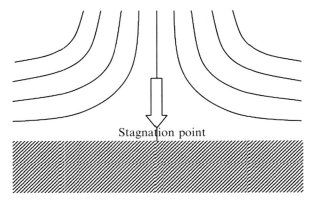

Fig. 3.68. Schematic illustration of the stagnation point flow in a vertical flow reactor

The problem of this design is the undesirable heating of the gas distributor head by radiation from the hot wafer susceptor, when the process temperature exceeds about 400°C. Since usual growth temperature of conventional III–V materials is more than 600°C, we need an effective cooling device of the gas distributor in order to prevent source gases from decomposing at the effusion nozzle in the gas distributor. Therefore, there were no commercial shower head reactors until Thomas Swan and SPIRE developed their shower head reactors with an efficient cooling system at around 2,000 [166, 167]. Before the evolution of the sophisticated cooling shower head reactor design, there was an attempt to realize a stagnation point flow without using a cool shower head design. The major obstacle of a cool shower head was condensation of arsenic or phosphorus on the surface of the cool head, which were derived from decomposition of AsH_3 or PH_3. In an attempt of avoiding this problem, Matsumoto et al. developed a vertical flow reactor by optimizing the shape of the reactor wall so that a stagnation point flow was realized at the diameter of 50 mm [168]. By using a tapered shape reactor wall, they realized a stagnation point flow over 50 mm diameter substrate. It is notable that they realized a laminar flow at atmospheric pressure when the total flow rate was between 3 and 7 slm. When the flow rate was less than 3 slm, natural convection occurred in the flow pattern. When the total flow rate was more than 7 slm, vortex by a forced convection flow appeared. This tapered flow design is very similar to the subflow of "two-flow reactor" invented by Shuji Nakamura [169]. Tapered wall design is essential to obtain a laminar flow in the reactor, which has a relatively distant space between the gas injection nozzle and the substrate. As it was difficult to scale up a single tapered flow design, the idea of multiple tapered nozzles was developed by Kondo et al. [170]. By using the multiple tapered nozzle design, they realized a stagnation point flow over the multiwafer susceptor, while avoiding a significant condensation of reaction byproducts on the nozzle.

There is another approach to realize a stagnation point flow. It is to use a high speed rotation disc. High speed rotation disc works like as a centrifugal pump. The flow over the rotating disc is attracted toward the rotating center so that a quasi-stagnation point flow is realized near the central region of the disc. By using this mechanism, we can set a distant space between gas distributor and the hot susceptor, so that we can avoid the heating up of the gas distributor [171, 172].

In Fig. 3.69, evolution of the stagnation point flow reactor is illustrated. A common problem of the stagnation point flow reactors is that the substrate faces a relatively cool gas distributor or space. Therefore, it is likely to occur that the temperature gradient across the wafer surface and the back-surface is large, which potentially causes a bowing of a large diameter substrate. At the same time, temperature gradient perpendicular to the substrate in vapor phase is likely to be large, which would result in condensation of reaction byproducts in case of high temperature grown materials, in which most of the precursors are decomposed in the vapor phase into reactive elemental species.

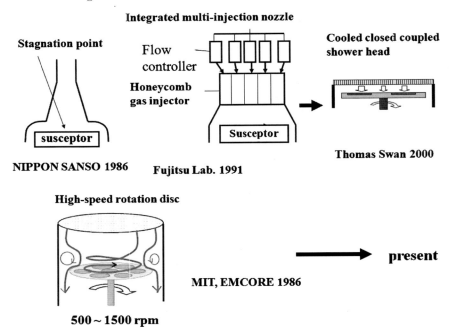

Fig. 3.69. Evolution of a stagnation point flow reactors

This would be a reason why stagnation point flow reactor is not very popular for mass-production of Si epitaxy or SiC.

Though stagnation point flow has a fundamental advantage of uniform precursor distribution, there are problems described above. For this reason, a horizontal flow design has also been developed. Horizontal flow reactor would be popular for HT grown materials, which are grown at higher than 1,000°C. In a horizontal flow reactor, precursors are consumed along the flow direction, which results in a growth rate and composition distribution over the substrate. Therefore, substrate rotation for averaging distribution of film properties is essential for obtaining a uniform film. In Fig. 3.70, a variation of the horizontal flow reactor is illustrated. Very uniform films are grown by using a two axis rotation of the substrate. Planetary rotation of the substrate to obtain a uniform film by CVD was first reported by RCA in 1968 for the application of borophosphosilicate glass (BPSG) thin film [173]. In early 1980s, centrifugal flow planetary rotation was also applied to Si epitaxy. As for the application of two axis rotation for III–Vs material growth by MOCVD, Komeno and his group developed a two axis rotation wafer holder for horizontal flow reactor in early 1980s [174]. Centrifugal flow planetary reactor for MOCVD was first developed by Frijlink, then, widely commercialized by AIXTRON [175]. Thanks to a geometrical symmetry, very uniform layers can be obtained by using planetary rotation reactor.

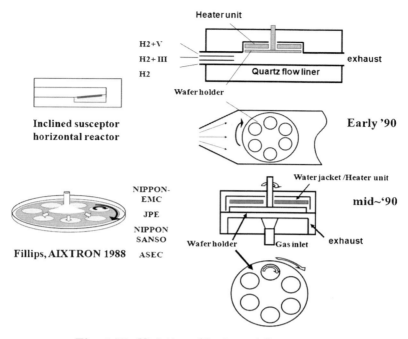

Fig. 3.70. Variation of horizontal flow reactors

A face down substrate holder was also developed for horizontal reactors, since the wall deposit on the flow restriction liner of the face-up configuration sometimes falls off onto the substrate surface. Both simple horizontal and centrifugal flow design face-down reactors were developed. In the case of high flow speed horizontal reactor for a large diameter substrate, a large entrance effect was a problem. The effect of the entrance effect is more significant in a centrifugal flow reactor, since the flow speed at the center is very high. Therefore, high flow speed centrifugal reactor design has been very difficult. In order to avoid a large gradient of film properties at the entrance region, three-layer-gas-injection was developed [176]. The detail of this design will be described in detail in the Sect. 3.2.5.

3.2.5 Atmospheric Pressure Reactor for Nitride

It still remains controversial that the optimization of the device structure or growth condition other than growth pressure would somehow replace growth pressure optimization, even though the growth pressure is a key parameter to obtain high quality nitride-semiconductors. Atmospheric pressure reactor would be important to grow high quality nitride-semiconductors all the more, provided that we can design the reactor with reasonable gas consumption. To be accurate, the control of a parasitic reaction is important.

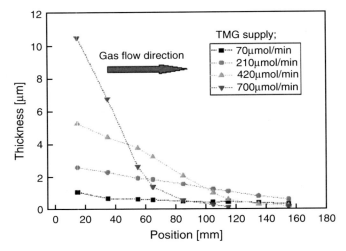

Fig. 3.71. Growth rate variation along the flow direction for a reactor with a forced flow vortex. Reprinted with permission from [161]. Copyright (2004), Elsevier

High flow speed is very important in atmospheric pressure MOCVD. However, high flow speed often brings about an undesirable recirculation in the flow of the reactor: forced flow vortex. When there is a vortex, particulates can grow even in a diluted growth condition of precursors. Figure 3.71 shows growth rate variation along the flow direction in a horizontal flow reactor [161].

All the precursors were mixed together at the upstream of the reactor. In addition to the premixing, it was calculated by a computer flow dynamic simulation where there was a forced flow vortex at the upstream region. This reactor is the same one that is used for experiments discussed in Fig. 3.58. By using this reactor, any AlGaN film could not be grown at an elevated pressure range as was shown in a previous chapter. When the TMG input flow rate was below $400\,\mu\mathrm{mol\,min^{-1}}$, the GaN growth rate variation along the flow direction was almost linear. The average growth rate with susceptor rotation was approximately $2\,\mu\mathrm{mol\,h^{-1}}$ at this flow rate. Note that the growth rate variation of GaN with TMG supply rate of $700\,\mu\mathrm{mol\,min^{-1}}$ is a nonlinear function of the wafer position. The GaN growth rate downstream was almost zero for an input TMG of $700\,\mu\mathrm{mol\,min^{-1}}$, which strongly suggests nano particle generation. Watwe et al. studied the gas phase chemistry of TMG and NH_3 by quantum chemical calculation, and showed that the $TMGa:(CH_3)_2GaNH_2:NH_3$ species has the lowest standard free energy activation barrier for methane elimination [177]. If we assume that $(CH_3)_2GaNH_2$ is back diffusing downstream, this catalytic reaction will easily occur and proceed to oligomer formation. Once an oligomer starts to grow, nano particles would result by an iterative process. In the downstream region, most of the precursors would be consumed in the nano

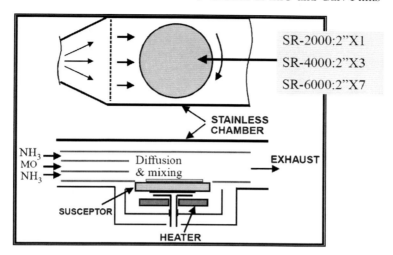

Fig. 3.72. Schematic for a reactor with a three-layered laminar flow gas injection. Reprinted with permission from [141]. Copyright (2004), Elsevier

particle growth by CVD on the particle. In this particular case, the growth rate of GaN showed saturation at approximately $2.5\,\mu m\,h^{-1}$.

By a careful inspection of the reactor design, laminar flow gas injection was developed by employing a three-layer laminar flow gas injection system (TAIYO NIPPON SANSO SR-6000) [178]. In Fig. 3.72, example of a laminar flow gas injection reactor is shown. This three-layer gas injection has been developed for high speed flow horizontal reactor in early 1990s in order to eliminate undesirable entrance effect for HEMT application [176, 179]. Then, the design was applied to the reactor for nitride-semiconductors because very high flow speed was required to reduce a parasitic reaction [178]. As shown in Fig. 3.72, organometallics and NH_3 are separately injected and mixed together by molecular diffusion.

Under typical flow conditions, the spacing between the gas separators and their positions are designed so that organometallics would diffuse laterally to the flow direction and reach the substrate surface just at the inlet of the growth region. By varying flow rate ratio between each carrier gases, the growth rate near the leading edge of the deposition zone is controlled. When the carrier gas flow off the bottom hydride the amount of gas is larger than those of the metal–alkyls, the growth starts at a downstream side of the flow. This is because metal–alkyls must travel through a thicker hydride flowing layer, since the large flow rate results in an expansion of the flow. On the contrary, when the bottom hydride gas flow rate is smaller, the growth starts at an upstream of the flow. The top hydride flow is used to vary the equivalent thickness height of the flow restriction liner. By increasing the carrier gas of the top flow, growth starts at the upstream side of the flow. This flow

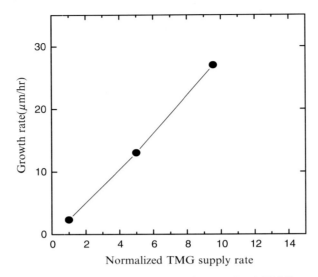

Fig. 3.73. Growth rate of GaN as a function of normalized TMG supply rate for a production scale MOCVD reactor. Reprinted with permission from [180]. Copyright (2008), Wiley

configuration is also useful for improving the growth efficiency because the unnecessary deposition at the upstream is eliminated [179]. In this reactor, maximum growth rate was improved to be more than $11\,\mu m\,h^{-1}$ [161]. The same gas injection method was applied to a production scale symmetric centrifugal flow planetary rotation reactor (TAIYO NIPPON SANSO SR-23 k). In this reactor, ten pieces of 2 in. diameter substrate can be treated at a time. GaN growth rate in this centrifugal reactor is shown as a function of the normalized input TMG supply rate in Fig. 3.73. In Fig. 3.73, NH_3 flow rate was fixed at 100SLM, and the TMG supply rate was changed. V/III ratio was 300–3,000 according to the growth rate. The growth experiment was conducted under atmospheric pressure. The maximum growth rate was more than $28\,\mu m\,h^{-1}$, because the present maximum growth rate in the figure was only limited by the available TMG supply [180].

Crystal quality of the GaN films grown with the high growth rate was characterized by XRD. Figure 3.74 shows FWHMs of ω-scan XRC for (0002) and (10–12) reflections for the GaN films as a function of growth rate [180]. Each HT grown GaN film was directly deposited on the LT buffer layer. The total thickness of the samples was fixed at about $4\,\mu m$. Lateral to vertical growth rate ratio seems smaller in the case of the highest growth rate of $28\,\mu m\,h^{-1}$, which resulted in broader line width for (10–12) XRC. It is notable that FWHM of XRC was not very much affected up to the growth rate of $12\,\mu m\,h^{-1}$, even though the HT GaN was directly grown on the LT buffer layer. This is probably due to an enhanced facet formation during the initial

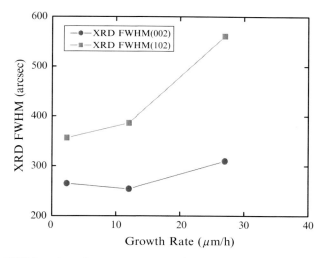

Fig. 3.74. FWHM of XRCs for symmetric (0002) and asymmetric (10–12) reflections of the GaN films grown with different growth rate. Reprinted with permission from [180]. Copyright (2008), Wiley

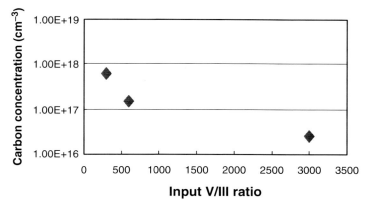

Fig. 3.75. Carbon concentration in the GaN films as a function of growth rate of GaN. Reprinted with permission from [180]. Copyright (2008), Wiley

growth stage by a relatively low V/III ratio. Surface morphology of all the samples was smooth regardless of the growth rate. The crystal quality would be very much improved if the HT GaN is grown after coalescence of the initial layer.

Since V/III ratio affects electrical property of GaN, carbon concentration and sheet resistance were measured for these samples. Figure 3.75 shows a carbon concentration of the GaN films as a function of V/III ratio. At the V/III ratio of 600 (growth rate of $12\,\mu\text{m}\,\text{h}^{-1}$), carbon concentration was

1.5×10^{17} cm^{-3} [180]. At the V/III ratio of 300 (growth rate of $28\,\mu\text{m}\,\text{h}^{-1}$), the carbon concentration was 6×10^{17} cm^{-3}. Due to the carbon contamination, sheet resistance of GaN was more than $15{,}000\,\Omega\,\text{sq}^{-1}$ for the GaN sample grown with the highest growth rate and with the V/III ratio of 300. From these results, it was concluded that the usable growth rate range applicable to the growth conditions for the doping of the films would be lower than $10\,\mu\text{m}\,\text{h}^{-1}$.

It would be interesting to investigate the intrinsic limit of growth rate of GaN by MOCVD. GaN growth rate of near $1\,\text{mm}\,\text{h}^{-1}$ has been realized by HVPE. A major difference in growth condition between MOCVD and HVPE is that V/III ratio is very small in HVPE compared to that in MOCVD. However, if the GaN growth rate is plotted as a function of V/III ratio, data of MOCVD and that of HVPE lie on the same line. For example, V/III ratio of $28\,\mu\text{m}\,\text{h}^{-1}$ was 300 in Fig. 3.73. If we increase Ga supply with constant NH$_3$ flow rate by ten times, we will obtain a growth rate of $280\,\mu\text{m}\,\text{h}^{-1}$ for HVPE provided that the growth rate is proportional to the supply rate of Ga species. Therefore, the difference between HVPE and MOCVD arises from a probability of generation of particulates in a vapor phase under high concentrations of Ga species.

At the moment, there is no intriguing demand for MOCVD of a very high speed growth of GaN. However, from an analogy of red LED, thick GaN layer would be attractive for high powered LED to improve light extraction efficiency. By employing a high flow speed laminar flow reactor, AlGaN can be grown even at an atmospheric pressure. Figure 3.76 shows solid Al composition

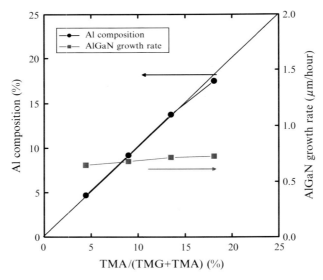

Fig. 3.76. Al composition in the AlGaN films as a function of input TMA/(TMG + TMA) ratio grown with the growth rate of about $0.7\,\mu\text{m}\,\text{h}^{-1}$. Reprinted with permission from [180]. Copyright (2008), Wiley

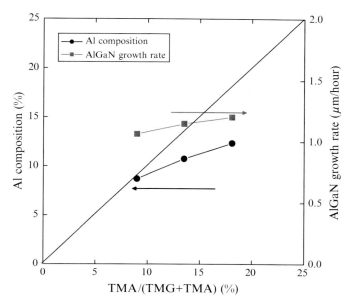

Fig. 3.77. Al composition in the AlGaN films as a function of input TMA/(TMG + TMA) ratio grown with the growth rate of more than $1.0\,\mu\text{m}\,\text{h}^{-1}$. Reprinted with permission from [180]. Copyright (2008), Wiley

of the AlGaN films as a function of input TMA/(TMG + TMA) ratio at the growth rate of about $0.7\,\mu\text{m}\,\text{h}^{-1}$ [180]. Al composition was linearly controlled up to ∼17%. Uniform AlGaN film was obtained for each Al composition. It would be concluded that a parasitic reaction was effectively suppressed for the growth rate of $0.7\,\mu\text{m}\,\text{h}^{-1}$ and Al composition of 17%. When the parasitic reaction takes place, the growth rate of AlGaN and the solid Al composition show saturation. Figure 3.77 shows a similar data of Fig. 3.76 but, the parasitic reaction starts to occur in this case.

In Fig. 3.77, growth rate was set to more than $1.0\,\mu\text{m}\,\text{h}^{-1}$. In the figure, TMA was increased with a constant TMG supply rate. It seems that a critical growth rate of AlGaN of Al composition of more than 8% would lie around $1.0\,\mu\text{m}\,\text{h}^{-1}$ for this particular flow speed condition [180].

As a summary, now we can suppress a parasitic reaction in a large scale production MOCVD. High flow speed as well as a careful design to keep a laminar flow is useful to realize atmospheric pressure MOCVD. By using atmospheric pressure MOCVD, in which a side-reaction is eliminated, we can grow sophisticated device structures under the controlled conditions.

3.3 HVPE of GaN and Related Materials
S.W. Lee and M.-W. Cho

3.3.1 Introduction

Hydride Vapor Phase Epitaxy (HVPE) is widely used to grow epitaxial GaN films in addition to MOCVD. The HVPE itself has a long history for semiconductor growth [81]. At the beginning of the 1960s, the advancement of the HVPE techniques applied to Si and Ge allowed great steps to be made in the spread of GaAs. In fact, the first single crystalline GaN film was obtained by using HVPE in 1969 [182]. HVPE has played an important role in the growth of III–V semiconductors and it was the singular method for growing epitaxial layers of GaN until the early 1980s [183, 184] However, this HVPE technique was substantially abandoned in the 1980s because of an evident inability for the p-type doping.

In the 1990s, HVPE has received attractions again because of its possibility to grow thick GaN films with low defects at a relatively low cost. Many efforts have shown good quality materials with promising characteristics. A. Usui et al. reported thick GaN layer with a dislocation density as low as $10^6 \, \mathrm{cm}^{-2}$ using selective growth techniques [185]. Now HVPE has become a promising method to grow high quality GaN crystal. Especially, combining of the HVPE technique with other epitaxy technique to produce high quality GaN templates or free-standing GaN substrates may become the key to fabricate GaN-based devices with superior performances.

In this chapter, firstly, HVPE of GaN in viewpoint of thermodynamics is discussed. Secondly, a conventional hot-wall horizontal system and a vertical system (boule-growth) are described. Thirdly, thick GaN film growth is briefly described. Next, topics for fabrication of free-standing (FS) GaN substrates are addressed by discussing several methods including laser lift-off (LLO), void assisted separation (VAS), and chemical lift-off (CLO) processes. Finally, HVPE of other III nitride materials including AlN, AlGaN, and InN are discussed.

3.3.2 Thermodynamic Considerations of HVPE of GaN

As an epitaxy process that works near thermodynamic equilibrium, it is of particular interest for applications such as selective area growth, overgrowth of buried structures, and planar growth. HVPE is a growth technique involving thermo–chemical reactions at atmospheric pressure. The HCl and carrier gases flowing through the reactors, the materials inside the reactor, and the temperature at the point of the reaction determine the reaction products. The final reaction leaves the desired material, which is deposited onto a substrate that is usually rotated for high uniformity. The temperatures of the reactions and deposition can be independently controlled. Generally, the heating system is constituted of a multiple zone furnace. Total pressure (mostly 1 atm), the partial pressures of reaction species, an input ratio of V/III, and the growth temperature are important parameters controlling the growth processes.

Koukitu et al. have investigated thermodynamics of various III nitrides grown by MOCVD, HVPE, and MBE. Details can be found in the comprehensive article published in 2001 [186] and specific results for each material can be found in [186]. In special, a detailed thermodynamic analysis of GaN in growing by HVPE is well reported in their article published in 1998 [187]. Equilibrium partial pressures of vapor species involving reactions as functions such as growth temperature, input partial pressure of GaCl, input V/III ratio etc. are well analyzed. In addition, the driving force as a function growth temperature and the calculated growth rates as a function of input partial pressure of GaCl can be found [187]. Here, we briefly describe the calculation procedures, definition of thermodynamic terms, and the results for the thermodynamic analysis related with the HVPE of GaN performed by Koukitu et al. [187].

They chose six species, i.e., GaCl, GaCl$_3$, NH$_3$, HCl, H$_2$ and inert gas (N$_2$, Ar, He etc.) as the necessary vapor species in analyzing the vapor growth of GaN. The chemical reactions which connect the species in the deposition zone can be expressed by following two reactions of (3.9) and (3.10).

$$\text{GaCl} + \text{NH}_3 = \text{GaN}(s) + \text{HCl} + \text{H}_2, \tag{3.9}$$

$$\text{GaCl} + 2\text{HCl} = \text{GaCl}_3 + \text{H}_2. \tag{3.10}$$

The equilibrium equations for these reactions are

$$K = \frac{P_{\text{HCl}} P_{\text{H}_2}}{P_{\text{GaCl}} P_{\text{NH}_3}} \tag{3.11}$$

and

$$K = \frac{P_{\text{GaCl}_3} P_{\text{H}_2}}{P_{\text{GaCl}} P_{\text{HCl}}^2}. \tag{3.12}$$

The total pressure of the system can be written as

$$\sum P_i = P_{\text{GaCl}} + P_{\text{GaCl}_3} + P_{\text{NH}_3} + P_{\text{HCl}} + P_{\text{H}_2} + P_{\text{IG}}, \tag{3.13}$$

where P_i's are the equilibrium partial pressures and IG means inert gas. Denoting the input partial pressures as P_i^0's, (3.14) can be obtained from conservation constant.

$$P_{\text{GaCl}}^0 - P_{\text{GaCl}} = P_{\text{NH}_3}^0 - P_{\text{NH}_3} \tag{3.14}$$

(3.14) shows the stoichiometric relationships for GaN deposition. In addition, they defined the parameters A and F, which can be expressed by (3.15) and (3.16), respectively.

$$A = \frac{1/2(P_{\text{GaCl}} + 3P_{\text{GaCl}_3} + P_{\text{HCl}})}{P_{\text{H}_2} + 3/2 P_{\text{NH}_3} + 1/2 P_{\text{HCl}} + P_{\text{IG}}}, \tag{3.15}$$

$$F = \frac{1/2(2P_{\text{H}_2} + 3P_{\text{NH}_3} + P_{\text{HCl}})}{P_{\text{H}_2} + 3/2 P_{\text{NH}_3} + 1/2 P_{\text{HCl}} + P_{\text{IG}}}. \tag{3.16}$$

Here, A is the ratio of the number of chlorine atoms to the number of hydrogen and inert gas atoms in the system, while F is the mole fraction of hydrogen relative to inert gas atoms. These parameters are kept invariant under a given growth condition, because A does not include any atoms which deposit in to a solid. Finally they introduced a parameter α, the mole fraction of decomposed NH_3, into the calculation because it is well known that the decomposition rate of NH_3 under typical growth conditions is slow without a catalyst, and the extent of the decomposition strongly depends on the growth conditions and equipment.

$$NH_3(g) \rightarrow (1-\alpha)NH_3(g) + \alpha/2 N_2(g) + 3\alpha/2 H_2(g). \tag{3.17}$$

Since the changes of α and F affects the change of H_2, N_2, and NH_3 partial pressures, they performed the calculation by changing the parameter F, and fixed α as 0.03. In order to do a calculation, information on the equilibrium constants for the reaction (3.11) and (3.12) are necessary. The values of equilibrium constants were obtained from the free energies of the formation of Ga(g), GaCl(g), $GaCl_3$(g), NH_3(g) and GaN(s). The values of the equilibrium constants for the reaction (3.11) and (3.12) are given by $-2.01 + 3.10 \times 10^3/T - 3.11 \times 10^{-1} Log_{10}(T)$ and $-5.54 + 8.93 \times 10^3/T - 5.70 \times 10^{-1} Log_{10}(T)$, respectively, where T is the temperature in K. In addition the activities should be determined.

Using these definitions on thermodynamic terms, parameters and constants, calculation for the variation of the equilibrium, partial pressures for each vapor species with several parameters were performed through the method they developed [188] and the results are shown in Fig. 3.78 [187]. As shown in Fig. 3.78a, the equilibrium partial pressures of $GaCl_3$ and HCl were decreased while increasing the temperature, while the partial pressure of GaCl

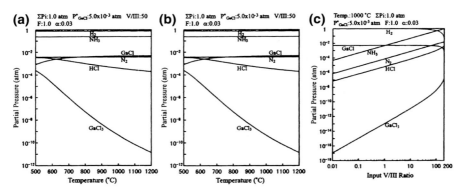

Fig. 3.78. Equilibrium partial pressures over GaN as functions of (**a**) growth temperature, (**b**) input partial pressure of GaCl at 1,000°C, (**c**) input V/III ratio at 1,000°C. Other fixed conditions and parameters are shown at each figure. Reprinted with permission from [187]. Copyright (1998), The Institute of Pure and Applied Physics

was increased with temperature in the temperature ranges of 500–700°C. However, the equilibrium partial pressures of H_2, N_2, and NH_3 were nearly not changed with the temperature. In the case of the parameter of input partial pressure of GaCl, the equilibrium partial pressures of NH_3, GaCl, N_2, HCl, and $GaCl_3$, except the H_2 were increased with the parameter as shown in Fig. 3.78b. The equilibrium partial pressures of NH_3, N_2, HCl, and $GaCl_3$, except H_2 and GaCl were also increased with the input V/III ratio as shown in Fig. 3.78c.

Here, it should be noted that the equilibrium partial pressure of GaCl is higher than that of $GaCl_3$ as shown in Fig. 3.78a and the difference increases rapidly with temperature, resulting in the equilibrium partial pressure of GaCl as large as seven orders larger than that of $GaCl_3$ at 1,000°C. It indicates that the reaction governing the growth at the substrate can be mostly described by (3.9). From the reaction of (3.9) governing the GaN growth, we can know that hydrogen is a reaction product not a reactant. Therefore, the reaction depositing GaN proceeds more effectively in the inert gas system and is prevented with an increase in H_2 [187]. This is opposite to the GaAs growth since hydrogen is a reactant not a reaction product, where the reaction governing the GaAs deposition can be described as $GaCl + 1/4As_4 + 1/2H_2 = GaAs + HCl$. It is also noteworthy that N_2 produced by the decomposition of NH_3 does not react effectively for the formation of GaN, while As_4 or As_2 produced from AsH_3 in the hydride system reacts effectively for the formation of GaAs [187].

Now, the driving force for the deposition is defined by following.

$$\Delta P_{Ga} = P^0_{GaCl} - (P_{GaCl} + P_{GaCl_3}) = P^0_{GaCl} - P_{Ga}. \quad (3.18)$$

Since the parameter F defined by (3.16) denotes the mole fraction of hydrogen relative to inert gas atoms as it was mentioned that the reaction depositing GaN is prevented with an increase in hydrogen, we can easily expect the decrease in the driving force for the deposition with an increase of F. Therefore, we can expect again that a higher growth rate is obtained in the inert gas system than in the H_2 system [187]. As mentioned by Koukitu et al. [187], this fact agrees well with Ban's experiments [189]. In his experiments, it was shown that the extent of GaCl consumption, i.e., the deposition of GaN, was significantly greater for He as the carrier gas than for H_2.

Finally, the growth rate under the mass transport type II or diffusion limit at constant pressure is expressed as [190]

$$r = K_g \Delta P_{Ga}, \quad (3.19)$$

where K_g is the mass transfer coefficient. Koukitu et al. calculated the growth rate as a function of input partial pressure of GaCl based on a series of their calculations [187].

Usui et al. [185] reported that the growth rate is increased linearly as the HCl partial pressure is increased and they obtained the growth rate of about 97 µm h^{-1}. Figure 3.79 shows the dependence of the GaN growth rate on the

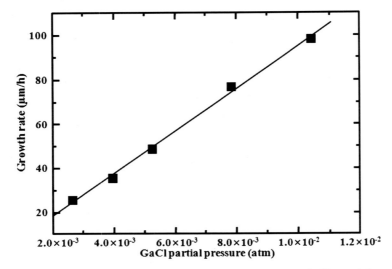

Fig. 3.79. The dependence of the GaN growth rate versus GaCl partial pressure. Reprinted with permission from [185]. Copyright (1997), The Institute of Pure and Applied Physics

GaCl partial pressure. The reactor was atmospheric pressured and the partial pressure of ammonia was 0.26 atm. In Fig. 3.79, the square points are experimental data points and solid linear line is the calculated growth rate using (3.19) with the Kg value of 1.18×10^5 mm h^{-1} atm^{-1}. The quite good agreement between the experimental data and calculation indicates that the growth of GaN using HVPE is thermodynamically controlled.

3.3.3 Growth System

HVPE system is usually a hot wall reactor made of quartz, which is placed in a multiple zone furnace. Figure 3.80 shows an illustration of horizontal HVPE system. The multiple zone furnace is designed for the optimum control of source and substrate temperatures, and the delivery of all reactants. The HCl and H$_2$ carrier gases flow through the tubes to the Ga source room with the metallic Ga inside the tube. The temperatures of the Ga source zone and GaN growth zone are nominally set to around 800°C and 1,000°C, respectively. The substrate is set on the susceptor perpendicular against the main stream of reaction gases. The susceptor is rotated during the growth to increase uniformity, where the growth is conducted at atmospheric pressure.

The reactions at the Ga source zone and at the GaN growth zone are mainly expressed by reactions (3.20) and (3.21), respectively.

$$\mathrm{Ga}(l) + \mathrm{HCl}(g) \to \mathrm{GaCl}(g) + 1/2\mathrm{H}_2(g), \tag{3.20}$$

$$\mathrm{GaCl}(g) + \mathrm{NH}_3(g) \to \mathrm{GaN}(s) + \mathrm{HCl}(g) + \mathrm{H}_2(g). \tag{3.21}$$

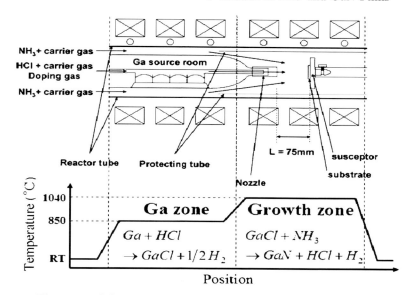

Fig. 3.80. Schematic illustration of a horizontal HVPE system

Generally, HVPE system is limited to a single wafer production. A multi-wafer system or the growth of GaN boules in a vertical system has been attempted for the mass production of free-standing GaN substrate. Recently, Schineller et al. obtained 7 cm-long GaN boules with 2 in. -diameter using a vertical reactor, where the optimal distance between the growth surface and the gas injector is continuously adjusted and maintained by pulling the susceptor up according to the speed of growth [191]. In order to obtain stable conditions for the boule growth, prereactions of GaCl and NH_3 and formation of NH_4Cl should be avoided. The prereactions at the end of the nozzles cause sticking and damaging of the nozzle. Figure 3.81 shows a schematic illustration of the vertical HVPE growth reactor with concentric reactor inlet geometry. Since the formation of NH_4Cl causes clogging of exhaust lines, they employed a sheath flow between the reactive species of GaCl and NH_3 to avoid the prereactions. In order to suppress the formation of the generally observed white powders, which nominally is the source for the growth defect generation in addition to act as the clogging material of the exhaust line, they removed the cold spots along the exhaust area of the reactor. A numerical simulation revealed that the temperature of the reactor exhaust is above 350°C. These three key technical points, i.e., in situ adjustment of the optimal distance between the growth surface and the gas injector, employment of the sheath flow nozzle, and managing the exhaust temperature to be high were critical factors to achieve the successful GaN boule growth.

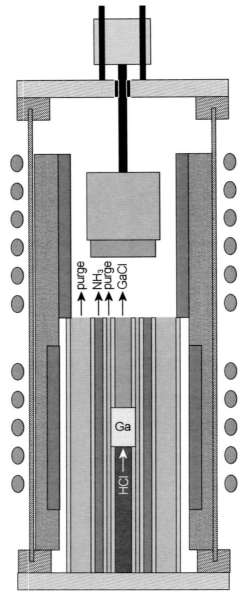

Fig. 3.81. Schematic illustration of the vertical HVPE growth reactor with concentric reactor inlet geometry. [191]

3.3.4 Thick GaN Film Growth

Nowadays, there are many approaches to produce GaN bulk crystal and substrates. Utsumi et al. reported that applying high pressure completely prevents the decomposition of GaN and allows the stoichiometric melting of GaN [192]. At pressures above 6.0 GPa, congruent melting of GaN occurred at about 2,220°C and by decreasing the temperature the GaN melt was to crystallize to the original structure. GaN single crystals were produced by cooling the melt slowly under the high pressures and were recovered at ambient conditions. Prowski et al. achieved the growth of GaN single crystals by using the high-pressure-solution-growth (1 GPa, 1,500°C) [193,194]. They obtained the 10 mm^2 sized plate-like GaN single crystal with good crystallographic quality by this method. However such an extreme condition is not suitable for industrial mass production. Yamane et al. reported the Na flux method for the growth of GaN single crystals [195]. In the growth of GaN using the Na flux method a continuous nitrogen dissolution was realized by applying the nitrogen gaseous pressure to the Ga–Na melt and the GaN single crystal were grown at about 1,000°C and at 10 MPa. On the other hand, GaN single crystal could be grown by ammonothermal method [196]. Reaction was carried out with supercritical ammonia at about temperatures of 500°C and with pressures of 100 MPa [197].

High quality GaN single crystals could be grown by these four different methods. However, the growth rate is in the order of several μm/h and such an extremely slow growth rate is not appropriate for industrial production of the GaN bulk crystal and the substrate. Contrary to this point HVPE technique has an added advantage of a high growth rate. Therefore, growing thick GaN films on foreign substrate by HVPE and removing out the substrate is a very attractive and desirable way to get the bulk-like GaN crystal and the substrate. However, there are important technical issues in growing the high quality GaN by HVPE for the application of bulk-like crystal or substrate. In this section, several issues are dealt with including cracking, bowing, and growth rate to achieve several hundred thick-GaN films.

High quality thick GaN films grown by HVPE have been used as either a template or a free-standing substrate. Strain accumulation, cracking and bending are critical problems in growing thick GaN films by HVPE. Hiramatsu et al. reported relaxation mechanisms of thermal stresses in GaN on c-plane sapphire substrate [198]. Cracks were not observed in the MOCVD grown GaN films of less than 20 μm thickness however, cracks were observed in MOCVD films thicker than 20 μm [198]. In the HVPE grown GaN films of thickness less than 20 μm, cracks were occasionally detectable, but cracks were regularly observed in films thicker than 20 μm [198]. Based on these observations they concluded that the occurrence of cracks is strongly dependent on GaN film thickness and relatively less dependent on the growth method. The strain in GaN was greater in films of less than a few microns' thickness. It was decreased in films of thickness from several to about a hundred microns, and was almost completely relaxed in those thicker than 100 µm. Relaxation of the stresses and strains in the heterostructure were thought as the results cracking in the sapphire. They proposed three relaxation mechanisms of the thermal strain for different film thicknesses as follows: (a) only lattice deformation (<4 µm), (b) enhancement of interface defects such as microcracks and/or dislocations (4–20 µm), and (c) generation of macrocracks in sapphire (>20 µm).

In 2001, Etzkorn and Clarke reexamined cracking of GaN films grown on sapphire substrates based on microstructural observation of cracking [199]. The appearance of mud-cracking patterns of cracks in GaN films indicates that it is indicative of cracking under biaxial tensile stress because such a pattern of cracks cannot be produced under compression. The tensile cracking in surprising and they pointed out three surprising features of the tensile cracking in GaN films as follows [199]. First, because GaN has a smaller thermal expansion coefficient than sapphire, GaN films grown on sapphire should be under biaxial compression on cooling. Second, considering the lattice mismatch between GaN and sapphire at the growth temperature and considering the epitaxial relationships, the GaN would be under compression at the growth temperature if the GaN were coherent and lattice planes were continuous across the interface. Above a critical thickness, this biaxial compression can be relaxed by a network of misfit dislocations at the GaN/sapphire interface. However, there is no dislocation mechanism that would switch the sign of the crystallographic mismatch from the compressive to the tensile. Third, thin GaN films, for instance those less than ∼5 µm thick, grown by MOCVD or HVPE on sapphire are generally not cracked. Furthermore, although cracking is common in the growth of thicker films and it is important to emphasize that it does not always occur [199]. However, the mud-cracking patterns indicate the tensile stress during the growth and the necessary tensile stress is possible, generated by an island coalescence mechanism. The tensile stress caused by island coalescence was proposed by early works of Hoffman and coworker, who proposed that the elastic strain energy introduced by this process was driven by the energy reduction associated with creating a grain boundary segment from the solid–vapor surfaces of two neighboring grains [200, 201] In fact, the most direct evidence that GaN films grown on sapphire are under tensile stress comes from in situ wafer curvature measurements, for instance, in the work by Hearne et al. on MOCVD growth [202].

Based on the microstructural observations, Etzkorn and Clarke concluded that the cracks lie below the surface of the film and rarely extend to the top surface [199]. This means that the cracks should be buried. Therefore, they proposed the mechanism governing the cracking and burying of cracks in GaN films on sapphire substrates as follows. Once the first cracks have been introduced and channeled a mixed state exists when additional material is deposited on the film. In the vicinity of the cracks, the newly deposited material grows on an almost stress-free surface since the cracks locally relieve a substantial proportion of the tensile strain in the film. Well away from the cracks the surface is still under tensile strain and so the additional material is also under stress. This continues until it becomes energetically favorable for another set of cracks, more closely spaced, to grow between the first set and relieve more of the elastic strain energy. As growth continues, more cracks are introduced, more of the tensile stress is relieved, and more of the growing surface is free of strain until, for all practical purposes, the additional material being deposited is growing essentially stress-free [199]. They also proposed the possibilities that the film grows across and buries the cracks as follows [199]. Although growth of GaN is normally strongly anisotropic with growth occurring primarily along the c axis, small growth fluctuations in the lateral direction would be sufficient for the crack faces to come together and touch. In doing so they would form a local region of grain boundary in preference to two separate GaN/gas interfaces provided the energy is thereby lowered. Once the crack faces touch newly deposited material will grow over the crack burying it. Touching of crack faces will presumably occur stochastically from place to place so there will be a period of growth before all the cracks are buried. Thus if growth is stopped before this closure process is complete, observations will show that some cracks intercept the surface whereas others do not. In addition to lateral growth fluctuations it is also possible that crack closure is promoted by the fact that as the film under stress thickens there is an increasing bending moment acting to decrease the length of the surface and bring the crack surfaces closer together [199].

Although cracking has happened inside the GaN films during the growth, the cracks formed during film growth can influence a subsequent fracture of the sapphire substrate on cooling. Substrate cracking has been a concern since first reported by Itoh et al. [203]. The most direct evidence that cracks can propagate on cooling is that if a sufficiently thick film is withdrawn from the growth chamber while it is still hot, for instance at 600°C, the substrate is uncracked [199]. As it is then cooled cracks abruptly form and the sound of them "popping-in" can often be heard. Observations show that the cracks formed in the GaN film propagate down into the underlying sapphire, albeit not very far, typically a few microns. The cracks initially propagate straight into the sapphire but then deflect onto one of the cleavage planes in the sapphire since the perpendicular direction does not lie along a cleavage plane [199].

Since the tensile stress in GaN films during the growth is generated by coalescence of islands, controlling the initial growth looks critical in controlling the strains and crack formation. In other words, as the magnitude of the tensile stresses is related to the island size at coalescence, control of the coalescence stage might involve controlling the nucleation density. Based on this idea, growth of crack-free thick GaN films on sapphire substrates was reported by Napierala et al. [204]. They reduced the tensile stress of GaN films grown by MOCVD through lowering the density of initial GaN crystallites. They controlled the grain size and their density during the LT nucleation step of GaN on sapphire. Subsequently, very thick (up to ∼200 μm) crack-free GaN layers were successfully grown by HVPE on the MOCVD GaN templates [204].

3.3.5 Fabrication of Free-Standing GaN Substrate

Fabrication of the Free-standing (FS) GaN substrates is strongly desired to be used as the substrates for the fabrication of high performance and reliable GaN-based devices. Although there are many approaches to obtain the free-standing GaN substrate using a HVPE method, the removing processes of the sapphire substrate from a thick GaN film is normally difficult because the sapphire has a greater hardness and a negligible etching rate with any etchant. Nevertheless, mechanical polishing or dry etching of substrates after GaN growth can be used to produce free-standing GaN substrates. However, this method is time-consuming and the GaN films often crack in the case of mechanical polishing. Because of these reasons, other ways including laser lift-off (LLO), void-assisted separation (VAS), and chemical lift-off (CLO) of the buffer layer were suggested as a debonding technique to get the FS GaN substrates. In this section various methods to produce the FS GaN substrates from the thick GaN films and related technical issues will be discussed. Properties of the FS GaN substrates are described briefly.

FS GaN Fabrication by Dry Etching of the Substrate

Melnik et al. reported FS GaN could be fabricated by HVPE in 1997 [205]. The FS GaN with a maximum size at $7 \times 6 \times 0.1$ mm^3 were obtained by HVPE growth of 100 µm thick GaN layers on SiC substrates and subsequent removal of the substrates by reactive ion etching (RIE). The layers were deposited directly on SiC substrates without any buffer layers. They used 6H–SiC and 4H–SiC wafers with 30 and/or 35 mm in diameter, and with thicknesses of 300, 200 and 40 µm. A thick GaN film was grown in a horizontal reactor with a growth rate of about 60 µm h^{-1}. After the GaN growth, the SiC substrate was removed by RIE using SF$_6$ containing a gas mixture. By etching the substrates, the GaN films were released from the substrates and FS GaN crystals were obtained as shown in Fig. 3.82. No cracks were observed on most of the crystals. The fracture of the thick GaN layer was the main factor limiting the size of the bulk crystal. The residual strains which cause the fracture are due to the lattice mismatch and by the difference in the thermal expansion coefficient between GaN and SiC. They observed the effect of these stresses at growth temperature. For example, the 40 µm-thick SiC substrates were bending during the deposition process and to avoid the bending, use of the thick SiC substrates were preferred [205].

FWHM of (0002) XRC from the FS GaN was about 150 arcsec. A residual stress has been detected in these crystals by X-ray analysis and Raman spectroscopy. HT annealing (at 830°C for 40 min in nitrogen gas ambient) was used to eliminate the residual strains in the crystal. The annealed FS GaN crystal was strain-free and had the lattice parameters of c = 5.18500 Å, and a = 3.1890 Å. The PL measurements performed at 14 K revealed the dominant peak at 3.474 eV having the FWHM of 3.5 meV [205].

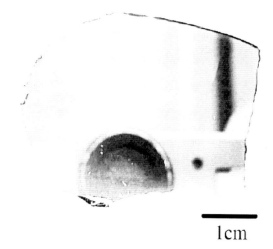

Fig. 3.82. Bulk GaN crystals obtained by dry etching the SiC substrate after ∼100 μm thick GaN growth by HVPE [205]

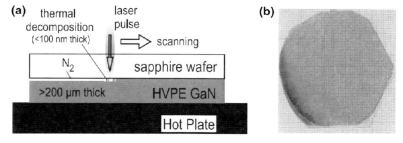

Fig. 3.83. (a) Schematic diagram of the laser process for removal of the sapphire substrate from thick HVPE-grown GaN films. (b) Image of a 275 μm thick freestanding GaN film, after removal from the 2 in. sapphire substrate. Reprinted with permission from [207]. Copyright (1999), The Institute of Pure and Applied Physics

FS GaN Substrate Fabrication by LLO

As mentioned, the LLO process for detaching of the GaN from the substrate was reported by Kelly et al. to separate the HVPE grown GaN film from the sapphire [206, 207]. The intense pulsed laser of appropriate wavelength can be used to locally decompose and split the interface between the nitride film and sapphire substrate as shown in Fig. 3.83a [207]. The LLO process was carried out above 600°C to reduce the bowing of the sample. The bowing was visibly reduced; almost 2 in. FS GaN films were separated from the sapphire wafer as shown in Fig. 3.83b.

Park et al. also demonstrated 2 in. FS GaN substrate by using the LLO process [208]. The 250–350 μm-thick GaN films were separated from the sapphire substrate using a KrF excimer laser. The FWHMs of the (0002) XRCs for the GaN/sapphire and FS GaN were 254 and 149 arcsec, respectively, which indicates a decrease of the FWHM by the relaxation of bowing after detaching [208]. The bowing of the GaN film changed from convex to concave due to the differences in the residual strain on the front and bottom surface.

Although FS GaN can be obtained as described by using the LLO process, due to the laser light irradiated during the LLO and the inhomogeneous temperature rise during the LLO process, the region of the GaN film close to the debonded position suffers from damages resulting in the degraded property [209].

FS GaN Fabrication by Void-Assisted Separation (VAS)

Oshima et al. developed a technique for preparing large-sized FS GaN substrate, where the thick GaN film was easily separated from the template after the cool-down with the assistance of many voids generated around the TiN film [210]. A thick GaN film was grown on a GaN template with a thin TiN film on the top. The GaN templates with a thickness of around 300 nm were prepared on sapphire substrates by MOCVD. A 20 nm-thick Ti layer was deposited on the GaN template by vacuum evaporation. Ti-deposited GaN template was annealed in a H_2 and NH_3 atmosphere at 1,060°C and the 300 µm-thick GaN film was grown on the annealed template by HVPE. Numerous small sized voids were generated at the GaN/TiN/GaN boundary during the HVPE growth. After the cool-down, the thick GaN film was easily separated from the template with the assistance of many voids generated around the TiN film [210]. A weak force was usually added at the growth interface for the separation after the cooling process of the HVPE growth. The wafer obtained had a diameter of 45 mm, and a mirror-like surface. The FWHMs of (0002) and (10$\bar{1}$0) XRCs were 60 and 92 arcsec, respectively. The dislocation density was evaluated at $5 \times 10^6 \, \text{cm}^{-3}$ by etch pit density measurement. Figure 3.84a shows a cross-sectional TEM image of the HVPE grown-GaN(VPE-GaN)/TiN/MOCVD grown-GaN(MO–GaN) boundary. Many threading dislocations (TDs) in the GaN template layer were effectively blocked by the TiN islands. The reduction mechanism of dislocation is probably due to dislocation bending by crystal facets of numerous GaN islands formed on the TiN nano-net. Figure 3.84b shows the FS GaN wafer obtained by the VAS method.

FS GaN Substrate Fabrication by Using the NdGaO$_3$ Substrate

Wakahara et al. reported FS GaN films grown on NdGaO$_3$ substrate by the HVPE process. After the cooling process, some GaN films were self-separated from the substrate [211]. Since the NdGaO$_3$ substrate was decomposed by the presence of H_2 and NH_3 at high temperature, N_2 was

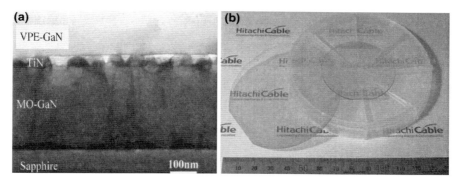

Fig. 3.84. (a) Cross-sectional TEM micrograph around the GaN/TiN/GaN boundary (b) Photograph of 2- and 3 in. – FS GaN wafers obtained by the VAS method. Reprinted with permission from [210]. Copyright (2003), The Institute of Pure and Applied Physics

used as the carrier gas. A LT GaN layer was adopted as a protective layer. The LT GaN protective layer was grown at 600°C, at which temperature the NdGaO$_3$ substrate does not react with NH$_3$. The epitaxial relationships of GaN films grown on (011) and (101) NdGaO$_3$ substrates were found to be GaN(0001)/NdGaO$_3$(011) with GaN[10$\bar{1}$0]//NdGaO$_3$[100] for the (011) NdGaO$_3$ and GaN(11$\bar{2}$4)/NdGaO$_3$(101) with GaN[11$\bar{2}$4]//NdGaO$_3$[10$\bar{1}$] for the (101) NdGaO$_3$. By optimizing the growth conditions, they obtained a FS GaN wafer. Both plan view TEM and CL images suggested that the dislocation density of FS GaN wafers could be expected to be as low as 10^6 cm^{-2}. The PL spectrum indicated strong near band edge emission without deep level related emission. Figure 3.85 shows a FS GaN wafer peeled off from the NdGaO$_3$ substrate [211]. The underlying mechanism for the automatic self-separation during the cooling process is believed to be the result of thermal stress during the cooling down process.

FS GaN Substrate Fabrication by Using the Evaporable Buffer Layer

The method getting the FS GaN substrate from the thick GaN films on (0001) sapphire substrates was reported by using the evaporable buffer layer (EBL) and LT GaN buffer as reported by Lee et al. [212]. The layer grown at around 500°C by HVPE consisted of a mixture of NH$_4$Cl and GaN seeds. By simply lowering the growth temperature to 500°C, NH$_4$Cl was easily synthesized in NH$_3$ and HCl atmosphere in addition to the GaN. The LT GaN film was grown at 600°C on the first layer (i.e., EBL) and annealing was performed. Then the HT GaN layer was grown at 1,040°C. The LT GaN buffer acted as a protective layer against severe evaporation of the NH$_4$Cl and acted as a seeding layer for the HT GaN. During the annealing of the LT GaN followed

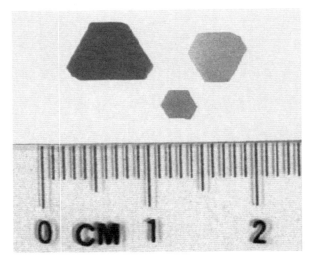

Fig. 3.85. A FS GaN wafer peeled off from the NdGaO$_3$ substrate. Reprinted with permission from [210]. Copyright (2000), The Institute of Pure and Applied Physics

by the HT GaN growth, the NH$_4$Cl was evaporated and many *voids* were formed at the buffer regions. By cooling down the thick HT GaN film, the self-separation of it mostly occurred, but sometimes a weak force was needed at the growth interface for the separation. Therefore, it is concluded that the separation was strongly assisted by the presence of the voids formed by the evaporating of the NH$_4$Cl in the EBL during the HVPE growth. The 200 μm thick FS GaN was obtained by this method [212]. The a-axis and c-axis lattice constants of FS GaN were 3.189 and 5.185 Å, respectively, which well agrees with those values for strain-free bulk GaN. The observed donor-bound exciton emission peak at 3.4718 eV agreed with the peak position of bulk GaN. All these features indicated that the obtained FS GaN is nearly strain-free.

FS GaN Substrate Fabrication by Dissolving the Substrate

Motoki et al. reported fabrication of 2 in. FS GaN substrate by using the (111) GaAs as a starting substrate [213]. They used the substrate with SiO$_2$ mask. Prior to GaN growth, SiO$_2$ layer with a 2 μm round pattern was defined directly on the GaAs (111) surface. After the growth of 500 μm-thick GaN by HVPE, they dissolved the GaAs substrate by chemical etching in an aqua regia, where the FS GaN substrate was obtained. First, a 60 nm-thick GaN buffer layer was selectively grown at 500°C on the patterned substrate. Subsequently, the substrate was raised to 1,030°C in NH$_3$ ambient, and then a thick GaN layer was grown. The FS GaN obtained by chemical etching of the GaAs substrate is shown in Fig. 3.86 [213].

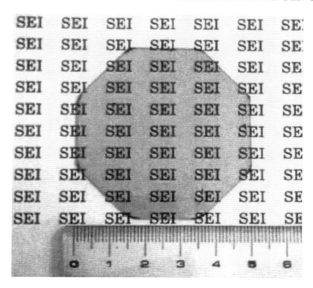

Fig. 3.86. Photograph of FS GaN substrate with a thickness of about 500 μm. Reprinted with permission from [213]. Copyright (2001), The Institute of Pure and Applied Physics

The dislocation density of their FS GaN substrate was determined to be as low as $2 \times 10^5\,\mathrm{cm}^{-2}$ characterized by plan–view transmission electron microscopy. Hall measurements revealed the n-type conductivity of the FS GaN substrate with carrier concentration of $5 \times 10^{18}\,\mathrm{cm}^{-3}$ and mobility of $170\,\mathrm{cm}^2\,\mathrm{V}^{-1}\,\mathrm{s}^{-1}$.

FS GaN Substrate Fabrication by CLO

Most recently, a chemical lift-off (CLO) technique has been developed, in which the CrN is used as a sacrificing layer and is chemically etched to detach the thick GaN film from the (0001) sapphire substrate [214]. High quality metallic Cr layer was deposited on a c-plane sapphire substrate by sputtering. Nitridation of the Cr layer was carried out in a HVPE reactor. After the nitridation of the Cr layer, the epitaxial GaN film was successively grown on the nitrided CrN surface in the same reactor. The CrN layer can be selectively chemical-etched to separate the GaN film from the sapphire substrate. They called this technique as chemical lift-off (CLO). The GaN layers on sapphire with CrN buffer layer were etched by CrN etchant, which was prepared by mixing 200 ml of deionized (DI) water, 50 g of diammonium cerium(IV) nitrate $(Ce(NH_4)_2(NO_3)_6)$, and 13 ml of perchloric acid $(HClO_4)$. Figure 3.87a, b and c show an as-grown GaN film on sapphire substrate with the CrN buffer, a fully-detached GaN film, and a partially-detached GaN film, respectively.

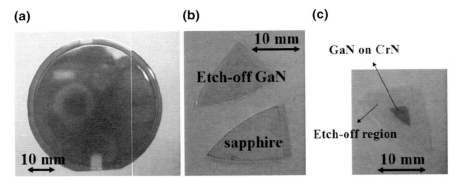

Fig. 3.87. Photographs showing the procedure to get FS GaN film separated by the CLO method. (**a**) An as-grown 40 μm thick-GaN film on sapphire substrate with the CrN buffer, (**b**) Detached GaN film and the sapphire substrate by the CLO method, and (**c**) The sample with a partial etch-off by the CLO method [214]

These figures show that the GaN layers with CrN buffer layers are successfully detached by chemical selective etching.

One of the promising applications of CLO technique is to use it as a device in fabrication of vertical LEDs. The world's first vertical GaN-based LED using the CLO technique was reported by Ha et al. in 2008 [215]. Details of the process and properties are discussed in Chap. 9.

FS GaN Substrate Fabrication by In Situ CLO Using the ZnO Buffer Layer

High quality GaN films can be grown on ZnO films because of the same crystal structure of Wurtzite and similar lattice constants of GaN ($a = 3.189$ Å, $c = 5.185$ Å) and ZnO ($a = 3.249$ Å, $c = 5.207$ Å). Lee et al. employed the ZnO buffer as a sacrificing layer to obtain virtually free-standing GaN buffer on which subsequent thick GaN films were deposited at HT [216]. The ZnO buffer can be easily etched by HCl and NH_3 which are commonly used gases in HVPE growth of GaN. Because of such chemical properties of ZnO, they did a strict control of the HVPE processes as well as a well needed proper protective layer for the ZnO buffer against deterioration of the ZnO surface by chemical etching prior to the GaN growth [216]. Figure 3.88 shows the schematics of the growth sequence, which they developed, to obtain the FS GaN substrates through the in situ lift-off process. A 200 nm thick single crystal ZnO layer with Zn-polarity were grown on a c-plane sapphire using an 8 nm-thick MgO buffer by PAMBE [43]. Then, a 1 μm-thick GaN layer was grown on the ZnO/c-sapphire sample by PAMBE, which enables the growth of Ga-polar GaN film [217] as shown in Fig. 3.88a. A thick GaN buffer film was grown on this template (i.e., PAMBE-GaN/ZnO/sapphire) by HVPE. In HVPE growth, HCl gas reacts with metallic Ga to form GaCl. The GaCl gas is transferred to the growth

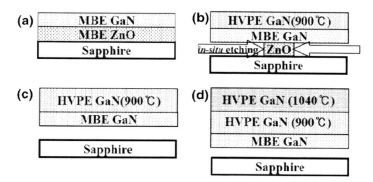

Fig. 3.88. Preparation sequence of the FS GaN substrate separated by in situ lift-off. (**a**) A 200 nm-thick single crystal ZnO layer and a 1 μm-thick GaN layer were grown on c-plane sapphire by PAMBE, (**b**) The ZnO layers were being etched by NH_3 and HCl gases on HVPE growth, (**c**) The first step thick GaN film was grown at 900°C. The ZnO layer between sapphire and GaN layer was all etched off, (**d**) The second step thick GaN film was grown at 1,040°C on the first GaN film. Reprinted with permission from [216]. Copyright (2007), American Institute of Physics

region to react with NH_3 to deposit GaN and HCl gas being produced as a result of the total reaction. The ZnO buffer was gradually etched from the side of the ZnO buffer by these NH_3 and HCl gases as the HVPE growth of GaN was started as shown in Fig. 3.88b. In order to suppress the possible thermal decomposition of the MBE-grown GaN protective layer, they limited the growth temperature of GaN from 700 to 900°C to at the initial stage of HVPE, since the GaN decomposes at temperatures above 900°C without NH_3 gas. The low temperature limit of 700°C was determined to grow the GaN film with a reasonable growth rate, since the growth rate of GaN was drastically decreased below 700°C. They controlled the in situ etching rate of the ZnO buffer layer by the growth temperature and the flow rates of NH_3 and HCl gases. The typical thickness of the LT GaN buffer was over 100 μm. The ZnO layer between the sapphire and the PAMBE-GaN layer were completely etched off during the growth of LT GaN buffer as schematically shown in Fig. 3.88c. They found that the LT GaN buffer should be thick enough to prevent the cracking of GaN buffer before the ZnO buffer is etched off. Finally, thick HT GaN film was successively grown at 1,040°C on the LT GaN buffer layer (i.e., on the LT-GaN/MBE-GaN), which was separated by the in situ lift-off process as shown in Fig. 3.88d [216].

The a-axis and c-axis lattice constants of the FS GaN were determined to be 3.189 and 5.185 Å, respectively, which agrees with those of strain-free GaN bulk. Photoluminescence results showed that the free exciton and donor bound exciton emissions emerge at the same energy positions as those from the bulk GaN. Extensive micro photoluminescence study revealed that strain-free states were extended throughout the HT grown GaN film [216].

3.3.6 Other III Nitrides Grown by HVPE

Aluminum nitride (AlN) and aluminum gallium nitride ($Al_xGa_{1-x}N$) are suitable materials as substrates for AlGaN-based UV LEDs. Indium nitride (InN) is expected to have the smallest effective mass, highest electron drift velocity and smallest band gap among group III nitrides. Therefore, InN will be a promising material for future high-speed, highfrequency electronic and long-wavelength LEDs. The bandgap energy of InN is determined to be around 0.75 eV [218, 219]. However, growth of high-quality InN layers remains difficult due to poor thermal stability of InN and the lack of a suitable substrate material for epitaxy. In this chapter, growth of other III nitrides, i.e., AlN, AlGaN, InN by HVPE is briefly described.

AlN and AlGaN

Nagashima et al. reported that a colorless and transparent AlN layer with a thickness of 83 μm was grown by HVPE [220]. Thermodynamic analysis for the HVPE of AlN using $AlCl_3$ was studied much earlier in 2003 by Kumagai et al. [221]. In the past, HVPE of AlN has a critical problem of reaction between the quartz (SiO_2) reactor and the source gases. The molten Al or hot AlCl gas generated at the source zone reacts violently with the quartz reactor. However, they succeeded in HVPE of AlN by employing the preferential generation of $AlCl_3$, which does not react with quartz at the source zone [222]. Figure 3.89 shows the equilibrium partial pressures of gaseous species as a function of growth temperature under the typical experimental conditions [222]. The equilibrium partial pressures of $AlCl_3$, $AlCl_2$, AlCl and $(AlCl_3)_2$ are very small as shown in the Fig. 3.89, while that of HCl is almost three times larger than that $AlCl_3$ at all growth temperatures. In addition, the equilibrium partial pressures of AlCl and $AlCl_2$ increase with increasing the growth temperature. The thermodynamic study and experiments provides stable growth conditions of AlN by HVPE. As a result, Al metal is typically placed at 500°C to generate $AlCl_3$ as far as possible with the minimized generation of AlCl. Based on the thermodynamic analysis, they obtained the AlN layer at a growth rate as high as 57 μm h^{-1}.

The FWHMs for the (0002) and (10$\bar{1}$1) XRCs were determined to be 295 and 432 arcsec, respectively, which represent the certainty of AlN as well as AlGaN growth by HVPE. Koukitu et al. reported HVPE growth of AlGaN ternary alloy using $AlCl_3$–$GaCl$–NH_3 system on sapphire without any buffer layers, where the typical growth rate was around 30 μm h^{-1} at 1,100°C [223]. They found that the solid composition could be controlled in all ranges from GaN to AlN by changing the input ratio of the group III precursors.

InN

InCl and $InCl_3$ have been used for HVPE of InN. InN epitaxial layer was not grown when InCl is used [224], while appreciable growth occurs when

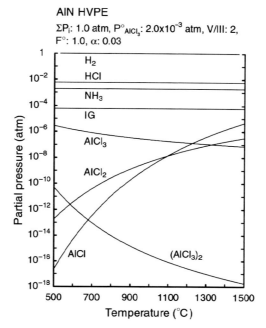

Fig. 3.89. Equilibrium partial pressures of gaseous species over AlN as a function of growth temperature. Reprinted with permission from [222]. Copyright (2006), Wiley

$InCl_3$ powder was used as In source [224, 225]. Kumagai et al reported that the equilibrium constant of the reaction between InCl and NH_3 is lower than, between $InCl_3$ and NH_3, and suppression of InN growth occurs under the presence of H_2 [226]. Although an appreciable growth of InN has been obtained using $InCl_3$ powder as a source material, $InCl_3$ usually contains water owing to its hygroscopic nature. In order to grow a high-quality InN layer without contamination, formation of $InCl_3$ in the source zone of the HVPE system is crucial. Kumagai et al. reported HVPE of InN using In metal, Cl_2 and NH_3 as source materials and N_2 carrier gas for the InN growth on sapphire (0001) substrates [227]. A thermodynamic analysis of the reaction between In metal and Cl_2 gas in the source zone of the HVPE system revealed that the equilibrium partial pressure of $InCl_3$ increased significantly with decreasing source zone temperature, while that of InCl remained almost constant. At a substrate temperature of 500°C, appreciable growth of InN occurred by decreasing the source zone temperature to 450°C. This means that InN growth occurs by the reaction between $InCl_3$ and NH_3. With an optimum surface nitridation time of the sapphire substrate prior to the InN growth, c-axis oriented single-crystalline layers of InN were obtained. Cathodoluminescence (CL) spectrum measured at RT showed a strong peak at 0.75 eV. FWHMs of the InN layer for the (0002) and ($10\bar{1}0$) XRCs were determined to be 13.2 and 94.2 arcmin, respectively [227].

References

1. M. Bagnall, Y.F. Chen, Z. Zhu, T. Yao, S. Koyama, M.Y. Shen, T. Goto, Appl. Phys. Lett. **70**, 2230 (1997)
2. Z.K. Tang, G.K.L. Wong, P. Yu, M. Kawasaki, A. Phtomo, H. Koinuma, Y. Segawa, Appl. Phys. Lett. **72**, 2446 (1998)
3. Y. Chen, H.J. Ko, S.K. Hong, T. Yao, Appl. Phys. Lett. **76**, 559 (2000)
4. Y. Chen, S.K. Hong, H.J. Ko, V. Kirshiner, H. Wenisch, T. Yao, K. Inaba, Y. Segawa, Appl. Phys. Lett. **78**, 3352 (2001)
5. T. Makino, C.H. Chia, N.T. Tuan, Y. Segawa, M. Kawasaki, A. Ohtomo, K. Tamura, H. Koinuma, Appl. Phys. Lett. **76**, 3549 (2000)
6. A. Valentini, A. Quirini, L. Vasanelli, Thin Solid Films **176**, L167 (1989)
7. M. Kadota, M. MIinakat, IEEE Trans. Ultrason. Ferroelectr. Freq. Control **42**, 345 (1995)
8. R. Cebulla, R. Wendt, K. Ellmer, J. Appl. Phys. **83**, 1087 (1998)
9. P. Yu, Z.K. Tang, G.K.L. Wong, M. Kawasaki, A. Ohtomo, H. Koinuma, Y. Segawa, J. Cryst. Growth **184/185**, 601 (1998)
10. R.D. Vispute, V. Tayansky, Z. Trajanovic, S. Choopun, M. Downs, R.P. Sharama, T. Venkatesan, M.C. Woods, R.T. Lareau, K.A. Jones, A.A. Iliadis, Appl. Phys. Lett. **70**, 2735 (1997)
11. S. Hayamizu, H. Tabata, H. Tanaka, T. Kawai, J. Appl. Phys. **80**, 787 (1996).
12. R.D Vispute, V. Tayansky, S. Choopun, R.P. Sharma, T. Venkatesan, M. He, X. Tang, J.B. Halpern, M.G. Spencer, Y.X. Li, L.G. Salamanca-Riba, A.A. Illiadis, K.A. Jones, Appl. Phys. Lett. **73**, 348 (1998)
13. K. Yamaya, Y. Yamichi, H. Nakanishi, S. Chichibu, Appl. Phys. Lett. **72**. 235 (1998)
14. F. Hamdani, A. Botchkarev, W. Kim, H. Morkoc, M. Yeadon, J.M. Gibson, S.C.Y. Tsen, D.J. Smith, D.C. Reynorlds, D.C. Look, K. Evans, C.W. Litton, W.C. Mitchel, P. Hemenger, Appl. Phys. Lett. **70**, 467 (1997)
15. Y. Chen, D.M. Bagnall, Z. Zhu, T. Sekiguchi, K. Park, K. Park, K. Hiraga, T. Yao, S. Koyama, M.Y. Shen, T. Goto, J. Cryst. Growth **181**, 165 (1997)
16. Y. Chen, D.M. Bagnall, H. Ko, K. Park, K. Hiraga, Z. Zhu, T. Yao, J. Appl. Phys. **84**, 3912 (1998)
17. K. Kobayashi, T. Matsubara, S. Matsushima, S. Shirrakata, S. Isomura, G. Okata, Thin Solid Films **266**, 106 (1995)
18. T. Minami, H. Nanto, S. Tanaka, Jpn. J. Appl. Phys. 2 **23**, L280 (1984)
19. C.K. Ryu, K. Kim, Appl. Phys. Lett. **67**, 3337 (1995)
20. P. Verardi, M. Dinescu, A. Andrei, Appl. Surf. Sci. **96–98**, 827 (1996)
21. Y. Chen, D. Bagnall, T. Yao, Master. Sci. Eng. B **75**, 190 (2000)
22. S.K. Hong, Y. Chen, H.J. Ko, H. Wenisch, T. Hanada, T. Yao, J. Electron. Mater. **30**, 647 (2001)
23. M.A.L. Jhonson, S. Fujita, W.H. Rowland Jr., W.C. Highes, J.W. Cook Jr., J.F. Schetzina, J. Electron. Mater. **25**, 855 (1996)
24. F. Vigue, P. Vennegues, S. Vezian, M. Laugt, J.P. Faurie, Appl. Phys. Lett. **79**, 194 (2001)
25. P. Fons, K. Iwata, S. Niki, A. Yamada, K. Matsubara, J. Cryst. Growth **201–202**, 627 (1999)
26. P. Fons, K. Iwata, S. Niki, A. Yamada, K. Matsubara, M. Watanabe, J. Cryst. Growth **209**, 532 (2000)

27. P. Fons, K. Iwata, A. Yamada, K. Matsubara, S. Niki, K. Nakahara, T. Tanabe, H. Takasu, Appl. Phys. Lett. **77**, 1801 (2000)
28. P. Fons, K. Iwata, A. Yamada, K. Matsubara, S. Niki, K. Nakahara, T. Tanabe, H. Takasu, J. Cryst. Growth **227–228**, 911 (2001)
29. P.F. Fewster, in *X-ray and Neutron Dynamical Diffraction*, ed. by A. Authier, S. Lagomarsion, B.K. Tanner (Plenum Press, New York, 1996), pp. 283
30. I. Ohkubo, A. Ohtomo, T. Ohnishi, Y. Matsumoto, H. Koinuma, M. Kawasaki, Surf. Sci. **443**, L1043 (1999)
31. K. Sakurai, M. Kanehiro, K. Nakahara, T. Tanabe, S. Fujita, S. Fujita, J. Cryst. Growth **214–215**, 92 (2000)
32. M. Sano, K. Miyamoto, H. Kato, T. Yao, Jpn. J. Appl. Phys. 2 **42**, L1050 (2003)
33. X. Wang, H. Iwaki, M. Murakami, X. Du, Y. Ishitani, A. Yoshikawa, Jpn. J. Appl. Phys. 2 **42**, L99 (2003)
34. H.J. Ko, Y. Chen, S.K. Hong, T. Yao, J. Cryst. Growth **209**, 816 (2000)
35. H. Kato, M. Sano, K. Miyamota, T. Yao, Jpn. J. Appl. Phys. 2 **42**, L1002 (2003)
36. H. Weinisch, V. Kirchner, S.K. Hong, Y.F. Chen, H.J. Ko, T. Yao, J. Cryst. Growth **227–228**, 944 (2001)
37. H.J. Ko, T. Yao, Y.F. Chen, S.K. Hong, J. Appl. Phys. **92**, 4354 (2002)
38. H. Kato, M. Sano, K. Miyamoto, T. Yao, Jpn. J. Appl. Phys. 2, **42**, L1002 (2003)
39. H. Kato, M. Sano, K. Miyamoto, T. Yao, Jpn. J. Appl. Phys. 1, **42**, 2241 (2003)
40. T. Ohnishi, A. Ohtomo, M. Kawasaki, K. Takahashi, M. Yoshimoto, H. Koinuma, Appl. Phys. Lett. **72**, 824 (1998)
41. S.K. Hong, T. Hanada, H.J. Ko, Y. Chen, T. Yao, D. Imai, K. Araki, M. Shinohara, Appl. Phys. Lett. **77**, 3571 (2000)
42. S.K. Hong, T. Hanada, H.J. Ko, Y. Chen, D. Imai, K. Araki, M. Shinohara, T. Yao, K. Saitoh, M. Terauchi, Phys. Rev. B **65**, 115331 (2002)
43. H. Kato, K. Miyamoto, M. Sano, T. Yao, Appl. Phys. Lett. **84**, 4562 (2004)
44. J.S. Park, S.K. Hong, T. Minegishi, S.H. Park, I.H. Im, T. Hanada, M.W. Cho, T. Yao, J.W. Lee, J.Y. Lee, Appl. Phys. Lett. **90**, 201907 (2007)
45. X.G. Wang, J.R. Smith, Phys. Rev. B **68**, 201402 (2003)
46. C. Rehbein, N.M. Harrison, A. Wander, Phys. Rev. B **54**, 14066 (1996)
47. Z.X. Mei, Y. Wang, X.L. Du, M.J. Ying, Z.Q. Zeng, H. Zheng, J.F. Jia, Q.K. Xue, Z. Zhang, J. Appl. Phys. **96**, 7108 (2004)
48. A. Setiawan, H.J. Ko, S.K. Hong, Y. Chen, T. Yao, Thin Solid Films **445**, 213 (2003)
49. Y. Kagamitani, A. Yoshikawa, T. Fukuda, T. Ono, Private communication
50. See http://global.kyocera.com/prdct/fc/product/pdf/s_c_sapphire.pdf for recent CTE value of sapphire
51. E.S. Shim, H.S. Kang, J.S. Kang, J.H. Kim, S.Y. Lee, Appl. Surf. Sci. **186**, 474 (2002)
52. S.L. Park, T.S. Cho, S.J. Doh, J.J. Lee, J.H. Je, Appl. Phys. Lett. **77**, 349 (2000)
53. P. Fons, K. Iwat, A. Yamada, K. Matsubara, S. Niki, K. Nakahara, T. Tanabe, H. Takasu, J. Cryst. Growth **227**, 911 (2001)
54. J.M. Myoung, W.H. Yoon, D.H. Lee, I. Yun, S.H. Bae, S.Y. Lee, Jpn. J. Appl. Phys. 1 **41**, 28 (2002)
55. S.H. Park, T. Hanada, D.C. Oh, T. Minegishi, H. Goto, G. Fujimoto, J.S. Park, I.H. Im, J.H. Chang, M.W. Cho, T. Yao, K. Inaba, Appl. Phys. Lett. **91**, 231904 (2007)

56. A. Ohtomo, K. Tamura, M. Kawasaki, T. Makino, Y. Segawa, Z.K. Tang, G.K.L. Wong, Appl. Phys. Lett. **77**, 2204 (2000)
57. E.R. Segnit, A.E. Holland, J. Am. Ceram. Soc. **48**, 412 (1965)
58. J.A. Van Vechten, T.K. Bergstresser, Phys. Rev. B **1**, 3351 (1970)
59. A. Ohtomo, R. Shiroki, I. Ohkubo, H. Koinuma, M. Kawasaki, Appl. Phys. Lett. **75**, 4088 (1999)
60. S. Choopun, R.D. Vispute, W. Yang, R.P. Sharma, T. Venkatesan, H. Shen, Appl. Phys. Lett. **80**, 1529 (2002)
61. T. Gruber, C. Kirchner, R. Kling, F. Reuss, A. Waag, Appl. Phys. Lett. **84**, 5359 (2004)
62. T. Makino, C.H. Chia, N.T. Tuan, Y. Segawa, M. Kawasaki, A. Ohtomo, K. Tamura, H. Koinuma, Appl. Phys. Lett. **77**, 1632 (2000)
63. W. Ma, Z.Z. Ye, L.L. Chen, Phys. Status Solidi A. **201**, 2929 (2004)
64. T. Makino, Y. Segawa, M. Kawasaki, A. Ohtomo, R. Shiroki, K. Tamura, T. Yasuda, H. Koinuma, Appl. Phys. Lett. **78**, 1237 (2001)
65. Y. Chen, H.-J. Ko, S.-k. Hong, T. Sekiuchi, T. Yao, Y. Segawa, J. Vac. Sci. Technol. B **18**, 1514 (2000)
66. A. El-Shaera, A. Bakina, M. Al-Suleimana, S. Ivanovb, A.C. Mofora, A. Waaga, Superlattice. Microst. **42**, 129–133 (2007)
67. H. Tanaka, S. Fujita, S. Fujita, Appl. Phys. Lett. **86**, 192911 (2005)
68. T. Takagi, H. Tanaka, S. Fujita, S. Fujita, Jpn. J. Appl. Phys. 2 **42**, L401 (2003)
69. H. Tampo, H. Shibata, K. Matsubara, A. Yamada, P. Fons, S. Niki, M. Yamagata H. Kanie, Appl. Phys. Lett. **89**, 132113 (2006)
70. H. Kato, K. Miyamoto, M. Sano, T. Yao, Appl. Phys. Lett. **84**, 4562 (2004)
71. J. Saito, K. Nanbu, T. Ishikawa, S. Hiyamizu, Jpn. J. Appl. Phys. 2 **22**, L79 (1983)
72. O. Ambacher, J. Smart, J.R. Shealy, N.G. Weimann, K. Chu, M. Murphy, W.J. Schaff, L.F. Eastman, R. Dimitrov, L. Wittmer, M. Stutzmann, W. Rieger, J. Hilsenbeck, J. Appl. Phys. **85**, 3222 (1999)
73. T. Wang, Y. Ohno, M. Lachab, D. Nakagawa, T. Shirahama, S. Sakai, H. Ohno, Appl. Phys. Lett. **74**, 3531 (1999)
74. A.D. Corso, M. Posternak, R. Resta, A. Baldereschi, Phys. Rev. B **50**, 10715 (1994)
75. F. Bernardini, V. Fiorentini, D. Vanderbilt, Phys. Rev. B **56**, R10024 (1997)
76. H. Shibata, H. Tampo, K. Matsubara, A. Yamada, K. Sakurai, S. Ishizuka, S. Niki, M. Sakai, Appl. Phys. Lett. **90**, 124104 (2007)
77. E. Cohen, M.D. Sturge, Phys. Rev. B **25**, 3828 (1982)
78. Y.R. Ryu, T.S. Lee, J.A. Lubguban, A.B. Corman, H.W. White, J.H. Leem, M.S. Han, Y.S. Park, C.J. Youn, W.J. Kim, Appl. Phys. Lett. **88**, 052103 (2006)
79. O. Madelung, *Semiconductors: Data Handbook*, 3rd edn. (Springer, New York, 2003)
80. D.B. Christey, G.K. Hubler, *Pulsed laser deposition of thin films* (Wiley, New York, 1994)
81. A. Ohtomo, A. Tsukazaki, Semicond. Sci. Technol. **20**, S1 (2005)
82. R. Triboulet, J. Perriier, *Progress in Crystal Growth and Characterization of Materials*, vol. 47 (Elsevier, Amsterdam, 2003), p. 65
83. D.P. Norton, Y.W. Heo, M.P. Ivill, K. Ip, S.J. Pearton, M.F. Chisholm, T. Steiner, Mater. Today, June 2004, pp. 34
84. S.J. Pearton, D.P. Norton, K. Ip, Y.W. Heo, J. Vac. Sci. Technol. B **22**, 932 (2004)

85. U. Ozgur, Y.I. Alivov, C. Liu, A. Teke, M.M. Reshchikov, S. Dogan, V. Avrutin, S.-J. Cho, H. Morkoc, J. Appl. Phys. **98**, 041401 (2005)
86. S.B. Ogale, Thin films and heterostructures for oxide electronics (Springer Science, New York, 2005)
87. A.V. Singh, R.M. Mehra, N. Buthrath, A. Wakahara, A. Yoshida, J. Appl. Phys. **90**, 5661 (2001)
88. T. Nakayama, Surf. Sci. **133**, 101 (1983)
89. H. Sankur, J.T. Cheung, J. Vac. Sci. Technol. A **1**, 1806 (1983)
90. V. Cracium, J. Elders, J.G.E. Gardeniers, I.W. Boyd, Appl. Phys. Lett. **65**, 2963 (1994)
91. M. Okoshi, K. Higashikawa, M. Hanabusa, Appl. Surf. Sci. **154**, 424 (2000)
92. L.N. Dinh, M.A. Schildbach, M. Balooch, W. Mclean II, J. Appl. Phys. **86**, 1149 (1999)
93. O.G. Noël, R.G. Roman, J. Perriere, J. Hermann, V. Craciun, C.-B. Leborgne, P. Barboux, J. Appl. Phys. **80**, 1803 (1996)
94. V. Craciun, S. Amirhaghi, D. Craciun, J. Elders, J.G.E. Gardeniers, I.W. Boyd, Appl. Surf. Sci. **86**, 275 (1995)
95. Y.R. Ryu, S. Zhu, J.D. Budai, H.R. Chandrasekhar, P.F. Miceli, H.W. White, Appl. Phys. Lett. **88**, 201 (2000)
96. A. Tsukazaki, A. Ohtomo, T. Onuma, M. Ohtani, T. Makino, M. Sukiya, K. Ohtani, S.F. Chichibu, S. Fuke, Y. Segawa, H. Ohno, H. Koinuma, M. Kawasaki, Nature Mater. **4**, 42 (2005)
97. W. Prellier, A. Fouchet, B. Mercey, C. Simon, B. Raveau, Appl. Phys. Lett. **82**, 3490 (2003)
98. A. Fouchet, W. Prellier, B. Mercey, L. Mechin, V.N. Kulkarni, T. Venktesan, J. Appl. Phys. **96**, 3228 (2004)
99. C. Belouet, Appl. Surf. Sci. **96**, 630 (1996)
100. V. Craciun, N. Basim, R.K. Singh, D. Craciun, J. Herman, C.B. Levorgne, Appl. Surf. Sci. **186**, 288 (2002)
101. V. Craciun, D. Craciun, M.C. Bunescu, C.B. Lebourgne, J. Hermann, Phys. Rev. B **58**, 6787 (1998)
102. R. Kelly, A. Miotello, Phys. Rev. E **60**, 2616 (1999)
103. P. Mukherjee, S. Chen, J.B. Cuff, P. Sakthival, S. Witanachci, J. Appl. Phys. **91**, 1828 (2002)
104. E. Millom, O. Albert, J.C. Loulergue, J. Etchepare, D. Hulin, W. Selier, J. Perriere, J. Appl. Phys. **88**, 06937 (2000)
105. J. Perriere, E. Millon, W. Seiler, C.B. Leborgne, V. Cracium, O. Albert, J.C. Loulergue, J. Etchepare, J. Appl. Phys. **91**, 690 (2002)
106. C. Jagadish, S.J. Pearton (eds.) Pulsed laser deposition of zinc oxide, Chap. 4. in *Zinc oxide bulk, thin films and nanostructures*. (Elsevier, Amsterdam, 2006)
107. P. Mukherjee, S. Chen, S. Witanachci, Appl. Phys. Lett. **74**, 1546 (1999)
108. P. Mukherjee, S. Chen, J.B. Cuff, P. Sakthivel, S. Witanachchi, J. Appl. Phys. **91**, 1837 (2002)
109. A. Ohtomo, M. Kawasakim, I. Ohkubo, H. Koinuma, T. Yasuda, Y. Segawa, Appl. Phys. Lett. **75**, 980 (1999)
110. E.M. Kaidashev, M. Lorentz, H.V. Wenckstern, A. Rahm, H.C. Semmelhack, K.H. Han, G. Benndorf, C. Bundesmann, H. Hockmuth, M. Grudmann, Appl. Phys. Lett. **82**, 3901 (2003)
111. M.Z. Lin, C.T. Su, H.C. Yan, M.Y. Chern, Jpn. J. Appl. Phys. **44**, L995 (2005)
112. B. Wesslerm A, Steinecker, W. Mader, J. Cryst. Growth **242**, 283 (2002)

113. K. Tamura, A. Ohtomo, K. Saikusa, Y. Osaka, T. Makino, Y. Segawa, M. Sumiya, S. Fuke, H. Koinuma, M. Kawasaki, J. Cryst. Growth **214/215**, 59 (2000)
114. M. Kawasaki, A. Ohtomo, I. Ohkubo, H. Koinuma, Z.K. Tang, P. Yu, G.K.L. Wong, B.P. Zhang, Y. Segawa, Mater. Sci. Eng. B **56**, 239 (1998)
115. D.C. Reynolds, D.C. Look, B. Jogai, C.W. Litton, T.C. Collins, W. Harsch, G. Cantwell, Phys. Rev. B **57**, 12151 (1998)
116. A. Tsukazakim, A. Ohtomo, M. Kawasaki, T. Makino, C.H. China, Y. Segawaand, H. Koinuma, Appl. Phys. Lett. **83**, 2784 (2003)
117. T. Koida, S.F. Chichibu, A. Uedono, A. Tsukazaki, M. Kawasaki, T. Sota, Y. Segawa, H. Koinuma, Appl. Phys. Lett. **82**, 532 (2003)
118. S.F. Chichibu, T. Sota, G. Cantwell, D.B. Eason, C.W. Litton, J. Appl. Phys. **93** 756 (2003)
119. E. Bellingeri, D. Marre, I. Pallecchi, L. Pellegrino, A.S. Siri, Appl. Phys. Lett. **86**, 012109 (2005)
120. M. Karger, M. Schilling, Phys. Rev. B **71**, 075304 (2005)
121. X.H. Wei, Y.R. Li, J. Zhu, W. Huang, Y. Zhang, W.B. Luo, H. Ji, Appl. Phys. Lett. **90**, 151918 (2007)
122. A. Ohtomo, M. Kawasaki, T. Koida, K. Masubuchi, H. Koinuma, Y. Sakurai, Y. Yoshida, T. Yasuda, Y. Segawa, Appl. Phys. Lett. **72**, 2466 (1998)
123. J.F. Sarver, F.L. Katnack, F.A. Hummel, J. Electrochem. Soc. **106**, 960 (1959)
124. L.A. Bendersky, I. Takeuchi, K.-S. Chang, W. Yang, S. Hullavarad, R.D. Vispute, J. Appl. Phys. **98**, 083526 (2005)
125. I. Takeuchi, W. Yang, K.S. Chang, M.A. Aronova, T. Ventkatesan, R.D. Vispute, L.A. Bendersky, J. Appl. Phys. **94**, 7336 (2003)
126. S. Heitsch, G. Zimmermann, D. Fritsch, C. Sturm, R. Schmidt-Grund, C. Schulz, H. Hochmuth, D. Spemann, G. Benndorf, B. Rheinländer, Th. Nobis, M. Lorenz, M. Grundmann, J. Appl. Phys. **101**, 083521 (2007)
127. R. Schmidt, B. Rheinländer, M. Schubert, D. Spemann, T. Butz, J. Lenzner, E.M. Kaidashev, M. Lorenz, A. Rahm, H.C. Semmelhack, M. Grundmann, Appl. Phys. Lett. **82**, 2260 (2003)
128. J. Christen, D. Bimberg, Phys. Rev. B **42**, 7213 (1990)
129. R.P. Koffyberg, Phys. Rev. B **13**, 4470 (1976)
130. F.K. Shan, G.X. Liu, W.J. Lee, B.C. Shin, J. Cryst. Growth **291**, 328 (2006)
131. T. Makino, Y. Segawa, M. Kawasaki, A. Ohtomo, R. Shiroki, K. Tamura, T. Yasuda, H. Koinuma, Appl. Phys. Lett. **78**, 1237 (2001)
132. H. Amano, N. Sawaki, I. Akasaki, T. Toyoda, Appl. Phys. Lett. **48**, 353 (1986)
133. S. Nakamura, M. Senoh, T. Mukai, Japanese J. Appl. Phys. **30**, 1708 (1991)
134. Y. Ohba, A. Hatano, Japanese J. Appl. Phys. **35**, L1013 (1996)
135. P. Finni, X. Wu, E.J. Tarsa, Y. Golan, V. Srikant, S. Keller, S.P. DenBaars, J.S. Speck, Jpn. J. Appl. Phys. **37**, 4460 (1998)
136. A. Watanabe, H. Takahashi, T. Tanaka, H. Ota, K. Chikuma, H. Amano, T. Kashima, R. Nakamura, I. Akasaki, Jpn. J. Appl. Phys. **38**, L1159 (1999)
137. H. Miyake, A. Motogaito, K. Hiramatsu, Jpn. J. Appl. Phys. **38**, L1000 (1999)
138. K. Hoshino, N. Yanagita, M. Araki, K. Tadatomo, J. Cryst. Growth **298**, 232 (2007)
139. A. Ubukata, K. Ikenaga, N. Akutsu, A. Yamaguchi, K. Matsumoto, T. Yamazaki, T. Egawa, J. Cryst. Growth **298**, 198 (2007)
140. M. Sakai, H. Ishikawa, T. Egawa, T. Jimbo, M. Umeno, T. Shibata, K. Asai, S. Sumiya, Y. Kuraoka, M. Tanaka, O. Oda, J. Cryst. Growth **244**, 6 (2002)

141. H. Tokunaga, A. Ubukata, Y. Yano, A. Yamaguchi, N. Akutsu, T. Yamasaki, K. Matsumoto, J. Cryst. Growth **272**, 3348 (2004)
142. M. Leroux, P. Venne'gue's, S. Dalmasso, M. Benaissa, E. Feltin, P. de Mierry, B. Beaumont, B. Damilano, N. Grandjean, P. Gibart, Phys. Stat. Sol. A **192**, 394 (2002)
143. D. Cherns, Y.Q. Wang, R. Liu, F.A. Ponce, Appl. Phys. Lett. **81**, 4541 (2002)
144. H. Xing, D.S. Green, H. Yu, T. Mates, P. Kozodoy, S. Keller, S.P. Denbaars, U.K. Mishra, Jpn. J. Appl. Phys. **42**, 50 (2003)
145. P. Kozodoy, S. Keller, S.P. DenBaars, U.K. Mishra, J. Cryst. Growth **195**, 265 (1998)
146. W. Lee, J. Limb, J.-H. Ryou, D. Yoo, T. Chung, R.D. Dupuis, J. Cryst. Growth **287**, 577 (2006)
147. C.H. Chen, H. Liu, D. Steigerwald, W. Imler, C.P. Kuo, M.G. Craford, Mater. Res. Soc. Symp. Proc. **395**, 103 (1993)
148. H. Tokunaga, H. Tan, Y. Inaishi, T. Arai, A. Yamaguchi, J. Hidaka, J. Cryst. Growth **221**, 616 (2000)
149. A.A. Allerman, M.H. Crawford, A.J. Fischer, K.H.A. Bogart, S.R. Lee, D.M. Follstaedt, P.P. Provencio, D.D. Koleske, J. Cryst. Growth **272**, 227 (2004)
150. A. Koukitu, T. Taki, N. Takahashi, H. Seki, J. Cryst. Growth **197**, 99 (1999)
151. T. Matsuoka, N. Yoshimoto, T. Sasaki, A. Katsui, J. Electron. Mater. **21**, 157 (1992)
152. Y. Kangawa, T. Ito, A. Mori, A. Koukitu, J. Cryst. Growth **220**, 401 (2000)
153. T. Akasaka, H. Gotoh, T. Saito, T. Makimoto, Appl. Phys. Lett. **85**, 3089 (2004)
154. J.K. Son, S.N. Lee, T. Sakong, H.S. Paek, O. Nam, Y. Park, J.S. Hwang, J.Y. Kim, Y.H. Cho, J. Cryst. Growth **287**, 558 (2006)
155. H. Yamada, K. Iso, M. Saito, K. Fujito, S.P. Denbaars, J.S. Speck, S. Nakamura, Jpn. J. Appl. Phys. **46**, L1117 (2007)
156. T.G. Mihopoulos, Dissertation, MIT, 1999
157. A. Thon, T.F. Kuech, Appl. Phys. Lett. **69**, 55 (1996)
158. D. Mazzaresse, A. Tripath, W.C. Conner, K.A. Jones, L. Calderon, D.W. Eckart, J. Electronic Mater. **18**, 369 (1989)
159. G.B. Stringfellow, *Organometallic Vapor-Phase Epitaxy: Theory and Practice*, 2nd edn. (Academic Press, New York, 1999)
160. J.R. Creighton, G.T. Wang, W.G. Breiland, M.E. Coltrin, J. Cryst. Growth **261**, 204 (2004)
161. K. Matsumoto, A. Tachibana, J. Cryst. Growth **272**, 360 (2004)
162. U. Bergmann, V. Reimer, B. Atakan, Phys. Chem. Chem. Phys. **1**, 5593 (1999)
163. A. Hirako, K. Kusabake, K. Ohkawa, Jpn. J. Appl. Phys. **44**, 874 (2005)
164. M. Dauelsberg, C. Martin, H. Protzmann, A.R. Boyd, E.J. Thrush, J. Käppeler, M. Heuken, R.A. Talalaev, E.V. Yakovlev, A.V. Kondratyev, J. Cryst. Growth **298**, 418 (2007)
165. A.V. Lobanova, K.M. Mazaev, R.A Talalaev, M. Leys, S. Boeykens, K. Cheng, S. Degroote, J. Cryst. Growth **287**, 601 (2006)
166. X. Zhang, I. Moerman, C. Sys, P. Demeester, J.A. Crawley, E.J. Thrush, J. Cryst. Growth **170**, 83 (1997)
167. D.W. Weyburne, B.S. Ahem, J. Cryst. Growth **170**, 77 (1997)
168. K. Matsumoto, K. Itoh, T. Tabuchi, R. Tsunoda, J. Cryst. Growth **77**, 151 (1986)

169. S. Nakamura, Jpn. J. Appl. Phys. **30**, L1705 (1991)
170. M. Kondo, J. Okazaki, H. Sekiguchi, T. Tanahashi, S. Yamazaki, K. Nakajima, J. Cryst. Growth **115**, 231 (1991)
171. C.A. Wang, S.H. Groves, S.C. Palmateer, D.W. Weyburne, R.A. Browwn, J. Cryst. Growth **77**, 136 (1986)
172. G.S. Tompa, M.A. McKee, C. Beckham, P.A. Zawadzki, J.M. Colabella, P.D. Reinert, K. Capuder, R.A. Stall, P.E. Norris, J. Cryst. Growth **93**, 220 (1988).
173. W. Kern, RCA Rev. 525 (1968)
174. H. Tanaka, H. Itoh, T. O'hori, M. Takikawa, K. Kasai, M. Takeuchi, M. Suzuki, J. Komeno, Jpn. J. Appl. Phys. **26**, L1456 (1987)
175. P.M. Frijlink, J. Cryst. Growth **93**, 207 (1988)
176. T. Arai, J. Hidaka, H. Tokunaga, K. Matsumoto, J. Cryst. Growth **170**, 88 (1997)
177. R.M. Watwe, J.A. Dumesic, T.F. Kuech, J. Cryst. Growth **221**, 751 (2000)
178. K. Uchida, H. Tokunaga, Y. Inaishi, N. Akutsu, K. Matsumoto, T. Itoh, T. Egawa, T. Jimbo, M. Umeno, Mat. Res. Soc. Symp. Proc. **449**, 129 (1997)
179. K. Matsumoto, T. Arai, H. Tokunaga, Vacuum **51**, 699 (1998)
180. H. Tokunaga, Y. Fukuda, A. Ubukata, K. Ikenaga, Y. Inaishi, T. Orita, S. Hasaka, Y. Kitamura, A. Yamaguchi, S. Koseki, K. Uematsu, N. Tomita, N. Akutsu, K. Matsumoto, Phys. Stat. Sol. C **5**, 3017 (2008)
181. P. Ruterana, J. Neugebauer, M. Albrecht, Nitride Semiconductors: handbook on materials and devices (Wiley, Dorderecht, 2003), pp. 193
182. H.P. Maruska, J.J. Tietjen, Appl. Phys. Lett. **15**, 327 (1969)
183. J.I. Pankove, J.E. Berkeyheiser, H.P. Maruska, J. Wittke, Solid State Commun. **8**, 1051 (1970)
184. J.I. Pankove, H.P. Maruska, J.E. Berkeyheiser, Appl. Phys. Lett. 7, **197** (1970)
185. A. Usui, H. Sunakawa, A. Sakai, A.A. Yamaguchi, Jpn. J. Appl. Phys. **36**, L899 (1997)
186. A. Koukitu, Y. kumagai, J. Phys. Condens. Matter. **13**, 6907 (2001)
187. A. Koukitu, S. Hama, T. Taki, H. Seki, Jpn. J. Appl. Phys. **37**, 762 (1998)
188. A. Koukitu, H. Seki, J. Cryst. Growth **49**, 325 (1980)
189. V.S. Ban, J. Electrochem. Soc. **119**, 762 (1972)
190. D.W. Shaw, in *Crystal Growth*, ed. by C.H.L. Goodman (Plenum, New York, 1978), p. 1
191. B. Schineller, J. Kaeppeler, M. Heuken, CS Mantech Conference, May 14–17 (Austin, Texas, USA, 2007), p. 122
192. W. Utsumi, H. Saitoh, H. Kaneko, T. Watanuki, K. Aoki, I. Shimomura, Nat. Mater. **2**, 735 (2003)
193. S. Porowski, Mater. Sci. Eng. B **44**, 407 (1997)
194. S. Porowski, I. Grzegory, J. Cryst. Growth **178**, 174 (1998)
195. H. Yamane, M. Shimada, S.J. Clarke, F.J. Disavo, Chem. Mater. **9**, 413 (1997)
196. R. Dwilinski, R. Doradzinski, J. Garczynski, L. Sierzputowski, M. Palczewska, A. Wysmolek, M. Kaminska, MRS Internet J. Nitride Semicond. Res. 3, **25** (1998)
197. A. Yoshikawa, E. Ohshimaa, T. Fukudaa, H. Tsujib, K. Oshimab, J. Cryst. Growth **260**, 67 (2004)
198. K. Hiramatsu, T. Detchprohm, I. Akasaki, Jpn. J. Appl. Phys. **32**, 1528 (1993)
199. E.V. Etzkorn, D.R. Clarke, J. Appl. Phys. **89**, 1025 (2001)
200. R.W. Hoffman, Phys. Thin Films **3**, 211 (1966)
201. F.A. Doljack, R.W. Hoffman, Thin Solid Films **12**, 71 (1972)

202. S. Hearne, E. Chason, J. Han, J.A. Floro, J. Hunter, H. Amano, I.S.T. Tsong, Appl. Phys. Lett. **74**, 356 (1999)
203. N. Itoh, J.C. Rhee, T. Kawabata, S. Koike, J. Appl. Phys. **58**, 1828 (1985)
204. J. Napierala, D. Martin, N. Grandjean, M. Ilegems, J. Cryst. Growth **289**, 445 (2006)
205. Y.V. Melnik, K.V. Vassilevski, I.P. Nikitina, A.I. Babanin, V. Yu. Davydov, V.A. Dmitriev, MRS Internet J. Nitride Semicond. Res. **2**, 39 (1997)
206. M.K. Kelly, O. Ambacher, B. Dahlheimer, G. Groos, R. Dimitrov, H. Angerer, M. Stutzmann, Appl. Phys. Lett. **69**, 1749 (1996)
207. M.K. Kelly, R.P. Vaudo, V.M. Phanse, J. Görgens, O. Ambacher, M. Stutzmann, Jpn. J. Appl. Phys. **38**, L217 (1999)
208. S.S. Park, I.W. Park, S.H. Choh, Jpn. J. Appl. Phys. **39**, L1141 (2000)
209. H.S. Kim, M.D. Dawson, G.Y. Yeom, J. Korean Phys. Soc. **40**, 567 (2002)
210. Y. Oshima, T. Eri, M. Shibata, H. Sunakawa, K. Kobayashi, T. Ichihashi, A. Usui, Jpn. J. Appl. Phys. **42**, L1 (2003)
211. A. Wakahara, T. Yamamoto, K. Ishio, A. Yoshida, Y. Seiki, K. Kainosho, O. Oda, Jpn. J. Appl. Phys. **39**, 2399 (2000).
212. H.J. Lee, S.W. Lee, H. Goto, S.H. Lee, H.J. Lee, J.S. Ha, T. Goto, S.K. Hong, M.W. Cho, T. Yao, Appl. Phys. Lett. **91**, 192108 (2007)
213. K. Motoki, T. Okahisa, N. Matsumoto, M. Matsushima, H. Kimura, H. Kasai, K. Takemoto, K. Uematsu, T. Hirano, M. Nakayama, S. Nakahata, M Ueno, D. Hara, Y. Kumagai, A. Koukitu, H. Seki, Jpn. J. Appl. Phys. **40**, L140 (2001)
214. S.W. Lee, Dissertation, Tohoku University (2007)
215. J.S. Ha, S.W. Lee, H.J. Lee, H.J. Lee, S.H. Lee, H. Goto, T. Kato, K. Hujii, M.W. Cho, T. Yao, IEEE Photon. Technol. Lett. **20**, 175 (2008)
216. S.W. Lee, T. Minegishi, W.H. Lee, H. Goto, H.J. Lee, S.H. Lee, H.-J. Lee, J.S. Ha, T. Goto, T. Hanada, M.W. Cho, T. Yao, Appl. Phys. Lett. **90**, 061907 (2007)
217. T. Suzuki, H.-J. Ko, A. Setiawan, J.-J. Kim, K. Saitoh, M. Terauchi T. Yao, Mat. Sci. Semicon. Proc. **6**, 519 (2003)
218. V. Yu, Davydov, A.A. Klochikhin, R.P. Seisyan, V.V. Emtsev, S.V. Ivanov, F. Bechstedt, J. Furthmü ller, H. Harima, A.V. Mudryi, J. Aderhold, O. Semchinova, J. Graul, Phys. Stat. Sol. B **229**, R1 (2002)
219. T. Matsuoka, H. Okamoto, M. Nakao, H. Harima, E. Kurimoto, Appl. Phys. Lett. **81**, 1246 (2002)
220. T. Nagashima, M. Harada, H. Yanagi, H. Fukuyama, Y. Kumagai, A. Koukitu, K. Takada, J. Cryst. Growth **305**, 355 (2007)
221. Y. Kumagai, T. Yamane, T. Miyaji, H. Murakami, Y. Kangawa, A. Koukitu, Phys. Stat. Sol. C **0**, 2498 (2003)
222. Y. Kumagai, K. Takemoto, J. Kikuchi, T. Hasegawa, H. Murakami, A. Koukitu, Phys. Stat. Sol. B **243**, 1431 (2006)
223. A. Koukitua, F. Satohb, T. Yamaneb, H. Murakamia, Y. Kumagai, J. Cryst. Growth **305**, 335 (2007)
224. N. Takahashi, J. Ogasawara, A. Koukitu, J. Cryst. Growth **172**, 298 (1997)
225. N. Takahashi, R. Matsumoto, A. Koukitu, H. Seki, Jpn. J. Appl. Phys. **36**, L743 (1997)
226. Y. Kumagai, K. Takemoto, A. Koukitu, H. Seki, J. Crystal Growth **222**, 118 (2001)
227. Y. Kumagai, J. Kikuchi, Y. Nishizawa, H. Murakami, A. Koukitu, J. Cryst. Growth **300**, 57 (2007)

4

Control of Polarity and Application to Devices

J.S. Park and S.-K. Hong

Abstract. Control of polarity in wurtzite nitride and oxide films and device applications of polarity are discussed. Controlling as well as evaluating the polarity of grown films is very important in exploring and investigating materials properties of polar wurtzite nitride and oxide films as the properties of films and devices have been strongly affected by the polarity of films. After describing various determination techniques of polarity of wurtzite GaN and ZnO films, polarity controlling methods based on interfacial engineering are presented. The field affects transistors and the periodically polarity inverted structures are presented for device applications of polarity as electronic and nonlinear optical applications, respectively.

4.1 Introduction

Gallium nitride (GaN) and related III-nitrides have been extensively explored for applications in optoelectronic devices including blue and ultraviolet light-emitting diodes and lasers [1], high-power and speed electronics [2,3], and field-emitter structures [4,5]. Most recently zinc oxide (ZnO) and related II–oxides have emerged as novel photonic materials for the UV region owing to their wide bandgap of 3.37 eV and very large exciton binding energy of 60 meV at room temperature [6]. GaN and ZnO exhibit a variety of material properties compared to conventional zinc blende compound III–V and II–VI semiconductors by virtue of their wurtzite crystal structure (space group P6mmc) with a high degree of ionicity.

Because of lack of center of symmetry along the $\langle 0001 \rangle$ direction of the wurtzite crystal, GaN and ZnO films grown along the $\langle 0001 \rangle$ direction have a polarity. The films with strains have piezoelectric and spontaneous polarizations, and properties of films and devices have been affected by the polarity of films. Therefore, controlling the polarity as well as evaluating the polarity of grown films is very important in exploring and investigating materials properties of polar wurtzite nitride and oxide films.

In this chapter control of polarity in polar GaN and ZnO films, and applications of polarity on devices are described. After introducing the basic physics underlying polarization phenomena, various determination techniques of polarity of wurtzite films are described. Among the many kinds of methods reported so far, we have selected an important technique which is categorized into the technique based on diffraction, spectroscopy, and microscopy. In addition, simple methods based on differences in wet etching and growth rates of different polar films are used to determine polarity. Next, controlling methods of polarity are described. Most of the methods are based on interface engineering, i.e., treatment of substrates before growing the films or the insertions of interfacial layers between the substrates and the films. Finally, device application of polarity is discussed. The field-effect transistors utilizing two-dimensional electron gas at the interfaces, which is affected by the directions of piezoelectric and spontaneous polarizations, are described for electronic applications. In addition, the periodically polarity-inverted structures are presented for application on nonlinear optical devices.

4.2 Polarity

The research for epitaxial growth of wurtzite structure films, especially the recent success in GaN and ZnO semiconductors, led the participants to recognize that such wurtzite structure semiconductors are very different from the traditional III–V and II–VI semiconductors. One important difference is the fact that the epitaxial growth of such semiconductors is typically done along a polar axis as shown in Fig. 4.1. The figure shows a schematic diagram of the wurtzite GaN and ZnO crystal structure. As seen in the figure, the wurtzite crystal structures lack inversion symmetry along the $\langle 0001 \rangle$ direction, which makes them piezoelectric active and gives rise to two

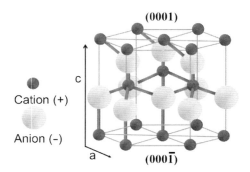

Fig. 4.1. The hexagonal wurtzite crystal structure. Anion atoms are shown as larger spheres, cation atoms as smaller spheres

possible polarities, i.e., (0001) cation-polar and (000$\bar{1}$) anion-polar surfaces. Hence, when epilayers are grown on lattice-mismatched substrates, a piezoelectric field is generated in the epilayers due to built-in strain. It should be mentioned that even when there is no strain, the wurtzite structure crystals possess spontaneous polarization owing to their characteristic atomic configuration [7, 8].

The upper (0001) surface in Fig. 4.1 is referred to as the cation-polar (Ga- or Zn-polar) face, and the lower (000$\bar{1}$) surface as the anion-polar (N- or O-polar) face. At an early stage of research on GaN and ZnO epitaxy, polarity determination of the films was not easy and accurate determination failed when surface sensitive spectroscopy such as X-ray photoelectron spectroscopy (XPS) or Auger electron spectroscopy (AES) was used to determine polarity. These surface sensitive methods are not suitable for polarity determination because lattice polarity is a bulk property determined by the bond sequence in the crystal not a surface property. Bulk property is different from surface termination which is determined by atoms on the top surface layer. In addition to experimental difficulties, a discussion of the polarity issue has been hindered by the use of confusing terminology: 'termination', 'polar', and 'face' [9]. One can imagine a situation where the anion face (N or O face) is covered with a monolayer of cation (Zn or Ga), but the orientation of the crystal is unchanged. In other words, polarity is different from termination because crystal polarity is a bulk property but termination is a surface property. Consequently, four different structures can be considered for wurtzite {0001} surfaces, as illustrated in Fig. 4.2: cation polar with a cation addlayer, cation polar with an anion addlayer, anion polar with an anion addlayer, and anion polar with a cation addlayer. Regarding polarity, the term "face" or "polar" (or "polarity") should be used instead

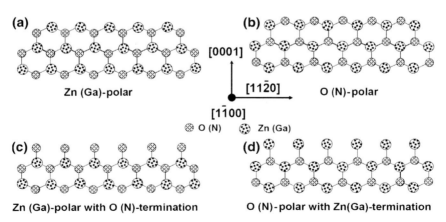

Fig. 4.2. Four kinds of structures for wurtzite {0001} surfaces. (**a**) Cation-polar with cation-termination. (**b**) Anion-polar with anion-termination. (**c**) Cation-polar with anion-termination, and (**d**) Anion-polar with cation-termination

of termination (or terminated). Hellman reported the relatively clear picture of the polarity of GaN with comparatively extensive results [9]. Specifically it has been found empirically that the orientation of high-quality GaN films grown on (0001) c-plane sapphire by metal-organic chemical vapor deposition (MOCVD) is typically (0001) polar (Ga-polar), whereas molecular beam epitaxy (MBE) grown GaN is typically (000$\bar{1}$) O-polar. Prolific investigations have also revealed the influence of the nucleation layer on the selection of crystal polarity in GaN films. (0001) and (000$\bar{1}$) polar films can be distinguished from each other by a variety of techniques including diffraction, spectroscopy, and characterization of physical properties which will be discussed in Sect. 4.4.

4.3 Spontaneous and Piezoelectric Polarization

Polarity in epitaxially grown wurtzite structured films is significant in the context of polarization effects because spontaneous (P_{SP}) and piezoelectric (P_{PE}) polarization fields have a well-defined orientation with respect to the (0001) and (000$\bar{1}$) crystal polarity. The spontaneous polarization field of nitride films is oriented in the [0001] direction which is indicated by theoretical calculations [7]. As a point of view of crystal structure, the wurtzite structured GaN and ZnO crystal is described schematically as a number of alternatively stacked planes composed of fourfold coordinated $O^{2-}(N^{3-})$ and $Zn^{2+}(Ga^{3+})$ ions along the [0001] axis. The spontaneous polarization P_{SP} field lies in the [0001] direction and oppositely charged ions produce positively charged (0001) plane and negatively charged (000$\bar{1}$) plane polar surfaces. P_{SP} constitutes nonzero dipole moments in the absence of external influence such as strain or applied electric field. Recent theoretical results have indicated that nitride and oxide semiconductors possess a large spontaneous polarization coefficient and it is estimated to be -0.057 C m^{-2} for ZnO [7, 10] and 0.029 C m^{-2} for GaN [7], which is much larger than the polarization due to piezoelectric effect, in general. The polarization induced surface charge is balanced by a bound charge of opposite sign at the films/substrate interface. While the predicted spontaneous polarization is independent of strain, piezoelectric polarization is strain-induced. In general, total polarization of nitride and oxide layer without external electric field is the sum of the spontaneous polarization in equilibrium lattice and the piezoelectric polarizations induced by strain. Ambacher et al. investigated the directions of P_{SP} and P_{PE} given for Ga-face, N-face, strained, unstrained AlGaN/GaN heterostructure as shown in the Fig. 4.3. For the Ga(Al)-face heterostructures P_{SP} is pointing toward the substrate. As consequence, the alignment of the P_{SP} and P_{PE} is parallel in the case of tensile strain, and antiparallel in the case of compressively strained top layer [11].

In addition, they calculated and demonstrated the amount of P_{SP}, P_{PE}, and total polarization of the AlGaN barrier layer, as well as the sheet charge density at the upper and lower GaN/AlGaN interfaces in N-face and Ga-face, respectively shown Fig. 4.4a, b. For N-face heterostructure, the sign

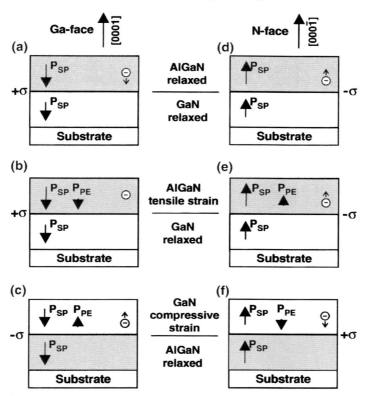

Fig. 4.3. Polarization induced sheet charge density and directions of the P_{SP} and P_{PE} in Ga- and N-polar strained and relaxed AlGaN/GaN heterostructures. Reprinted with permission from [11]. Copyright (1999), American Institute of Physics

of polarization induced sheet charge is determined to be negative for the lower AlGaN/GaN and positive for the upper GaN/AlGaN interface. In contrast to the N-face, Ga-face GaN/AlGaN/GaN heterostructures, the amount of sheet charge remains the same, but positive sheet charge causing a two-dimensional electron gas (2DEG) is located at the lower AlGaN/GaN interface. According to the results, the expected diagram of the band profiles is shown in Fig. 4.5.

The result for the direction of P_{SP} and P_{PE} in O-polar ZnO/ZnMgO is similarly reported by Yano et al. [12]. The existence of a piezoelectric polarization P_{PE} field along the [0001] axis, which is associated with electrostatic charge densities in strained material, has an influence on carrier distributions, electric fields, and consequently a wide range of optical and electronic properties of materials and devices [13].

The P_{PE} is determined by piezoelectric coefficients e_{ij} and the strain tensor ε_j by the following formulae:

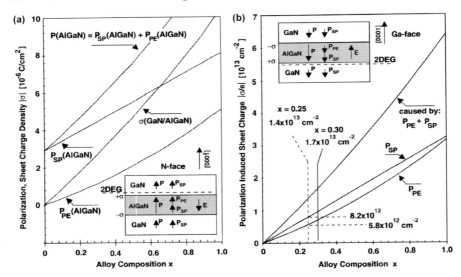

Fig. 4.4. Calculated sheet charge density caused by P_{SP} and P_{PE} at the upper interface of a N-polar (**a**) and low interface of a Ga-polar (**b**) GaN/AlGaN/GaN heterostructures with variation of alloy composition of the barrier. Reprinted with permission from [11]. Copyright (1999), American Institute of Physics

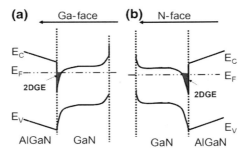

Fig. 4.5. Energy band diagram of (**a**) Ga-polar and (**b**) N-polar GaN/AlGaN/GaN heterostructures

$$P_{\mathrm{PE}} = e_{ij}\varepsilon_j = d_{ij}c_{jk}\varepsilon_k, \quad j,k = \mathrm{xx, yy, zz, yz, zx, xy}, \qquad (4.1)$$

where d_{ij} is the piezoelectric moduli relating to polarization, the stress tensor $\sigma_j = c_{jk}\varepsilon_k$, and elastic tensor c_{jk} respectively. For the wurtzite crystal, the piezoelectric tensor has three independent nonzero components. Therefore, the piezoelectric coefficients e_{ij} can be given by

Table 4.1. Comparison of selected physical constants for wurtzite GaN, AlN, and ZnO

	GaN	AlN	ZnO
a_0 (nm)	0.3189 [14]	0.3112 [14]	0.3250 [15]
c_0 (nm)	0.5185 [14]	0.4982 [14]	0.5204 [15]
C_{11} [Gpa]	367 [11]	396 [11]	209.6 [16]
C_{12} [Gpa]	135 [11]	137 [11]	121.1 [16]
C_{13} [Gpa]	103 [11]	108 [11]	105.1 [16]
C_{33} [Gpa]	405 [11]	373 [11]	210.9 [16]
C_{44} [Gpa]	95 [11]	116 [11]	42.5 [16]
e_{33} [C m^{-2}]	0.73 [7]	1.46 [7]	0.96 [17]
e_{31} [C m^{-2}]	−0.49 [7]	−0.60 [7]	−0.62 [17]
e_{15} [C m^{-2}]	−0.30 [11]	−0.48 [11]	−0.37 [17]
P_{SP} [C m^{-2}]	−0.029 [7]	−0.081 [7]	−0.057 [10]

C elastic constant; e piezoelectric constant

$$e_{ij} = \begin{bmatrix} 0 & 0 & 0 & 0 & e_{15} & 0 \\ 0 & 0 & 0 & e_{15} & 0 & 0 \\ e_{31} & e_{31} & e_{33} & 0 & 0 & 0 \end{bmatrix}. \quad (4.2)$$

Two of these components measure piezoelectric polarization P_{PE} induced along the c axis or in the basal plane. The relevant relationship is

$$P_{PE} = e_{33}\varepsilon_z + e_{31}(\varepsilon_x + \varepsilon_y), \quad (4.3)$$

where $\varepsilon_z = (c - c_0)/c_0$ is the strain along the c axis, in-plane strain $\varepsilon_x = \varepsilon_y = (a - a_0)/a_0$ is assumed to be isotropic, e_{33}, e_{31} are the piezoelectric coefficients, and a and c are the lattice constants of the strained layer. The relation between the lattice constants in the hexagonal system is given by

$$\frac{c - c_0}{c_0} = -2\frac{C_{13}}{C_{33}}\frac{a - a_0}{a_0}, \quad (4.4)$$

where, C_{13} and C_{33} are elastic constants. Using (4.3) and (4.4), piezoelectric polarization in the direction of the c axis can be determined by

$$P_{PE} = 2\frac{a - a_0}{a_0}\left(e_{31} - e_{33}\frac{C_{13}}{C_{33}}\right). \quad (4.5)$$

Calculations on numerical values of spontaneous polarization, piezoelectric polarization, and elastic constants along the (0001) axis have been reported for the GaN, AlN, and ZnO semiconductor as summarized in Table 4.1.

4.4 Determination of Polarity

Determination of polarity of the films is important and basic, in addition to polarity control, to study growth behaviors, to investigate material properties, to apply polarity related phenomena to devices. In this section, various techniques to determine polarity are described, based on their underlying principles.

4.4.1 Determination Based on Diffraction

Characterization by transmission electron microscopy (TEM) using convergent beam electron diffraction (CBED) has been widely used to determine the crystal polarity of GaN [18–20] and ZnO [21, 22] single crystal or films. The CBED method basically decides the crystal orientations and requires comparison of experimental and simulated CBED patterns to determine absolute polarity. The simulated results are very sensitive to input values, which constitute the main source of error in polarity determination by CBED. Furthermore, the symmetry of the CBED pattern is strongly affected by the existence of defects, which sometimes make CBED difficult to determine polarity. Even with these difficulties, the absolute decision of the polarity is possible by CBED, which is regarded as one of the most reliable techniques to determine the polarity in epitaxial GaN and ZnO films. In addition, since CBED measurements are performed by cross-sectional TEM observations, we can determine the change of polarities along the growth direction or along the interface in one sample. The polarity settlement is conducted by studying asymmetric diffraction spots of (0002) and (000$\bar{2}$) from the cross-sectional observation of GaN or ZnO films, where the zone axis is set to $\langle 10\bar{1}0 \rangle$ and $\langle 11\bar{2}0 \rangle$ of ZnO or GaN. The first attempts to determine the polarity by using CBED were made by Liliental-Weber et al. and Ponce et al. [18, 23] for the GaN single crystals. The smooth face is indicated as Ga-face by Ponce et al., while Liliental–Weber indicates it as N-face.

In the case of ZnO films, Hong et al. reported the comparison of the experimental and simulated CBED pattern of the ZnO films grown on different surface-treated MOCVD grown Ga-polar GaN [21]. Figure 4.6a shows the experimental CBED patterns of the upper ZnO and the lower GaN film, where Zn preexposure was performed on the GaN surface before the ZnO growth. The experimental CBED patterns from the upper ZnO and the lower GaN films are nearly the same (same symmetry). This indicates that the polarity of the upper ZnO film is Zn on Ga-polar GaN. In fact, the simulated patterns obtained from the Zn-polar ZnO and Ga-polar GaN in Fig. 4.6c show that the experimental pattern observed in Fig. 4.6a does correspond to the Zn-polar. Figure 4.6b shows the experimental CBED patterns of the upper ZnO and the lower GaN films, where O-plasma preexposure were performed on the GaN surface before the ZnO growth. The experimental CBED pattern from the upper ZnO and the lower GaN films shows opposite symmetry, which indicates that the polarity of the upper ZnO film may be opposite to that of the lower GaN film, considering the fact that the crystal structure and symmetry of ZnO and GaN are the same. In fact, the experimental pattern from the upper ZnO film in Fig. 4.6b shows opposite symmetry to the simulated pattern for the Zn-polar ZnO (left side pattern in Fig. 4.6c), which indicates that the upper ZnO film in Fig. 4.6b has the O polarity. The results shown in Fig. 4.6 clearly show that we can determine the polarity of single and multilayered samples at the same time through the CBED analysis of cross-sectional observation.

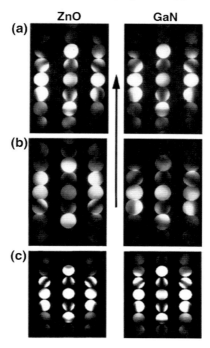

Fig. 4.6. CBED patterns of the upper ZnO and the lower GaN films (**a**) Zn-polar ZnO/GaN (**b**) O-polar ZnO/GaN (**c**) Simulated CBED patterns of ZnO and GaN with the zone axis of the $[2\bar{1}\bar{1}0]$ direction. The arrow indicates growth direction. Reprinted with permission from [21]. Copyright (2002), American Physical Society

Another representative method based on diffraction technique is reflection high-energy electron diffraction (RHEED). The surface reconstructs to optimize the surface energy, and different surface stoichiometry and crystallographic orientation give rise to different stable surface structure, which can be a fingerprint of film polarity. In situ observation of RHEED is a useful method to monitor the surface change of films in real time. Smith et al. have reported on the surface reconstructions according to the polarity of GaN films experimentally and computationally [24, 25]. A (1×1) RHEED pattern is generally observed during GaN growth regardless of polarity. However, various RHEED patterns with different reconstructions (1×1), (3×3), (6×6), and $c(6 \times 6)$ are observed, depending on temperature and Ga coverage from the surface of GaN films that they are believed to be N face (N-polar). In the case of Ga face (Ga-polar) GaN films, they observed (2×2), (5×5), (6×4), and (1×1) reconstructions at various temperatures. The details of RHEED reconstruction in Ga- and N-polarity is summarized in Fig. 4.7 [26].

In the case of ZnO films, two reconstructions (1×1) and (3×3) were reported by many groups [27–29] in the growth of the O-polar ZnO films. For the

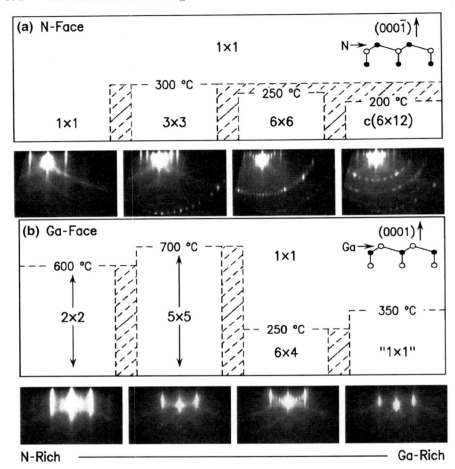

Fig. 4.7. Schematic phase diagrams of the reconstructions existing on the N face (*upper*), and Ga face (*lower*). Cross-hatched regions indicate either mixed or intermediate phases. RHEED patterns for both the Ga and N face, as viewed along the [11$\bar{2}$0] azimuth. Reprinted with permission from [26]. Copyright (1998), American Institute of Physics

Zn-polar ZnO films, reconstructions were seldom reported. Mei et al. reported the (4 × 4) surface reconstruction for the uniform Zn polar ZnO films on nitrided sapphire substrate [30]. They did not give the theoretical consideration of the mechanism about the (4×4) reconstruction from the Zn-polar ZnO film. Therefore, more investigations of reconstruction of ZnO surface depending on polarity are needed. Here it should be noted that the absolute determination of polarity from the polarity-unknown sample by RHEED reconstructions is not an easy task and it needs additional studies on surface science.

X-ray diffraction (XRD) method is often employed only for the evaluation of crystal quality. Little attention has been paid to determine the polarity of wurtzite films with XRD, although the first attempt to determine the crystallographic polarity of ZnO using XRD was reported by Mariano and Hanneman [31]. They compared the diffraction contours from the continuous spectrum produced by (0002) and (000$\bar{2}$) planes of ZnO which is correlated to the etching figures and morphology. The application of high-resolution (HR) XRD and X-ray standing wave (XSW) techniques to study the structure and polarity of GaN films was reported by Kazimirov et al. [32]. The analysis of the Ga-K edge yield modulations unambiguously reveal the polarity of GaN films as shown in Fig. 4.8. The Ga-K XSW fluorescence yields calculated for the Ga- and N-polar GaN films were fitted to the experimental data with the static Debye–Waller factor as the only fitting parameter in Fig 4.8a. Typical χ^2 value of the Ga-polar films was 4∼6 times lower than those of N-polar films as shown in Fig. 4.8b.

Recently, Inaba et al. reported the polarity of polar and nonpolar (11$\bar{2}$0) GaN films with a prototype high-resolution XRD system with an Au rotating-anode X-ray generation [33]. Tampo et al. have demonstrated a simple and nondestructive method for the settlement of the crystallographic polarity of ZnO epilayers using XRD with anomalous dispersion near the Zn K edge [34].

4.4.2 Determination Based on Spectroscopy

The atomic structures of the surface can be nondestructively analyzed in real space with coaxial impact collision ion scattering spectroscopy (CAICISS), which has been developed and modified from the impact collision ion scattering spectroscopy (ICISS). It is a simple way of quantitatively analyzing the atomic arrangement of atoms on the surface, including the distance and angle between the neighboring atoms. In principle, the CAICISS takes only ions being impact-collided with target atoms into account in the analysis based on the focusing and the shadowing effects in ion scattering. The ion source and energy analyzer are on almost the same axis in the CAICISS, i.e., coaxial. However, it is impossible to put them on the same axis in the case of the ICISS system because a turntable-type electrostatic energy analyzer is used. In case of CAIXISS, by using a time-of-flight (TOF) energy analyzer rather than an electrostatic energy analyzer and by placing the TOF analyzer on the same axis to the ion source, the experimental scattering angle could be increased up to about 180°. In addition to realization of nearly 180°-experimental scattering angle, not only ions but also neutrals can be detected by using a TOF analyzer, which means that a much higher intensity of scattered particles is obtainable.

The CAICISS analysis for determining surface structure including atomic arrangement has been demonstrated for nitride [35] and oxide films [36]. Moreover, polarity can be determined by the positions of the parabolic shadowing dip and the focusing peak of the He$^+$ beam in CAICISS analysis.

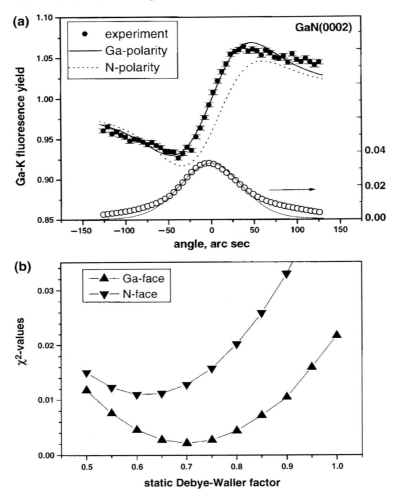

Fig. 4.8. (a) One of the experimental XSW data sets. The experimental X-ray reflectivity curve (*bottom*) is shown as *open circles*, the solid line is the convolution of the theoretical X-ray rocking curve with a Gaussian function with $\sigma = 31.5$ arcsec. The experimental Ga–K fluorescence yield is shown at the *top* as *solid circles*. The theoretical curve for the N face is shown as a dotted line for comparison; (b) the χ^2 values as a function of the Dw factor. The Ga-face model gives five times lower χ^2 values than the N-face model. Reprinted with permission from [32]. Copyright (2001), American Institute of Physics

CAICISS signals are obtained by incident angle dependence of a He^+ beam which is irradiated along the azimuth direction of $[11\bar{2}0]$ and azimuth angle dependence on an incident angle. Ohnishi et al. have reported the experimental determination of polarity of ZnO films from the angular dependence measured by CAICISS for bulk ZnO with Zn and O polarity [37]. Hong et al. reported different polarity of ZnO films grown on Ga polar GaN template with different pretreatment, in which polarity was determined by using the CAICISS [38]. They obtained the CAICISS signals from both by changing the polar angle and by changing azimuth angle, which showed different spectra depending on the polarity as shown in Fig. 4.9. Polarity of the ZnO films was determined by comparison of experimental CAICISS spectra with simulated spectra. Figures 4.9a, b show experimental (closed squares) and simulated (open circles) polar angle dependent Zn signals from (a) Zn and (b) O-plasma preexposed ZnO films. In the case of Zn preexposed ZnO films, the observed polar angle dependence is characterized by three dominant peaks at 24, 50, and 74°. This feature agrees well with the simulated spectrum obtained for Zn-polar ZnO. In the case of O-plasma preexposed ZnO films, the experimental polar angle dependence is characterized by four dominant peaks at 22, 36, 54, and a broad peak 72–76°. This feature agrees well with the simulated spectrum obtained for O-polar ZnO. These spectra shapes agreed well with reported data on the Zn-face and O-face bulk ZnO [39]. Figures 4.9c, d show the azimuth angle dependent Zn signal at an incident angle of 58.5° for the Zn preexposed and the O-plasma preexposed samples, respectively. The experimental results agree well with simulated spectra for Zn-polar and

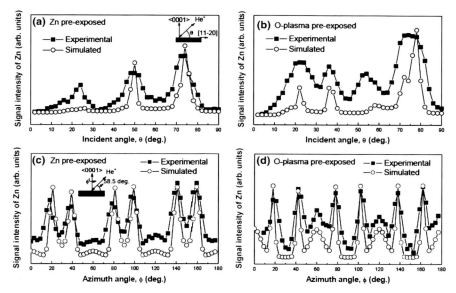

Fig. 4.9. Incidence-angle and Azimuth angle dependence of Zn signal from ZnO films measured by CAICISS on (**a**),(**c**) Zn preexposure and (**b**),(**d**) O-plasma preexposure, respectively. The set-up of the sample for measurements is the same for all samples. Reprinted with permission from [38]. Copyright (2000), American Institute of Physics

O-polar ZnO. Later they confirmed the polarity of the samples by CBED and obtained the same results on polarity determination [21], which clearly shows the validity of the CACISS in determining polarity and surface structure.

X-ray photo electron spectroscopy (XPS) and Auger electron spectroscopy (AES) are sensitive to surface properties. The earliest attempt to study GaN crystal polarity using the XPS was reported by Sasaki and Matsuoka in 1998 [39]. They used XPS to probe the surfaces of GaN films grown on two faces of SiC, i.e., Si- and C-face. After studying the energy shift of the Ga core levels, they proposed that the film grown on the C face was more easily oxidized, and thus likely to be Ga-face GaN. They concluded that Ga-polar GaN grows on the C-face SiC, while N-polar GaN grows on the Si-face SiC. It should be noted that the measurements of Sasaki et al. are not a polarity characterization, but rather surface chemistry measurements, therefore their polarity determinations need reconsideration based on another direct determination method like CBED observations. On the other hand, Harada et al. reported clear differences in binding energy and intensity ratio O/N or N/Al in the XPS spectra from the AlN films grown on the Si and C faces of SiC substrate [40]. They concluded that the AlN film grown on Si and C faces have Al and N polarity, respectively.

Khan et al. attempted to use AES to decide the polarity of GaN films grown on AlN buffer layer in 1993 [42]. This technique measures the composition of the top ($\sim$1 nm) of the film weighted by the escape probabilities of Auger electrons. Their calculations indicated that N and Ga terminated surfaces should have $I(N_{KLL})/I(Ga_{KLL})$ peak ratios of 0.82 and 1.14, respectively. Based on their measured N/Ga ratio of 1.05, they concluded that their films were N terminated. However, it should be noted again that surface stoichiometry might not be a reliable indicator of film polarity because polarity is bulk property. Recently, Niebelschutz et al. demonstrated that a normalized Auger peak shift of approximately 0.25 ± 0.01 eV from the N-face to the Ga-face domain points towards higher energies for both the nitrogen and gallium peaks as shown in Fig. 4.10 [43]. They calculated the Auger kinetic energy dependence on the work function of the material and explained the possible band model of lateral polarity heterostructures which showed upward bowing at the Ga-face surface and downward bowing at the N-face surface. Therefore, the LMM peak of Ga has higher kinetic energy than the N KLL Auger peak. They suggested that the AES method is suitable to identify the domains of lateral polarity heterostructures.

4.4.3 Determination Based on Microscopy

The piezoresponse force microscopy (PFM) offers a significant advantage compared to other techniques in studying the piezoelectric properties of films. In addition to measuring the piezoelectric displacement, a two-dimensional image with information on the magnitude and the sign of piezoresponse signal makes it possible to see the polarity change with nanoscale lateral resolution. Use of

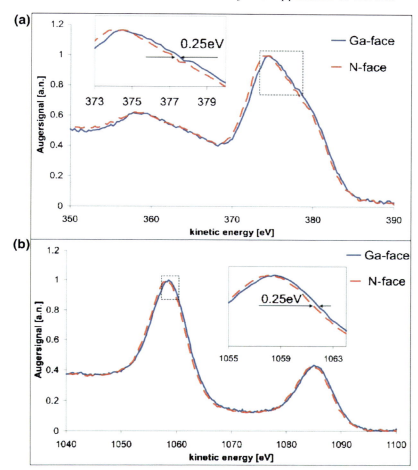

Fig. 4.10. Normalized Auger peak of (**a**) nitrogen [KLL] and (**b**) gallium [LMM]. Reprinted with permission from [43]. Copyright (2006), American Institute of Physics

PFM to determine the polarity of nitride films was conducted by Rodriguez et al. [44]. They reported the polarity distributions of GaN based lateral polarity heterostructures from the phase images of the piezoresponse as shown in Fig. 4.11 [44]. In PFM, when a modulation voltage is applied to a piezoelectric material, the vertical displacement of the probing tip, which is in mechanical contact with the sample accurately follows the piezoelectric motion of the sample surface. The piezoelectric material is expanded when a positive electric field is applied along the [0001] axis, while it is contracted when reversing the direction of electric field, which results in different PFM image contrast.

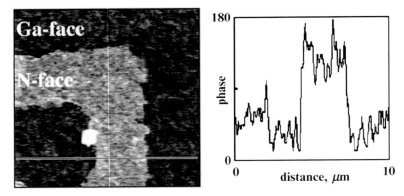

Fig. 4.11. PFM phase image of Ga- and N-polar regions with corresponding cross-sectional profile. Reprinted with permission from [44]. Copyright (2002), American Institute of Physics

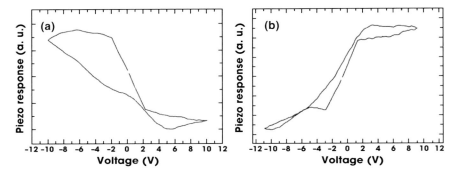

Fig. 4.12. Voltage–piezoresponse curve (V–Z) of ZnO film grown on patterned CrN and Cr_2O_3 (**a**) Zn-polar ZnO and (**b**) O-polar ZnO film, respectively. Reprinted with permission from [45]. Copyright (2008), American Vacuum Society

Park et al. showed the image contrast of piezoelectric response for the periodically polarity inverted (PPI) ZnO heterostructures [45]. The details of growth of PPI ZnO will be discussed in Sect. 4.7.2. They reported different brightness coming from different piezoelectric response behavior according to the polarity, i.e., the brighter regions and darker regions correspond to O-polar and Zn-polar, respectively. In addition to the PFM image, surface voltage–piezoresponse distance (V–Z) curves were obtained from each region as shown in Fig. 4.12. It clearly shows the opposite hysteresis piezoresponse curves with input voltage variation. The hysteresis curve for the ZnO film on the CrN patterns implies Zn-polar property (Fig. 4.12a), while the hysteresis curve for the ZnO films on the Cr_2O_3 area indicates O polarity (Fig. 4.12b). Placing a PFM tip on the top surface of Zn-polar ZnO film and applying a positive tip bias to the top (0001) face of ZnO resulting in a negative field with respect to the [0001] direction and a contraction of the Zn-polar ZnO film. This situation is reversed for the O-polar ZnO film. That is, the positive tip bias produces an expansion and the negative tip bias produces a contraction [46]. Therefore, the PFM images and the V–Z curves can be applied to determine the polarity.

Kelvin force microscopy (KFM) can be used to compare the surface potential distribution. Katayama et al. reported that the average potential of N-face domain is 1.17 eV higher than that of Ga-face domain, which is related to surface polarity [47].

Electrostatic force microscopy (EFM) has been employed to perform characterization of the local electronic properties of GaN-based lateral polarity heterostructures. EFM of the polar surface should respond to the net surface charge density, which is equal to the sum of polarization charge and screening charge [48]. Rodriguez et al.have shown the EFM phase contrast indicating that the net surface charge is positive for the N-face surface and negative for the Ga-face surface.

Photoelectron emission microscopy (PEEM) is an emission microscopy technique in which images of material surface are formed by photoexcited electron. Yang et al. using the UV–PEEM method to compare the polarity contrast and inversion domain formed at the boundary of Ga-face and N-face [49]. Polarity contrast was observed from different polar regions and the enhanced emission of N-face surface was attributed to photoemission from electrons in the conduction band at the surface induced by band bending.

As another microscopy based technique, electron holography has been used to determine the polarity [50, 51]. The charges induced by the polarization in the films will influence the potential in the vacuum near the films. The potential distribution around the charges depends on the sign of bounded charges and can be used to determine the polarity of the films [50]. As electron holography can precisely detect the electrical potential distribution in both vacuum and films. Wang et al. demonstrated the polarity determination of ZnO films grown on cubic nitridation layers by electron holography as shown in Fig. 4.13 [51]. Figure 4.13a–c correspond to the result for the Zn-polar and Fig. 4.13d–f correspond to the result for the O-polar ZnO films, respectively.

4.4.4 Determination Based on Differences in Etching and Growth Rates

Different polarities result in different chemical etching characteristics, i.e. different etching rate and etched surface morphology. However, it should be noted that etched morphology can be affected also by the initial morphology of as-grown sample, which was affected by growth methods and modes. Therefore, a decision on the polarity from the etched morphology can be a hint for the polarity and should additionally check from the reference sample of which polarity was determined by another direct method.

It is well known that characteristics of GaN surface depending on the polarity are smooth surfaces for the Ga polar GaN and surfaces with

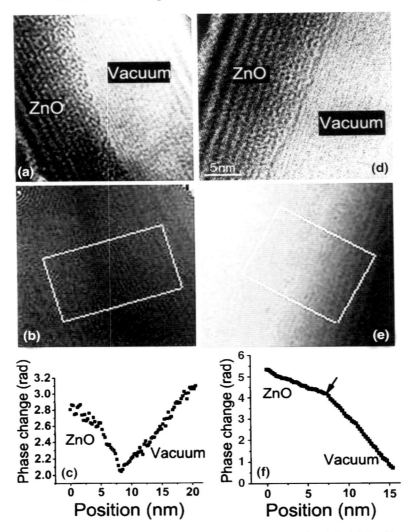

Fig. 4.13. The polarity determination by electron holography (**a**)–(**c**) for Zn–polar and (**d**)–(**f**) for O-polar ZnO. The hologram of ZnO films (**a** and **d**), the reconstructed phase Image (**b, e**), the averaged 1D phase change profile (**c** and **f**) obtained from the boxed area in (**b, c**). Reprinted with permission from [51]. Copyright (2005), American Institute of Physics

hexagonal-shaped pyramidal pits for the N-polar GaN. However, surface morphology is strongly affected by growth conditions. Attempts for improvement of surface morphology and crystal quality of N-polar films were reported [52, 53]. In case of ZnO, the surface morphology is strongly affected by growth conditions, too. The surface morphology contains hexagonal pits which is observed for the Zn-polar ZnO, while the hexagonal hillock can be observed from the O-polar one. The surface morphology features of as grown ZnO films are also polarity sensitive. The Zn-polar ZnO with a smooth surface can be obtained under the O-rich growth condition, while hexagonal pits are usually observed under the stoichiometric and the Zn-rich growth conditions. On the other hand, the hexagonal hillock was usually found in the O-polar ZnO, even under the Zn-rich growth conditions [54].

It is expected that the cation-face and anion-face surfaces of GaN and ZnO should have quite different chemical properties. Therefore, polarity determination based on the wet etching rate is highly reliable and reflects a response related to the polarity itself, with less effect from the initial morphology. Mariano and Hanneman reported the different etching rates of opposite polar direction based on surface bonding model for A_{II}–B_{VI} compounds [31]. The layer of zinc surface atoms has a positive charge and the layer of oxygen surface atoms has a negative charge with the oxygen atoms having two dangling electrons due to the transfer of electrons of zinc atoms to the oxygen atoms [31]. The dangling electrons on the O-polar surface account for the high etching rate owing to their susceptibility to reaction with electron-seeking agents in the etchant than the Zn-polar surface.

Such differences in chemical wet etching rates, depending on the polarity of the wurtzite structure materials, have been reported in III- nitride [26,55,56] and ZnO [34,54]. Figures 4.14a, c show the surface morphology of the as grown Zn-polar and O-polar ZnO films grown on nitrided sapphire substrate, respectively. The surface morphologies after being etched by using a 0.05% HCl solution for 1 min are shown in Fig. 4.14b, d. Large hexagonal pits can be observed on the Zn-polar surface, while hexagonal islands are clearly found on the O-polar surface from the as grown samples. The Zn-polar ZnO surface almost kept the initial surface with less etching, while the O-polar surface changed drastically during the etching. The etching rate of O-polar sample was much higher than that of the Zn-polar surface.

Similar results have been reported in GaN bulk crystal. The N face of the GaN single crystal can be etched more easily than the Ga-face. The etching solutions used normally for GaN are alkaline solutions, including aqueous KOH and NaOH [57].

In addition to the differences in wet etching behaviors depending on the polarities, differences in the growth rate for different polar ZnO films have been reported [58, 59]. The growth rate of Zn-polar ZnO films is higher than that of O-polar films. The higher growth rate for Zn-polar ZnO film can be understood by considering the sticking of atoms on the growth surfaces under the growth conditions. Growth of the ZnO film is mostly limited by the sticking of the Zn atoms than that of the O atoms [59]. Each oxygen atom on a Zn-polar surface has three dangling bonds, while each oxygen atom on an O-polar surface has only one dangling bond. Therefore, Zn sticking coefficient on the oxygen atom plane of the Zn-polar ZnO is larger than that on the O-polar ZnO, which results in a higher growth rate for the Zn-polar ZnO film. As

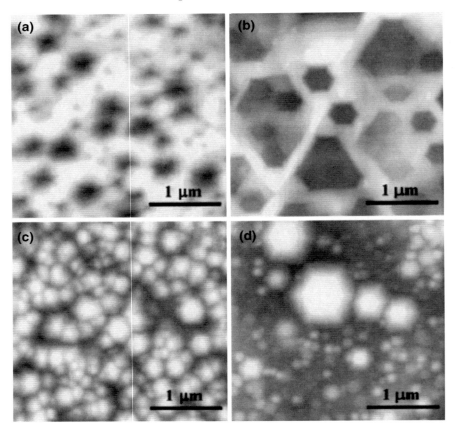

Fig. 4.14. AFM images of the Zn-polar and O-polar ZnO films before and after etching. (**a**) and (**c**) are the as-grown, (**b**) and (**d**) are the etched surface of Zn-polar and O-polar ZnO, respectively. Reprinted with permission from [54]. Copyright (2005), American Institute of Physics

shown in Fig. 4.15, the growth rate of the Zn-polar ZnO films showed higher growth rate than the O-polar ZnO films independent of the II/VI ratio, which means comparison of the growth rate at the same growth condition can be used for the polarity determination for the ZnO films [59].

In the case of the GaN films, the Ga-polar (0001) surface has a slightly higher growth rate than that of the N-polar (000$\bar{1}$) surface under the stoichiometric flux conditions [60]. However, growing under more Ga rich conditions, the growth rates for (0001) and (000$\bar{1}$) surfaces have similar values, which makes it difficult to determine the polarity of GaN films based on the growth rate.

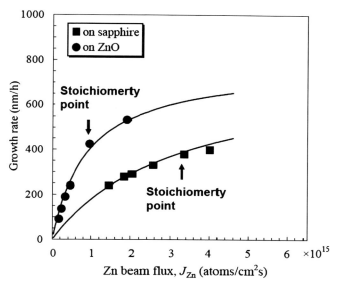

Fig. 4.15. Growth rate of ZnO films grown on Zn-polar ZnO and (0001) sapphire substrates versus Zn beam flux, J_{zn}. Reprinted with permission from [59]. Copyright (2004), Elsevier

4.5 Control of Polarity

III-Nitride and II-Oxide films are deposited by various growth techniques; MBE, pulsed laser deposition (PLD), hydride vapor phase epitaxy (HVPE), and MOCVD. Each growth method claims the polarity control of films by own methods. Some methods can be applied to control of polarity without consideration of growth methods while others are dependent. In this section, some interesting polarity control methods of GaN and ZnO films are reviewed. The results on the polarities of AlN and InN films are also briefly described.

4.5.1 Polarity Control of GaN Films

The polarity of GaN films grown on $c-Al_2O_3$ substrate under optimal conditions by MOCVD has been reported to be Ga-polar [61]. The growth mechanism of Ga-polar GaN films grown by MOCVD was explained in terms of chemical bond strength between the first GaN layer and the $c-Al_2O_3$ substrate. Since the bonding energy of Al–N which is larger than that of O–Ga and Al_2O_3 tends to be terminated by Al atoms in a reductive atmosphere, a Ga–N=Al–O sequence at the interface is expected to dominate, resulting in Ga polar. The key factors for the polarity control of GaN grown on (0001) Al_2O_3 substrates can be the following two points: (a) Surface treatment condition of sapphire substrate with selected buffer layer and (b) use of intentional or unintentional interfacial layers between films and substrate. One of the surface treatment methods, the nitridation of sapphire substrate is usually used to obtain high quality GaN films and control crystal polarity [54, 62–67]. However, the mechanism for polarity control has not been fully understood yet due to difficulty in getting the well-defined microstructures of nitrided layers. The theoretical and experimental research to find out the truth of underlying mechanism of polarity for the GaN films with surface nitridation of the Al_2O_3 substrates including the microstructures of nitridation and polarity selection are reported [68–70].

Mikroulis et al. reported that different nitridation temperatures led to distinct Ga-polar and N-polar GaN films using the MBE method [67]. The polarity of GaN overgrown on high temperature (HT) nitrided sapphire at 750°C was found to be Ga-polar and that of GaN grown on low temperature (LT) nitrided substrate was found to be N-polar GaN films, respectively. For the growth of Ga-polar GaN films independent of the nitridation temperatures, the intentional AlN buffer layer was essential [66, 71].

As a different approach, insertion of an Mg or Al metal layer makes it possible to select crystal polarity and control the orientation [72]. Ramachandran et al. reported the surfactant effect of Mg on GaN and the polarity invert from the Ga-polar to N-polar GaN by Mg incorporation during the MBE growth of the GaN films on SiC substrate [73]. The effect of the Mg layer on the GaN template with the PAMBE on the GaN-polarity was further confirmed by Romano et al., who observed a polarity inversion from the GaN films with sufficiently high Mg doping [74]. Grandjean et al. reported that a polarity inversion was achieved by formation of Mg_3N_2 interfacial layer by simultaneous surface exposition to Mg and NH_3 fluxes using the MBE system on the GaN template [75]. Here, it should be noted that Mg_3N_2 has inversion symmetry.

Xu et al. reported polarity conversion by the insertion of Al metal layer using the model of two monolayers of Al based on experiments in PAMBE [76] and Yoshikawa et al. observed the polarity manipulation of GaN using same method as in MOCVD [77]. Lim et al. demonstrated the polarity inversion of GaN by introducing the preflow of trimethyl aluminum (TMA) before the GaN buffer growth on the nitrided sapphire substrate, in which the insertion of more than two monolayers of Al were carried out [78]. As TMA preflow time increased, the polarities of GaN films changed from the N polarity to a mixed polarity, and finally to a pure Ga polarity when the TMA preflow time was over than 5 s in the MOCVD system [78]. On the other hand, the direct deposition of an Al metal layer on a nitrided sapphire substrate before the growth of an AlN buffer layer showed the successful change of the polarity from N- to Ga-polar GaN films grown by PAMBE [79].

Suzuki et al. showed the Ga-polar GaN films grown on O-polar ZnO templates [80]. They grew O-polar ZnO on sapphire substrate by PAMBE and performed NH_3 preexposure before the GaN growth. Figure 4.16a shows a SIMS depth profile for the sample across GaN/ZnO interface. It is evident that

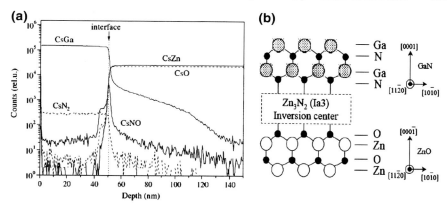

Fig. 4.16. (a) SIMS profile from the GaN/ZnO sample and (b) schematic diagram for bond sequence at a Ga-polar GaN/O-polar ZnO interface. Reprinted with permission from [80]. Copyright (2005), Elsevier

Zn and O diffused into the GaN epilayer from ZnO substrate. Furthermore, Ga and N diffused into the ZnO substrate from the GaN epilayer. Therefore it is likely that compounds such as Zn_3N_2 were formed between the GaN and ZnO layers as illustrated in Fig. 4.16b. By forming the Zn_3N_2 layer, which has inversion symmetry, the upper GaN film showed the Ga-polar from the CBED patterns to determine the polarity [80]. The method to control and invert the polarity by using the intentional interlayer with inversion symmetry reported by Suzuki et al. is basically the same approach to the method reported by Hong et al., who succeeded in the O-polar ZnO growth on the Ga-polar GaN template by using the Ga_2O_3 interfacial layer with inversion symmetry [21].

On the other hand, Namkoong et al. studied PAMBE of GaN on Zn-polar and O-polar ZnO substrates and reported that the Ga-polar GaN could be grown on the Zn-polar substrate, while mixed polar GaN was grown on the O-polar ZnO substrate [81]. However, Kobayashi et al. reported the Ga-polar GaN growth on the atomically flat O-polar ZnO substrate by PLD at room temperature (RT), while the growth of N-polar GaN by PLD at 700°C [82]. In addition, they could grow Ga-polar GaN at 700°C by using the RT buffer layer.

4.5.2 Polarity Control of ZnO Films

The first attempt to investigate the polarity and in-plane orientation of ZnO films on Al_2O_3 substrates grown by PLD was conducted by Ohkubo et al. [83]. The twisting of the in-plane orientation by 30° was accompanied by the flipping of polarity from the O-polar to Zn-polar which was confirmed by CAICISS. ZnO films grown at low (400–450°C) and high (800–835°C) temperatures were rotated by 30° which was measured by XRD Φ-scan results

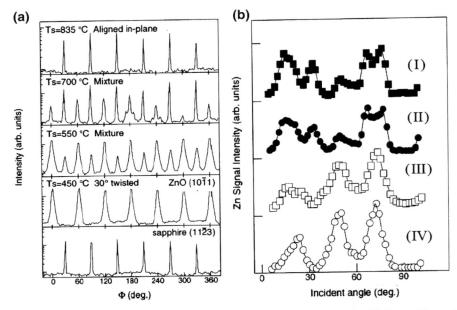

Fig. 4.17. (a) XRD Φ-scans of a sapphire and ZnO films grown by PLD at different temperatures. (b) Incident angle dependence of the Zn signal intensity in CAICISS time-of flight spectra: (I) single crystal O-polar ZnO; (II) thin film grown at 835°C; (III) single crystal Zn-polar ZnO; (IV) thin film grown at 450°C. Reprinted with permission from [83]. Copyright (2005), Elsevier

as shown in Fig. 4.17a. Furthermore, at that time, the polarity of ZnO films was changed as shown in Fig. 4.17b. The fact of change of polarity can be explained by two possible interface structures, the Zn–O = Al–O and the O–Zn = O–Al sequence which are Zn-polar and O-polar, respectively.

However, it should be noted that most of the epitaxial ZnO films grown by PAMBE, PLD, and MOCVD on (0001) Al_2O_3 substrates have crystallographic 30° rotation to reduce the lattice mismatch, which inevitably expect the O-polar ZnO films according to the results of Ohkubo et al. [83].

Polarity control by using the buffer or the interfacial layer has been reported for the ZnO films grown on (0001) Al_2O_3 substrate by PAMBE [51,84]. Kato et al. used the MgO buffer layer to control the polarity in 2004. The polarity control is achieved by changing the thickness of the MgO buffer layer [84]. They suggested the possible atomic arrangements between ZnO and MgO for polarity selection as shown in Fig. 4.18. When the thickness of MgO is lesser than 3 nm, wurtzite–MgO is formed. The polarity of wurtzite MgO is considered to be O-polar, which resulted in the O-polar ZnO growth on the O-polar wurtzite MgO. When the thickness of MgO is more than 3 nm, it has rocksalt structure. The ZnO film on the rocksalt MgO showed Zn polarity. Here, it is notable that the (111) rocksalt MgO has inversion symmetry and

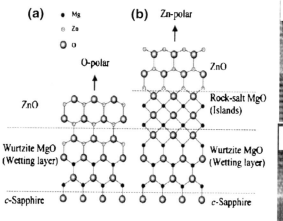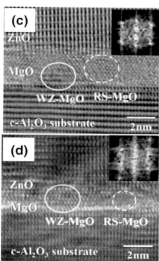

Fig. 4.18. Schematics of atomic arrangement of ZnO on c-sapphire with (**a**) 1 nm-thick MgO, and (**b**) an MgO buffer layer thicker than 3 nm. HRTEM images with the zone axis of ZnO ($11\bar{2}0$) grown on (**c**) 3 nm-thick MgO and (**d**) the 1.5 nm-thick MgO buffer. Digital diffraction patterns obtained from the MgO buffer layer were shown in the insets. Reprinted with permission from [58]. Copyright (2005), American Vacuum Society and Reprinted with permission from [84]. Copyright (2004), American Institute of Physics

the Zn-polar ZnO was grown on the O-polar wurtzite MgO with the rocksalt MgO in between, as shown in Fig. 4.18a, b, respectively. To precisely determine the crystal structure and thickness of the MgO, cross-sectional TEM observations were carried out [61]. Figure 4.18c shows that the 1.5 nm-thick MgO buffer is consist of wurtzite structure layer with a small mixture of rocksalt structure as evidenced by the digital diffraction pattern from the MgO buffer obtained by Fast Fourier Transform (FFT). Figure 4.18d shows that the 3 nm-thick MgO buffer consists mostly of rocksalt structure with a small mixture of wurtzite structure.

On the other hand, Wang et al. used well-defined ultra-thin cubic AlN layers formed by nitridation of the sapphire substrates to control the polarity of upper ZnO films in 2005 [54]. Nitridation and ZnO films growth were carried out by PAMBE. They controlled the polarity of cubic AlN layer by controlling the nitridation temperatures and then the polarity of ZnO films were changed depending on the polarity of underlying AlN layer [54].

Mei et al. reported the role of nitridation temperature in the polarity selection of ZnO films for the growth of uniform Zn-polar ZnO films in 2005 [30]. Similar results were reported by Wang et al. in 2005, using the nitrided sapphire substrate. They showed that the Zn-polar ZnO film grown on the amorphous interfacial layer formed by LT nitridation and the O-polar ZnO film on the Ga predeposited nitrided substrate [54]. On the other hand, Roh et al. have shown that ZnO films grown on LT GaN buffer layers typically exhibited the Zn-polar, while those grown on HT GaN buffer layers typically exhibited the O-polar [85].

On the other hand, Park et al. reported the selective growth of polarity-controlled ZnO films using the Cr-compound intermediate layers in 2007. The polarity of ZnO films grown on rocksalt CrN was determined to be the Zn-polar, while O-polar ZnO films were grown on rhombohedral Cr_2O_3 layers [86]. They suggested that the key factors for the selective growth of Zn- and O-polar ZnO films on Al_2O_3 are the use of intermediate layers of CrN with cubic N sublattice and Cr_2O_3 with hexagonal O sublattice, respectively.

Although the films were not grown on the sapphire substrates, the first reliable polarity controlled ZnO films by PAMBE were reported by Hong et al. in 2000. They succeeded in selective growth of Zn-polar and O-polar ZnO films on Ga-polar GaN template by interface engineering [38]. Zn preexposure on the GaN prior to the ZnO growth results in Zn-polar ZnO films, in which there is no interfacial layer in between the ZnO and the GaN. On the other hand, O-plasma preexposure leads to the growth of O-polar ZnO films, in which the monoclinic Ga_2O_3 interfacial layer is formed between the ZnO and the GaN. The key of polarity control that they suggested is the insertion of an interface layer having inversion symmetry because the polarity comes from lack of inversion symmetry [21]. Here it should be noted that they have grown the O-polar ZnO (anion-polar) film on the Ga-polar GaN (cation-polar) film by using the Ga_2O_3 interfacial layer with inversion symmetry.

4.5.3 Polarity of AlN and InN Films

The polarity of epitaxial films grown on polar substrates is expected to be the same as that of the substrate if there is no interfacial layer between the film and the substrate. When growing GaN films on 6H–SiC substrates, in general, the Ga-polar GaN on the Si-face SiC and N-polar GaN on C-face SiC were reported [9]. However, Keller et al. observed the polarity conversion from the N polar AlN to Al polar AlN on C-face SiC substrate when the films were deposited at a low V/III ratio using MOCVD in 2006 [87]. Takeuchi et al. reported the growth of high quality N-polar AlN layers on 0.15°-off (0001) sapphire substrates by MOCVD [88].

Xu and Yoshikawa reported the polarity control of InN films grown on nitrided sapphire substrates by PAMBE just based on the polarity of GaN on the substrates, i.e., the N-polar InN growth on the N-polar GaN template, while the In-polar InN growth on the Ga-polar GaN template [89]. The effects of growth temperature on the polarity of the InN grown on Al_2O_3 substrate were investigated by Saito et al. [90]. The polarity of the LT InN grown at 300°C showed mainly N polarity, while that of the HT InN grown at 550°C showed mainly In polarity.

4.6 Effects of Polarity on Material Properties

The wurtzite crystal structure with polarity makes the difference of structural, chemical, optical and electrical properties depending on the polarity [9,91,92]. Crystal polarization also affects growth kinetics [93], thermal stability [94], and impurity incorporation [95, 96] in epitaxial growth processes. Difference in impurity doping efficiency depending on the polarity is especially interesting and has been reported for GaN [97] and ZnO [98]. Ptak et al. reported a dramatic difference of Mg incorporation rates between the N- and Ga-polar surfaces during the Mg doping of GaN [97]. They showed the evidence of surface segregation and accumulation of Mg by showing the incorporation of Mg after the shutter is closed. It was found that there was approximately 15–20 times less Mg doping level for the N-polar GaN as compared with the Ga-polar as shown in Fig. 4.19. On the other hand, a higher N doping level for the Zn-polar ZnO compared with the O-polar ZnO was found [98]. Polarity dependent doping efficiency is a very important factor for the success of reliable p-type conductivity in ZnO epitaxy since p-type conductivity is still favored for commercialization of the ZnO-based optoelectronic devices.

Chichibu et al. investigated the enhanced incorporation of impurity in N-polar GaN which was connected to the incorporation of vacancy-type defects [99]. The effect of polarity on the incorporation of impurities and formation of vacancies in the growth of GaN was investigated by the positron annihilation method. The polarity had a critical impact on the formation of vacancies and vacancy clusters, which was more abundant in N-polar [99]. These results indicate that N-polar GaN contains a higher density of vacancy-type defects or defect-complexes than Ga-polar GaN.

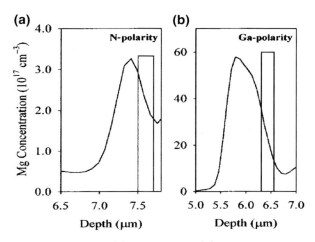

Fig. 4.19. Mg concentration for (**a**) N-polar and (**b**) Ga-polar GaN. Reprinted with permission from [97]. Copyright (2001), American Institute of Physics

Sasaki et al. reported the stronger and narrower emission of Zn-polar ZnO single crystal than O-polar one [100]. The emission from neutral-donor-bound excitons (D^0X) was dominant on all ZnO substrates, but emission from ionized-donor-bound excitons (D^+X) was only observed from the O-face [100]. The difference in PL spectra for ZnO faces was explained by the surface states due to oxygen causing the formation of ionized donors. Allen et al. investigated the electrical and optical properties of bulk ZnO [101]. They reported that the emissions from both longitudinal and transverse free A excitons, A_T(3.3759 eV) and A_L(3.3772 eV), are stronger from the Zn-polar ZnO as shown in Fig. 4.20a. The increased free exciton emission from the Zn-face substrate is caused by recombination from extra excitons generated in an inversion layer near the surface. The inversion layer was expected to contain a high density of holes affected by spontaneous polarization as shown in Fig. 4.20b.

Contact properties on different polar GaN and ZnO surfaces have been studied. The Schottky barrier height (SBH) on the Ga-polar surface is reported to be higher and the associated leakage of current to be lower than that of N-polar surface [102, 103]. Karrier et al. measured an effective SBH between Pt/n-GaN using the I–V characteristics as a function of the ideality factor [102]. The value of the SBH was determined to be 1.1 and 0.9 eV for devices on Ga-polar and N-polar GaN, respectively. The similar result of SBH is also reported by Jang et al. [104]. The specific interpretation of experimental results about the SBH on N-polar GaN was given by Rizzi et al. based on the contribution of spontaneous polarization charges to the interface charge balance [105]. Rickert et al. studied the SBH for six thin metal overlayers of Au, Al, Ni, Ti, Pt and Pd on n- and p-type Ga-polar GaN samples using synchrotron radiation-based XPS [106]. Liu et al. studied the SBH of Ni/Au contacts on Ga-polar and N-polar GaN under hydrostatic pressure and

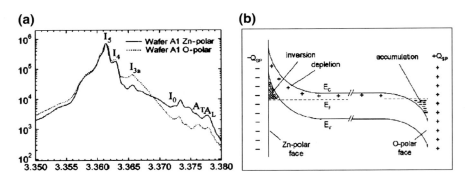

Fig. 4.20. (a) 4.4 K PL spectra taken on the Zn-polar and O-polar faces of bulk ZnO in the bound and free exciton range. (b) Band bending and charge redistribution in undoped bulk ZnO. Reprinted with permission from [101]. Copyright (2007), American Institute of Physics

applied in-plane uniaxial stress [107]. Under hydrostatic pressure two different polarities of GaN yielded significantly different rates of SBH increase with increasing pressure [107].

In spite of the fact that high quality Schottky contacts are critical for ZnO device applications, there is little information about the Schottky contacts on ZnO. Oh et al. reported that the Schottky characteristics are dependent on the growth temperature and polar direction of ZnO:N layers [108]. The SBH for Au contact to a Zn:N layer is estimated to be 0.66 and 0.69 eV by I–V and C–V measurements [108]. The SBH is proportional to the resistivity and incorporated N concentration of ZnO:N layers. Consequently, they suggested that the lower growth temperature and a Zn-polar ZnO were favorable for the incorporation of N into the ZnO films. Besides, Allen et al. investigated the Pd, Pt, Au, and Ag Schottky diodes with low ideality factors in which the diodes were fabricated on the Zn-polar and O-polar ZnO [109]. A polarity effect was observed for Pt and Pd diodes, where the higher quality barriers were achieved on the O-polar ZnO, while significant effect was not observed for Au or Ag diode.

The effects of crystal polarity on ohmic contacts to GaN and ZnO have been investigated. Jang et al. compared the characteristics of an ohmic contact using Ti/Al/Ni/Au metal on Ga- and N-polar GaN [104]. The contact resistivity on the Ga face was lower by two orders of magnitude than that on N-face GaN after annealing at 700°C. The lower contact resistivity on the Ga face after annealing could be attributed to the formation of polarization-induced 2DEG at the AlN/GaN interface [104].

Contrary to the dependence of ohmic contact characteristics on the GaN polarity, clear experimental or theoretical results are hardly reported in ZnO. Murphy et al. experimentally suggested that the differences in contact resistivity could also be attributed to the difference between O-polar and Zn-polar ZnO films [110]. However, their results did not include properties in detail. Moreover, Yang et al. reported no significant polarity effects on the Ti/Au n-type ohmic contacts to bulk ZnO substrates [111].

4.7 Device Applications of Polarization Induced Properties

As discussed in previous sections, the polarity of wurtzite structured GaN and ZnO affects the physical and chemical properties, which is applicable to device fabrication and design. In addition, the appropriate utilization of the polarization induced properties in designing a device can be an important point to maximize the device performances. In this section, the field effects transistors and the periodically polarity-inverted structures are discussed for the electronic and nonlinear optical applications, respectively.

4.7.1 Electronic Devices

A large polarization coming from natural crystal polarity will induce an electric field to modify the band structures and to affect optical [112] and electrical properties [113]. The piezoelectric and spontaneous polarization effects have been revealed as the one of the significantly important factors in designing the nitride and the oxide heterostructures for the electronic devices applications. For example, polarization induced electric fields have strong effects on the formation of 2DEG at the heterointerfaces for GaN/InGaN [114] and ZnO/MgZnO [115] heterostructures.

The systematic variations of the polarization induced 2DEG density and the band profile for an AlGaN/GaN interfaces were theoretically calculated by Ambacher et al. [113] They predicted the sheet carrier charge by considering various piezoelectric constants, stain, bound polarization, composition of the ternary alloy, and free carriers. Morkoc et al. have reviewed the Al(In)GaN/GaN band profile according to the polarity and the characteristics of modulation-doped field effect transistor (FET) considering spontaneous and piezoelectric polarizations [116]. In general, the general GaN-based FET heterostructures have been fabricated using Ga-polar material because the Ga-polar $Al_xGa_{1-x}N$/GaN interface has an enhanced 2DEG, which resulted from the net effect of piezoelectric and spontaneous polarizations. Besides, high electrical quality N-polar AlGaN/GaN FET characteristics were demonstrated by Rajan et al. in 2007 [117].

In the case of ZnO-based heterostructures for FET applications, Koike et al. demonstrated the 2DEG formation at the O-polar $Zn_{0.6}Mg_{0.4}O$/ZnO heterointerface with a sheet carrier density of $\sim 1.2 \times 10^{13}$ cm^{-2} by the piezoelectric polarization in a ZnO well [118]. In 2005, they reported the characteristics of the $Zn_{0.7}Mg_{0.3}O$/ZnO heterostructure FET (HFET) [119]. The schematic illustration of HFET structure is shown in Fig. 4.21a and the top view image of the fabricated device is shown in Fig. 4.21b [119]. They designed the gate electrode with a dimension of 50-μm long and 50-μm width. The Au/In metal on ZnO and Au/Ti bilayer metal system on MgO are used for ohmic and Schottky contacts, respectively.

The characteristics of the fabricated HFET at RT are shown in Fig. 4.22 [119]. Figure 4.22a shows a typical drain-source current dependence on drain-source voltage ($I_{DS} - V_{DS}$) for the HFET in the dark. The operation of HFET device was n-channel depletion mode, which is in agreement with the expectation from the 2DEG in single quantum well as shown in the Fig. 4.23. In addition, a clear pinch-off and a current saturation were observed. Figure 4.22b shows the $I_{DS} - V_{GS}$ relationship at $V_{DS} = 3.0$ V. The saturation is about 150 μA at around $V_{DS} = 2.0$ V. The calculated μ_{FE} is 140 cm^2 Vs^{-1} from the experimental results, and maximum transconductance g_m is estimated to be 0.7 mS mm^{-1} at $V_{GS} = -4.6$ V, as shown by the $g_m - V_{GS}$ curve in Fig. 4.22b.

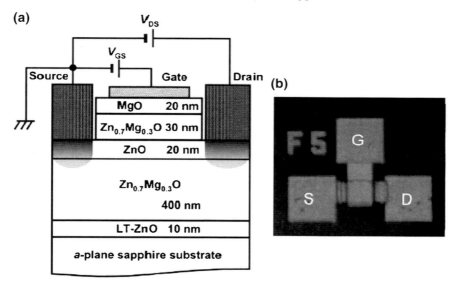

Fig. 4.21. (a) Schematic illustration of a HFET and (b) top view photograph of the fabricated device. Reprinted with permission from [119]. Copyright (2005), American Institute of Physics

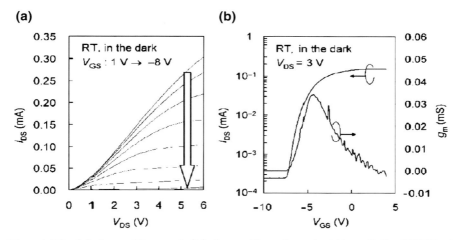

Fig. 4.22. (a) $I_{DS} - V_{DS}$ and (b) $I_{DS} - V_{DS}$ characteristics of the HFET at RT. Reprinted with permission from [119]. Copyright (2005), American Institute of Physics

On the other hand, Tampo et al. have investigated the formation of 2DEG in the Zn-polar ZnMgO/ZnO heterostructure in 2006 [115]. A sheet electron concentration of $\sim 1 \times 10^{13}$ cm^{-2} and strong electron confinement for low Mg compositions were confirmed as O-polar ZnO based FET.

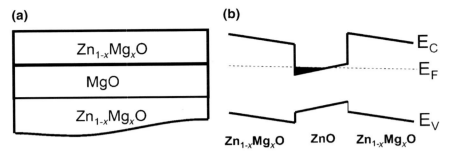

Fig. 4.23. (a) Schematic pictures of $Zn_{1-x}Mg_xO/ZnO$ single quantum well and (b) band structure

4.7.2 Nonlinear Optical Devices

The first experimental result of nonlinear optics is the second harmonic generation (SHG) ($\lambda = 0.347\,\mu m$) through quartz crystal using ruby laser ($\lambda = 0.694\,\mu m$) [120]. Within a few months from the first publication, birefringently phase-matched SHG were demonstrated in potassium dihydrogen phosphate (KDP) [121] as well as sum frequency generation in triglycine sulfate [122]. One of the main applications of nonlinear optics is spreading out the wavelength range of laser sources through optical frequency conversion.

Due to their nonlinear optical properties, ZnO and GaN films show second and third order nonlinear optical features [123–125], which make them promising material for wide application in the nonlinear optical devices. Previously, we described various polarity control methods for GaN and ZnO films. Periodically polarity inverted (PPI) GaN or ZnO heterostructures, in which polarity is periodically inverted parallel to the film/substrate interface, can be applied to the devices for frequency conversion. There has been much attention in the development of new nonlinear optical materials for potential applications in integrated optics. Although the periodical poling of $LiNbO_3$ and $LiTaO_3$ has paved the way for nonlinear optical devices [126, 127], low cost for device fabrication and flexibility are needed. Therefore, the periodically polarity-controlled ZnO and GaN films in one or two dimensions are expected to open new application fields of ZnO and GaN in nonlinear optical devices [45, 123, 128].

The first attempt of using the PPI GaN structures to SHG was conducted by Chowdhury et al. [128]. They showed the SHG at $\lambda = 1658.6\,nm$ in one dimensional PPI GaN structure with $1.7\,\mu m$-periodicity based on the polarity control achieved by growing the GaN films with and without the AlN buffer layers, i.e., Ga-polar GaN on the AlN buffer and N-polar GaN on the Al_2O_3 substrate [128]. Two dimensional structures [129] and applications on the SHG were demonstrated using the photonic crystal of GaN [130].

In case of PPI ZnO structure, the MgO buffer [131] or the Cr compound buffer [45, 132] was employed to control the polarity and to fabricate the one dimensional PPI ZnO structures on (0001) Al_2O_3 substrates. The PPI ZnO structure fabricated through pattering and regrowth is schematically shown in the Fig. 4.24a. Figure 4.24b shows optical microscope image for the fabricated

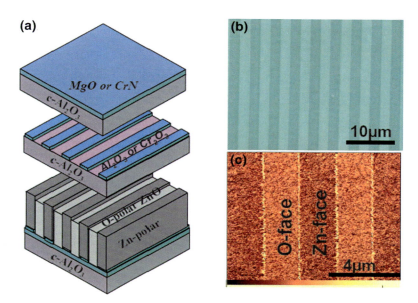

Fig. 4.24. (a) Schematic illustration of the PPI ZnO structure fabricated through patterning and regrowth processes, (b) Optical microscope image, and (c) PFM image of the fabricated PPI ZnO structure. Reprinted with permission from [45]. Copyright (2008), American Vacuum Society

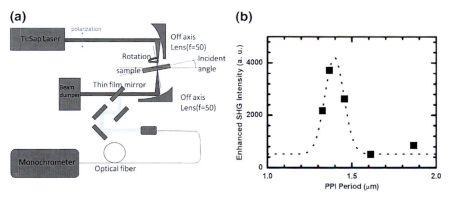

Fig. 4.25. (a) Schematic set-up of the optics used in the measurements of SHG from the PPI ZnO structures and (b) intensity of SHG as a function of periodic distance in the PPI structures [131]

one dimensional PPI ZnO structure and Fig. 4.24c shows the PFM image indicating the periodical change of polarity [45].

For the characterization of nonlinear optical properties from the PPI ZnO structure, SHG experiments were carried out [131]. The schematic configuration of experimental setup is shown in Fig. 4.25a. The Ti: sapphire laser with about 10 fs pulse duration was used for the input fundamental wave ($\lambda = 800$ nm) and clear SHG peak was observed from the ZnO PPI structure with 1.4 µm-periodicity as shown in Fig. 4.25b, which agreed well with calculated quasi phase matching conditions [131].

References

1. S. Nakamura, G. Fasol, *The Blue Laser Diode: GaN Based Light Emitters and Lasers* (Springer, Berlin, 1997)
2. M.A. Khan, Q. Chen, M.S. Shur, B.T. McDermott, J.A. Higgins, J. Burm, W.J. Schaff, L. F. Eastrman, IEEE Electr. Device L. **17**, 584 (1996)
3. Y.F. Wu, B.P. Keller, S. Keller, D. Kapolnek, P. Kozodiy, S.P. DenBaars, U.K. Mishra, Appl. Phy. Lett. **69**, 1438 (1996)
4. R.D. Underwook, S. Keller, U.K. Mishra, D. Kapolnek, B.P. Keller, S.P. Denbaars, J. Vac. Sci. Technol. B **16**, 822 (1998)
5. T. Kozawa, M. Suzuki, Y. Taga, Y. Gotoh, J. Ishikaa, J. Vac. Sci. Technol. B **16**, 833 (1998)
6. D.M. Bagnall, Y.F. Chen, Z. Zhu, S. Koyama, M.Y. Shen, T. Goto, T. Yao, Appl. Phys. Lett. **70**, 2230 (1997)
7. F. Bernardini, V. Fiorentini, D. Vanderbilt, Phys. Rev. B **56**, R10024 (1997)
8. F. Bernardini, V. Fiorentini, Phys. Rev. B **58**, 15292 (1998)
9. E.S. Hellman, MRS Internet J. N. S. R. **3**, 11 (1998)
10. Y. Noel, C.M. Zocovich-Wilson, B. Civalleri, Ph. D'Arco, R. Doversi, Phys. Rev. B **65**, 014111 (2001)
11. O. Ambacher, J. Smart, J.R. Shealy, N.G. Weimann, K. Chu, M. Murphy, W.J. Schaff, L.F. Easterman, R. Dimitrov, L. Wittmer, M. Stutzmann, W. Rieger, J. Hilsenbeck, J. Appl Phys. **85**, 3222 (1999)
12. M. Yano, K. Hashimoto, K. Fujimoto, K. Koike, S. Sasa, M. Inoue, Y. Uetuji, T. Ohnishi, K. Inaba, J. Cryst. Growth **301–302**, 353 (2007)
13. E.T. Yu, X.Z. Dang, P.M. Asbeck, S.S. Lau, G.J. Sullivan, J. Vac. Sci. Technol. B **17(4)**, 1742 (1999)
14. H. Morkoc, S. Strite, G.B. Gao, M.E. Lin, B. Sverdlow, M. Burns, J. Appl. Phys. **76**, 1363 (1994)
15. H. Karzel, W. Potzel, M. Kofferlein, W. Schiessl, M. Steiner, U. Hiller, G.M. Kalvius, D.W. Mitchell, T.P. Das, P. Blaha, K. Schwarz, M.P. Pasternak, Phys. Rev. B **53**, 11425 (1996)
16. T.B. Bateman, J. Appl. Phys. **33**, 3309 (1962)
17. I.B. Kobiakov, Solid State Commun. **35**, 305 (1980)
18. Z. Liliental-Weber, S. Ruvimov, Ch. Kisielowaki, Y. Chen, W. Swider, J. Washburn, N. Newman, A. Gassmann, X. Liu, L. Schloss, E.R. Weber, I. Grzegory, M. Bockowaki, J. Jun, T. Suski, K. Pakula, J. Baranowski, S. Porowski, H. Amano, I. Akasaki, Mater. Res. Soc. Symp. Proc. **395**, 351 (1996)

19. P. Vermaut, P. Ruterana, Gl. Nouet, Philos. Mag. A. **76**, 1215 (1997)
20. H. Morkoc, Mater. Sci. Eng. R. **33**, 135 (2001)
21. S.K. Hong, T. Hanada, H.J. Ko, Y. Chen, T. Yao, D. Imai, K. Araki, M. Shinohara, Phys. Rev. B **65**, 115331 (2002)
22. Z.H. Zhang, H.H. Liu, J.K. Jian, K. Zou, X.F. Duan, Appl. Phys. Lett. **88**, 193101 (2006)
23. F.A. Ponce, D.P. Bour, W.T. Young, M. Saumders, J.W. Steeds, Appl. Phys. Lett. **69**, 337 (1996)
24. A.R. Smith, R.M. Feenstar, D.W. Gerve, J. Neugebauer, J.E. Northrup, Phys. Rev. Lett. **79**, 3934 (1997)
25. J.E. Northrup, J. Neugebauer, R.M. Feenstar, A.R. Smith, Phys. Rev. B **61**, 9932 (2000)
26. A.R. Smith, R.M. Feenstra, D.W. Greve, M.S. Shin, M. Skorwringski, J. Neugebauer, J.E. Northrup, Appl. Phys. Lett. **72**, 2114 (1998)
27. Y.F. Chen, H.J. Ko, S.K. Hong, T. Yao, Appl. Phys. Lett. **76**, 559 (2000)
28. Y.F. Chen, H.J. Ko, S.K. Hong, T. Yao, Y. Segawa, J. Cryst. Growth **214**, 87 (2000)
29. K. Iwata, P. Fons, S. Niki, A. Yamada, K. Matsubara, K. Nakahara, H. Takasu, Phys. Status Solid. A **180**, 287 (2000)
30. Z.X. Mei, X.L. Du, Y. Wang, M.J. Ying, Z.Q. Zeng, H. Zheng, J.F. Jia, Q.K. Xue, Z. Zhang, Appl. Phys. Lett. **86**, 112111 (2005)
31. A.N. Mariano, R.E. Hanneman, J. Appl. Phys. **34**, 384 (1963)
32. A. Kazimirov, N. Faleev, H. Temkin, M.J. Bedzyk, V. Dmiriev, Y. Melnik, J. Appl. Phys. 89, 6092 (2001)
33. K. Inaba, H. Amano, Phys. Stat. Solidi B **244**, 1775 (2007)
34. H. Tampo, P. Fons, A. Yamada, K.-K. Kim, H. Shibata, K. Matsubara, S. Niki, H. Yoshikawa, H. Kanie, Appl. Phys. Lett. **87**, 141904 (2005)
35. M. Sumiya, S. Nakamura, S.F. Chichibu, Appl. Phys. Lett. **77**, 2512 (2000)
36. H. Maki, N. Ichinose, S. Sekiguchi, N. Ohashi, T. Nichihara, H. Haneda, J. Tanaka, Jpn. J. Appl. Phys. **38**, 2741 (1999)
37. T. Ohnishi, A. Ohtomo, M. Kawasaki, K. Takahashi, M. Yoshimoto, H. Koinuma, Appl. Phys. Lett. **72**, 824 (1998)
38. S.K. Hong, T. Hanada, H.J. Ko, Y. Chen, T. Yao, D. Imai, K. Araki, K. Shinohara, Appl. Phys. Lett. **77**, 3571 (2000)
39. T. Sasaki, T. Matsuoka, J. Appl. Phys. **64**, 4531 (1998)
40. M. Harada, Y. Ishikawa, T. Saito, N. Shibata, Jpn. J. Appl. Phys. **42**, 2829 (2003)
41. L. Zhang, D. Wett, R. Szargan, T. Chasse, Surf. Interf. Anal. **36**, 1479 (2004)
42. M. Asif Khan, J.N. Kuznia, D.T. Olson, R. Kaplan, J. Appl. Phys. **73**, 3108 (1993)
43. M. Niebelshutz, G. Ecke, V. Climalla, K. Tonisch, O. Ambacher, J. Appl. Phys. **100**, 074909 (2006)
44. B.J. Rodriguez, A. Gruverman, A.I. Kingon, R.J. Nemanich, O. Ambacher, Appl. Phys. Lett. **80**, 4166 (2002)
45. J.S. Park, T. Minegishi, S.H. Lee, I.H. Im, S.H. Park, T. Hanada, T. Goto, M.W. Cho, T. Yao, S.K. Hong, J.H. Chang, J. Vac. Sci. Technol. A **26**, 90 (2008)
46. D.A. Scrymgeour, T.L. Sounart, N.C. Simmons, J.W.P. Hsu, J. Appl. Phys. **101**, 014316 (2007)

47. R. Katayama, Y. Kuge, K. Onabe, T. Matushita, T. Kondo, Appl. Phys. Lett. **89**, 231910 (2006)
48. B.J. Rodriguez, W.C. Yang, R.J. Nemanich, A. Gruverman, Appl. Phys. Lett. **86**, 112115 (2005)
49. W.C. Yang, B.J. Rodriguez, M. Park, R.J. Nemanich, O. Ambacher, V. Cimalla, J. Appl. Phys. **94**, 5720 (2003)
50. Q.Y. Xu, Y. Wang, Y.G. Wang, X.L. Du, Q.K. Xue, Z. Zhang, Appl. Phys. Lett. **84**, 2067 (2004)
51. Y. Wang, X.L. Du, Z.X. Mei, Z.Z. Zeng, M.J. Ying, H.T. Yuan, J.F. Jia, Q.K. Xue, Z. Zhnag, Appl. Phys. Lett. **87**, 051901 (2005)
52. M. Sumiya, K. Yoshimura, T. Ito, K. Ohtsuka, S. Fuke, K. Mizuno, M. Yoshimoto, H. Koinuma, A. Ohtomo, M. Kawasaki, J. Appl. Phys. **88**, 1158 (2000)
53. X. Wang, S.B. Che, Y. Ishitani, A. Yoshikwa, J. Appl. Phys. **99**, 073512 (2006)
54. X. Wang, Y. Tomita, O.H. Roh, M. Ohsugi, S.B. Che, Y. Ishitani, A. Yoshikawa, Appl. Phys. Lett. **86**, 011921 (2005)
55. D. Zhuang, J.H. Edgar, Mater. Sci. Eng. R. **48**, 1 (2005)
56. J.L. Rouviere, J.L. Weyher, M.S. Eggebert, S. Porowski, Appl. Phys. Lett. **73**, 668 (1998)
57. J.L. Weyher, S. Muller, I. Grzwgory, S. Porowski, J. Cryst. Growth **182**, 17 (1997)
58. T. Minegishi, J.H. Yoo, H. Suzuki, Z. Vashaei, K. Inaba, K Shim, T. Yao, J. Vac. Sci. Technol. B **23(3)**, 1286 (2005)
59. H. Kato, M. Sano, K. Miyamoto, T. Yao, J. Cryst. Growth **265**, 375 (2004)
60. E.C. Piquette, P.M. Bridger, Z.Z. Bandic, T.C. McGill, J. Vac. Sci. Technol. B **17(3)**, 1241 (1999)
61. M. Sumiya, M. Tanaka, K. Ohtsuka, S. Fuku, T. Ohnishi, I. Ohkubo, M. Yoshimoto, H. Koinuma, M. Kawasaki, Appl. Phys. Lett. **75**, 674 (1999)
62. T.D. Moustakas, R.J. Molnar, T. Lei, G. Menon, C.R. Eddy Jr., Mater. Res. Soc. Symp. Proc. **242**, 427 (1992)
63. X.Q. Wang, H. Iwaki, M. Murakami, X.L. Du, Y. Ishitani, A. Yoshikawa, Jpn. J. Appl. Phys. B **42**, L99 (2003)
64. C. Heinlein, J. Grepstad, T. Berge, H. Riechert, Appl. Phys. Lett. **71**, 341 (1997)
65. K. Uchida, A. Watanabe, F. Yano, M. Kouguchi, T. Tanaka, S. Minagawa, J. Appl. Phys. **79**, 3487 (1996)
66. F. Widmann, G. Feuillet, B. Daudin, J.L. Rouviere, J. Appl. Phys. **85**, 1550 (1999)
67. S. Mikroulis, A. Georgakilas, A. Kostopoulos, V. Cimalla, E. Dimakis, Appl. Phys. Lett. **80**, 2886 (2002)
68. R.D. Felice, J.E. Nrothrup, Appl. Phys. Lett. **73**, 936 (1998)
69. Y. Wang, A.S. Ozcan, G. Ozaydin, K.F. Ludwig, A. Bhattacharyya, T.D. Moustakas, H. Zhou, R.L. Headrick, D. Peter Siddons, Phys. Rev. B **74**, 235304 (2006)
70. J.E. Nrothrup, R. Di Felice, Phys. Rev. B **55**, 13878 (1997)
71. M.J. Murphy, K. Chu, H. Wu, W. Yeo, W.J. Scahff, O. Ambacher, J. Smart, J.R. Shealy, L.F. Eastmann, T.J. Eustis, J. Vac. Sci. Technol. B **17**, 1252 (1999)
72. M. Sumiya, S. Fuke, MRS Internet J. N. S. R. **9**, 1 (2004)
73. V. Ramachandran, R.M Feenstra, W.L. Sarney, L. Salamanca-Riba, J.E. Northrup, L.T. Romano, D.W. Greve, Appl. Phys. Lett. **75**, 808 (1999)

74. L.T. Romano, J.E. Northrup, A.J. Ptak, T.H. Myers. Appl. Phys. Lett. **77**, 2479 (2000)
75. N. Grandjean, A. Dussaigne, S. Pezzagna, P. Vennéguès. J. Cryst. Growth **251**, 460 (2003)
76. K. Xu, N. Yano, A.W. Jia, A. Yoshikawa, K. Takahashi, J. Cryst. Growth **237–239**, 1003 (2002)
77. A. Yoshikawa, K. Xu, Thin solid Films **412**, 38 (2002)
78. D.H. Lim, K. Xu, S. Arima, A. Yoshikawa, K. Takahashi, J. Appl. Phys. **91**, 6461 (2002)
79. Y.S. Park, H.S. Lee, J.H. Na, H.J. Kim, S.M. Si, H.M. Kim, J.E. Oh, J. Appl. Phys. **94**, 800 (2003)
80. T. Suzuki, H.J. Ko, A. Setiawan, J.J. Kim, H. Sitoh, M. Terauchi, T. Yao, Mater. Sci. Semicon. Proc. **6**, 519 (2003)
81. G. Namkoong, S. Burnham, K.K. Lee, M. Losurdo, P. Capezzuto, G. Bruno, B. Nemeth, J. Nause, Appl. Phys. Lett. **87**, 184104 (2005)
82. A. Kobayahsi, Y. Kawaguchi, J. Ohta, H. Fujioka, K. Fujiwara, A. Ishii, Appl. Phys. Lett. **88**, 181907 (2006)
83. I. Ohkubo, A. Ohtomo, T. Ohnishi, Y. Mastumoto, H. Konuma, M. Kawasaki, Surf. Sci. **443**, L1043 (1999)
84. H. Kato, K. Miyamoto, M. Sano, T. Yao, Appl. Phys. Lett. **84**, 4562 (2004)
85. O.H. Roh, Y. Tomita, M. Ohsugi, X. Wang, Y. Ishitani, A. Yosikawa, Phys. Stat. Sol. B **241**, 2835 (2004)
86. J.S. Park, S.K. Hong, T. Minegishi, S.H. Park, I.H. Im, T. Hanada, M.W. Cho, T. Yao, J.W. Lee, J.Y. Lee, Appl. Phys. Lett. **90**, 201907 (2007)
87. S. Keller, N. Fichtenbaum, F. Wu, G. Lee, S.P. Denbaars, J.S. Speck, M.K. Mishra, Jpn. J. Appl. Phys. **45**, L322 (2006)
88. M. Takeuchi, H. Shimizu, R. Kajitani, K. Kawasaki, Kumagai, A. Koukitu, Y. Aoyagi, J. Cryst. Growth **298**, 336 (2007)
89. K. Xu, A. Yoshikawa, Appl. Phys. Lett. **83**, 251 (2003)
90. Y. Saito, Y. Tanabe, T. Yamaguchi, N. Teraguchi, A. Suzuki, T. Araki, Y. Nanishi, Phys. Stat. Solidi. B **228**, 13 (2001)
91. M. Losurdo, M.M. Giangregorio, Appl. Phys. Lett. **86**, 091901 (2005)
92. D. Huang, M.A. Reshchikov, P. Visconti, F. Yun, A.A. Baski, T. King, H. Morkoc, J. Jashinski, Z. Liliental-wever, C.W. Litton, J. Vac. Sci. Technol. B **20(6)**, 2256 (2002)
93. H. Kato, M. Sano, K. Miyamoto, T. Yao, Jpn. J. Appl. Phys. **42**, 2241 (2003)
94. M.A. Mastro, O.M. Kryliouk, T.J. Anderson, A. Davydov, A. Shapiro, J. Cryst. Growth **274**, 38 (2005)
95. H. Kato, M. Sano, K. Miyamoto, T. Yao, Jpn. J. Appl. Phys. **42**, L1002 (2003)
96. F. Tuomisto, K. Saarinen, B. Lucznik, I. Grzegrory, H. Teisseyre, T. Suski, S. Porowski, P.R. Hageman, J. Likonen, Appl. Phys. Lett. **86**, 031915 (2005)
97. A.J. Ptak, Th.H. Myers, L.T. Romano, C.G. Van de Walle, J.E. Northup, Appl. Phys. Lett. **78**, 285 (2001)
98. K. Nakahara, H. Ken, USA Patent 7002179: ZnO system semiconductor device
99. S F. Chichibu, A. Setoguchi, A. Uedono, K. Yoshimura, M. Sumiya, Appl. Phys. Lett. **78**, 28 (2001)
100. H. Sasaki, H. Kato, F. Izumida, H. Endo, K. Maeda, M. Ikeda, Y. Kashiwaba, I. Niikura, Y. Kashiwaba, Phys. Stat. Sol. C **3**, 1034 (2006)
101. M.W. Allen, P. Miller, R.J. Reeves, S.M. Durbin, Appl. Phys. Lett. **90**, 062104 (2007)

102. U. Karrer, O. Ambacher, M. stutzmann, Appl. Phys. Lett. **77**, 2012 (2000)
103. Z.Q. Fang, D.C. Look, R. Visonti, D.F. Wang, C.Z. Lu, F. Yun, H. Morkoc, S.S. Park, K.Y. Lee, Appl. Phys. Lett. **78**, 2178 (2001)
104. H.W. Jang, J.H. Lee, J.L. Lee, Appl. Phys. Lett. **80**, 3955 (2002)
105. A. Rizzi, H. Luth, Appl. Phys. Lett. **80**, 530 (2002)
106. K.A. Rickert, A.B. Ellis, J.K. Kim, J.L. Lee, F.J. Himpsel, F. Dwikusuma, T.F. Kuech, J. Appl. Phys. **92**, 6671 (2002)
107. Y. Liu, M.Z. Kauser, M.I. Nathan, P.P. Ruden, S. Dogan, H. Morkoc, S.S. Park, K.Y. Lee, Appl. Phys. Lett. **84**. 2112 (2004)
108. D.C. Oh, J.J. Kim, H. Makino, T. Hanada, M.W. Cho, T. Yao, H.J. Ko, Appl. Phys. Lett. **86**, 042110 (2005)
109. M.W. Allen, M.M. Alkaisi, S.M. Durbin, Appl. Phys. Lett. **89**, 103520 (2006)
110. T.E. Murphy, J.O. Blaszczak, K. Moazzami, W.E. Bowen, J.D. Philips, J. Electron. Mater. **34**, 389 (2005)
111. H.S. Yang, D.P. Norton, S.J. Pearton, F. Ren, Appl. Phys. Lett. **87**, 212106 (2005)
112. N. Grandjean, B. Damilano, S. Dalmasso, M. Leroux, M. Laugt, J. Massies, J. Appl. Phys. **86**, 3714 (1999)
113. O. Ambacher, B. Foutz, J. Smart, J.R. Shealy, N.G. Weimann, K. Chu, M. Murphy, A.J. Sierakowski, W.J. Schaff, L.F. Eastman, R. Dimitrov, A. Mitchell, M. Stutzmann, J. Appl. Phys. **87**, 334 (2000)
114. J. Cai, F.A. Ponce, J. Appl. Phys. **91**, 9856 (2002)
115. H. Tampo, H. Shibat, K. Matsubara, A. Yamada, P. Fons, S. Niki, M. Yamagata, H. Kanie, Appl. Phys. Lett. **89**, 132113 (2006)
116. H. Morkoc, A. Di Carlo, R. Cingolani, Sol. State. Electron. **46**, 157 (2002)
117. S. Rajan, A. Chini, M.H. Wong, J.S. Speck, U.K. Mishara, J. Appl. Phys. Lett. **102**, 044501 (2007)
118. K. Koike, K. Hama, I. Nakashima, G.Y. Takada, M. Ozaki, K. Ogata, S. Sasa, M. Inoue, M. Yano, Jpn. J. Appl. Phys. **43**, L1372 (2004)
119. K. Koike, I. Nakashima, K. Hashimoto, S. Sasa, M. Inoue, M. Yano, Appl. Phys. Lett. **87**, 112106 (2005)
120. P.A. Franken, A.E. Hill, C.W. Peters, G. Weinreich, Phys. Rev. Lett. **7**, 118 (1961)
121. P.D. Maker, R.W. Terhune, M. Nisenoff, C.M. Savage, Phys. Rev. Lett. **8**, 21 (1962)
122. M. Bass, P.A. Franken, A.E. Hill, C.W. Peters, G. Weinreich, Phys. Rev. Lett. **8**, 18 (1962)
123. G.I. Petrov, V. Shcheslavskiy, V.V. Yakovlev, I. Ozerov, E. Chelnokov, W. Marine, Appl. Phys. Lett. **83**, 3993 (2003)
124. U. Neumann, R. Grunwal, U. Griebner, G. Steinmeyer, M. Schmidbauer, W. Seeber, Appl. Phys. Lett. **87**, 171108 (2005)
125. M.C. Larciprete, D. Haertle, A. Belardini, M. Bertolotti, F. Sarto, P. Gunter, Appl. Phys. B Lasers Opt. **82**, 431 (2006)
126. S.N. Zhu, H.Y. Zhu, N.B. Ming, Science **278**, 843–846 (1997)
127. A. Bruner, P. Shaier, D. Eger, Opt. Express **14**, 9371 (2006)
128. A. Chowdhury, H.M. Ng, M. Bhardwaj, N.G. Weimann, Appl. Phys. Lett. **83**, 1077 (2003)
129. D. Coquillat, J. Torres, D. Peyrade, R. Legros, J.P. Lascaray, M. Le Vassor d'Yerville, E. Centeno, D. Cassagne, J.P. Albert, Opt. Express **12**, 1097 (2004)

130. R. Katayama, Y. Kuge, T. Kondo, K. Onabe, J. Cryst. Growth **301–302**, 447 (2007)
131. T. Minegishi, Dissertation, Tohoku University, 2008
132. J.S. Park, J.H. Chang, T. Minegishi, H.J. Lee, S.H. Park, I.H. Im, T. Hanada, S.K. Hong, M.W. Cho, T. Yao, J. Electron. Mater. **37**, 736 (2008)

5
Growth of Nonpolar GaN and ZnO Films

S.-K. Hong and H.-J. Lee

Abstract. Growing high quality nonpolar films is difficult compared to polar films. Pure a-plane nonpolar films have been grown without difficulty; however, growing pure m-plane films is not easy and other planes coexist parallel to the interface. In this chapter, growth of nonpolar GaN and ZnO films is described. Growth characteristics and properties of nonpolar (a-plane and m-plane) and semipolar GaN films, and nonpolar ZnO films are discussed. The emphasis is on the typical features of growth, structural properties, and procedures to grow nonpolar films with improved crystal quality. The results of lateral epitaxial overgrowth of nonpolar GaN films are also discussed.

5.1 Introduction

The GaN and ZnO generally have a wurtzite, or unstable zincblend structure, which atomically has four tetrahedral bonds for each atom. As a result of the lack of inversion symmetry, wurtzite crystals have crystallographic polarity [1]. Most epitaxial films of GaN, ZnO, and their heterostructures such as InGaN/GaN, AlGaN/GaN, MgZnO/ZnO have been grown along the polar axis, i.e., $\langle 0001 \rangle$ direction in general, which results in the generation of polarization-induced built-in electric fields due to spontaneous and piezoelectric polarization [2]. The polarization induced electric field causes negative effects on device properties, including a decrease in the overlapping of the electron and hole wave function in the quantum well and consequently a decrease in the quantum efficiency of the emitting devices [3,4]. In order to eliminate polarization effects, growing films without polarity along the growth direction, i.e., growth of nonpolar films, is needed.

Research on nonpolar growth of GaN and ZnO films and device demonstrations has been accelerated because fabrication of high performance light emitting diodes (LEDs) free from polarization induced electric fields is expected. Most of the nonpolar films studied correspond to films with $(10\bar{1}0)$

m-plane and $(11\bar{2}0)$ a-plane of the wurtzite crystal. In this chapter, the growth features of nonpolar GaN and ZnO films, including epitaxial lateral overgrowth (ELO), are described. The focus is given on the growth issues of nonpolar films. Growth characteristics and properties of nonpolar (a-plane and m-plane) and semipolar GaN films and nonpolar ZnO films are discussed. The emphasis is on typical features of growth, structural properties, and procedures to grow nonpolar films with improved crystal quality. Results of lateral epitaxial overgrowth of nonpolar GaN films are also discussed.

5.2 Polar Surface, Nonpolar Surface, and Heterostructures

Figure 5.1 shows schematic views of polar, nonpolar, and semipolar surfaces of wurtzite GaN [5]. The inclined planes of $(10\bar{1}3)$, $(10\bar{1}1)$, $(11\bar{2}2)$ etc. are semipolar planes because these planes have lower polarization fields.

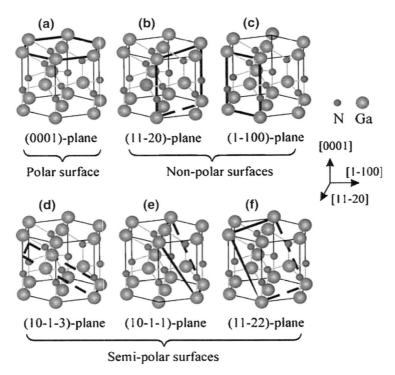

Fig. 5.1. Schematic view of (**a**) polar (0001) plane, (**b**) nonpolar $(11\bar{2}0)$ plane, (**c**) nonpolar $(1\bar{1}00)$ plane, and (**d–f**) semipolar $(10\bar{1}3)$, $(10\bar{1}1)$ and $(11\bar{2}2)$ planes in the wurtzite GaN crystal. Reprinted with permission from [5]. Copyright (2008), Wiley

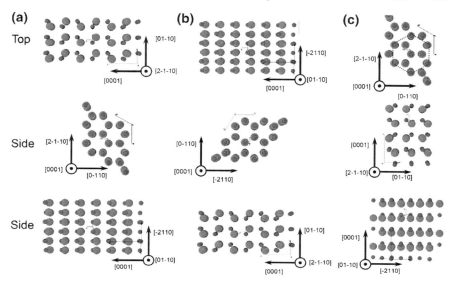

Fig. 5.2. Schematic atomic configurations of *top* and *side views* of the (**a**) a-plane, (**b**) m-plane, and (**c**) c-plane ZnO. *Large* and *small spheres* are Zn and O atoms, respectively

Figure 5.2 shows schematic atomic configurations for nonpolar ($11\bar{2}0$), nonpolar ($1\bar{1}00$) plane, and polar (0001) plane of the ZnO, which has the same wurtzite crystal structure as the GaN. The polar (0001), nonpolar ($11\bar{2}0$), and nonpolar ($1\bar{1}00$) planes are called c-plane, a-plane, and m-plane, respectively. The polar (0001) plane has only cations or anions, while ($11\bar{2}0$) and ($1\bar{1}00$) nonpolar planes have equivalence with cations and anions as shown in Fig. 5.2.

In single crystalline film and heterostructure growths by epitaxy technique, the surface energies of the growing surface have strong effects on the sticking of adatoms, growth mode, surface evolution, surface reconstruction, heterointerfaces, etc. Therefore, figuring out the surface energies of polar and nonpolar surfaces is valuable. Table 5.1 summarizes the calculated surface energies of polar and nonpolar GaN and ZnO surfaces [6–8]. Here, in order to make a comparison, the surface energy values that are determined by local density functional theory are collected, as the reported values are a little different, depending on the calculation methods [8]. From Table 5.1, we can extract several conclusions: (1) the surface energies of polar surfaces are likely to be higher than those of nonpolar surfaces, (2) the surface energies of nonpolar a-planes are higher than those of m-planes for both GaN and ZnO, (3) the surface energies of nonpolar GaN surfaces are higher than those of ZnO.

As shown in Fig. 5.2, nonpolar a-plane and m-plane have $\langle 0001 \rangle$ direction parallel to the surface. Therefore, there is no electric field due to the spontaneous polarization along the growth direction. In c-axis-oriented quantum wells for polar films, polarization-related charges at the heterointerfaces

Table 5.1. Surface energies (in J m^{-2}) of polar and nonpolar GaN and ZnO surfaces

Surface	Surface energy	Ref.
GaN {0001}[a]	1.97	[6]
GaN (11–20)	1.97	[7]
GaN (10–10)	1.89	[7]
ZnO {0001}[b]	2.15	[8]
ZnO (11–20)	1.25	[8]
ZnO (10–10)	1.15	[8]

[a] The value is based on $1/2\{(0001)+(000\bar{1})\}$
[b] The value is given by assuming half of the cleavage energy given in [8]

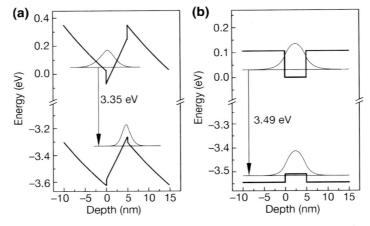

Fig. 5.3. Schematic illustration of energy band discontinuities in AlGaN/GaN quantum wells grown with (**a**) polar (0001) c-plane, where large polarization field is applied and (**b**) nonpolar ($10\bar{1}0$) m-plane free from the polarization induced electric field. Reprinted with permission from [10]. Copyright (2000), Macmillan

induce large electric fields in the quantum wells, which cause the quantum-confined Stark effect (QCSE) [9]. The polarization-induced electric field leads to spatial separation of the electron and hole wavefunctions, i.e., their reduced overlap [10]. Figure 5.3 shows the schematic illustration of energy band discontinuities in c-plane and nonpolar AlGaN/GaN heterostructures with polar and nonpolar quantum wells [10]. As shown in Fig. 5.3, the LEDs based on polar surfaces have inherently reduced recombination probability and hence decreased device efficiency, compared with the LEDs based on nonpolar surfaces. This is the main reason why nonpolar growth and fabrication of devices based on it are important and attractive, although nonpolar growth with high film quality has been difficult compared to polar growth.

5.3 Growth of Nonpolar GaN Films

The first nonpolar GaN film (a-plane) was grown by Sano and Aoki in 1976, more than 30 years ago, using the hydride vapor phase epitaxy (HVPE) [11]. Since the report by Sano et al., reports on nonpolar GaN films have continued but most results were not promising for device applications because of their low crystal quality [5]. Most of all, one of the most critical problems of nonpolar GaN films was their very rough surface, which made it difficult to be used for device fabrication. Nonpolar GaN films that seemed usable for device applications were available in 2000, and had surface roughness comparable to polar GaN films [10]. This work was regarded as a milestone in nonpolar GaN research. In this section, the growth of nonpolar GaN films is described. Planar growth of m-plane and a-plane GaN films, epitaxial lateral overgrowth (ELO), and growth of semipolar films are also addressed.

5.3.1 M-plane GaN Films

Most heteroepitaxial m-plane GaN film growths were performed using the (100) γ-LiAlO$_2$ substrate. However, all GaN films on (100) γ-LiAlO$_2$ substrates are not m-plane GaN. Hellman et al. reported that the c-plane (0001) GaN grew on the (100) γ-LiAlO$_2$ substrate [12]. However, Ke et al. reported m-plane GaN growth using metal organic chemical vapour deposition (MOCVD) [13]. Waltereit et al. also reported m-plane GaN growth using plasma-assisted molecular beam epitaxy (PAMBE) [10]. Sun et al. reported that the nitridation of the substrate and/or immediate N-rich nucleation conditions invariably induced mixed m- and c-planes, while immediate Ga-rich nucleation without the nitridation treatment resulted in the growth of pure m-plane GaN films using PAMBE [14]. Figure 5.4 shows the epitaxial relationship between m-plane GaN and (100) γ-LiAlO$_2$ [15].

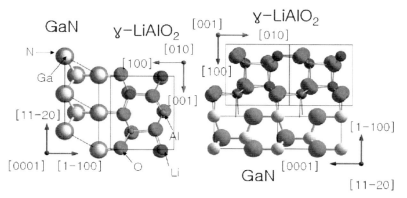

Fig. 5.4. Ball and stick models illustrating epitaxial relationship and lattice matching in the m-plane GaN (100) γ-LiAlO$_2$ system. Reprinted with permission from [15]. Copyright (2007), Wiley

The lattice mismatch between GaN [0001] and LiAlO$_2$ [100] direction is −0.2%, and that between GaN [11$\bar{2}$0] and LiAlO$_2$ [001] is 1.7%.

As the m-plane GaN is not familiar compared with the c-plane GaN, the appearance of reflection high energy electron diffraction (RHEED) patterns for the m-plane GaN is interesting. Figure 5.5 shows typical RHEED patterns from an m-plane GaN film recorded along various azimuths [16].

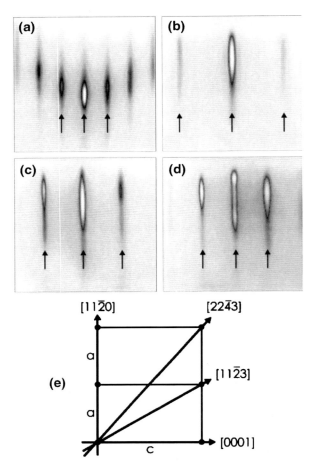

Fig. 5.5. Typical RHEED patterns recorded along azimuths of (**a**) 0°, (**b**) 38°, (**c**) 58° and (**d**) 90°. *Arrows* denote bulk streaks of zeroth and first order which corresponds to estimated lattice rod spacings of (**a**) 5.18 Å, (**b**) 2.01 Å, (**c**) 2.70 Å and (**d**) 3.15 Å. *Lower part* (**e**) shows a schematic of the (1$\bar{1}$00) plane with its in-plane lattice constants a and c. Azimuths of (**a**)–(**d**) were assigned to be the [11$\bar{2}$0], [22$\bar{4}$3], [11$\bar{2}$3] and [0001] directions, based on calculating angles and spacings. Reprinted with permission from [16]. Copyright (2000), Elsevier

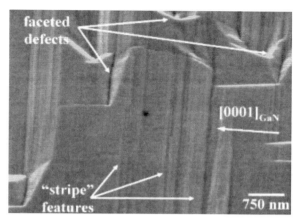

Fig. 5.6. Scanning electron microscope image of the pure m-plane GaN film on m-plane sapphire substrate grown by MOCVD. Image taken with sample tilted 45° toward GaN a-axis. Stripe features and faceted defects are shown. Reprinted with permission from [17]. Copyright (2008), American Institute of Physics

There are very few reports of the achievement of pure m-plane GaN growth on sapphire substrate, which is the most used substrate for c-plane GaN. In general, (0002) and/or (10$\bar{1}$3) and/or (11$\bar{2}$2) planes have been detected in addition to the main (1$\bar{1}$00) m-plane [17]. Armitage and Hirayama reported pure m-plane GaN growth on m-plane sapphire substrates by MOCVD [17]. They found that both substrate preparation (annealing) and initial nitridation conditions of the sapphire substrate strongly affected the purity of the GaN films, and succeeded in growing pure m-plane GaN films under the limited growth conditions. However, the detailed mechanism governing the formation of mixed planes or pure m-planes has not been fully understood. The surface morphology of pure m-plane GaN films showed stripe features running perpendicular to the c-axis of GaN and faceted defects were also found on some regions as shown in Fig. 5.6 [17].

The crystal quality of GaN film was addressed by X-ray rocking curves (XRCs). Figure 5.7 shows symmetric and asymmetric XRD ω scans for m-GaN film on m-plane sapphire [17]. Two symmetric scans for (1$\bar{1}$00) reflection were performed with x-rays incident, perpendicular or parallel to the GaN c-axis, which showed significant anisotropy. Full width at half maximums (FWHMs) of the XRCs were 0.22° and 0.56° for the symmetric (1$\bar{1}$00) reflections with x-rays incident, perpendicular or parallel to the GaN c-axis, respectively. Looking at the results, we find that there is more tilt mosaic parallel to the c-axis of GaN and less tilt mosaic perpendicular to the c-axis of GaN, which is believed to be mostly related to the stripe features perpendicular to the c-axis of GaN. That is, there is lower tilt mosaic along the stripe direction and larger tilt mosaic across the stripe direction. FWHMs of the XRCs were 0.41° and 0.61° for the asymmetric (1$\bar{1}$01) and (11$\bar{2}$0) reflections, respectively.

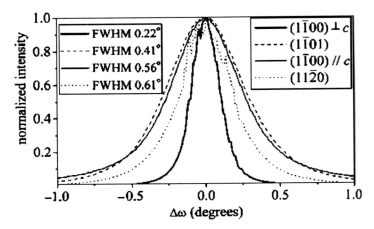

Fig. 5.7. Symmetric and asymmetric XRD ω scans for m-plane GaN film on m-plane sapphire. The two symmetric scans shown correspond to data for x-rays incident perpendicular or parallel to the GaN c-axis. Reprinted with permission from [17]. Copyright (2008), American Institute of Physics

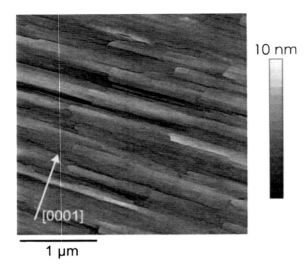

Fig. 5.8. Atomic force micrograph showing morphology of a 1.5 μm thick GaN film grown on (100) $LiAlO_2$ by PAMBE, where stripes perpendicular to the GaN c-axis are clearly shown. Reprinted with permission from [16]. Copyright (2000), Elsevier

Almost all the reports on m-plane GaN, directly grown on foreign substrates, including m-plane sapphire (Fig. 5.6) and $LiAlO_2$ (Fig. 5.8) showed surface morphologies with a commonly observed striated or slate surface, which was oriented parallel to the [11$\bar{2}$0] direction. Figure 5.8 shows the surface morphology of m-plane GaN film grown on (100) LiAlO$_2$ by PAMBE [16], where the stripes perpendicular to the c-axis of GaN are shown.

Hirai et al. reported that slate surface morphology was attributed to instabilities initiated at stacking faults (SFs), because the spacing of the slate features in the [0001] direction was essentially the same as the spacing between stacking faults [18]. They concluded that slate morphology disappeared only when stacking faults were completely eliminated from the films.

Because of difficulties in growing pure m-plane GaN films on sapphire substrates, free standing (FS) m-plane GaN substrate [19–21], m-plane SiC substrate [22, 23], and m-plane ZnO substrate [24] are used for the growth of m-plane GaN films. In the case of growth on FS m-plane GaN substrates, the FS substrates are obtained from HVPE grown GaN crystals. That is, the FS m-plane GaN are obtained by slicing an m-plane GaN crystal grown on (100) LiAlO$_2$ substrates [19] or by slicing a c-plane GaN crystal grown by HVPE [20,21]. In SiC and ZnO substrates, the m-plane substrates are available from bulk SiC or bulk ZnO material.

Surface morphology of pure m-plane GaN films on m-plane FS GaN substrates showed striated surfaces, but could be controlled by changing the growth conditions. Chichibu et al. observed atomically flat surfaces with monolayer steps from the GaN films grown under optimized growth conditions [21]. Figure 5.9 shows surface morphologies of m-plane FS GaN substrates and m-plane GaN film. The m-plane GaN film showed very small FWHMs for the (10$\bar{1}$0) XRCs. The FWHMs of the (10$\bar{1}$0) on-axis XRCs taken along the $\langle 0001 \rangle$ and $\langle 11\bar{2}0 \rangle$ azimuths were 31 and 91 arcsec, respectively [21].

Kobayashi et al. used m-plane ZnO substrate in growing the m-plane GaN film by pulsed laser deposition (PLD) [24]. Interestingly, they grew the GaN films at room temperature (RT) and observed the RHEED intensity oscillations indicating a monolayer height (0.276 nm) layer-by-layer growth as shown in Fig. 5.10 [24]. The XRCs for the (1$\bar{1}$00) on-axis reflections with the x-ray incidence perpendicular and parallel to the c-axis showed the same FWHMs of 252 arcsec [24].

5.3.2 A-plane GaN Films

In contrast to m-plane GaN films, a-plane GaN films have been easily grown on sapphire substrates without the mixed planes [5]. That is, nonpolar (11$\bar{2}$0) a-plane GaN films are grown on (1$\bar{1}$02) r-plane sapphire substrates through various epitaxy techniques including HVPE, MOCVD, and PAMBE. Historical events in nonpolar GaN film growths, including a-plane GaN film growth, have been reviewed by Paskova [5].

The a-plane GaN film on r-plane sapphire substrate has the epitaxial relationship of [0001]GaN//[$\bar{1}$101]Sapphire and [$\bar{1}$100]GaN//[11$\bar{2}$0]Sapphire [25]. The films had defects – high density structural defects including threading dislocations (TDs) and basal SFs [25, 26]. In the case of thick HVPE grown

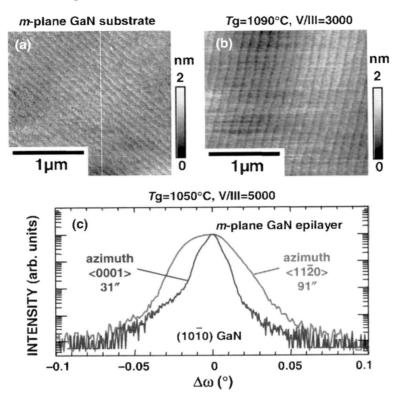

Fig. 5.9. AFM images of (**a**) m-plane FS GaN substrate and (**b**) m-plane homoepitaxial film grown by MOCVD at 1,090°C with V/III = 3,000. (**c**) Representative (10$\bar{1}$0) XRCs of the GaN film (grown at 1,050°C with V/III = 5,000) taken along the (0001) and (11$\bar{2}$0) azimuths. Reprinted with permission from [21]. Copyright (2008), American Institute of Physics

film, internal cracks were also observed [26]. The defect densities for TDs were in the order of $10^9 \sim 10^{10}$ cm^{-2} and for SFs in the order of $10^5 \sim 10^6$ cm^{-1}.

In order to reduce defect density and improve crystal quality, various approaches including employment of SiN$_X$ nanomask [27], two-step growth [28], and employment of a buffer layer [29] were tried. Chakraborty et al. used SiN$_X$ nanomask for defect reduction in MOCVD grown a-plane GaN films. The layer was grown in situ using disilane (Si$_2$H$_6$) and ammonia (NH$_3$) gases before growing the GaN films. As shown in Fig. 5.11, FWHMs of on-axis and off-axis XRCs were decreased with increasing SiN$_X$ deposition time [27]. By employing the SiN$_X$ nanomask, surface morphology (Fig. 5.12) and crystal quality were improved. However, the TD and SF densities were still as high as 9×10^9 cm^{-2} and 3×10^5 cm^{-1}, respectively [27]. From Figs. 5.11a, 5.12, it is evident that there is anisotropy in structural quality and surface morphology. The FWHMs

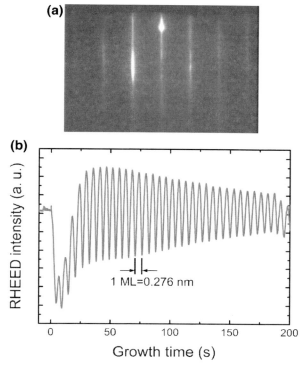

Fig. 5.10. (a) RHEED pattern of an m-plane GaN grown on ZnO at room temperature. Incidence of electron beam is parallel to [11$\bar{2}$0]. (b) Intensity profile for the RHEED specular spot during growth of GaN. Reprinted with permission from [24]. Copyright (2007), American Institute of Physics

of on-axis (11$\bar{2}$0) XRCs showed smaller values when the x-ray beam was incident along the GaN ⟨0001⟩ direction compared to cases where the x-ray beam was incident perpendicular to the GaN ⟨0001⟩ direction. This meant that the tilt mosaic component along the ⟨0001⟩ direction was smaller than that along the [$\bar{1}$100] direction.

The anisotropic structural property of a-plane GaN films, which was clearly shown in XRC for the on-axis (11$\bar{2}$0), was improved by employing the two-step growth in the MOCVD process [28]. Hollander et al. reported that lateral growth was favored using a low V/III ratio resulting in films with a smooth surface, while pitted films were grown at a high V/III ratio indicating preferential on-axis growth. Both film types showed strong anisotropy in the peak XRC FWHMs for the on-axis (11$\bar{2}$0) with respect to the in-plane phi angles. They achieved in-plane isotropic behavior of crystallinity with overall reduced XRC FWHMs by starting growth with a high V/III ratio initially before reducing the V/III ratio for film coalescence [28]. Figure 5.13

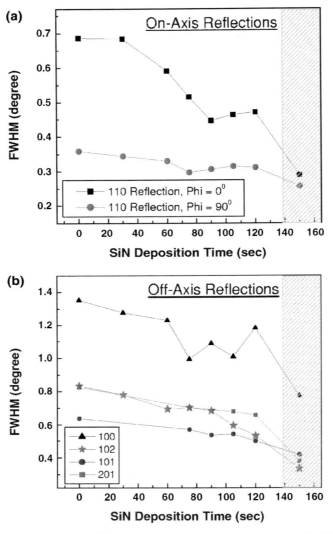

Fig. 5.11. (a) On-axis and (b) off-axis XRC FWHMs of a-plane GaN films grown by MOCVD on r-plane sapphire with different SiNx deposition times. Reprinted with permission from [27]. Copyright (2003), American Institute of Physics

shows XRC FWHMs with respect to in-plane phi angles for the a-plane GaN films grown with different initial growth times at high V/III ratio [28].

Improving crystal quality by using a buffer layer was tried. Wu et al. used AlGaN buffer layer for growing the a-plane GaN films on r-plane sapphire through MOCVD [29]. Depending on the buffer layer thickness, defect density was changed as shown in Fig. 5.14 [29]. These results indicated that studies on

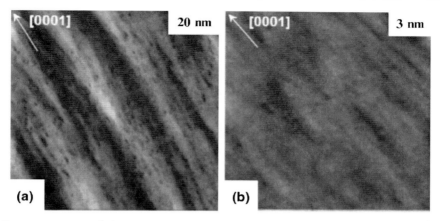

Fig. 5.12. $5 \times 5\,\mu m^2$ AFM micrographs of $2\,\mu m$ thick a-plane GaN film on r-sapphire substrate grown by MOCVD. (**a**) Without the SiN_x interlayer and (**b**) with the 120 s of SiN_x interlayer. Reprinted with permission from [27]. Copyright (2003), American Institute of Physics

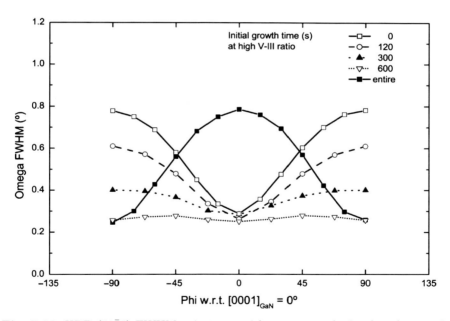

Fig. 5.13. XRD (11$\bar{2}$0) FWHM anisotropy with respect to the in-plane beam orientation. Films are grown at low V/III ratio with an initial 0, 120, 300, and 600 s of high V/III ratio growth. Additionally, a film is shown that was grown entirely at high V/III ratio. Reprinted with permission from [28]. Copyright (2008), American Institute of Physics

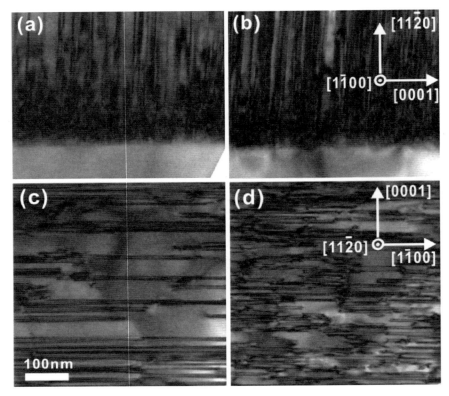

Fig. 5.14. Two-beam bright-field TEM images of a-plane GaN films grown on r-sapphire substrates taken under $g = (1\bar{1}00)$. Cross-sectional images for films grown via (**a**) a 30 nm-thick AlGaN buffer layer, and (**b**) a 90 nm-thick buffer layer. Plan-view images for films grown via (**c**) the 30 nm buffer layer, and (**d**) the 90 nm buffer layer. Magnification is the same for all four images, and corresponds to the scale in (**c**). Reprinted with permission from [29]. Copyright (2008), American Institute of Physics

searching the buffer layer and its optimization are further needed in nonpolar film growth also, as in polar film growth.

5.3.3 Semipolar GaN Films

Semipolar planes are those with a nonzero h, k or i index and a nonzero l index in the (hkil) Miller–Bravais indexing convention. Semipolar planes extend diagonally across the hexagonal unit cell and form an angle with the c-plane other than 90°. Compared to polar (0001) films, semipolar films exhibit reduced polarization effects because the polarization vector is inclined with respect to the growth direction. In addition, the possibility of zero polarization along the growth direction is expected for specific strain states on specific semipolar planes [30]. Baker et al. reported semipolar GaN film growth on m-plane sapphire substrate through HVPE [31]. (10$\bar{1}$3) and (11$\bar{2}$2) semipolar GaN films were grown with in-plane epitaxial relationship of [30$\bar{3}\bar{2}$]GaN//[1$\bar{2}$10]Sapphire and [1$\bar{2}$10]GaN//[0001]Sapphire for the (10$\bar{1}$3) GaN films, and of [11$\bar{2}$1]GaN//[0001]Sapphire and [1$\bar{1}$00]GaN// [1$\bar{2}$10]Sapphire for the (11$\bar{2}$2) GaN films [31]. The (10$\bar{1}$3) films were determined to have N-polarity, while the (11$\bar{2}$2) films had Ga-polarity. On the other hand, (11$\bar{2}$2) semipolar GaN films were grown using the (11$\bar{2}$2) GaN substrates not a sapphire substrate [32]. Although it depends on the planes, growth of semipolar films with improved surface morphology and crystal qualities was possible.

5.4 Lateral Epitaxial Overgrowth of Nonpolar GaN Films

In order to improve the crystal quality of nonpolar GaN films, various overgrowth techniques, similar to those for polar GaN films, have been reported. Lateral epitaxial overgrowth (LEO) was mostly employed for the growth of nonpolar films by overgrowth. As a modified method, sidewall lateral epitaxial overgrowth (SLEO) was also developed. In this section, various LEO techniques for nonpolar GaN film growth to achieve significant reduction of defects are described.

5.4.1 LEO of A-plane GaN

Defect reductions by using the LEO techniques have been reported on nonpolar a-plane GaN film growth, similar to those for polar films. LEO of a-plane GaN films was reported by both MOCVD [32] and HVPE [33]. LEO of a-plane GaN through the wing regions showed significant reduction of extended defects compared to the planar films grown without the LEO technique.

As the crystallographic orientation of the mask stripe openings dictates the facets that form, and hence, the characteristics of lateral overgrowth, as has been shown for laterally overgrown c-plane GaN, Craven et al. investigated orientation dependence of laterally overgrown a-GaN using the SiO_2 mask patterned with an array of rectangular mask openings (windows) which formed a "wagon wheel" design [32]. Figure 5.15 shows SEM and panchromatic cathodoluminescence (CL) images for GaN films grown through the window regions between the SiO_2 masks, where the stripe masks with a width of 5 μm were fabricated along the [0001], [$\bar{1}$101], and [$\bar{1}$100] directions of GaN. The facet formation of the overgrown GaN films was different depending on the stripe directions, as shown in Fig. 5.15a. It was found that the overgrown GaN [0001] stripe had vertical and inclined facets with {1$\bar{1}$00} planes. In the case of the GaN [1$\bar{1}$01] stripe, the overgrown GaN had asymmetric morphologies with one microfaceted vertical (1$\bar{1}$02) sidewall and one inclined (10$\bar{1}$2) sidewall. In

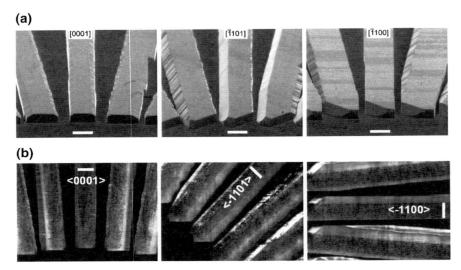

Fig. 5.15. (a) SEM images of stripes oriented parallel to [0001], [$\bar{1}$101], and [$\bar{1}$100] of GaN. (b) Plan-view panchromatic CL images of stripes oriented parallel to [0001], [$\bar{1}$101], and [$\bar{1}$100] of GaN. Images directly correspond to stripes imaged by SEM in (a). Reprinted with permission from [32]. Copyright (2002), American Institute of Physics

contrast, the overgrown GaN [$\bar{1}$100] stripe had rectangular cross-sections with vertical (0001) basal plane sidewalls. Distribution of defects in the overgrown GaN stripes was investigated from the CL images as shown in Fig. 5.15b, where the width and position of window regions are marked by a white bar. In Fig. 5.15b, we can see mottled regions on each GaN overgrown stripe. The mottled regions indicate the region where TDs were extended unimpeded to the top mask of the LEO stripe. The region of lateral epitaxial growth in the [1$\bar{1}$00] stripe was revealed to be relatively free of TDs, while the entire width of the [0001] stripe was covered by the mottled area. The results indicate the efficiency of the stripe mask along the [$\bar{1}$100] direction in defect reduction compared to the stripe mask along the [0001] direction. Additional information obtained from Fig. 5.15b included the lateral overgrowth rates along the $\langle 0001 \rangle$ direction depending on the polarity. The Ga-face shows a higher growth rate than the N-face [32].

The difference in defect reduction, depending on stripe directions, as shown in Fig. 5.15b, was investigated by TEM observations [32]. Figure 5.16 shows cross-sectional TEM images for a-plane LEO GaN with different stripe directions [32]. TD reduction was observed for LEO stripes aligned along [$\bar{1}$100], as shown in the cross-sectional TEM image in Figs. 5.16a, b. Mask blocking was the primary dislocation reduction mechanism because no dislocations were observed to bend in the direction of the lateral overgrowth as shown in

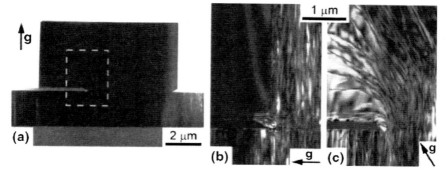

Fig. 5.16. Cross-sectional TEM image of a [1̄100] stripe (**a**) which shows threading dislocation reduction in the asymmetric overgrowth regions. Magnified views of the mask edge region defined by the dashed box on part (**a**) are shown for (**b**) [1̄100] and (**c**) [0001] stripes. Dislocation lines bend from the window region into the overgrowth region for stripes aligned along [0001] while no dislocation bending is observed for [1̄100] stripes. Parts (**a**) through (**c**) are bright-field images with various diffraction conditions of (**a**) **g** = 112̄0, (**b**) **g** = 0006, and (**c**) **g** = 011̄0, respectively. Reprinted with permission from [32]. Copyright (2002), American Institute of Physics

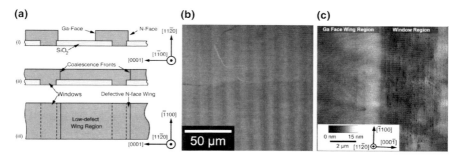

Fig. 5.17. (**a**) Schematics of LEO growth using mask bars oriented along the GaN [1̄100] direction. (i) Cross section of uncoalesced stripes. (ii) Cross section of coalesced stripes showing the offset coalescence front. (iii) Plan view of coalesced stripes identifying large low-defect wing region. (**b**) Nomarski optical contrast micrograph of a coalesced LEO a-plane GaN film. (**c**) 10-μm AFM image of coalesced [1̄100]-oriented GaN stripes. Reprinted with permission from [33]. Copyright (2003), American Institute of Physics

Fig. 5.16b. Unlike the [1̄100] stripes, dislocations propagated into overgrown regions of stripes aligned along the [0001], as shown in Fig. 5.16c.

Similar to the MOCVD grown LEO a-plane GaN, significant defect reduction was reported in the LEO of a-plane GaN prepared by HVPE [33]. Figure 5.17 shows the schematic diagram and surface images of the LEO of a-plane GaN sample prepared by HVPE [33]. As illustrated in Fig. 5.17a,

Ga-face shows higher growth rate than N-face. As in the LEO GaN prepared by MOCVD, the [$\bar{1}$100] stripes were effective in reducing defects. Whereas non LEO a-plane GaN exhibited basal plane SF and TD densities of 10^5 cm^{-1} and 10^9 cm^{-2}, respectively, the overgrown LEO material grown with the [$\bar{1}$100] stripes had essentially less extended defects. The TD and basal plane SF densities in the wing regions were below the $\sim 5 \times 10^6$ cm^{-2} and 3×10^3 cm^{-1}, respectively.

SLEO is proposed as a modified technique [34]. The most important difference between LEO and SLEO is that low defect regions are limited only to the wing regions in the LEO, while the low defect region is possible throughout the material in the SLEO. Figure 5.18 shows the entire procedures of SLEO, including the processing of the patterned template fabrication (Fig. 5.18a) and the growth stages (Fig. 5.18b) [34].

Figure 5.18a schematically shows the main processing steps. The SiO$_2$ mask patterns with stripes along the $\langle 1\bar{1}00 \rangle$ GaN direction are formed on the a-plane GaN template layer through conventional photolithography techniques. The GaN layer is etched through the mask openings, which expose Ga-face (0001) and N-face (000$\bar{1}$) planes on the sidewalls and (11$\bar{2}$0) plane at the bottom of the trenches. Here, it is not necessary that the GaN etching proceeds down to the r-plane sapphire substrate, but the etch depth is chosen such that it is greater than the window width so that there is sufficient space for the material to grow laterally from the sidewalls. The next is the growth stage. There are three main stages during GaN SLEO as shown in Fig. 5.18b [34]. The first, "SIDE" stage involves growth predominantly from the exposed Ga-face sidewall until coalescence with the opposite N-face sidewall. This stage suppresses growth from the bottom of the trench by starving the bottom of precursor species. The N-face forms straight sidewalls with hexagonal hillock features, and the Ga-face grows about 10 times faster than the N-face, confining the SFs only to these slow growing N-face regions. The Ga-face sidewall forms an arrowhead shape with inclined {11$\bar{2}$n} facets. The high lateral growth rates of the Ga-face are favored by growth parameters (high growth temperatures, low pressure, and low V/III ratio). Both the second, "UP" growth mode and the third, "OVER" growth mode require high temperature, low pressure growth, and concurrently progressively decreasing V/III ratio, as the material coalesces and forms a continuous film. For all three stages, high lateral to vertical growth rate ratios are preferred [34]. By employing the SLEO technique in a-plane GaN, they achieved low TD density of $\sim 10^6$–10^7 cm^{-2} throughout the material, which clearly shows the strong advantage of the SLEO compared with the LEO. Note that the LEO techniques can get a TD density of similar or a little higher order but low defect regions are limited to wing regions and not the entire material.

Maskless LEO of a-plane GaN was reported by Lee et al. [35]. They etched a 2 μm-thick a-plane GaN/r-sapphire template to form 4 μm wide seed GaN with a periodicity of 14 μm stripe pattern in the direction of GaN $\langle 1\bar{1}00 \rangle$ by the inductive coupled plasma etcher. Then, a LEO–GaN layer was grown on

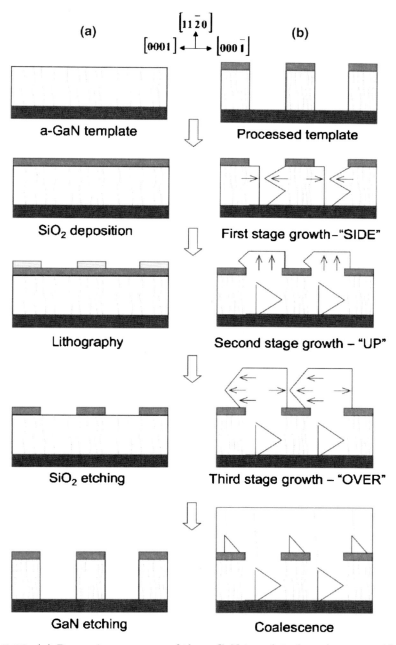

Fig. 5.18. (a) Processing sequence of the a-GaN template for subsequent sidewall lateral overgrowth. (b) Schematic of growth stages of sidewall lateral epitaxial overgrowth (SLEO). Reprinted with permission from [34]. Copyright (2006), American Institute of Physics

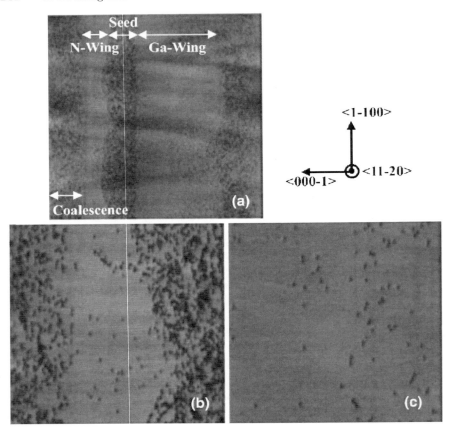

Fig. 5.19. AFM images with $20 \times 20\,\mu m^2$ (**a**) for surface of a-plane InGaN MQWs grown on a-plane maskless LEO GaN template. AFM images ($5 \times 5\,\mu m^2$) of N-face wing region with coalescence boundary (**b**) and Ga-face wing region (**c**). Reprinted with permission from [35]. Copyright (2008), American Institute of Physics

the patterned substrate by using MOCVD. Fully coalesced a-plane GaN with a 10 µm-thickness without the triangle shape surface pits were grown without any mask material, by optimizing growth parameters. Although defect configurations in the GaN films were not reported, the InGaN multi-quantum well (MQW) active layer, which was composed of five period 3 µm-thick InGaN wells and 13 µm-thick GaN barriers, showed effective defect reductions as could be expected in the AFM images shown in Fig. 5.19 [35]. Periodic distribution of surface defects was clearly observed on the surface of a-plane InGaN MQWs/LEO–GaN template as shown in Fig. 5.19. The surface defect density of a-GaN seed region was $\sim 5 \times 10^{10}\,cm^{-2}$, which was similar to that of a-plane GaN/r-sapphire template. The surface defect density of the two low defective wing regions was $\sim 1 \times 10^7\,cm^{-2}$, which was much lower than that of the seed GaN region. Surface defects are believed to be closely related mainly with TDs and SFs.

Here it should be noted that there are different reports on the ratio of growth rates for Ga-face and N-face growth in the $\langle 0001\rangle$ direction of LEO a-plane GaN, which could have resulted from differences in growth parameters, although higher growth rates for the Ga-face direction were always observed. Faster growth rates of Ga-face compared to N-face with factors as large as ~ 10 [32, 34] and ~ 4 [35] were reported for the MOCVD, and a factor of ~ 6 [33] for the HVPE was reported. Slower growth rate in N-face and faster growth rate in Ga-face effectively confined the basal plane SFs in the N-face wing.

5.4.2 LEO of M-plane GaN

The LEO technique was also applied to m-plane GaN growth by HVPE and successful reduction of defect density was achieved [36]. Figure 5.20 shows schematic diagrams and SEM images for LEO m-plane GaN with $\langle 0001\rangle$ and $\langle 11\bar{2}0\rangle$ stripes [36]. Figures 5.20a, b show the LEO GaN with the $\langle 0001\rangle$ stripes. Cross-sectional SEM images revealed that there were five m-plane faces rotated about the $\langle 0001\rangle$ axis. Figures 5.20c, d show the LEO GaN with the $\langle 11\bar{2}0\rangle$ stripes. In this case, the LEO GaN showed vertical (0001) and (000$\bar{1}$) sidewalls and rectangular cross-section. The ratio of the lateral growth rates of the (0001) and (000$\bar{1}$) wings in the m-plane GaN LEO stripe showed about $2 \sim 4$, while it also changed from sample to sample. Such a non-uniform growth rate is believed to have resulted from local variations in the V/III ratio [36].

Defect configurations of LEO m-plane GaN with different stripe orientations were very different. At first, in the case of TDs, TD densities of window regions were $\sim 6 \times 10^8 \, \text{cm}^{-2}$ and $\sim 4 \times 10^9 \, \text{cm}^{-2}$, respectively, for the $\langle 0001\rangle$ stripes and $\langle 11\bar{2}0\rangle$ stripes. The lower TD density for the $\langle 0001\rangle$ stripes is understood to be the result of the bending of TDs from window regions to wing regions, which was promoted by the presence of m-plane facets. Such bending of TDs to the wing regions from the window regions is not favorable for the vertical sidewalls of the LEO GaN with the $\langle 11\bar{2}0\rangle$ stripes, which resulted in the higher TD density in the window regions. In the lateral overgrown wing regions, the TD densities were $\sim 5 \times 10^6 \, \text{cm}^{-2}$ for both stripes [36]. However, the SF density for the overgrown wing regions for both stripes was significantly different [36]. In the case of $\langle 0001\rangle$ stripes, both overgrown wing and window regions were faulted throughout. As the lateral growth direction was within the basal plane in the case of the LEO GaN with $\langle 0001\rangle$ stripes, the basal SFs could propagate unimpeded into the overgrown wing regions, which resulted in comparable average SF densities for wing and window regions, as illustrated in Fig. 5.20a. On the other hand, in the case of the LEO GaN with the $\langle 11\bar{2}0\rangle$ stripes, the windows and N-face (000$\bar{1}$) wings exhibited basal plane SFs with an average density of $\sim 1 \times 10^5 \, \text{cm}^{-1}$, while Ga-face (0001) wings had an average SF density below $3 \times 10^3 \, \text{cm}^{-1}$. As lateral

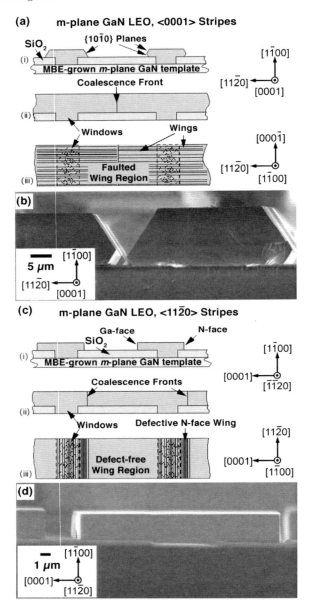

Fig. 5.20. Schematic representation and corresponding cross-sectional scanning electron micrographs of growth evolution for (**a**), (**b**) ⟨0001⟩- and (**c**), (**d**) ⟨11$\bar{2}$0⟩-oriented LEO stripes. Reprinted with permission from [36]. Copyright (2005), American Institute of Physics

growth proceeded normally to the basal planes in the case of the LEO with the ⟨11$\bar{2}$0⟩ stripes, SFs in the overgrown GaN could be present only in these regions [36].

5.5 Growth of Nonpolar ZnO Films

A-plane ZnO films are mostly grown on r-plane sapphire substrates without the mixture of other planes. Compared with the pure a-plane ZnO films, which are easily grown on r-plane sapphire substrates, m-plane ZnO films on m-plane sapphire substrates showed mixed planes of (0002) and (01$\bar{1}$3) in addition to (10$\bar{1}$0) m-plane depending on the growth conditions. Pure (10$\bar{1}$0) m-plane ZnO films have been grown on LiAlO$_2$ substrates similar to the m-plane GaN films. LEO ZnO films have not been reported yet. In this section, the growth of nonpolar ZnO films is briefly described.

5.5.1 A-plane ZnO Films

A-plane ZnO films have been grown mostly on r-plane sapphire substrates using pulsed laser deposition (PLD) [37], MOCVD [38], and PAMBE [39, 40]. Figure 5.21 shows RHEED patterns for r-plane sapphire and a-plane ZnO film on it [39]. From the RHEED patterns, the in-plane orientation relationship (epitaxial relationship) was determined to be [$\bar{1}$101]Al$_2$O$_3$//[0001]ZnO and [11$\bar{2}$0]Al$_2$O$_3$//[$\bar{1}$100]ZnO. The RHEED pattern of Fig. 5.21c shows a roughly v-shaped spot pattern, which implies a faceted surface. The chevron-like spot pattern is more clear in the initial stage of growth [39].

The epitaxial relationship can also be determined by XRD or by TEM diffraction patterns. Figures 5.22a and b shows XRD results for θ−2θ and ϕ scan of the ZnO film and substrate, respectively. As shown in Fig. 5.22a, ZnO (11$\bar{2}$0) reflection at 56.602°, and sapphire (1$\bar{1}$02), (2$\bar{2}$04), (3$\bar{3}$06) reflections are detected, which means that the grown ZnO film is nonpolar a-plane ZnO, and the epitaxial relationship is (1$\bar{1}$02)Al$_2$O$_3$//(11$\bar{2}$0)ZnO. In order to determine in-plane orientation relationship by XRD, off-axis diffraction peaks are measured. Figure 5.22b shows ϕ scans for ZnO (10$\bar{1}$0), (10$\bar{1}$1), and sapphire (2$\bar{1}$$\bar{1}$0) reflections. The ϕ scan for ZnO (10$\bar{1}$0) shows peaks separated by 180° as seen in Fig. 5.22b, which indicates the twofold symmetry of (10$\bar{1}$0) planes in a-plane (11$\bar{2}$0). The observed twofold symmetry of the (10$\bar{1}$0) planes indicates that there is no mixed domain. In Fig. 5.22b, the zone axis of the ZnO (10$\bar{1}$1) plane and (11$\bar{2}$0) is [$\bar{1}$101]ZnO and that of the Al$_2$O$_3$ (2$\bar{1}$$\bar{1}$0) plane and (1$\bar{1}$02) is [02$\bar{2}$1] Al$_2O_3$. Therefore, the epitaxial relationship is (1$\bar{1}$02) Al$_2$O$_3$//(11$\bar{2}$0)ZnO and [02$\bar{2}$1]Al$_2$O$_3$//[$\bar{1}$101]ZnO, which is consistent with the epitaxial relationship determined by RHEED [39].

Figures 5.23a and b show selected area diffraction (SAD) patterns at interfacial regions obtained by TEM along the ZnO[0001]//Al$_2$O$_3$[$\bar{1}$101] zone axis and the ZnO[$\bar{1}$100]//Al$_2$O$_3$[11$\bar{2}$0] zone axis [39]. From this, we know

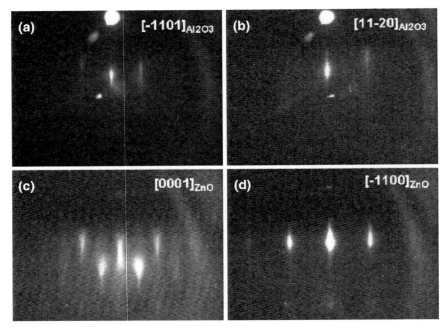

Fig. 5.21. RHEED patterns for r-plane Al$_2$O$_3$ substrates (**a, b**) after thermal cleaning and ZnO films grown for 1 h (**c, d**). Patterns (**a**) and (**c**) were acquired from the same sample position to the incident electron beam and patterns (**b**) and (**d**) were acquired by rotating the position 90° from (**a**) and (**c**). Determined azimuth directions are shown on each pattern, which shows the in-plane orientation relationship (epitaxial relationship) of $[\bar{1}101]$Al$_2$O$_3$//$[0001]$ZnO and $[11\bar{2}0]$Al$_2$O$_3$//$[\bar{1}100]$ZnO. Reprinted with permission from [39]. Copyright (2007), Elsevier

the epitaxial relationship of ZnO$[0001]$//Al$_2$O$_3$$[\bar{1}101]$ and ZnO$[\bar{1}100]$//Al$_2$O$_3$ $[11\bar{2}0]$ in addition to $(11\bar{2}0)$ZnO//$(1\bar{1}02)$Al$_2$O$_3$ as can be seen in Fig. 5.23a. Figure 5.23c shows a schematic drawing of the relative crystallographic directions between the a-plane ZnO and the r-plane Al$_2$O$_3$ [39].

The surface morphology of the a-plane ZnO films on r-plane sapphire substrates is anisotropic with the slate surface morphology as shown in Fig. 5.24 [39]. The average width of the slates is smaller for the thinner film. Here it should be noted that the direction of the slates is parallel to the $\langle 0001 \rangle$ ZnO direction similar to the a-plane GaN film [27], while the slate direction is perpendicular to the $\langle 0001 \rangle$ direction in the case of m-plane GaN films [16,17].

Defects in a-plane ZnO films on r-plane sapphire substrates are revealed to be misfit dislocations at the interface, basal SFs, and TDs in the films by comprehensive investigations through TEM [41]. Figure 5.25 shows high resolution TEM (HRTEM) micrographs. Zone axes of Fig. 5.25a and c are ZnO $\langle 0001 \rangle$ and ZnO $\langle \bar{1}100 \rangle$, respectively. Regularly spaced misfit dislocations are clearly visible as marked by arrows in Fig. 5.25b, which is a

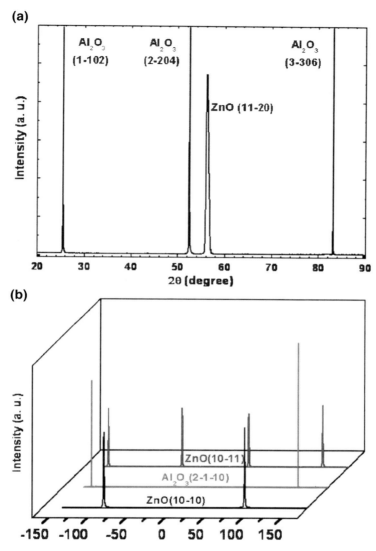

Fig. 5.22. (a) XRD θ–2θ scan and (b) φ scans for the ZnO (10$\bar{1}$0), (10$\bar{1}$1), and Al$_2$O$_3$ (2$\bar{1}\bar{1}$0) reflections from the a-plane ZnO film grown on r-plane sapphire substrate by PAMBE, which shows single crystalline pure a-plane ZnO film and the epitaxial relationship of (1$\bar{1}$02) Al$_2$O$_3$//(11$\bar{2}$0)ZnO and [02$\bar{2}$1]Al$_2$O$_3$//[$\bar{1}$101]ZnO. Reprinted with permission from [39]. Copyright (2007), Elsevier

Fourier-filtered image of Fig. 5.25a. The in-plane translational period of ZnO along the [1$\bar{1}$00] is $\sqrt{3}a_{ZnO} = 5.629$ Å, while the one for the Al$_2$O$_3$ [$\bar{1}$1$\bar{2}$0] direction is $a_{Al_2O_3} = 4.758$ Å. Therefore, the lattice misfit (δ) along the ZnO ⟨1$\bar{1}$00⟩ direction is calculated to be 0.183 (i.e., 18.3%), which give us the

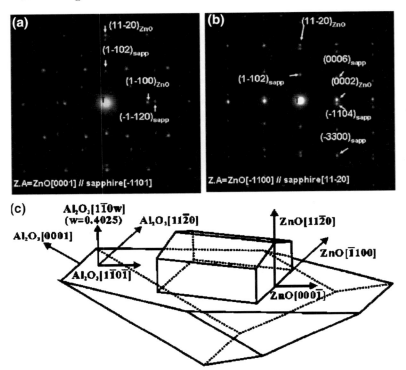

Fig. 5.23. TEM SAD patterns at the ZnO/Al$_2$O$_3$ interfaces viewed with the zone axes of (a) ZnO[0001]//Al$_2$O$_3$ [$\bar{1}\bar{1}01$] and (b) ZnO[$\bar{1}$100]//Al$_2$O$_3$[11$\bar{2}$0]. (c) Schematic drawing for the relative crystallographic directions of a-plane ZnO and r-plane Al$_2$O$_3$ based on determined epitaxial relationship. Reprinted with permission from [39]. Copyright (2007), Elsevier

spacing (D) of misfit dislocations as 1.3 nm using the equation that describes geometrically expected spacing of misfit dislocations D = |**b**| /δ, where |**b**| is the magnitude of the Burgers vector component parallel to the interface, i.e., 2.379 Å. This value is almost similar to the observed spacing of misfit dislocations in Fig. 5.25b. On the other hand, in the case of orthogonal direction, the HRTEM micrograph of Fig. 5.25c, which is taken from the ZnO ⟨1$\bar{1}$00⟩ zone axis, shows that ZnO planes are consecutively connected with sapphire planes at the interface. The translational period of ZnO along the [0001] direction is c$_{ZnO}$ = 5.207 Å, while that for the Al$_2$O$_3$ [$\bar{1}$101] direction is $\sqrt{3a^2_{Al_2O_3} + c^2_{Al_2O_3}}$ = 15.384 Å, which gives us the lattice misfit of 1.54%, considering the domain matching epitaxy [42]. Due to such small lattice misfits along the ZnO ⟨0001⟩ direction, the misfit dislocations are rarely observed from this zone axis as shown in Fig. 5.25d.

In addition to misfit dislocations at the a-ZnO/r-sapphire interfaces, SFs and TDs are observed in the films. Figure 5.26 shows plan-view TEM bright

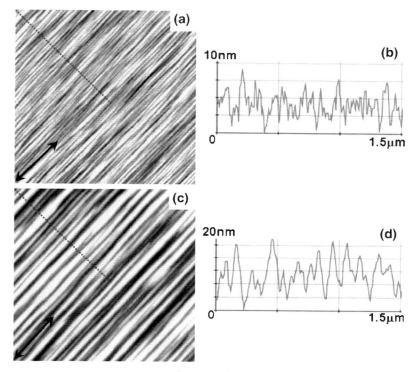

Fig. 5.24. AFM images ($2 \times 2\,\mu m^2$) for ZnO films with different film thicknesses of (**a**) 90 and (**c**) 300 nm. Surface morphology is anisotropic with the slate surface morphology. *Arrow* on the image indicates the $\langle 0001 \rangle$ ZnO direction. Line scan results for the (**a**) and (**c**) performed on *dotted lines* on the images are displayed in (**b**) and (**d**), respectively. Average width of slates are smaller for the thinner film. Reprinted with permission from [39]. Copyright (2007), Elsevier

field (BF) micrographs of the same regions but taken under different two-beam diffraction conditions [41]. Depending on the **g** vectors, the observed defects are very different. In Fig. 5.26a with **g** = 0002, TDs are observed without the appearance of SFs. Mostly SFs and a few TDs are observed with the **g** = $1\bar{1}00$ two-beam condition as shown in Fig. 5.26b. Most of the TDs in Fig. 5.26a have the Burgers vector of $\langle 0001 \rangle$, while TDs in Fig. 5.26b have the Burgers vector of $1/3\langle 11\bar{2}0 \rangle$ [41]. Dislocation densities in Figs. 5.26a, b are estimated to be $\sim 7.3 \times 10^{10}$ and $\sim 6.1 \times 10^9$ cm^{-2}, which correspond to the dislocations with the Burgers vectors of $\langle 0001 \rangle$ and of $1/3\langle 11\bar{2}0 \rangle$, respectively [41]. SFs in Fig. 5.26b is determined to be a type-I1 intrinsic stacking fault with a displacement vector of $1/6\langle 0\bar{2}\bar{2}3 \rangle$ based on additional HRTEM observation (Fig. 5.27). The stacking sequence of ZnO (0002) planes in the faulted region are determined to be AB˙ABC'BCBC as shown in Fig. 5.27b. This SF is the type-I1 intrinsic

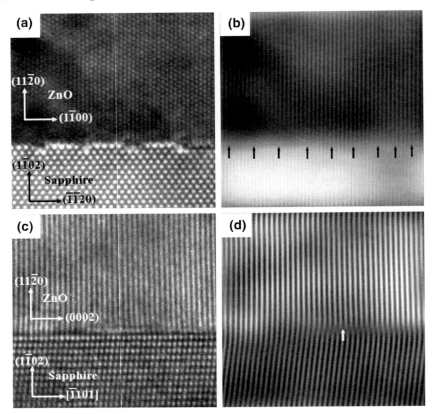

Fig. 5.25. HRTEM micrographs at the interfaces between nonpolar a-plane ZnO (11$\bar{2}$0) films and r-plane Al$_2$O$_3$ (1$\bar{1}$02) substrates. (**a**) ZnO [0 0 01] zone axis HRTEM micrograph, (**b**) Fourier-filtered image corresponding to the image in (**a**), (**c**) ZnO [$\bar{1}$100] zone axis HRTEM micrograph, and (**d**) Fourier-filtered image corresponding to image in (**c**). Misfit dislocations at interfaces are marked by arrows. Reprinted with permission from [41]. Copyright (2008), Elsevier

stacking fault with bounding Frank partial dislocations at the end of the SF. From Fig. 5.26b, SF density is estimated to be $\sim 1.2 \times 10^5\,\mathrm{cm}^{-1}$ [41].

Cross-sectional TEM observations clearly show TDs configurations in the a-plane ZnO film on r-plane sapphire. Figure 5.28a is a cross-sectional TEM BF micrograph of the a-plane ZnO film taken from the ZnO [$\bar{1}$100] zone axis, while Figs. 5.28b and c reveal the same regions of Fig. 5.28a but imaged under two-beam conditions with diffraction **g** vectors of $\mathbf{g} = \langle 0002 \rangle$ and $\langle 11\bar{2}0 \rangle$, respectively [41]. The TDs observed in Fig. 5.28b are determined to be the ones with Burgers vector of $\langle 0001 \rangle$ or $1/3\langle 11\bar{2}3 \rangle$, while the TDs in Fig. 5.28c should have Burgers vector of $1/3\langle 11\bar{2}0 \rangle$ or $1/3\langle 11\bar{2}3 \rangle$ based on visible and invisible criteria of dislocations [41]. Therefore, the

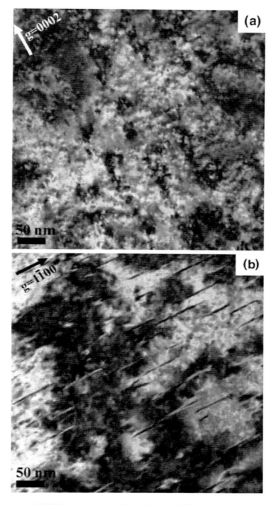

Fig. 5.26. Plan-view TEM micrographs of the 240 nm-thick a-plane ZnO film on r-plane sapphire. (**a**) g = 0002 and (**b**) g = 1–100 two-beam bright field micrographs. Perfect threading dislocations with a Burgers vector of $\langle 0001 \rangle$ were dominantly observed in (**a**), while stacking faults were observed in (**b**). Reprinted with permission from [41]. Copyright (2008), Elsevier

dislocations appearing at the same positions in Figs. 5.28b and c correspond to TDs with Burgers vector of $1/3\langle 11\bar{2}3\rangle$. These dislocations are marked by arrows in Fig. 5.28c; unmarked dislocations observed in Fig. 5.28c correspond to dislocations with Burgers vector of $1/3\langle 11\bar{2}0\rangle$. Here it should be noted that a great number of TDs are shown in Fig. 5.28b compared with Fig. 5.28c,

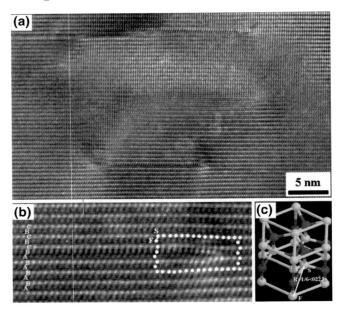

Fig. 5.27. (a) Plan-view HRTEM micrograph of the a-plane ZnO film on r-plane sapphire. (b) Enlarged HRTEM micrograph of one of the stacking faults in (a). The stacking fault was determined to be a type-I1 intrinsic stacking fault having stacking sequence of (AB'ABC'BC) with bounding Frank partial dislocation with the Burgers vector of $1/6\,[02\bar{2}3]$. (c) Magnitude and direction of displacement vector of $1/6[02\bar{2}3]$ in a ZnO crystal are shown. Reprinted with permission from [41]. Copyright (2008), Elsevier

and only a few TDs are observed at the same locations in both Figs. 5.28b and c. This means that almost all TDs in Fig. 5.28b, i.e., almost all TDs in the a-plane ZnO film have Burgers vector of $\langle 0001 \rangle$ [41].

5.5.2 M-plane ZnO Films

M-plane ZnO films have been grown by using MOCVD [43–45], PLD [46], and PAMBE [47], where m-plane sapphire [43,44], (100) $LiAlO_2$ [45], m-ZnO [46], and (001) MgO [47] substrates were used.

Moriyama and Fujita reported m-plane ZnO growth on m-plane sapphire substrates [43]. They found that both higher growth temperature (800°C) and higher VI/II ratio (2.8×10^4) were needed in the growth of pure m-plane ZnO, while lower growth temperature (500°C) or lower VI/II ratio (5.4×10^3) resulted in the formation of $(01\bar{1}3)$ planes in addition to the $(01\bar{1}0)$ m-planes. The in-plane orientation relationship (epitaxial relationship) was found to be $[0001]ZnO//[2\bar{1}\bar{1}0]Al_2O_3$ and $[\bar{2}110]ZnO//[0001]Al_2O_3$ [43], which gives us the lattice misfit of 7.8 and 9.5%, respectively.

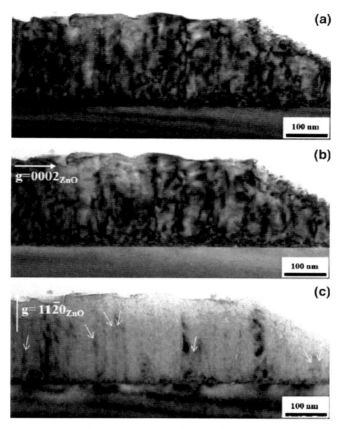

Fig. 5.28. Cross-sectional TEM micrographs of the a-plane ZnO film on r-plane sapphire taken near the ZnO [−1100] zone axis. (**a**) Zone axis BF micrograph, (**b**) **g** = 0002, and (**c**) **g** = 11$\bar{2}$0 two-beam BF micrographs. Marked threading dislocations by arrows and unmarked ones in (**c**) are dislocations with the Burgers vector of 1/3⟨11$\bar{2}$3⟩ and 1/3⟨11$\bar{2}$0⟩, respectively. A great many number of dislocations were observed in (**b**). Almost all threading dislocations in (**b**) have the Burgers vector of ⟨0001⟩. Reprinted with permission from [41]. Copyright (2008), Elsevier

Similar to the m-plane GaN growth, a (100) LiAlO$_2$ (γ-LiAlO$_2$) substrate is used to grow pure m-plane ZnO films [45]. γ-LiAlO$_2$ has a tetragonal structure where the a–c (100) plane has the same atomic arrangement as the (10$\bar{1}$0) plane of wurtzite structure [45]. The reported in-plane orientation relationship (epitaxial relationship) between ZnO and LiAlO$_2$ is [001]LiAlO$_2$//[11$\bar{2}$0]ZnO and [010]LiAlO$_2$//[0001]ZnO, which gives us the lattice misfit of 3.47 and 2.71%, respectively [45].

Matsui and Tobata reported homoepitaxial growth of m-plane ZnO films on m-plane ZnO substrates by using PLD [46]. Surface morphology showed the slate morphology, where they called as the slates nanostripes. Figure 5.29a

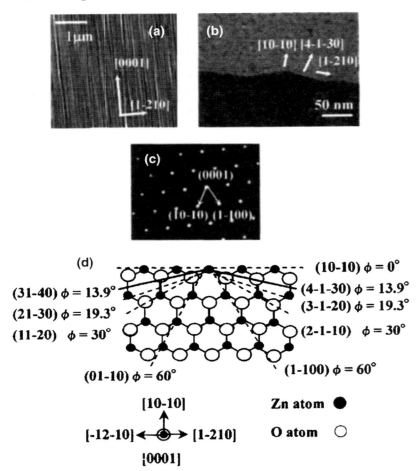

Fig. 5.29. (a) AFM image (5 × 5 μm²) of the m-plane ZnO film on m-plane ZnO (10$\bar{1}$0) substrate showing nanostripe arrays, (b) Cross-sectional TEM image of nanostripe arrays with the ⟨0001⟩ zone axis, (c) Electron diffraction pattern of nanostripe arrays, and (d) schematic cross section of the ZnO lattice viewed along the [0001] direction. Angle between each plane with (10$\bar{1}$0) plane is indicated by the index of the plane. Reprinted with permission from [46]. Copyright (2008), American Institute of Physics

shows the AFM image of the top surface of the m-ZnO film (220 nm-thick) on ZnO (10$\bar{1}$0) substrate [46]. The image revealed a nanostripe array running along the ⟨0001⟩ direction. This direction is orthogonal to the slates observed from m-GaN films, where the slates runs along the ⟨11$\bar{2}$0⟩ direction [16, 17]. Figures 5.29b and c show cross-sectional TEM image and electron diffraction pattern, revealing that nanostripe arrays are elongated along the ⟨0001⟩ direction of hexagonal ZnO. The plane angle of the side facets and (10$\bar{1}$0) planes

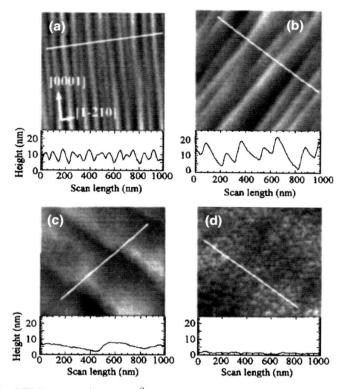

Fig. 5.30. AFM images ($1 \times 1\,\mu m^2$) and cross-sectional profiles of ZnO layers on vicinal ZnO ($10\bar{1}0$) substrates with (**a**) 0°-, (**b**) 5°-, (**c**) 10°-, and (**d**) 15°-off degrees from the ($10\bar{1}0$)-oriented surface toward the ($2\bar{1}\bar{1}0$)-oriented surface. All stripe arrays were elongated along the $\langle 0001 \rangle$ ZnO direction. Reprinted with permission from [46]. Copyright (2008), American Institute of Physics

is 14°, which is consistent with the ($4\bar{1}30$) and ($31\bar{4}0$) equivalent planes having a misorientation of 13.9° along the [$1\bar{2}10$] and [$\bar{1}2\bar{1}0$] directions from the ($10\bar{1}0$) surface, respectively, as shown in Fig. 5.29d [46]. Nanostripe arrays are triangular-shaped in cross-section and the two side-bonding facets are composed of high-index planes.

The facets observed in Fig. 5.29a are weakened by using vicinal substrates as shown in Fig. 5.30 [46]. The m-plane ZnO film shows a smooth surface with roughness below 2 nm when the off-angle of the substrate reaches 15°. The off-angle of 15° is close to that of inclination of nanostripe arrays consisting of the high index ($4\bar{1}30$) plane and suppressing the development of self-faceting in relation to the formation of the nanostripe arrays [46].

M-plane ZnO film growth on (001) MgO substrate has been reported by using PAMBE [47]. Although a small mixture of (0002) planes is detected from the θ−2θ XRD, the growth of nonpolar ZnO films on cubic substrate is interesting. Furthermore, another interesting finding is that such m-plane ZnO films on MgO substrates were grown by the PAMBE, while the PLD leads the polar c-plane ZnO growth.

References

1. S.K. Hong, T. Hanada, H.J. Ko, Y. Chen, T. Yao, D. Imai, K. Araki, M. Shinohara, Phys. Rev. B **65**, 115331 (2002)
2. F. Bernardini, V. Fiorentini, D. Vanderbilt, Phys. Rev. B **56**, R10024 (1997)
3. T. Takeuchi, S. Sota, M. Katsuragawa, M. Komori, H. Takeuchi, H. Amano, I. Akasaki, Jpn. J. Appl. Phys. **36**, L382 (1997)
4. C. Wetzel, T. Takeuchi, H. Amano, I. Akasaki, Phys. Rev. B **62**, R13302 (2000)
5. T. Paskova, Phys. Stat. Sol. B, 245, 1011 (2008)
6. J.E. Northrup, L.T. Romano, J. Neugebauer, Appl. Phys. Lett. **74**, 2319 (1999)
7. J.E. Northrup, J. Neugebauer, Phys. Rev. B **53**, R10477 (1996)
8. B. Meyer, D. Mark, Phys. Rev. B. **67**, 035403 (2003)
9. T. Deguchi, K. Sekiguchi, A. Nakamura, T. Sota, R. Matsuo, S. Chichibu, S. Nakamura, Jpn. J. Appl. Phys. **38**, L914 (1999)
10. P. Waltereit, O. Brandt, A. Trampert, H.T. Grahn, J. Menniger, M. Ramstiner, M. Reiche, K.H. Ploog, Nature **406**, 865 (2000)
11. M. Sano, M. Aoki, Jpn. J. Appl. Phys. **15**, 1943 (1976)
12. E.S. Hellman, Z. Liliental-Weber, D.N.E. Buchanan, MRS Internet J. Nitride Res. **2**, Article 30 (1997)
13. X. Ke, X. Jun, D. Peizhen, Z. Yongzong, Z. Guoqing, Q. Rongsheng, F. Zujie, J. Cryst. Growth **193**, 127 (1998)
14. Y.J. Sun, O. Brandt, K.H. Ploog, J. Vac. Sci. Technol. B **21**, 1350 (2003)
15. B.A. Haskell, S. Nakamura, S.P. DenBaars, J.S. Speck, Phys. Stat. Sol. B **244**, 2847 (2007)
16. P. Waltereit, O. Brandt, M. Ramsteiner, R. Uecker, P. Reiche, K.H. Ploog, J. Cryst. Growth **218**, 143 (2000)
17. R. Armitage, H. Hirayama, Appl. Phys. Lett. **92**, 092121 (2008)
18. A. Hirai, B.A. Haskell, M.B. McLaurin, F. Wu, M.C. Schmidt, K.C. Kim, T.J. Baker, S.P. DenBaars, S. Nakamura, J.S. Speck, Appl. Phys. Lett. **90**, 121119 (2007)
19. C.Q. Chen, M.E. Gaevski, W.H. Sun, E. Kuokstis, J.P. Zhang, R.S.Q. Fareed, H.M. Wang, J.W. Yang, G. Simin, M.A. Khan, H.P. Maruska, D.W. Hill, M.M.C. Chou, B. Chai, Appl. Phys. Lett. **81**, 3149 (2002)
20. A. Hirai, Z. Jia, M.C. Schmidt, R.M. Farrell, S.P. DenBaars, S. Nakamura, J.S. Speck, K. Fujito, Appl. Phys. Lett. **91**, 191906 (2007)
21. S.F. Chichibu, H. Yamaguchi, L. Zhao, M. Kubota, K. Okamoto, H. Ohta, Appl. Phys. Lett. **92**, 091912 (2008)
22. M. McLaurin, T.E. Mates, F. Wu, J.S. Speck, J. Appl. Phys. **100**, 063707 (2006)
23. Q. Sun, S.Y. Kwon, Z. Ren, J. Han, T. Onuma, S.F. Chichibu, S. Wang, Appl. Phys. Lett. **92**, 051122 (2008)
24. A. Kobayashi, S. Kawano, Y. Kawaguchi, J. Ohta, H. Fujioka, Appl. Phys. Lett. **90**, 041908 (2007)
25. M.D. Craven, S.H. Lim, F. Wu, J.S. Speck, S.P. DenBaars, Appl. Phys. Lett. **81**, 469 (2002)

26. B.A. Haskell, F. Wu, S. Matsuda, M.D. Craven, P.T. Fini, S.P. DenBaars, J.S. Speck, S. Nakamura, Appl. Phys. Lett. **83**, 1554 (2003)
27. A. Chakraborty, K.C. Kim, F. Wu, J.S. Speck, S.P. DenBaars, U.K. Mishra, Appl. Phys. Lett. **89**, 041903 (2003)
28. J.L. Hollander, M.J. Kappers, C. McAleese, C.J. Humphreys, Appl. Phys. Lett. **92**, 101104 (2008)
29. Z.H. Wu, A.M. Fischer, F.A. Ponce, T. Yokogawa, S. Yoshida, R. Kato, Appl. Phys. Lett. **93**, 011901 (2008)
30. T. Takeuchi, H. Amano, I. Akasaki, Jpn. J. Appl. Phys. **39**, 413 (2000)
31. T.J. Baker, B.A. Haskell, F. Wu, J.S. Speck, S. Nakamura, Jpn. J. Appl. phys. **45**, L154 (2006)
32. M.D. Craven, S.H. Lim, F. Wu, J.S. Speck, S.P. DenBaars, Appl. Phys. Lett. **81**, 1201 (2002)
33. B.A. Haskell, F. Wu, M.D. Craven, S. Matsuda, P.T. Fini, T. Fujii, K. Fujito, S.P. DenBaars, J.S. Speck, S. Nakamura, Appl. Phys. Lett. **83**, 644 (2003)
34. B.M. Imer, F. Wu, S.P. DenBaars, J.S. Speck, Appl. Phys. Lett. **88**, 061908 (2006)
35. S.N. Lee, H.S. Paek, H. Kim, Y.M. Park, T. Jang, Y. Park, Appl. Phys. Lett. **92**, 111106 (2008)
36. B.A. Haskell, T.J. Baker, M.B. McLaurin, F. Wu, P.T. Fini, S.P. DenBaars, J.S. Speck, S. Nakamura, Appl. Phys. Lett. **86**, 111917 (2005)
37. T. Koida, S.F. Chichibu, A. Uedono, T. Sota, A. Tsukazaki, M. Kawasaki, Appl. Phys. Lett. **84**, 1079 (2004)
38. Y. Zhang, G. Du, H. Zhu, C. Hou, K. Hunag, S. Yang, Opt Mater, **27**, 399 (2004)
39. S.K. Han, S.K. Hong, J.W. Lee, J.Y. Lee, J.H. Song, Y.S. Nam, S.K. Chang, T. Minegishi, T. Yao, J. Cryst. Growth **309**, 121 (2007)
40. Y.S. Nam, S.W. Lee, K.S. Baek, S.K. Chang, J.H. Song, J.H. Song, S.K.Han, S.K. Hong, T. Yao, Appl. Phys. Lett. **92**, 201907 (2008)
41. J.W. Lee, S.K. Han, S.K. Hong, J.Y. Lee, T. Yao, J. Cryst. Growth, **310**, 4102 (2008)
42. T. Zheleva, K. Jagannadham, J. Narayan, J. Appl. Phys. **75**, 860 (1994)
43. T. Moriyama, S. Fujita, Jpn. J. Appl. Phys. **44**, 7919 (2005)
44. J.Z. Perez, V.M. Sanjose, E.P. Lidon, J. Colchero, Appl. Phys. Lett. **88**, 261912 (2006)
45. M.M.C. Chou, L. Chang, H.Y. Chung, T.H. Huang, J.J. Wu, C.W. Chen, J. Cryst. Growth **308**, 412 (2007)
46. H. Matsui, H. Tabata, Appl. Phys. Lett. **87**, 143109 (2005)
47. E. Cagin, J. Yang, W. Wang, J.D. Phillips, S.K. Hong, J.W. Lee, J.Y. Lee, Appl. Phys. Lett. **92**, 233505 (2008)

6

Structural Defects in GaN and ZnO

S.-K. Hong and H.K. Cho

Abstract. Characteristics of the various kinds of structural defects in GaN and ZnO films and advances in techniques to reduce a dislocation density are presented. Heteroepitaxial GaN and ZnO films have various kinds of structural defects including misfit dislocations, threading dislocations, stacking faults, nanopipes, and inversion domain boundary, which inevitably affect the properties of the films and devices. Transmission electron microscopy of these defects is described in addition to discussions on basic characteristics of the defects. Many different technical approaches to reduce a dislocation density have been reported. Detailed techniques including epitaxial lateral over growth and employment of buffers that lead to the growth of high quality films with a low dislocation density are discussed.

6.1 Introduction

After the first single crystalline, colorless GaN growth by vapor–phase growth, which is very similar to the current hydride vapor phase epitaxy (HVPE), in 1969 [1], growth of high quality GaN films by metal organic chemical vapor deposition (MOCVD) were succeeded by using the low temperature AlN nucleation layer in 1986 [2] or by using the low temperature (LT) GaN nucleation layer in 1991 [3]. Although there is a report on defect evolution from a single crystalline GaN film by the etch pit formation in 1976 [4], results on direct observation and characterization of various defects in epitaxial GaN films were explosively reported in 1995 [5–11]. In 1995, various kinds of defects such as threading dislocation (TD), stacking fault (SF), inversion domain boundary (IDB), and nanopipe were found and/or characterized in details.

GaAs-based light emitting diodes (LEDs) and laser diodes (LDs) achieved their long operating lifetimes by reducing the dislocation density of crystal. Most optoelectronic devices malfunction with dislocations because these defects cause rapid recombination of holes with electrons without conversion of their available energy into photons, i.e., nonradiative recombination, which causes heating up of the crystal. However, the blue and green InGaN quantum-well structure LEDs with luminous efficiencies of 5 and 30 lumens per watt, respectively, could be made despite the large number of TDs (1×10^8–1×10^{12} cm^{-2}) until 1997 [12]. In fact, at that time the research society on compound semiconductors were surprised at the high density of dislocations of the order of 10^{10} cm^{-2} in operating high brightness GaN based LEDs and thought that the TDs do not act as efficient nonradiative recombination sites in nitride materials [13]. However, the dislocation was a nonradiative recombination site but it was just inefficient. Necessity of GaN films with low dislocation density was clearly demonstrated by Nakamura et al. [14]. Nakamura et al. demonstrated the LD with a lifetime of 3000 h and an estimated lifetime of over 10,000 h, which was surprisingly improved from their first continuous-output GaN blue light semiconductor laser with a lifetime of only 27 h. Such a dramatic improvement of lifetime was attributed by two technological advances at that time: the one is employment of modulation-doped superlattices instead of the thick AlGaN layers and the other was fabricating the LDs onto epitaxially laterally overgrown (ELOG) GaN substrates, which have significantly low dislocation density compared with the GaN films on normal substrates. In fact, although the dislocations in GaN materials are inefficient nonradiative centers compared with the traditional compound semiconductors, various kinds of processes and buffers have been developed to obtain high quality GaN and related materials with low dislocation density, which are inevitably necessary to fabricate optoelectronic devices with long lifetimes and high performances.

In case of ZnO materials, after the reports on lasing operations from single crystalline ZnO films [15, 16], characterizations of various defects in epitaxial ZnO films were reported [17–20]. Because of the similarity in the structural properties of ZnO with GaN, the same kinds of structural defects and behaviors that were observed in the GaN films have been reported from the ZnO and related materials, too.

In this chapter, characteristics of structural defects in GaN and ZnO films are addressed, extensively. At first, possible dislocations and SFs in wurtzite structure and their characteristics are described. Next, as a practical viewpoint, detailed characterization of various kinds of defects including TD, misfit dislocation, SF, nanopipe, and IDB by transmission electron microscopy (TEM) is given. At last, various kinds of technological approaches in reducing the dislocation density are addressed with the emphasis on the key results of those techniques.

6.2 Dislocation and Stacking Faults

6.2.1 Dislocations

Dislocation is a lattice defect where a discontinuity in the lattice exists on the lattice plane. The discontinuity on a specific area of plane forms many

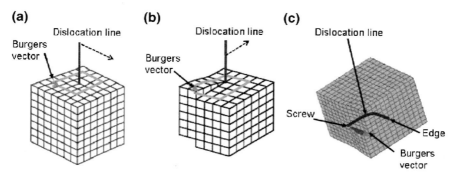

Fig. 6.1. Three basic types of dislocations. (**a**) the edge, (**b**) the screw and (**c**) the mixed dislocations

unsatisfied bonds, i.e., dangling bonds. The edge of the discontinuous plane can be thought as a line with dangling bonds. For this reason, the dislocation is classified as a typical line defect and normally 'dislocation' means that line. There are three basic types of dislocations: the edge dislocation, the screw dislocation, and the mixed dislocation as illustrated in Fig. 6.1.

By moving the dislocation along the direction of the dashed arrows in Fig. 6.1a and Fig. 6.1b, each part of the lattice is displaced by the magnitude of one atomic plane and this displacement can be defined by a Burgers vector. Here, we can see that for the edge dislocation, the Burger vector is perpendicular to the direction of the dislocation line while it is parallel to the direction of the dislocation line for the screw dislocation. This is the typical feature of edge and screw dislocations, characterizing each dislocation. The mixed dislocation has mixed characteristic of both edge and screw dislocations as expected from its name. At each end point of the mixed dislocation, the Burgers vector is perpendicular and parallel, respectively, to the direction of the dislocation line. Except the end point, the Burgers vector and the dislocation line are neither perpendicular nor parallel. The two vectors, i.e., the direction of the dislocation line and the Burgers vector can define one plane, where both vectors exist for the edge dislocation. On the other hand, several planes can be defined for the screw dislocations because the Burgers vector and the direction of the line are parallel. The plane defined by the Burgers vector and the direction of the dislocation line is called a slip plane, on which normally, the dislocations are moving.

With another view point, any dislocation can be classified into the perfect and the partial dislocations. The perfect dislocation is a dislocation of which the Burgers vector has a magnitude of the lattice translation vector of a lattice, while the partial dislocation is a dislocation of which Burgers vector has a magnitude smaller than the lattice translation vector of a lattice.

6.2.2 Misfit and Threading Dislocations

Epitaxy is a specialized thin film growth technique that grows a single crystalline film on a crystalline substrate. Here, the film grown in such manner is called an epitaxial film or an epitaxial layer or an epilayer. Epitaxy comes from the Greek words – a prefix "epi" and a suffix "taxis." The prefix "epi" means "on" or "upon" and the suffix "taxy" means "order" or "arrangement." Two basic types of epitaxy are homoepitaxy and heteroepitaxy. Homoepitaxy is a kind of epitaxy performed with only one material; that is, a film is grown on a substrate of the same material in homoepitaxy. Heteroepitaxy is a kind of epitaxy performed with materials that are different from each other. That is, a film grows on a crystalline substrate of another material in heteroepitaxy. Therefore, there are mismatches in the lattice constants and thermal expansion coefficients in most of heteroepitaxy. Typically, the mismatch in the lattice constant is called as a lattice misfit or simply a misfit. Normally the misfit is expressed by f and it is defined by the following equation, where a_s and a_f mean in-plane lattice constant of the substrate and the film, respectively.

$$f = \frac{a_s - a_f}{a_s}, \qquad (6.1)$$

In heteroepitaxy, the epilayer is grown initially by adjusting the in-plane lattice constant to the substrate in the manner that the planes of the epilayer are connected one by one to those of the substrate. In order to do that, the lattice of the epilayer should be distorted under in-plane tensile strain or compressive strain depending on the differences in in-plane interplanar spacing of the epilayer and the substrate. The film at this stage is a strained film and is called 'psedomorphic'. With increase in the film thickness, the strains in the epilayer are accumulated and in order to reduce the strain energy in the film, the one by one matching of planes between the epilayer and the substrate is destroyed and dislocations are introduced at the interface between the epilayer and the substrate. The film at this stage is a relaxed film. Here, the dislocation introduced to relax the strains caused by the lattice misfit is called as a misfit dislocation and the film thickness at which the misfit dislocations start to be formed is called the critical thickness. Figure 6.2 illustrates the strained, psedomorphic film without the misfit dislocations and the relaxed film with the misfit dislocations.

In addition to the misfit dislocations another type dislocation in the film is a threading dislocation (TD). The TD is a dislocation thread into the film. The TD will be a perfect dislocation or a partial dislocation. Also, the types of the TDs can be the edge, the screw, or the mixed dislocation depending on the Burgers vector and the line direction of dislocations. The misfit dislocation is normally the edge or the mixed dislocation. Here, it should be noted that the misfit dislocations exist at the interface while, the TDs are in the inside of the film as illustrated in Fig. 6.3.

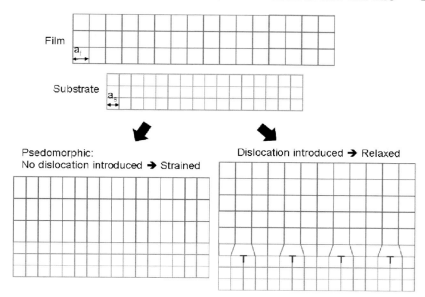

Fig. 6.2. Schematic illustration of the strained, psedomorphic film without the misfit dislocations and the relaxed film with the misfit dislocations

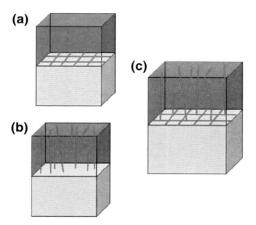

Fig. 6.3. Illustration of (**a**) misfit dislocations at a interface, (**b**) threading dislocations in a film and (**c**) misfit and threading dislocations in a heteroepitaxy system

6.2.3 Dislocation in Wurtzite Structure

There are three kinds of perfect dislocations and three kinds of partial dislocations in the wurtzite structure as summarized in Table 6.1. Assuming the c/a ratio to be the ideal value of $c/a = (8/3)^{1/2}$, the relative energy of the dislocations which is proportional to the $|\mathbf{b}|^2$ is shown. Here, **b** is the Burger

Table 6.1. Dislocations in wurtzite structure

Burgers vector, b	Type	Relative energy $\approx |b|^2$
$1/3[11\bar{2}0]$	Perfect	a^2
$[0001]$	Perfect	$c^2 = (8/3)a^2$
$1/3[11\bar{2}3]$	Perfect	$a^2 + c^2 = (11/3)a^2$
$1/3[01\bar{1}0]$	Partial (Shockley)	$(1/3)a^2$
$1/2[0001]$	Partial (Frank)	$(1/4)c^2 = (2/3)a^2$
$1/6[02\bar{2}3]$	Partial (Frank)	a^2

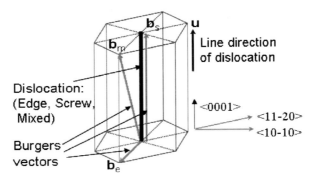

Fig. 6.4. Illustration of the perfect dislocations in an hcp unit cell showing the possible Burger vectors and the line direction. Assuming the line direction of [0001], the dislocations can be an edge, a screw, and a mixed type dislocations depending on their Burgers vector

vector of dislocation and a and c are the lattice constants of the hexagonal close packed (hcp) lattice. One of the three partial dislocations is Shockley partial, while two of the three partial dislocations are Frank partials. As mentioned, types of dislocations can be determined by the Burger vector and the direction of the dislocation line. Assuming the line direction to be [0001], the perfect dislocation with a Burgers vector of $1/3[11\bar{2}0]$ is an edge, the dislocation with a Burgers vector of $[0001]$ is a screw, and the dislocation with a Burgers vector of $1/3[11\bar{2}3]$ is a mixed type as illustrated in Fig. 6.4.

6.2.4 Stacking Fault in Wurtzite Structure

Stacking fault (SF) means staking errors in close packed lattice resulting from the removal or insertion of one close packed layer. Such stacking errors often happened in epitaxial growth results in simple formation of SFs. Also, the SFs can be formed by the propagation of dislocations. SFs are basically related with partial dislocations in the manner that the stacking faulted area is bounded by two partial dislocations. For these reasons, the SF can be named as the Shockley type or the Frank type depending on the partial dislocations that are related with the SFs.

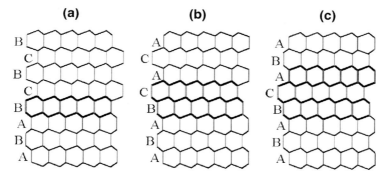

Fig. 6.5. Schematic illustration of three types of stacking faults in an hcp lattice. (**a**) type I, single intrinsic I_1 (**b**) type II, double intrinsic I_2 and (**c**) type III, extrinsic stacking faults

There are three types of SFs in the hcp structure, where the stacking rules are different from the normal stacking rule of hcp lattice, i.e., ...ababab.... These are single, double, and triple faults as illustrated in Fig. 6.5. In the single fault, one violation of the stacking rule is involved resulting in the stacking sequence of ...abab**abc**bcbc... with a displacement vector $\mathbf{R} = 1/6[02\bar{2}3]$. The double fault has two violations of the stacking rule giving a stacking sequence of ...abab**abca**caca... with a displacement vector $\mathbf{R} = 1/3[01\bar{1}0]$. The triple fault contains three violations of the stacking rule resulting in the stacking sequence of ...abab**abcab**abab... with a displacement vector $\mathbf{R} = 1/2[0001]$. The single and double faults are called single intrinsic and double intrinsic, respectively, or type I and type II, while the triple fault is called extrinsic or type III stacking fault. The stacking fault energy increases with increase in the number of stacking errors. Therefore, the stacking fault energy of the single fault is the lowest and that of the triple fault is the largest. Figure 6.6 shows the magnitude and direction of the displacements in a wurtzite structure for the three types of SFs.

6.3 TEM of Defects in GaN and ZnO Films

6.3.1 Defect Contrast in TEM

Defects can be described by translational vectors which represent displacement of atoms from their regular positions in the lattice. Amplitude of electron wave from the perfect crystal can be expressed by the following equations [21].

$$\frac{d\Phi_o}{dz} = \frac{\pi i}{\xi_o}\Phi_o(z) + \frac{\pi i}{\xi_g}\Phi_g(z)\exp(2\pi i s z) \quad (6.2)$$

$$\frac{d\Phi_g}{dz} = \frac{\pi i}{\xi_o}\Phi_g(z) + \frac{\pi i}{\xi_g}\Phi_o(z)\exp(-2\pi i s z) \quad (6.3)$$

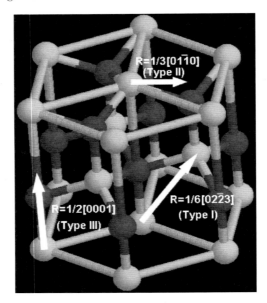

Fig. 6.6. Displacement vectors for the type I, II, III stacking faults

In the case of the imperfect crystal with defect and when its general displacement vector is **R**, the amplitude of electron wave from the imperfect crystal can be expressed by the following equations [21].

$$\frac{d\Phi o}{dz} = \frac{\pi i}{\xi_o}\Phi o(z) + \frac{\pi i}{\xi_g}\Phi g(z)\exp(2\pi i s z + 2\pi i \bar{g} \cdot \bar{R}) \tag{6.4}$$

$$\frac{d\Phi g}{dz} = \frac{\pi i}{\xi_o}\Phi g(z) + \frac{\pi i}{\xi_g}\Phi o(z)\exp(-2\pi i s z - 2\pi i \bar{g} \cdot \bar{R}) \tag{6.5}$$

Here, Φ_o and Φ_g are amplitudes of the direct and diffracted beams for reflection g, respectively, dz is the thickness of a diffraction slice, ξ means a reflection length. Intensities of the direct and diffracted beams are $|\Phi_o|^2$ and $|\Phi_g|^2$. Comparing (6.2)–(6.5), we can see that the amplitude scattered by the perfect crystal is modified by the factor $2\pi \mathbf{g} \cdot \mathbf{R} = 2\pi \mathbf{n}$ and this factor is called the phase factor. Here, n can be integral, zero, or fractional. The magnitude of $\mathbf{g} \cdot \mathbf{R}$ should be sufficient to change the intensity from the background so that the defect contrast is detectable. For example, in case of dislocations, the values of $\mathbf{g} \cdot \mathbf{R}$ or g·b should be greater than 1/3 [21]. Bragg diffraction is controlled by the crystal structure and orientation of the specimen in TEM and we use this diffraction to create a diffraction contrast image, which is just a special form of amplitude contrast. In order to get good strong diffraction contrast in images we tilt the TEM specimen to a so called two-beam condition, in which one diffracted beam is much stronger than other diffracted beam. Since the direct beam is another strong beam in the pattern in addition to

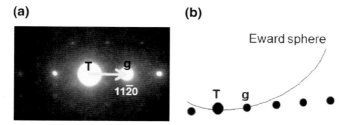

Fig. 6.7. (a) Two-beam condition obtained from the diffraction pattern of a hcp crystal with a zone axis of ⟨0001⟩, in which the strongly excited diffracted spot is 11$\bar{2}$0 and so the **g** vector is [11$\bar{2}$0]. (b) Eward sphere showing the two-beam diffraction condition

the strong diffracted beam, these two strong beams make the main feature of the two-beam condition. Figure 6.7a shows an example of a diffraction pattern under the two-beam condition, in which T indicates the direct beam and the 11$\bar{2}$0 diffracted spot is strongly excited. Figure 6.7b shows the Eward sphere for the diffraction condition and the vector from the directed beam T to the 11$\bar{2}$0 diffracted beam is the vector **g** for this two-beam condition.

By tilting the crystalline TEM specimen we can get many different two-beam conditions, in which the strongly excited diffracted beams are different. This means we can change the vector **g** in the two-beam conditions, which results in the setting up of different two-beam conditions. Depending on the g vectors that get excited in the two-beam condition, the specific dislocation is being imaged. This is because, although the Burgers vector **b** of the specific dislocation is fixed, the values of $2\pi\mathbf{g} \cdot \mathbf{b}$ changes depending on the **g** values, which results in change of contrast of the dislocation line: strong, weak, or none. Since the Burgers vectors of dislocations has specific values that are limited by the crystal structure of the material, we can determine the Burgers vectors of specific dislocations by observing the dislocations under several different two-beam conditions with considerations of the **g** · **b** criteria.

6.3.2 Analysis of Threading Dislocation by Cross-sectional TEM

High quality epitaxial GaN films have been grown by MOCVD or by HVPE, while high quality epitaxial ZnO films have been grown by molecular beam epitaxy (MBE) or pulsed laser deposition (PLD). Independent of the growth methods, TDs are one of the main defects in epitaxial GaN or ZnO films grown on sapphire substrates, which are the most used substrates for the epitaxial growth of these materials. Dislocations in materials can be observed and characterized by TEM. Figure 6.8 shows a typical feature of TDs in an epitaxial (0001) GaN film grown on (0001) Al_2O_3 substrate, in which most of the dislocations runs along the GaN ⟨0001⟩ direction, i.e., along the growth direction. In order to know the types of dislocations observed in Fig. 6.8, we have to know the Burgers vector and the line direction of the dislocations.

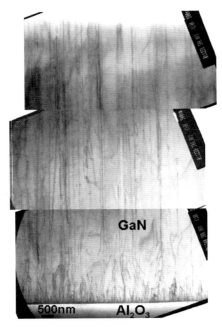

Fig. 6.8. Typical feature of threading dislocations in an epitaxial (0001) GaN film grown on (0001) Al$_2$O$_3$ substrate

The best way to characterize the Burgers vector of dislocations is by analyzing the Burgers vector from several sets of observations performed under the two-beam conditions. This is because we can determine the Burgers vector of dislocations based on the basic rule for dislocation visibility depending on the values of **g** · **b**., i.e., **g** · **b** criteria.

Practically, dislocations in ZnO or GaN films can be analyzed by cross-sectional or plan-view TEM observations. For the single crystalline GaN or ZnO films grown along the ⟨0001⟩ directions, TDs can be analyzed easily from the cross-sectional views with the zone axis of ⟨2$\bar{1}$$\bar{1}$0⟩ or ⟨01$\bar{1}$0⟩. Figure 6.9 shows illustrated single crystalline diffraction patterns of hcp structure with the zone axes of [2$\bar{1}$$\bar{1}$0] and [01$\bar{1}$0].

Figures 6.10 and 6.11 show electron diffraction pattern obtained from the interfacial regions of the epitaxial (0001) ZnO/Al$_2$O$_3$ with the zone axes of ZnO [2$\bar{1}$$\bar{1}$0] and [01$\bar{1}$0], respectively. In Fig. 6.10, a diffraction spot from the ZnO (0001) is shown due to the double diffraction although it has not appeared as a result of forbidden diffraction in Fig. 6.9a.

Since the TDs in epitaxial (0001) GaN or (0001) ZnO films are almost perfect dislocations with the Burgers vector of [0001], 1/3[11$\bar{2}$0], and 1/3[11$\bar{2}$3], the determination of the Burgers vector for these TDs can be performed by observing two sets of two-beam conditions with the cross-sectional samples under

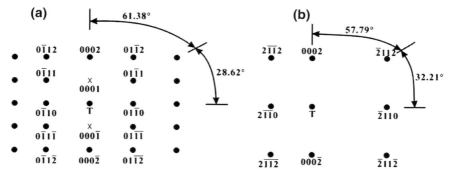

Fig. 6.9. Illustration of the single crystalline diffraction patterns of hcp structure with zone axes of (a) $[2\bar{1}\bar{1}0]$ and (b) $[01\bar{1}0]$

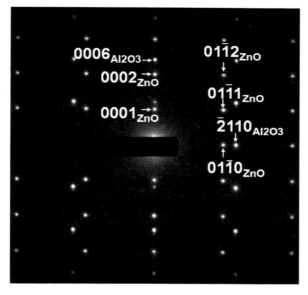

Fig. 6.10. Electron diffraction pattern from the epitaxial (0001) ZnO/Al$_2$O$_3$ with the zone axis of ZnO $[2\bar{1}\bar{1}0]$

the $[2\bar{1}\bar{1}0]$ or the $[01\bar{1}0]$ zone axis. For example, from the two-beam condition with $\mathbf{g} = 0002$, the dislocations with Burgers vector of $[0001]$ and $1/3[11\bar{2}3]$ can be imaged, while the dislocations with Burgers vector of $1/3[11\bar{2}0]$ cannot be imaged considering the $\mathbf{g} \cdot \mathbf{b}$ values. On the other hand, in the case of the two-beam condition with $\mathbf{g} = 01\bar{1}0$ (for the $[2\bar{1}\bar{1}0]$ zone axis) or with the two-beam condition with $\mathbf{g} = \bar{2}110$ (for the $[01\bar{1}0]$ zone axis), the dislocations with Burgers vector of $1/3[11\bar{2}0]$ and $1/3[11\bar{2}3]$ can be imaged, while the dislocations with Burgers vector of $[0001]$ cannot be imaged considering the

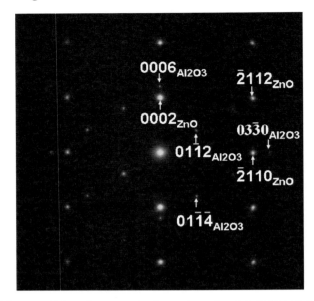

Fig. 6.11. Electron diffraction pattern from the epitaxial (0001) ZnO/Al$_2$O$_3$ with the zone axis of ZnO [01$\bar{1}$0]

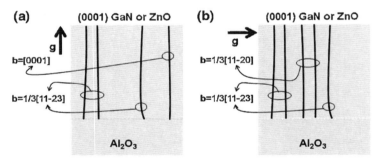

Fig. 6.12. Illustration of the Burgers vector determination from the images taken with the two two-beam conditions by cross-sectional views of (0001) GaN or ZnO film on Al$_2$O$_3$ substrate. (**a**) With the **g** vector of ⟨0001⟩ and (**b**) with the **g** vector of [01$\bar{1}$0] (for the [2$\bar{1}\bar{1}$0] zone axis) or with the **g** vector of [$\bar{2}$110] (for the [01$\bar{1}$0] zone axis)

g·b values. Therefore, by observing these two sets of the two-beam conditions the Burgers vectors of these perfect TDs can be determined as illustrated in Fig. 6.12.

In addition, if the line direction of the threading dislocation is ⟨0001⟩, then we can determine the types of dislocations: the dislocations with Burgers vectors of [0001], 1/3[11$\bar{2}$0], and 1/3[11$\bar{2}$3] are screw, edge, and mixed type dislocations, respectively. Table 6.2 summarizes the visible and invisible criteria

Table 6.2. Summary of the visible and invisible criteria for the dislocations in a wurtzite-structured materials by using the different **g** vectors, which are normally used for the dislocation analyses for the cross-sectional views of the (0001) GaN or ZnO films

	$\mathbf{g} = 0001$	$\mathbf{g} = \bar{2}110$	$\mathbf{g} = 01\bar{1}0$
$\mathbf{b} = 1/3[11\bar{2}0]$ (Perfect)	Invisible	Visible	Visible
$\mathbf{b} = [0001]$ (Perfect)	Visible	Invisible	Invisible
$\mathbf{b} = 1/3[11\bar{2}3]$ (Perfect)	Visible	Visible	Visible
$\mathbf{b} = 1/3[01\bar{1}0]$ (Shockley Partial)	Invisible	Visible	Visible
$\mathbf{b} = 1/2[0001]$ (Frank Partial)	Visible	Invisible	Invisible
$\mathbf{b} = 1/6[02\bar{2}3]$ (Frank Partial)	Visible	Visible	Visible

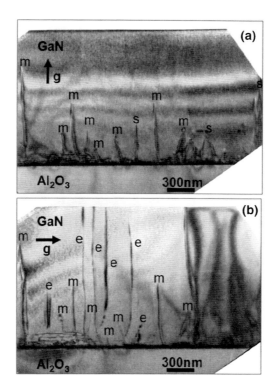

Fig. 6.13. Cross-sectional TEM bright field micrograph of (0001) GaN on Al$_2$O$_3$ substrate under the two-beam conditions with the **g** vectors of (**a**) <0002> and (**b**) <01$\bar{1}$0>. e, s, and m means an edge, a screw, and a mixed dislocations, respectively

for the dislocations using the different **g** vectors, which are normally used for the dislocation analyses from the cross-sectional views of the (0001) GaN or ZnO films. Practically, this simple method is applied to the MOCVD grown (0001) GaN film on Al$_2$O$_3$ substrate as shown in Fig. 6.13. Here, the notation e, s, and m mean an edge, a screw, and a mixed dislocation, respectively.

For the nonpolar films such as a-plane (11$\bar{2}$0) GaN or ZnO films, typical zone axes for the cross-sectional views are [0001] and [01$\bar{1}$0]. In this case by observing under the two-beam conditions with the zone axis of [01$\bar{1}$0] we can determine the Burgers vector of TDs as the case for the (0001) GaN or ZnO films. On the other hand, for the nonpolar m-plane (01$\bar{1}$0) GaN or ZnO films, typical zone axes for the cross-sectional views are [0001] and [2$\bar{1}\bar{1}$0]. In this case by observing under the two-beam conditions with the zone axis of [2$\bar{1}\bar{1}$0] we can determine the Burgers vector of TDs, too.

6.3.3 Analysis of Threading Dislocation by Plan-View TEM

In general, plan-view observations are frequently performed in order to determine a TD density. However, TD analysis of (0001) GaN or ZnO films by plan-view observations is not so simple compared with the cross-sectional view. As shown in Fig. 6.8, since most of the TDs thread along the ⟨0001⟩ direction, the dislocations are mostly observed as the so called end on dislocations and it means that the dislocations are observed as a point. However, in order to clearly observe the dislocations, the sample is tilted to satisfy a specific two-beam condition and it makes the dislocations be inclined relative to the direction of the electron beam. Therefore, the dislocations appear as a line not a point. Additionally, the dislocation line is not straight but shows a zigzag contrast [22]. Figure 6.14 show a plan-view TEM micrograph of the (0001) GaN film taken under the two-beam condition with the **g** vector of [11$\bar{2}$0].

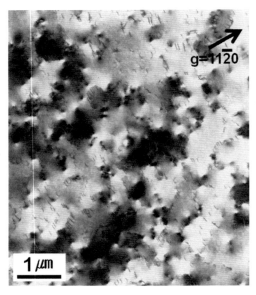

Fig. 6.14. Plan-view TEM micrograph of the (0001) GaN film taken under the two beam condition with the **g** vector of [11$\bar{2}$0]

Determination of the Burgers vectors of the threading perfect dislocations in (0001) GaN or ZnO films by plan-view observations is not a simple task, especially for the screw dislocation with the Burgers vector of $\langle 0001 \rangle$. Since all the diffracted spots under the $\langle 0001 \rangle$ zone axis have (hki0) indices as shown in Fig. 6.15a, the $\mathbf{g} \cdot \mathbf{b}$ values are zero with any $\mathbf{g}$ vectors for the dislocation with Burgers vector of $\langle 0001 \rangle$ from the [0001] zone axis, although the dislocations with Burgers vector of $1/3[11\bar{2}0]$ or $1/3[11\bar{2}3]$ are easily observable. Therefore, in order to analyze the dislocation with the Burgers vector of $\langle 0001 \rangle$ from the plan-view observation, we have to change the zone axis in which diffraction spots of (hkil) with nonzero l index are appeared. For example, the zone axis of $[1\bar{2}1\bar{3}]$ is one of the diffraction patterns in which we can find the diffraction spots of (hkil) with nonzero l index as shown in Fig. 6.15b. Figure 6.15c shows electron diffraction pattern with the zone axis of [0001] obtained from the (0001) GaN film on Al_2O_3 substrate, while Fig. 6.15d shows electron diffraction pattern with the zone axis of $[1\bar{2}1\bar{3}]$, which is obtained by the tilting of 31.4° from the [0001] zone axis. Alternatively, lots of different two-beam

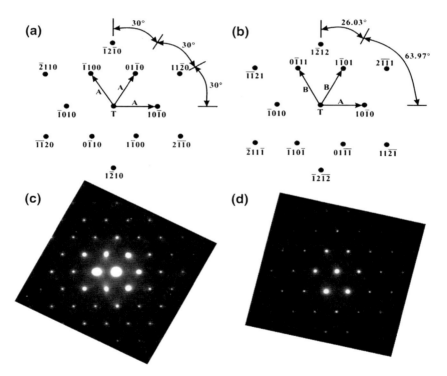

Fig. 6.15. Illustration of the single crystalline diffraction patterns of hcp structure with zone axes of (a) [0001] and (b) $[1\bar{2}1\bar{3}]$, and electron diffraction patterns from the (0001) GaN film under the zone axes of (c) [0001] and (d) $[1\bar{2}1\bar{3}]$

Table 6.3. Summary of the **g** · **b** values for the perfect dislocations in an hcp crystal using the various **g** vectors [23]

Reflection		±[11$\bar{2}$0]	±[$\bar{1}$2$\bar{1}$0]	±[$\bar{2}$110]	±[11$\bar{2}$3]	±[$\bar{1}$2$\bar{1}$3]	±[$\bar{2}$113]	±[11$\bar{2}\bar{3}$]	=[$\bar{1}$2$\bar{1}\bar{3}$]	±[$\bar{2}$11$\bar{3}$]	±[0003]
1	10$\bar{1}$0	±1	0	$\mp$1	±1	0	$\mp$1	±1	0	$\mp$1	0
	01$\bar{1}$0	±1	±1	0	±1	±1	0	±1	±1	0	0
	$\bar{1}$100	0	±1	±1	0	±1	±1	0	±1	±1	0
2	0002	0	0	0	±2	±2	±2	$\mp$2	$\mp$2	$\mp$2	±2
3	10$\bar{1}$1	±1	0	$\mp$1	±2	±1	0	0	$\mp$1	$\mp$2	±1
	10$\bar{1}\bar{1}$	±1	0	$\mp$1	0	$\mp$1	$\mp$2	±2	±1	0	$\mp$1
	01$\bar{1}$1	±1	±1	0	±2	±2	±1	0	0	$\mp$1	±1
	01$\bar{1}\bar{1}$	±1	±1	0	0	0	$\mp$1	±2	±2	±1	$\mp$1
	$\bar{1}$101	0	±1	±1	±1	±2	±2	$\mp$1	0	0	±1
	$\bar{1}$10$\bar{1}$	0	±1	±1	$\mp$1	0	0	±1	±2	±2	$\mp$1
4	10$\bar{1}$2	±1	0	±1	±3	±2	±1	$\mp$1	$\mp$2	$\mp$3	±2
	10$\bar{1}\bar{2}$	±1	0	±1	$\mp$1	$\mp$2	±3	±3	±2	±1	$\mp$2
	01$\bar{1}$2	±1	±1	0	±3	±3	±2	$\mp$1	$\mp$1	$\mp$2	±2
	01$\bar{1}\bar{2}$	±1	±1	0	$\mp$1	$\mp$1	$\mp$2	±3	±3	±2	$\mp$2
	$\bar{1}$102	0	±1	±1	±2	±3	±3	$\mp$2	$\mp$1	$\mp$1	±2
	$\bar{1}$10$\bar{2}$	0	±1	±1	$\mp$2	$\mp$1	$\mp$1	±2	±3	±3	$\mp$2
5	11$\bar{2}$0	±2	±1	$\mp$1	$\mp$2	±1	$\mp$1	±2	±1	$\mp$1	0
	$\bar{1}$2$\bar{1}$0	±1	±2	±1	±1	±2	±1	±1	±2	±1	0
	$\bar{2}$100	$\mp$1	±1	±2	$\mp$1	±1	±2	$\mp$1	±1	±2	0
6	10$\bar{1}$3	±1	0	$\mp$1	±4	±3	±2	$\mp$2	$\mp$3	$\mp$4	±3
	10$\bar{1}\bar{3}$	±1	0	$\mp$1	$\mp$2	$\mp$3	$\mp$4	±4	±3	±2	$\mp$3
	01$\bar{1}$3	±1	±1	0	±4	±4	±3	$\mp$2	±2	$\mp$3	±3
	01$\bar{1}\bar{3}$	±1	±1	0	$\mp$2	$\mp$2	$\mp$3	±3	±4	±3	$\mp$3
	$\bar{1}$103	0	$\mp$1	±1	±3	±4	±4	$\mp$3	$\mp$2	$\mp$2	±3
	$\bar{1}$10$\bar{3}$	0	$\mp$1	±1	$\mp$3	$\mp$2	$\mp$2	±3	±4	±4	$\mp$3
7	11$\bar{2}$2	±2	±1	$\mp$1	±4	±3	±1	0	$\mp$1	$\mp$3	±2
	11$\bar{2}\bar{2}$	±2	$\mp$1	$\mp$1	0	$\mp$1	$\mp$3	±4	±3	±1	$\mp$2
	$\bar{1}$2$\bar{1}$2	±1	±2	±1	±3	±4	±3	$\mp$1	0	$\mp$1	±2
	$\bar{1}$2$\bar{1}\bar{2}$	±1	±2	±1	$\mp$1	0	$\mp$1	±3	±4	±3	$\mp$2
	$\bar{2}$112	$\mp$1	±1	±1	±2	±1	±3	±4	$\mp$3	0	±2
	$\bar{2}$11$\bar{2}$	$\mp$1	±1	±1	±2	$\mp$3	$\mp$1	0	±1	±4	$\mp$2

conditions can be applied under the appropriate zone axes in order to analyze the Burgers vectors of dislocations. Table 6.3 summarizes the **g** · **b** values for the perfect dislocations in an hcp crystal using the various **g** vectors [23].

As reported by Datta et al. observing all types of dislocations to determine the dislocation density by plan-view is possible by taking the image using the multibeam diffractions under the [1$\bar{2}$13] zone axis [24]. However, to determine the Burgers vectors of TDs exactly we have to observe several images under several sets of two-beam conditions from both [0001] and [1$\bar{2}$13] zone axes, especially if we want to distinguish the dislocations between the Burgers vector of [0001] and 1/3[11$\bar{2}$3]. Observing the dislocations even with the zero **g** · **b** values is sometimes possible by using the so called 'black and white lobe contrast' from the end on dislocation [25], and this method was applied for analysis of screw dislocations in (0001) GaN film [26]. However, this method needs careful interpretations in determining the dislocation types because obtaining a clear lobe contrast is not easy.

Follstaedt et al. reported a new plan-view image contrast technique using the two-beam condition under the [0001] zone axis, in which screw dislocations can be identified in addition to edge and mixed dislocations in the (0001) GaN film [27]. The method is to tilt the sample to about 18° to the $\mathbf{g} = (11\bar{2}0)$ two-beam condition. Edge and mixed dislocations are visible using this reflection but the dislocation line itself is invisible for the screw dislocation because $\mathbf{g} \cdot \mathbf{b} = 0$ for this reflection. However, a screw dislocation is revealed in the image as a pair of black–white spots, one pair at each surface of the specimen, because of the surface strain relaxation effects where the dislocation line intersects the surface [24]. An edge dislocation is imaged as a general zigzag line, and a mixed dislocation is a combination of a screw and an edge, and hence is imaged as a pair of black–white spots plus a connecting zigzag line between these spots. Fig. 6.16 shows TEM bright field images taken (a) on [0001] orientation, (b) tilted 3° to $\mathbf{g} = (11\bar{2}0)$ two-beam condition, and (c) $\mathbf{g} = (11\bar{2}0)$ but tilted 18°. In Fig. 6.16c, arrows indicate edge, boxes indicate mixed, and circles indicate screw dislocations [27]. The limit of this method is that all of the dislocations are not detected but ~90% or more of the total dislocations are detected [27]. Notwithstanding this limit, the method is very useful from the viewpoint that we can determine the Burgers vectors with the tilting less than 20°.

6.3.4 Misfit Dislocation

Misfit dislocations are the dislocations which are generated at the heterointerface in order to relax the misfit between two films forming the interface. Basically, observing the misfit dislocations by TEM are possible through the plan-view and cross-sectional view observations. The lattice misfit directly affects the critical thickness and the number of misfit dislocations at the interface. If the misfit is large, the critical thickness is thin and the number of misfit dislocation is large resulting in a narrow spacing between the misfit dislocations. On the other hand, if the misfit is not large, the reverse happens, i.e., there is a thick critical thickness, small number of misfit dislocations and wide spacing between the misfit dislocations. For example, in the case of epitaxial (001) ZnSe film on GaAs substrate which has the lattice misfit of 0.27%, the critical thickness is about 150 nm and the misfit dislocations are formed on the interface with the relatively large spacing of an order of micrometers as shown in Fig. 6.17.

However, in the cases of (0001) GaN and ZnO films on Al_2O_3 substrates, the lattice misfits are ~14 and ~16%, respectively, which are extremely large compared with the previous ZnSe/GaAs system. For the (0001) GaN and ZnO films on Al_2O_3 substrates, both the critical thicknesses and the spacing between the misfit dislocations are in the order of a few nanometer or less,

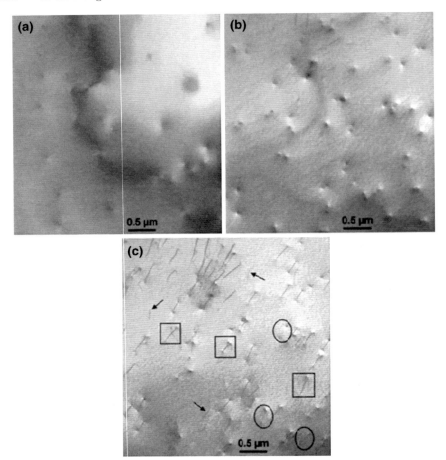

Fig. 6.16. TEM bright field images taken (**a**) on [0001] orientation, (**b**) tilted 3° to **g** = (11$\bar{2}$0) two-beam condition, and (**c**) **g** = (11$\bar{2}$0) but tilted 18°. *Arrows* indicate edge, *boxes* indicate mixed, and *circles* indicate screw dislocations. Reprinted with permission from [27]. Copyright (2003), American Institute of Physics

hence the plan-view observation of misfit dislocations are basically impossible from the conventional two-beam conditions. Therefore, observing the misfit dislocations in hetero epitaxial system with large lattice misfit is performed from cross-sectional view with a high-resolution TEM (HRTEM) imaging condition. Figure 6.18 shows a HRTEM micrograph of the misfit dislocations at the ZnO/GaN heterointerface, in which the lattice misfit is about 1.8%. In Fig. 6.18a we can find extra half planes in the GaN layer, which represent the misfit dislocations. These misfit dislocations are clearly visualized in Fig. 6.18b in which the image of Fig. 6.18a is Fourier filtered using the reflections of {01$\bar{1}$0} planes.

6 Structural Defects in GaN and ZnO   279

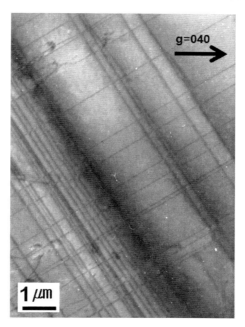

Fig. 6.17. Plan-view bright field TEM micrograph under the two-beam condition with the **g** vector of [040], showing array of misfit dislocations at the interface in **a** (001) ZnSe/GaAs hetero epitaxial system

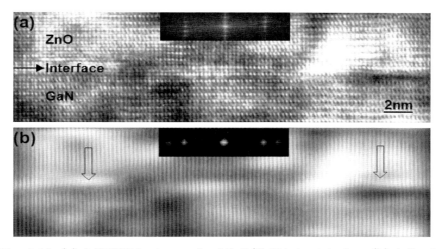

Fig. 6.18. (a) A HRTEM micrograph of ZnO/GaN heterointerface. (b) A Fourier filtered image where planes perpendicular to the interface are visualized in order to clearly show the location of the misfit dislocations (*extra half planes*). The insets show the diffraction spots which contributed in forming the HRTEM images

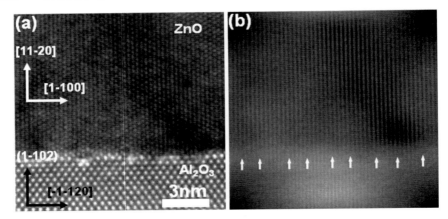

Fig. 6.19. (a) A High-resolution TEM micrograph at the interfaces in a (11$\bar{2}$0) ZnO/(1$\bar{1}$02) Al$_2$O$_3$ heteroepitaxial system observed along the [0001]ZnO//[$\bar{1}$101]Al$_2$O$_3$ zone axes and (b) the Fourier-filtered image of (a) using the ($\bar{1}$1$\bar{2}$0) and (11$\bar{2}$0) Al$_2$O$_3$ reflections and ($\bar{1}$100) and (1$\bar{1}$00) Zno reflections. The position of misfit dislocations is marked by *arrows*. Reprinted with permission from [28]. Copyright (2007), Elsevier

Figure 6.19a shows the HRTEM micrograph at the interfaces for the a-plane (11$\bar{2}$0) ZnO on r-plane (1$\bar{1}$02) Al$_2$O$_3$ substrate, while Fig. 6.19b shows the Fourier-filtered image of Fig. 6.19a using the ($\bar{1}$1$\bar{2}$0) and (11$\bar{2}$0) Al$_2$O$_3$ reflections and the (1$\bar{1}$00) and ($\bar{1}$100) ZnO reflections [28]. Here, the position of the misfit dislocations is marked by the arrows and we can see that the misfit dislocations are formed with a relatively regular spacing, i.e., the misfit dislocations constitute the array that every fifth or sixth (11$\bar{2}$0) plane of Al$_2$O$_3$ terminating at the interface.

Figure 6.20 shows the schematic drawing for the (11$\bar{2}$0) ZnO on (1$\bar{1}$02) Al$_2$O$_3$ substrate with a projection on the [0001] direction for ZnO and on the [$\bar{1}$101] direction for Al$_2$O$_3$, where the in-plane translational periods for the ZnO and the Al$_2$O$_3$ are shown and denoted by T [28].

The in-plane translational period of ZnO along the [1$\bar{1}$00] direction is $\sqrt{3}$ a$_{ZnO}$ = 5.629 Å, while the one for the Al$_2$O$_3$ [$\bar{1}$1$\bar{2}$0] direction is a$_{Al2O3}$ = 4.758 Å as mentioned in Fig. 6.20, which gives the lattice misfit of 18.31%. Therefore, the magnitude of the Burgers vector component parallel to the [$\bar{1}$1$\bar{2}$0] Al$_2$O$_3$ is 4.758 Å/2 = 2.379 Å, assuming that the lattice misfit is fully accommodated by the Burger vector of the misfit dislocations. The spacing of the misfit dislocations can be calculated to be 1.3 nm using the geometrically expected spacing of the misfit dislocations using the following equation.

$$D = |\mathbf{b}|/\delta \tag{6.6}$$

Here, |**b**| is the magnitude of the Burgers vector component, i.e., 2.379 Å and δ is the misfit strain, i.e., 0.183. Note that the calculated D value of 1.3 nm agrees well with the observed spacing of the misfit dislocations in Fig. 6.19 [28].

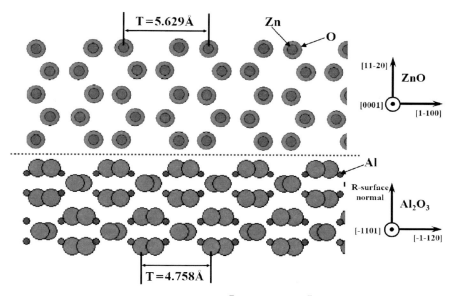

Fig. 6.20. Schematic drawing for the $(11\bar{2}0)$ ZnO on $(1\bar{1}02)$ Al$_2$O$_3$ substrate with a projection on the [0001] direction for ZnO and on the $[\bar{1}101]$ direction for Al$_2$O$_3$, where the in-plane translational periods for the ZnO and the Al$_2$O$_3$ are shown and denoted by T. Reprinted with permission from [28]. Copyright (2007), Elsevier

6.3.5 Nanopipe

The hollow pipe in a material was first predicted by Frank [29]. He predicted that a dislocation with the Burgers vector exceeding a certain value should have an empty core forming a tube to satisfy a state of local equilibrium [29]. Long hollow pipes have been observed in SiC [30] and ZnS crystals [31]. The pipes observed in SiC and ZnS were named to micropipes because their diameters were of the order of several tens of micrometers. However, the hollow pipe observed in GaN has a diameter of the order of a nanometer and is called as the nanopipe after its first observation in (0001) GaN by Qian et al. in 1995 [9,32]. The main origin of a leakage current in GaN based light emitting diodes has been attributed to the nanopipes [33]. As shown in Fig. 6.21, nanopipes generally have hexagonal cross-section with facet planes of $\{10\bar{1}0\}$ [34].

The formation of a nanopipe is strongly related with the segregation of oxygen [35, 36] as shown in Fig. 6.22, which shows clear increase of oxygen content at the side wall of the nanopipe.

6.3.6 Stacking Fault

In GaN and ZnO films, type I stacking faults (SFs) have been mostly observed compared to type II and type III. This is because the type I stacking fault

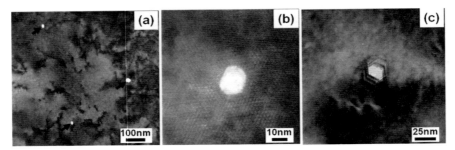

Fig. 6.21. Nanopipes observed by TEM form the (0001) GaN film on Al_2O_3 substrate grown by MOCVD. *Arrows* indicates the nanopipes. (**a**) Low and (**b**) high magnification end on view and (**c**) tilted view showing the contour of the Nanopipes running through the film. Reprinted with permission from [34]. Copyright (1998), Elsevier

has the lowest formation energy [37,38]. Furthermore, it can be formed during crystal growth by the deposition of one layer in a wrong position. The stacking faults on (0001) basal planes in GaN films on Al_2O_3 substrates frequently observed at interfacial regions or in the low temperature buffer layer, are mostly caused by stacking errors during the growth. Figure 6.23a shows the stacking faults in (0001) film. As shown in Fig. 6.23 many stacking faults on basal (0001) planes appeared as horizontal lines parallel to the (0001) plane. The inset in Fig. 6.23a shows a HRTEM images for the stacking faults. The presence of such high density of stacking faults can be detected even from the diffraction pattern. Figures 6.23b–d are the selected area electron diffraction pattern from the GaN film and the Al_2O_3 substrate, the GaN film only, and the Al_2O_3 substrate only, respectively. Here it should be noted that vertical lines connecting the diffraction spots appeared in addition to the substrate and GaN diffraction spots. Note that the additional vertical diffraction lines are connecting the GaN diffraction spots not the diffraction spots of Al_2O_3 substrate as shown in Fig. 6.23b, c, which indicates that the vertical additional diffraction lines come from the GaN film. In fact these vertical diffraction lines were caused from the horizontal high density of stacking faults, similar to the appearance of satellite diffractions from a superlattice structure.

Figure 6.24 shows a HRTEM micrograph for the SF in a (0001) ZnO film grown on Al_2O_3 substrate. The stacking sequence at the faulted region is ...ABABABCBC..., which means the stacking fault is the type I stacking fault.

Here, it should be noted that every stacking fault is bounded by partial dislocations. The partial dislocations in an hcp crystal are summarized in Table 6.1 and the displacement vectors of the three types of SFs are shown in Fig. 6.6. These displacement vectors are the same to the Burgers vector of partial dislocations bounding the stacking fault. That is, the type I stacking

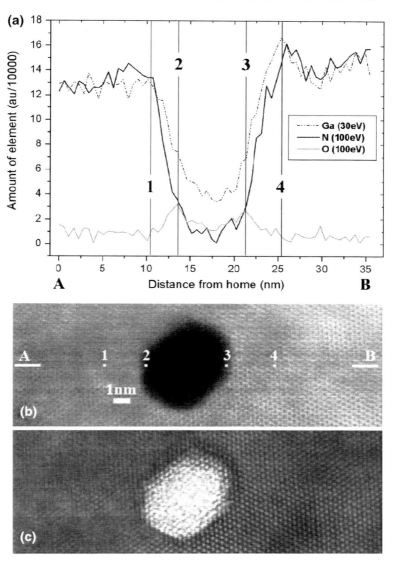

Fig. 6.22. Scanning TEM (STEM) analysis of an end-on nanopipe in a (0001) GaN/sapphire sample. (a) Ga–$M_{2,3}$, O–K, and N–K profiles from an Electron energy loss spectroscopy (EELS) line scan, (b) high angle annular dark field (HAADF) image, and (c) bright field image of the nanopipe. Points 1–4 in. (b) show corresponding points in (a). Reprinted with permission from [35]. Copyright (2005), American Institute of Physics

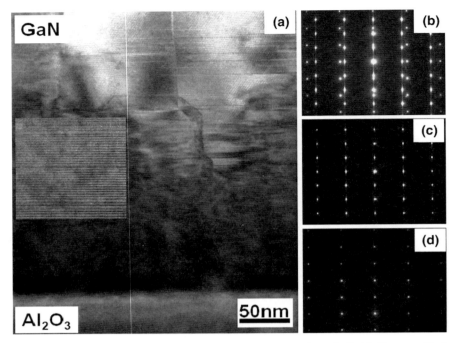

Fig. 6.23. (a) Bright field TEM micrograph for the (0001) GaN film on Al_2O_3 substrate in which many number of stacking faults are shown as *horizontal lines* in addition to threading dislocations. Selected area diffraction patterns from (b) both GaN and Al_2O_3, (c) GaN film only, and (d) Al_2O_3 substrate only are shown. Here it should be noted that *vertical lines* connecting the diffraction spots appeared in addition to the substrate and GaN diffraction spots. Note that the additional *vertical diffraction lines* are connecting the GaN diffraction spots not the diffraction spots of Al_2O_3 substrate

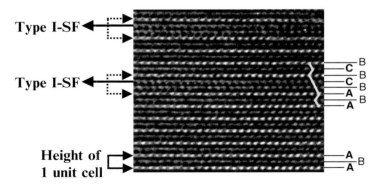

Fig. 6.24. HRTEM micrograph showing the type I stacking fault observed in the ZnO film

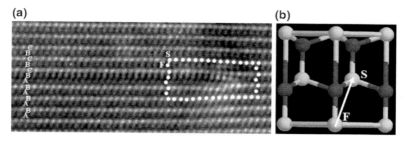

Fig. 6.25. (a) HRTEM micrograph showing the type I stacking fault. The stacking sequence ...ABABABCBC... is mentioned and the Burgers circuit is made around the partial dislocation at the end of stacking fault position. The vector from the final point F to the start point S of the Burgers circuit is just the Burgers vector of the partial dislocation, which is correspond to the vector $1/6[02\bar{2}3]$ shown in (b)

fault has the Frank partial dislocation with a Burgers vector of $1/6[02\bar{2}3]$, the type II stacking fault has the Shockley partial dislocation with a Burgers vector of $1/3[01\bar{1}0]$, and finally the type III stacking fault has the Frank partial dislocation with a Burgers vector of $1/2\,[0001]$ bounding each of the stacking fault. Figure 6.25a shows another HRTEM micrograph showing type I stacking fault. In Fig. 6.25a, the stacking sequence ...ABABABCBC... is mentioned and the Burgers circuit is made around the partial dislocation at the end of stacking fault position. The vector from the final point F to the start point S of the Burgers circuit is just the Burgers vector of the partial dislocation, which corresponds to the vector $1/6\,[02\bar{2}3]$ as shown in Fig. 6.25b.

6.3.7 Inversion Domain Boundary

An inversion domain boundary (IDB) is one kind of planar defect, which has been observed in GaN or ZnO film. The IDB is the boundary of the inversion domain, which has the characteristic of opposite polarity. That is, across this boundary, the stacking of ABAB changes to BABA. However, just forming a simple switching in the stacking sequence across the IDB results in the formation of Zn–Zn and O–O bonding at the boundary. Such bonding configurations are highly unstable. By shifting one side of an IDB by $c/2$ along the [0001] direction, an energetically ideal IDB with a significantly reduced energy can be formed [39] as shown in Fig. 6.26. In Fig. 6.26, a schematic representation of a simple inversion domain boundary in wurtzite has wrong bonds between like atoms as denoted by the dashed lines, while a schematic representation of IDB* structure formed by translating one side of an IDB by $c/2$ along the [0001] direction contains fourfold and eightfold rings of bonds, but there are no Ga–Ga or N–N bonds [39]. The real IDB* in GaN was observed and identified by atomic-resolution HRTEM [40].

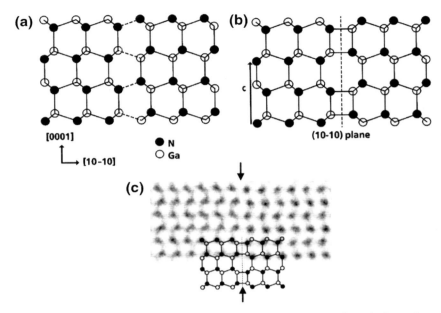

Fig. 6.26. (a) Schematic representation of a simple inversion domain boundary in wurtzite. The *dashed lines* denote wrong bonds between like atoms. Atom positions are projected onto the (1$\bar{2}$10) plane. (b) Schematic representation of IDB*. This structure can be formed by translating one side of an IDB by $c/2$ along the [0001] direction. The IDB* structure contains fourfold and eightfold rings of bonds, but there are no Ga–Ga or N–N bonds [39]. (c) Atomic-resolution HRTEM micrograph showing the IDB* in GaN. Reprinted with permission from [39] and [40]. Copyright (1996), American Physical Society and Copyright (2004), American Institute of Physics

In addition to the vertical IDB lines on (10$\bar{1}$0) plane, the faceted IDBs have been reported in Mg doped GaN films as shown in Fig. 6.27 [41]. The faceted IDB formation by increasing the Mg doping concentrations in the MOCVD grown GaN films have been studied [42].

The IDBs with various facet planes observed near the top surface regions in the highly Mg doped GaN film as shown in Fig. 6.28, support the possibility of the Mg_3N_2 phase formation at the IDB regions [43].

6.4 Dislocation Reduction of Epitaxial Films by Process

6.4.1 Defects in Epitaxial Lateral Overgrowth (ELOG)

The improvement in high efficiency blue and green LEDs and LDs and the development of ultraviolet emitters as a pump source for the realization of

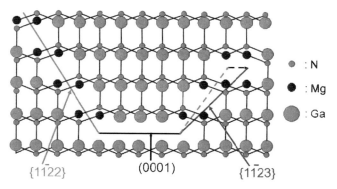

Fig. 6.27. Schematic representations of various possible inversion domain boundaries in GaN. A (11$\bar{2}$2) boundary is shown on the *left*. The shaded atoms are threefold coordinated Mg or Ga atoms. Shown on the *right* is a (11$\bar{2}$3) boundary comprised of a nanometer scale mixture of (11$\bar{2}$2) segments and (0001) segments. Reprinted with permission from [41]. Copyright (2000), American Institute of Physics

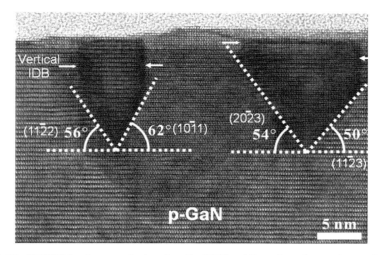

Fig. 6.28. HRTEM micrograph showing the IDBs with various facet planes observed near the top surface regions in the highly Mg doped GaN film [42]

white LEDs have been the main focus of the research in GaN based III-nitrides [44–46]. Conventional GaN epilayers grown on sapphire substrates show a considerably high threading dislocation density due to a large lattice mismatch (∼14%) and the difference in thermal coefficient difference between GaN and sapphire. Successful GaN growth on sapphire substrate is achieved through the two-step deposition using LT buffer layers [2, 3, 13, 47–50]. These defects including TDs of mid $10^8 \sim 10^9$ cm^{-2} act as nonradiative recombination centers that finally limit the output power of light emission [51,52]. Therefore, reducing the number of TDs is essential for fabricating low reverse bias leakage current LEDs, high power/long lifetime bluish violet LDs, and low dark current photodetectors [53–55].

Up to now, several approaches to the growth of low dislocation density GaN have been proposed and demonstrated. Among them, the most promising methods are epitaxial lateral overgrowth (ELOG) and pendeo-epitaxial (PE) growth, which have mainly been performed by MOCVD and HVPE. LDs consisting of InGaN/GaN MQWs and Al$_{0.14}$Ga$_{0.86}$N/GaN modulation doped strained-layer superlattice cladding layers grown on an ELOG GaN with a lower dislocation density showed a lifetime >1,150 h under room temperature continuous-wave operation [56].

As shown in Fig. 6.29, the standard ELOG process consists fundamentally of growth, lithography, and regrowth procedures. Initially, GaN nucleation layers grown at LT or GaN epitaxial films grown on sapphire substrate with a thin nucleation layer are deposited as a template. Amorphous dielectric layers are then deposited and stripes separated with a periodic distance are patterned using a conventional photolithography process. A silicon dioxide (SiO$_2$) or silicon nitride (SiN$_x$) dielectric layer as a mask is deposited by chemical vapor deposition or sputtering [57–60]. The directions of the stripes are mostly the $\langle 11\bar{2}0 \rangle$ and $\langle 1\bar{1}00 \rangle$. The choice of the stripe direction results in a different surface morphology at the initial stage and a different behavior of dislocation movement [61–63]. The morphological evolution of the resulting film is strongly affected by various growth parameters including temperature, reactor pressure, V/III flow ratio, the width and distance of the dielectric stripes, etc.

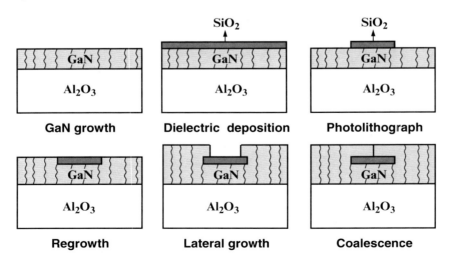

Fig. 6.29. Schematic diagrams showing the growth evolution of epitaxial lateral overgrowth (ELOG) process

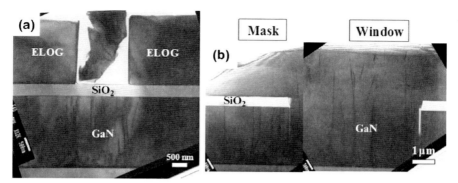

Fig. 6.30. Cross-sectional TEM images showing the dislocation distribution of (**a**) before and (**b**) after complete coalescence. The direct dislocation blocking in the mask region is observed

There is no GaN deposition on any region coated with the dielectric layer when the degree of supersaturation is controlled by the appropriate variations in growth parameters, and the regions opened with the template GaN begin to grow along the vertical direction during the initial stages. This indicates selective overgrowth screened by amorphous dielectric layers. Further growth causes deposition along both the lateral and vertical directions simultaneously and the lateral overgrowth on stripe mask is initiated. Finally, fully coalescenced GaN films with a flat surface can be obtained under appropriate growth conditions. The opened and the covered areas are called "window" and "mask" regions, respectively [64,65].

Most TDs observed in the GaN layers below the mask are blocked completely by dielectric layers, while the window regions regrown on the template GaN show a similar threading dislocation density with the GaN seed layer due to the continuous growth of dislocations as shown in Fig. 6.30. This method was first introduced by Kato et al. [66] who suggested the selective epitaxy of GaN on a sapphire substrate by MOCVD. This method allows epitaxial films to be grown selectively on the regions confined by a dielectric mask, which effectively prevents the propagation of threading dislocations into the GaN surface. Extensive studies have been carried out with the aim of obtaining high quality GaN films with a low dislocation density.

Hiramatsu et al. and Zheleva et al. carried out comparative studies of the dislocation distribution and the evolution of surface morphology [61,63,67,68]. According to Hiramatsu et al. [61,67], the growth behavior of the overgrown GaN epilayers is strongly dependent on the stripe direction. For a dielectric stripe along the $\langle 1\bar{1}00 \rangle$ direction of the underlying GaN, an increase in growth temperature or a decrease in reactor pressure results in a morphological change in the overgrown GaN from a pyramidal to rectangular shape, while the GaN overgrown on the dielectric stripe of the $\langle 11\bar{2}0 \rangle$ direction showed only pyramidal shape regardless of the growth temperature and reactor pressure.

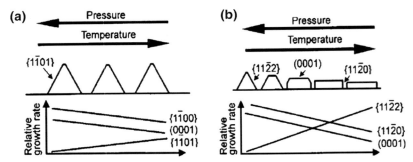

Fig. 6.31. Morphological change in ELOG GaN on the (a) ⟨11$\bar{2}$0⟩ stripe and (b) ⟨1$\bar{1}$00⟩ stripe for different reactor pressure and growth temperatures. Reprinted with permission from [67]. Copyright (1999), Wiley

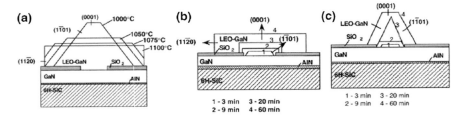

Fig. 6.32. Schematic diagram showing (a) the temperature dependence of the side facet morphologies of selectively grown GaN stripes along the ⟨1$\bar{1}$00⟩ direction. The lateral dimensions of the top (0001) planes are increased, as the lengths of the slanted side walls are reduced until the vertical (11$\bar{2}$0) side walls are developed as the growth temperature is increased from 1,000 to 1,100°C. Schematic diagrams showing the dependence of the side surface morphology of selectively grown GaN stripes as a function of the growth time along (b) ⟨1$\bar{1}$00⟩ and (c) ⟨11$\bar{2}$0⟩ orientations of the stripes. Reprinted with permission from [63]. Copyright (2001), Elsevier

When a stripe of the dielectric layers is formed along the ⟨1$\bar{1}$00⟩ direction, an increase in growth temperature leads to an enhancement in the surface diffusivity of adatoms and lateral growth over vertical growth is promoted for the formation of a flat surface as illustrated in Fig. 6.32a. This causes a decrease in the surface area of the {1$\bar{1}$01} facet planes and an increase in the (0001) surface area, which finally forms GaN stripes with a rectangular shape consisting of {11$\bar{2}$0} side facets in the window regions as illustrated in Figs. 6.31 and 6.32.

If the GaN epilayers grown on the dielectric stripes patterned along the ⟨1$\bar{1}$00⟩ and ⟨11$\bar{2}$0⟩ directions show rectangular and triangular shapes, respectively, during the initial growth stage, their morphological shapes are expected to follow the evolution shown in Fig. 6.32. In the case of a stripe along the ⟨1$\bar{1}$00⟩ direction, Zheleva et al. reported that the {1$\bar{1}$01} planes first disappear with increasing growth time [63, 68], and the (0001) top surface and {11$\bar{2}$0} side facets become dominant, as shown in Fig. 6.32b. This results in rapid merging between growth fronts deposited from the neighboring window regions due to the fast lateral growth rate. In addition, the surface of the overgrown GaN after the complete coalescence is expected to be relatively flat, i.e., the improved surface roughness. Microstructural characterization using TEM reveals a considerable decrease in threading dislocation density in the mask region as shown in Fig. 6.30b. However, a high defect density GaN film is still observed in the GaN layers grown in the window region, as shown in Figs. 6.30 and 6.34b. A special treatment should be considered for the development of GaN films with a uniformly reduced defect density, which will be explained in the following sections.

In contrast, overgrowth on the dielectric patterned along the $\langle 11\bar{2}0 \rangle$ direction shows a rapid growth rate along the vertical direction. Therefore, the film morphology is initially triangular in shape with the $\{1\bar{1}01\}$ side facets, and further growth results in a trapezoidal shape as illustrated in Fig. 6.32c. Figure 6.33b shows the distribution of dislocations in the resulting film. In addition, the coalescence in the growth fronts of the GaN domains is delayed due to the rapid vertical growth rate resulting in a rough film surface. Therefore, most studies on the initial stages of the ELOG process have focused on regrowth using the dielectric layers patterned along the $\langle 1\bar{1}00 \rangle$ direction.

The respective GaN domains overgrown from the GaN template in the window regions are merged in the center of the dielectric mask during lateral growth. However, several types of defects are often observed near the coalescence boundaries and dielectric/GaN interfaces [69–71]. The width and distance of these dielectric patterns determine when the coalescence events occur and the distribution of the defects generated in the coalescence boundaries. The representative defects observed in this region are voids, a few nm $\sim$ a few

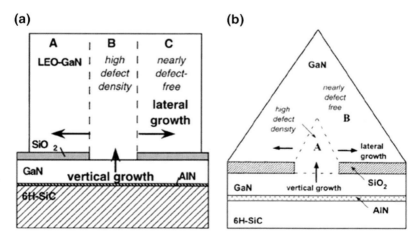

Fig. 6.33. (a) Schematic diagrams of the ELOG in a selectively grown (a) GaN stripe and (b) GaN hexagonal pyramid. Reprinted with permission from [63]. Copyright (2001), Elsevier

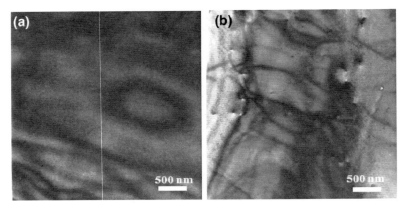

Fig. 6.34. Plan-view TEM images obtained from the top surfaces of (**a**) mask and (**b**) window regions. The ELOG GaN corresponding to the mask region is nearly dislocation free

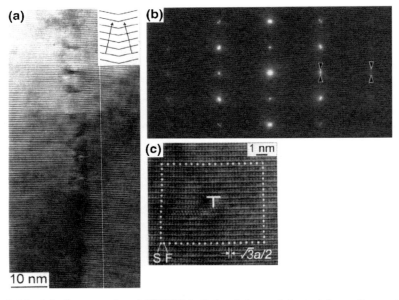

Fig. 6.35. (**a**) Cross-sectional HRTEM of the defect originated from the center of the SiO$_2$ mask. The tilting of the c-axis directions in two adjacent GaN crystals is illustrated in the inset. (**b**) A corresponding diffraction pattern. (**c**) Burgers circuit around one of the dislocations forming the defect in (**a**). Reprinted with permission from [69]. Copyright (1998), American Institute of Physics

μm size, a high density of horizontal dislocations, and stacking faults aligned parallel to the (0001) plane as shown in Figs. 6.35 and 6.36 [72]. The possible origins of the formation of these defects are as follows: (1) low angle grain

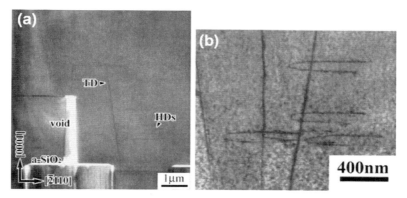

Fig. 6.36. (a) Cross-sectional TEM image showing the formation of internal voids on the dielectric and (b) Magnified cross-sectional image of (a) showing horizontal dislocations with a loop or semiloop shape on the c-plane. Reprinted with permission from [72]. Copyright (2001), Wiley

boundary due to the small tilting of each grain produced in two windows as shown in Fig. 6.35a, (2) the formation of internal voids on the dielectric under non optimized growth conditions during coalescence as shown in Fig. 6.36a, and (3) the increased thermal stress induced by large mismatch in the thermal expansion coefficients between GaN and the dielectric layer during the cooling process after the growth.

According to Horibuchi's report [72], the ELOG GaN layer grown on a stripe with the narrow mask window width leads to an increase in the horizontal dislocations aligned on the (0001) plane, which are evenly distributed over the entire area of the mask and window. On the other hand, horizontal dislocations are mainly observed in the edge region of the mask for a wide mask window. Unlike the general TDs showing a linear shape, horizontal dislocations have a loop or semiloop shape on the c-plane as shown in Fig. 6.36b [72, 73].

Tomiya et al. reported that the amount of crystallographic tilt in the overgrown films can be reduced by introducing PECVD SiO_2 and SiN_x dielectric layers instead of electron-beam evaporated SiO_2 [74]. Wagner et al. examined the effect of the carrier gas composition on the morphology of the overgrown GaN layers [75]. They reported that an increase in the hydrogen content leads to morphological changes from trapezoidal to triangular shapes due to the reduced lateral growth velocity.

6.4.2 Defects in PENDEO Epitaxy (PE)

The ELOG process using a dielectric layer is effective in decreasing the dislocation density to $<10^7\,\mathrm{cm}^{-2}$, as explained in Sect. 6.4.1. Nevertheless,

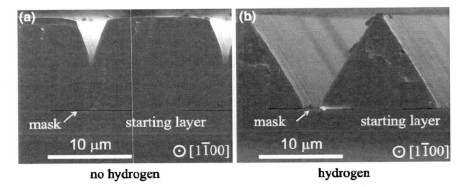

Fig. 6.37. Morphology of stripes along the [1$\bar{1}$00] grown using (**a**) pure nitrogen as the carrier gas, resulting in a trapezoidal cross section, and (**b**) hydrogen/nitrogen mixture as the carrier gas, resulting in a triangular cross section. Reprinted with permission from [75]. Copyright (2002), Applied Institute of Physics

the dislocation density in the window region is $10^8 \sim 10^9\,\mathrm{cm}^{-2}$ and the films with low defect density can be obtained in limited areas. A double ELOG process repeating the deposition of dielectric layers on the existing ELOG GaN was suggested to resolve this problem [76, 77]. In the second step, the GaN overgrown from the window region in the first step was covered with a second dielectric layer to suppress the propagation of dislocations. However, this method is quite complex and expensive because it requires additional fabrication processes as well as precise alignment of the photolithography process. In addition, there are many defects in the coalescence boundaries and dielectric/GaN interfaces. Therefore, Davis et al. proposed the pendeo-epitaxy (PE) method to obtain nitride samples with a uniform distribution and low dislocation density [78–81]. The terminology of "Pendeo" is derived from Latin meaning "to hang" or "to be suspended".

The PE process also deposits a dielectric layer as in the ELOG process. The detail sequences are as follows. First, a thick GaN seed layer is grown at a high temperature with low temperature nucleation layer followed by the deposition of the dielectric layer. After forming the patterns using photolithography, plasma etching is carried out to selectively remove the GaN seed layer. The GaN layer is completely removed until the substrate surface is exposed. The GaN epilayers are then regrown. Initially, no nitride films are deposited on the dielectric layer and the propagation of threading dislocations is suppressed. The remarkable difference between ELOG and PE is in the initial stage when the planes start the regrowth process. While the ELOG begins growing on the seed layer, the regrowth of GaN in the PE method begins along the side facets of the GaN below the dielectric layers because all GaN seed layers in the window region have been removed, as shown in Fig. 6.38. In particular, regrowth is performed by MOCVD at relatively high temperatures

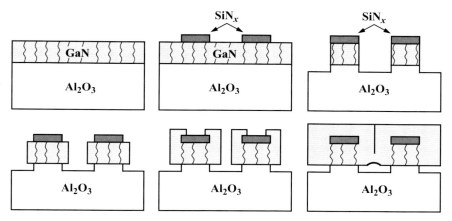

Fig. 6.38. Schematic diagrams showing the growth evolution of pendeo-epitaxy (PE) process

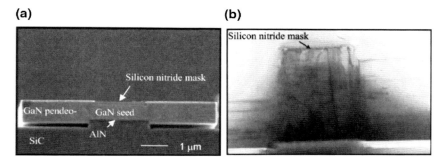

Fig. 6.39. (a) Cross-sectional SEM image of the initial stage of GaN PE grown on a 2 μm wide GaN seed stripe oriented along the $\langle 1\bar{1}00\rangle$ direction. Cross-sectional TEM image of a 1 μm wide GaN seed form oriented along the $\langle 1\bar{1}00\rangle$ direction, and the GaN PE layer which has coalesced over the silicon nitride seed mask. Reprinted with permission from [78]. Copyright (1999), American Institute of Physics

to promote lateral growth. As shown in Fig. 6.38, the (0001) surface is exposed in the initial step of regrowth, which leads to growth along the [0001] direction. Continuous growth results in the deposition of GaN on the dielectric layer, which makes the morphology of the suspended films appear similar to that of a bird's wing, as shown in Figs. 6.38 and 6.39a. Further growth leads to the coalescence of each GaN domain suspended in the side of the seed layer resulting in a flat surface.

If the microstructure of the resulting film is examined, the propagation of the dislocations in the mask regions are blocked by the dielectric layer and the wing region also shows a reduced dislocation density. The generation of TDs originates from the large mismatch in the lattice constant and thermal expansion coefficient between the nitride film and sapphire substrate.

However, in the PE process the overgrown GaN layers in the wing region do not come in contact with the substrate because the GaN on the substrate is only formed by the lateral growth of the seed layer. Therefore, the surface of the overgrown GaN layer would have a low and uniform dislocation density, as shown in Fig. 6.39b. In an analogy with ELOG, newly generated defects such as voids, horizontal dislocations, and stacking faults are observed near the dielectric/nitride interface as a result of thermal expansion coefficient mismatch between the dielectric layer and nitride. Crystallographic tilting between wings also occurs. However, PE can be applied to various substrates such as Si, SiC, and sapphire, due to the advantage of having no contact with the substrates.

A maskless PE process was also proposed by the group who first introduced PE using a dielectric mask [82–84]. Coating with dielectric materials in the ELOG and PE processes leads to several problems including defect generation induced by thermal stress, crystallographic tilting between wings, and the incorporation of impurities through the diffusion of dielectric materials [74,85]. Meanwhile, lateral and vertical growth occur simultaneously during regrowth if the dielectric layer is removed after forming the GaN patterns, as shown in Fig. 6.40. In particular, maskless PE GaN suspended on a seed layer (no contact with substrate) can be grown if suitable growth conditions are selected to promote relatively lateral growth. Before regrowth, longtime etching resulting in a deep trench on the substrate is more effective in removing contact with the substrate [86]. In this case, a portion of the threading dislocations appear on the GaN regrown from the seed layer, while in the etched regions, the source of dislocation formation is removed due to noncontact with the substrate [82–84]. In addition, the crystallographic tilting observed in the PE process with a dielectric mask is significantly reduced due to the reduced thermal stress by eliminating the dielectric film.

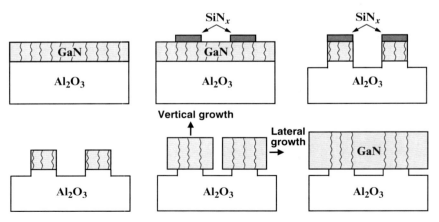

Fig. 6.40. Schematic diagrams showing the growth evolution of maskless pendeoepitaxy process

6.4.3 Defects in Facet-Controlled Epitaxial Lateral Overgrowth (FACELO)

In the Sect. 6.4.1, it is shown that the ELOG process is effective in reducing the number of TDs in the mask region. Cross-sectional TEM clearly shows the propagation of TDs along the [0001] direction in the films overgrown from the window region and crystallographic tilting in the center of the mask region, even though the propagation of dislocations is blocked in the mask region. Therefore, using the ELOG process, the suppression of dislocation propagation is limited only in the mask region. As explained in the Sect. 6.4.1, the morphology of the overgrown GaN strongly depends on the stripe direction of the dielectric layer. In particular, in the case of the $\langle 1\bar{1}00 \rangle$ direction, the changes in growth temperature and reactor pressure allow for control of the growth morphology from a rectangular to trapezoidal and triangular shape. For the trapezoidal and triangular shapes, the coalescence of GaN domains overgrown from the window is delayed due to the increased growth along the vertical direction and a thicker film is needed to make a flat surface. Interestingly, some researchers reported that many of the TDs originally propagating along the c-axis were bent in the GaN stripes with a triangular and trapezoidal shape during the initial regrowth stage [86–89]. In the GaN and ZnO material systems with the wurtzite structure, TDs were found to originate at the film/substrate interface and propagate into the surface through the entire layer along the vertical direction. In addition, the mobility of dislocations in the GaN is a factor of $10^{11} \sim 10^{14}$ lower than AlGaAs due to the high bond strength. Therefore, it is difficult to reduce dislocation density through a reaction of dislocations in nitride semiconductors. Defects such as TDs generate a stress field that can be relaxed if the defects meet the surface and disappear [90–92]. Therefore, defects generally propagate toward the surface during crystal growth. According to the traditional defect behavior, if a GaN film with a triangular or trapezoidal shape is formed, the direction of dislocation lines can be changed and redirected toward the surface, which increases the probability of reaction between dislocations. In extreme cases, the formation of horizontal dislocations induced by bending dislocations prevents the propagation of dislocations into the active layer, consisting of InGaN/GaN MQWs [12].

As suggested by Vennegues et al. and Hiramatsu et al. [89, 91], this principle was applied to the two-step overgrowth technique, which is known as two-step ELOG or facet-controlled epitaxial lateral overgrowth (FACELO). In the first step of FACELO, GaN with a triangular shape covering the window region bends the TDs to the horizontal direction. In the second step, the growth conditions enhancing lateral growth are selected for the rapid coalescence of GaN domains. Consequently, the propagation of TDs in the mask and window regions is suppressed by blocking through the dielectric layer and bending dislocations, respectively, as shown in Fig. 6.41.

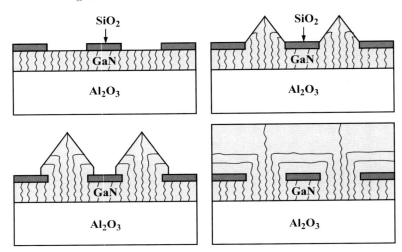

Fig. 6.41. Schematic diagrams showing the growth evolution of two-step ELOG or facet-controlled epitaxial lateral overgrowth (FACELO) process

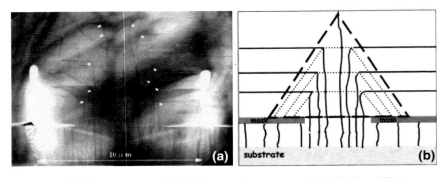

Fig. 6.42. (a) Cross-sectional TEM image of a two-step ELOG film. *White arrows* indicate the bending points of dislocations. (b) Diagram of two-step ELOG. *Dotted lines* represent the shape of the ELOG material at different stages of the first part of the two-step ELOG process. The *dashed triangle* represents the shape at the end of this first part of the process. Reprinted with permission from [93]. Copyright (1999), Wiley

Figures 6.41 and 6.42 show details of the growth procedure and microstructural properties of the GaN films using FACELO. First, regrowth of the GaN film is performed on dielectric layers patterned along the $\langle 1\bar{1}00 \rangle$ direction. In the first step, a GaN film with a triangular or trapezoidal shape can be obtained by deposition at a relatively low temperature and high reactor pressure, where the $\{11\bar{2}2\}$ side facets were exposed. The formation of a GaN stripe showing a perfect triangular shape in the initial stage leads to the inclination of almost all dislocations to the horizontal (0001) plane. In the case of the triangular shape, some dislocations originating from the center area still propagate along the c-axis, as shown in Fig. 6.42 [93]. In the next step, the growth temperature and reactor pressure are increased and decreased, respectively, which promote lateral growth for fast coalescence. The surface morphology of the resulting film is changed from a triangular to rectangular shape. The alternative method facilitating lateral growth is the addition of impurities such as $(MeCp)_2Mg$ [91]. The main purpose of this step is to form coalescence boundaries with a low defect density and a GaN film with the improved surface roughness. In contrast, there is no change in the propagation direction of dislocations. However, there should be a decrease in the number of defects generated in the coalescence boundaries by optimizing the growth conditions. Finally, the FACELO process leads to the growth of high-quality GaN films with a low and uniform distribution of dislocations over the entire surface, where the dislocation density ranges from 10^6 to 10^7 cm^{-2}. This method is cost-effective and technically simple because no additional process is required and only the growth conditions are changed in situ.

The inclined side facets are $\{11\bar{2}2\}$ planes, which show a 58.4° angle with the (0001) basal plane. In order to approach the surface, the dislocations should be inclined at an angle of 58.4° in the triangular stripe. However, most TDs are aligned towards the (0001) plane with a bending angle of 90°. Gradecak et al. [87] reported that the edge- and mixed-type dislocations that propagate along the $\langle 0001 \rangle$ direction are bent at the angles of 90° and $44 \pm 7°$, and $\sim$90°, 62°, and $41 \pm 7°$, respectively, with respect to the $\langle 0001 \rangle$ direction, while screw dislocations tend to propagate vertically. In addition, they predicted the inclination angle quantitatively using anisotropic elasticity theory. The results show that while screw dislocations propagate along the c-axis without bending, edge and mixed dislocations tend to redirect their line direction close to the surface, which is in agreement with their TEM observations [87].

6.4.4 Defects in Other Overgrowth Techniques

The two-step ELOG process explained in the Sect. 6.4.3 can be applied to various underlying layers. In the Maskless PE process using rectangular shaped GaN patterns, regrowth in the seed region leads to the continuous propagation of TDs. If the two-step regrowth technique showing different surface morphologies is utilized in the maskless PE, the formation of GaN stripes with side facets subsequently results in the bending of dislocations [93, 94]. The primary growth parameters affecting the surface morphology of overgrown GaN are growth temperature and reactor pressure. Figure 6.43 shows the cross-sectional SEM images of the GaN films overgrown at different growth temperatures on the GaN seed layers to confirm the effect of the growth temperature. The stripe patterns were formed on a GaN seed layer etched with stripe period of 4 μm using a dielectric film mask. The GaN overgrowth was performed at a reactor pressure of 100 mbar and a V/III ratio of 1965. By

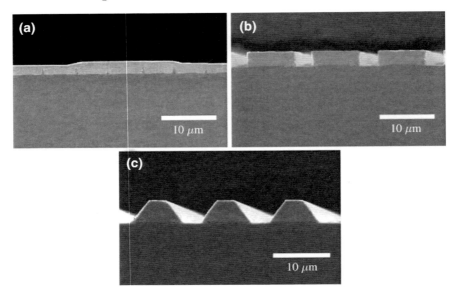

Fig. 6.43. Cross-sectional SEM images of GaN layers grown at (**a**) 1,160, (**b**) 1,110, and (**c**) 1,060°C on the patterned GaN. Reprinted with permission from [94]. Copyright (2004), Elsevier

decreasing the growth temperature, the surface morphology of GaN films changes gradually from a rectangular to a trapezoidal shape, which is due to the rapid growth of the $\{11\bar{2}2\}$ planes, as shown in Fig. 6.43c. It is thus clear that vertical growth is dominant at a lower growth temperature and the ratio of lateral to vertical growth rate is less than 1. Meanwhile, by increasing the growth temperature, the ratio of lateral to vertical growth rate became enhanced, which is believed to be due to the increase in lateral growth from the $\{11\bar{2}0\}$ sidewall facet, as shown in Figs. 6.43a and b. As mentioned in Sect. 6.4.3, the formation of trapezoidal shape at reduced growth temperature retard the coalescence observed in wing regions (window region), which indicated that thicker film is needed to obtain the flat surface. A decrease in the ratio of lateral to vertical growth rate is found to be another important parameter enhancing the degree of crystallographic tilt [95]. Therefore, the regrowth step in maskless PE was normally performed at higher temperatures and lower reactor pressures in order to facilitate lateral growth after the bending of dislocations.

As shown in Fig. 6.44, the cross-sectional TEM images showing the distribution of dislocations observed in the overgrown GaN films deposited at high temperature indicate that the window region has considerably fewer TDs, because there are no interfaces between the GaN film and sapphire substrate. In contrast, a similar dislocation density to the underlying GaN layer in the seed regions (mask area) is observed.

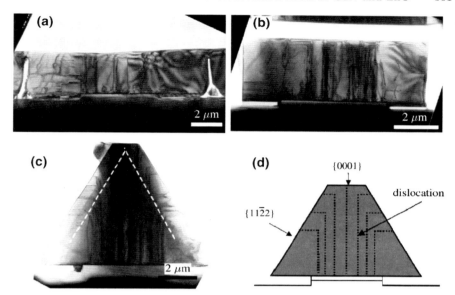

Fig. 6.44. Cross-sectional TEM images of GaN layers grown at (**a**) 1,160, (**b**) 1,110, and (**c**) 1,060°C on the patterned GaN. The *dotted line* in (**c**) indicates the starting point of the dislocation bending. (**d**) Schematic of the dislocation bending observed in the GaN layers of *triangular shape*. Reprinted with permission from [94]. Copyright (2004), Elsevier

For the GaN layer overgrown at the reduced temperature, the trapezoidal-shape was observed as shown in Fig. 6.44c. The dotted line indicates the position that the propagation of dislocations changes from vertical to horizontal direction. Thus, for the second step performed at lower temperature, the propagation of dislocations into the surface becomes diminutive due to the bending of dislocations by the formation of the $\{11\bar{2}2\}$ facet planes. However, a GaN film with a triangular shape without a (0001) plane results in the delay of complete coalescence of the wing regions and crystallographic tilting from each GaN domains. Thus, the two-step growth is introduced to simultaneously utilize the advantages of these two growth modes, because the bending of TDs is mostly accomplished at the initial stage showing triangular shape, as shown in Figs. 6.44c and d.

For the first step, a reduced growth temperature and increased reactor pressure is chosen to enhance vertical growth with a triangular shape resulting in the bending of TDs. Then, the growth at high temperatures and low reactor pressures is performed to obtain the flat surface of the final sample. The growth of the films with a low dislocation density could be achieved in the overgrown GaN deposited on the patterned seed layer by adopting a two-step growth controlling growth temperature and reactor pressure, as shown in Fig. 6.45b. The dislocation density observed in the surfaces of the window

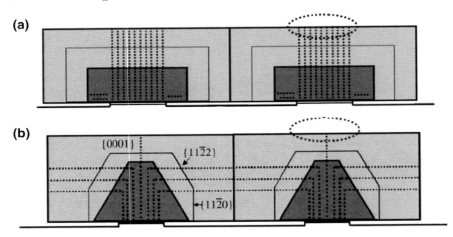

Fig. 6.45. Schematic models of dislocation behavior for (**a**) continuous growth and (**b**) two-step growth. Reprinted with permission from [94]. Copyright (2004), Elsevier

region as well as the seed region was significantly reduced, compared to the sample using continuous growth at single temperature. The essential reasons for the reduction of dislocations are the bending of dislocations and noncontact between substrate and film, as shown in Fig. 6.46c.

Both ELOG and PE require the deposition of a GaN seed layer and a dielectric or photoresist prior to the growth of overgrown GaN. Some techniques to reduce the process step in epitaxial overgrowth have been suggested. Among them, DHELO (direct heteroepitaxial lateral overgrowth) grows the patterned dielectric layers on the sapphire substrate directly followed by the epitaxial growth of GaN [96,97], as shown in Fig. 6.47. This process does not require the etching of GaN and sapphire. A two-step process was also used during epitaxial growth to redirect the TDs originating from the GaN/sapphire interfaces. In this process, the first step showing a triangular shape was designed to bend TDs in the window regions and the second step was planed to enhance lateral growth for the propagation of dislocations into the horizontal direction in the mask regions. As shown in Fig. 6.48, the TDs in the opening region near the dielectric mask are bent toward the lateral direction over the mask. In contrast, no TDs were observed near the surface over the dielectric mask due to noncontact between the GaN and sapphire. This result indicates a significant decrease in the density of TDs in the mask regions through two-step overgrowth.

6.4.5 Other Growth and Process Techniques for Defect Reduction

ELOG and PE procedures offer an interesting opportunity for significantly reducing the dislocation density through the blocking and bending of dislocations.

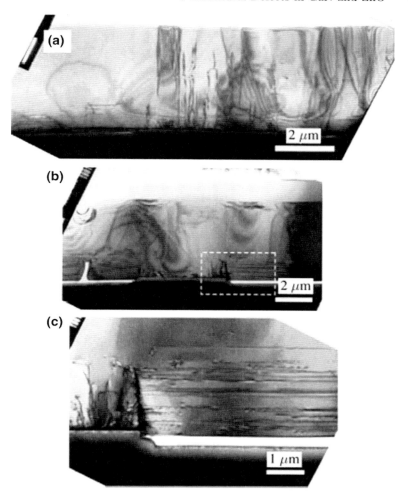

Fig. 6.46. Cross-sectional TEM images of GaN films deposited by (**a**) continuous growth and (**b**) two-step growth on the patterned GaN. (**c**) High magnification image obtained from the *dashed rectangle* in (**b**). Reprinted with permission from [94]. Copyright (2004), Elsevier

These require complex ex situ processes, such as lithography and etching, and finally result in low production yields and high cost compared with the direct growth of the GaN and ZnO on sapphire substrates. Recently, a variety of methods using a single step were presented focusing mainly on the introduction of new concepts of buffer layer [98–100]. These results demonstrated a reduced dislocation density, etch pit density, and small asymmetrical reflection, based on TEM and high-resolution X-ray diffraction (HRXRD).

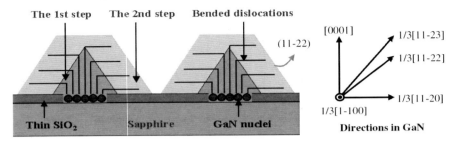

Fig. 6.47. Schematic diagram of the DHELO–GaN growth by the two-step condition. Reprinted with permission from [97]. Copyright (2004), Wiley

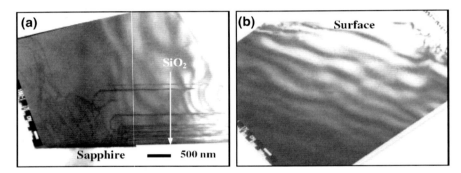

Fig. 6.48. Cross-sectional bright-field TEM images of the DHELO–GaN film grown by the two-step condition: (**a**) a region near the SiO_2 mask, (**b**) an almost dislocation-free region near the surface. Reprinted with permission from [97]. Copyright (2004), Wiley

The inserting layers applied in the buffer layer were SiN_x, Mg_xN_y, and ultra thin MgO. Wang et al. [98] deposited a thin SiN_x layer on a sapphire substrate before the growth of a nucleation layer. The introduction of this layer plays a role in promoting the lateral growth of nucleation islands. Moreover, an increase in island size leads to a decrease in TDs generated in the boundaries between the islands. Chen et al. [99] introduced an extremely thin MgO buffer, which was mainly used in ZnO film growth with a plasma assisted MBE system. The coverage of the strain related MgO buffer on the sapphire substrate allows the 2D epitaxial growth due to reduced surface energy through a wetting process. Based on the observation that dislocations decrease rapidly beyond 20 nm, it is expected that they would interact with each other and become quickly annihilated in the initial stages of ZnO growth. The multiple Mg_xN_y/GaN buffer layers with 12 pairs grown by MOCVD suggested by Tun et al. [100] prolong the 3D–2D transition time and result in a low nuclei density. Because the coalescence of island–island results in the pure edge dislocations with low angle grain boundary, the lower density of the islands in the buffer layer indicates a decrease in the density of edge-type dislocations (or low angle grain boundaries). This method resulted in improved GaN properties, such as high mobility, low background concentration, low etch pit density.

In similar methods, various interlayers were inserted during the deposition of the GaN and ZnO to obtain low defect density films. The interlayers used were SiN_x, porous TiN, ScN, and CrN nanoislands [101–104]. Kappers et al. [101] reported a process for reducing the number of dislocations by inserting SiN_x interlayers during the high temperature GaN growth. After depositing the SiN_x dielectric layer on high temperature GaN, the regrowth showed a triangular morphology with side facets in the initial stage producing TDs redirected to the horizontal direction. Therefore, there was a decrease in the number of dislocations in the coalescence boundaries. Fu et al. [102] reported a method using porous TiN templates, which were formed after the deposition of an e-beam evaporated Ti layer followed by thermal annealing in a NH_3 and H_2 gas mixture. The OMVPE GaN grown on TiN template showed a lower dislocation density by up to one order of magnitude. Based on the observation in the voids of the TiN layer, the GaN film is expected to show lateral overgrowth behavior, which prevents the propagation of dislocations into the surface due to the elimination of GaN/TiN interfaces. Moram et al. [103] suggested that a polycrystalline ScN layer made by depositing Sc metal is quite effective in reducing the number of dislocations by acting as a dislocation blocking layer without lithography. The reasons for the decrease in dislocations are closely related to void formation during annealing and the matching of the lattice spacing between the ScN interlayer and GaN. On the other hand, Ha et al. [104] decreased the dislocation density in GaN films on AlN/sapphire templates using CrN nanoislands. The Cr metal layers coated on the templates were first nitrided. The CrN layers partially covered the AlN layer and showed nanosized tetrahedral shapes. These CrN nanoislands block the propagation of dislocations from the AlN templates and allow the dislocations formed in the GaN overgrown on the surface covered with CrN to be bent horizontally, which significantly reduces the number of dislocations observed on the surface, as shown in Fig. 6.49.

With the exception of overgrowth based on ELOG and PE, various techniques have been proposed to reduce the number of defects. Among them, Benamara et al. [105] reported a method to suppress the propagation of TDs towards the active region using an intermediate layer. The growth of epitaxial GaN at high temperatures is interrupted and the growth temperature is reduced to allow the growth of an intermediate layer at low temperatures, which is a similar temperature to the growth of a nucleation layer. The resulting film shows quite a low TD density $\leq 8 \times 10^7 \, cm^{-2}$. The morphology of the intermediate layer exhibits 3D islands. Subsequently, the materials redeposited at high temperature grew both laterally and vertically over the islands with inclined side facets until the growth fronts from each island coalesced. The bending of TDs was observed during the initial stages of the regrowth, which is similar to that observed using overgrowth techniques.

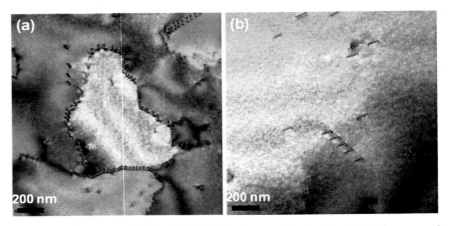

Fig. 6.49. Plan view bright-field TEM micrographs for the GaN films on the AlN/sapphire templates without (**a**) and with (**b**) the CrN. Reprinted with permission from [104]. Copyright (2008), American Institute of Physics

Consequently, it is expected that the GaN surface has a lower dislocation density due to the bending of dislocations. However, small voids were observed at the boundaries of islands.

References

1. H.P. Maruska, J.J. Tietjen, Appl. Phys. Lett. **15**, 327 (1969)
2. H. Amano, N. Sawaki, I. Akasaki, T. Toyoda, Appl. Phys. Lett. **48**, 353 (1986)
3. S. Nakamura, M. Senoh, T. Mukai, Jap. J. Appl. Phys. **30**, 1708 (1991)
4. A. Shintani, S. Minagawa, J. Electrochem. Soc. **123**, 706 (1976)
5. W. Qian, M. Skowronski, M.D. Graef, K. Doverspike, L.B. Rowland, D.K. Gaskell, Appl. Phys. Lett. **66**, 1252 (1995)
6. Z.L. Weber, H. Sohn, N. Newman, J. Washburn, J. Vac. Sci. Technol. B **13**, 1578 (1995)
7. F.A. Ponce, B.S. Krusor, J.S. Major Jr., W.E. Plano, D.F. Welch, Appl. Phys. Lett. **67**, 410 (1995)
8. D.J. Smith, D. Chandrasekhar, B. Sverdlov, A. Botchkrev, A. Salvador, H. Morkoc, Appl. Phys. Lett. **67**, 1830 (1995)
9. W. Qian, G.S. Rohrer, M. Skowronski, K. Doverspike, L.B. Rowland, D.K. Gaskill, Appl. Phys. Lett. **67**, 2284 (1995)
10. B.N. Sverdlov, G.A. Martin, H. Morkoc, D.J. Smith, Appl. Phys. Lett. **67**, 2063 (1995)
11. N.E. Lee, R.C. Powell, Y.W. Kim, J.E. Greene, J. Vac. Sci. Technol. A **13**, 2293 (1995)
12. S. Nakamura, Science **281**, 956 (1998)
13. S.D. Lester, F.A. Ponce, M.G. Craford, D.A. Steigerwald, Appl. Phys. Lett. **66**, 1249 (1995)

14. S. Nakamura, M. Senoh, S. Nagahama, N. Iwasa, T. Yamada, T. Matsushita, H. Kiyoku, Y. Sugimoto, T. Kozaki, H. Umemoto, M. Sano, K. Chocho, Jpn. J. Appl. Phys. **36**, L1568 (1997)
15. D.M. Bagnall, Y.F. Chen, Z. Zhu, T. Yao, S. Koyama, M.Y. Shen, T. Goto, Appl. Phys. Lett. **70**, 2230 (1997)
16. Z.K. Tang, G.K.L. Wong, P. Yu, M. Kawasaki, A. Ohtomo, H. Koinuma, Y. Segawa, App. Phys. Lett. **72**, 3270 (1998)
17. S.K. Hong, H.J. Ko, Y. Chen, T. Yao, J. Cryst. Growth **209**, 537 (2000)
18. S.H. Lim, D. Shindo, H.B. Kang, K. Nakamura, J. Vac. Sci. Technol. B **19**, 506 (2001)
19. S.H. Lim, J. Washburn, Z.L. Weber, D. Shindo, J. Vac. Sci. Technol. A **19**, 2601 (2001)
20. F. Vigue, P. Vennegues, S. Vezian, M. Laugt, J.P. Faurie. Appl. Phys. Lett. **79**, 194 (2001)
21. D.B. Williams, C.B. Carter, Transmission Electron Microscopy (Springer, New York, 1996)
22. T. Goringe, Transmission Electron Microscopy of Materials (Wiley, New York, 1979)
23. P.G. Partridge, Metallurgical Rev. **12**, 169 (1967)
24. R. Datta, M.J. Kappers, J.S. Barnard, C.J. Humphreys, Appl. Phys. Lett. **85**, 3411 (2004)
25. W.J. Tunstall, P.B. Hirsch, J.W. Steeds, Phil. Mag. **9**, 99 (1964)
26. S.K. Hong, T. Yao, B.J. Kim, S.Y. Yoon, T.I. Kim, Appl. Phys. Lett. **77**, 82 (2000)
27. D.M. Follstaedt, N.A. Missert, D.D. Koleske, C.C. Mitchell, K.C. Cross, Appl. Phys. Lett. **83**, 4797 (2003)
28. S.K. Han, S.K. Hong, J.W. Lee, J.Y. Lee, J.H. Song, Y.S. Nam, S.K. Chang, T. Minegishi, T. Yao, J. Cryst. Growth **309**, 121 (2007)
29. F.C. Frank, Acta Crystallogr. **4**, 497 (1951)
30. A.R. Verma, Phil. Mag. **42**, 1005 (1951)
31. S. Mardix, A.R. Lang, Phil. Mag. A **24**, 683 (1971)
32. W. Qian, M. Skowronski, K. Doverspike, L.B. Rowland, D.K. Gaskill, J. Cryst. Growth **151**, 396 (1995)
33. S.W. Lee, D.C. Oh, H. Goto, J.S. Ha, H.J. Lee, T. Hanada, M.W. Cho, T. Yao, S.K. Hong, H.Y. Lee, S.R. Cho, J.W. Choi, J.H. Choi, J.H. Jang, J.E. Shin, J.S. Lee, Appl. Phys. Lett. **89**, 132117 (2006)
34. S.K. Hong, B.J. Kim, H.S. Park, Y. Park, S.Y. Yoon, T.I. Kim, J. Cryst. Growth 191, **275** (1998)
35. M.E. Hawkridge, D. Cherns, Appl. Phys. Lett. **87**, 221903 (2005)
36. I. Arslan, N.D. Browning, Phys. Rev. Lett. **91**, 165501 (2003)
37. A.F. Wright, J. Appl. Phys. **82**, 5259 (1997)
38. C. Stampfl, C.G. Van de Walle, Phys. Rev. B, **57**, R15052 (1998)
39. J.E. Northrup, J. Neugebayer, L.T. Romano, Phys. Rev. Lett. **77**, 103 (1996)
40. C. Iwamoto, X.Q. Shen, H. Okumura, H. Matuhata, Y. Ikuhara, Appl. Phys. Lett. **79**, 3941 (2004)
41. L.T. Romano, J.E. Northrup, A.J. Ptak, T.H. Myers, Appl. Phys. Lett. **77**, 2479 (2000)
42. J.W. Lee, Dissertation, Korea Advanced Institute of Science and Technology, Korea (2005)

43. V. Ramachandran, R.M. Feenstra, W.L. Sarney, L. Salamanca-Riba, J.E. Northrup, L.T. Romano, D.W. Greve, Appl. Phys. Lett. **75**, 808 (1999)
44. D.J.H. Lambert, M.M. Wong, U. Chowdhury, C. Collins, T. Li, H.K. Kwon, B.S. Shelton, T.G. Zhu, J.C. Campbell, R.D. Dupuis, Appl. Phys. Lett. **77**, 1900 (2000)
45. P. Sandvik, K. Mi, F. Shahedipour, R. McClintock, A. Yasan, P. Kung, M. Razeghi, J. Cryst. Growth **231**, 366 (2001)
46. D.J.H. Lambert, M.M. Wong, U. Chowdhury, C. Collins, T. Li, H.K. Kwon, B.S. Shelton, T.G. Zhu, J.C. Campbell, R.D. Dupuis, Appl. Phys. Lett. **88**, 113505 (2006)
47. D. Kapolnek, X.H. Wu, B. Heying, S. Keller, B.P. Keller, U.K. Mishra, S.P. DenBaars, J.S. Speck, Appl. Phys. Lett. **67**, 1541 (1995)
48. C.F. Shin, N.C. Chen, S.Y. Lyn, K.S. Liu, Appl. Phys. Lett. **86**, 211103 (2005)
49. J. Han, T.-B. Ng, R.M. Biefeld, M.H. Crawford, D.M. Follstaedt, Appl. Phys. Lett. **71**, 3114 (1997)
50. K.S. Kim, C.S. Oh, K.J. Lee, G.M. Yang, C.H. Hong, K.Y. Lim, H.J. Lee, A. Yoshikawa, J. Appl. Phys. **85**, 8441 (1999)
51. S.J. Rosner, E.C. Carr, M.J. Ludowise, G. Girolami, H.I. Erikson, Appl. Phys. Lett. **70**, 420 (1997)
52. F.A. Ponce, D.P. Bour, W. Gotz, N.M. Johnson, H.I. Helava, I. Grzegory, J. Jun, S. Porowski, Appl. Phys. Lett. **68**, 917 (1995)
53. C.J. Collins, T. Li, D.J.H. Lambert, M.M. Wong, R.D. Dupuis, J.C. Campbell, Appl. Phys. Lett. **77**, 2810 (2000)
54. P. Kozodoy, J.P. Ibbetson, H. Marchand, P.T. Fini, S. Keller, J.S. Speck, S.P. DenBaars, U.K. Mishra, Appl. Phys. Lett. **73**, 975 (1998)
55. V. Narayanan, K. Lorenz, W. Kim, S. Mahajan, Appl. Phys. Lett. **78**, 1544 (2001)
56. S. Nakamura, M. Senoh, S. Nagahama, N. Iwasa, T. Yamada, T. Matsushita, H. Kiyoku, Y. Sugimoto, T. Kozaki, H. Umemoto, M. Sano, K. Chocho, Appl. Phys. Lett. **72**, 211 (1998)
57. F. Bertram, T. Riemann, J. Christen, A. Kaschner, A. Hoffmann, C. Thomsen, K. Hiramatsu, Appl. Phys. Lett. **74**, 359 (1999)
58. T. Paskova, E.M. Goldys, P.P. Paskov, Q. Wahab, L. Wilzen, M.P. de Jong, B. Monemar, Appl. Phys. Lett. **78**, 4130 (2001)
59. Z. Liliental-Weber, D. Cherns, J. Appl. Phys. **89**, 7833 (2001)
60. S. Nakamura, J. Cryst. Growth **195**, 242 (1998)
61. K. Hiramatsu, H. Matsushima, T. Shibata, Y. Kawagachi, N. Sawaki, Mater. Sci. Eng. B **59**, 104 (1999)
62. Y. Honda, Y. Iyechika, T. Maeda, H. Miyake, K. Hiramatsu, H. Sone, N. Sawaki, Jpn. J. Appl. Phys. **38**, L1299 (1999)
63. T.S. Zheleva, O.H. Nam, W.M. Ashmawi, J.D. Griffin, R.F. Davis, J. Cryst. Growth **222**, 706 (2001)
64. A. Sakai, H. Sunakawa, A. Usui, Appl. Phys. Lett. **71**, 2259 (1997)
65. O.H. Nam, M.D. Bremser, T.S. Zheleva, R.F. Davis, Appl. Phys. Lett. **71**, 2638 (1997)
66. Y. Kato, S. Kitamura, K. Hiramatsu, N. Sawaki, J. Cryst. Growth **144**, 133 (1994)
67. K. Hiramatsu, K. Nishiyama, A. Motogaito, H. Miyake, Y. Iyechika, T. Maeda, Phys. Stat. Sol. A **176**, 535 (1999)

68. T.S. Zheleva, O.H. Nam, M.D. Bremser, R.F. Davis, Appl. Phys. Lett. **71**, 2472 (1997)
69. A. Sakai, H. Sunakawa, A. Usui, Appl. Phys. Lett. **73**, 481 (1998)
70. H. Marchand, X.H. Wu, J.P. Ibbetson, P.T. Fini, P. Kozodoy, S. Keller, J.S. Speck, S.P. DenBaars, U.K. Mishra, Appl. Phys. Lett. **73**, 747 (1998)
71. S. Tanaka, Y. Honda, N. Sawaki, M. Hibino, Appl. Phys. Lett. **79**, 955 (2001)
72. K. Horibuchi, S. Nishimoto, M. Sueyoshi, N. Kuwano, H. Miyake, K. Hiramatsu, Phys. Stat. Sol. A **192**, 360 (2002)
73. Z.L. Weber, D. Cherns, J. Appl. Phys. **89**, 7833 (2001)
74. S. Tomiya, K. Funato, T. Asatsuma, T. Hino, S. Kijima, T. Asano, M. Ikeda, Appl. Phys. Lett. **77**, 636 (2000)
75. V. Wagner, O. Parillaud, H.J. Buhlmann, M. Ilegems, S. Gradecak, P. Stadelmann, T. Riemann, J. Christen, J. Appl. Phys. **92**, 1307 (2002)
76. O.H. Nam, T.S. Zheleva, D.B. Thomson, R.F. Davis, Mater. Res. Soc. Symp. Proc. **482**, 301 (1998)
77. M. Benyoucef, M. Kuball, B. Beaumont, V. Bousquet, Appl. Phys. Lett. **81**, 2370 (2002)
78. K. Linthicum, T. Gehrke, D. Thomson, E. Carlson, P. Rajagopal, T. Smith, D. Batchelor, R. Davis, Appl. Phys. Lett. **75**, 196 (1999)
79. T. Zheleva, S. Smith, D. Thomson, K. Linthicum, P. Rajagopal, R.F. Davis, J. Electr. Mater. **28**, L5 (1999)
80. T. Zheleva, S. Smith, D. Thomson, T. Gehrke, K. Linthicum, P. Rajagopal, E. Carlson, W. Ashmawi, R.F. Davis, MRS Internet J. Nitride Semicond. Res. **4S1**, G338 (1999)
81. K.J. Linthicum, T. Gehrke, D.B. Thomson, K.M. Tracy, E.P. Carlson, T.P. Smith, T.S. Zheleva, C.A. Zorman, M. Mehregany, R.F. Davis, MRS Internet J. Nitride Semicond. Res. **4S1**, G49 (1999)
82. J.B. Li, J.C. Tedenac, J. Electron. Mater. **31**, 421 (2002)
83. R.I. Barabash, G.E. Ice, W. Liu, S. Einfeldt, A.M. Roskowski, R.F. Davis, J. Appl. Phys. **97**, 013504 (2005)
84. A.M. Roskowski, E.A. Preble, S. Einfeldt, P.M. Miraglia, J. Schuck, R. Grober, R.F. Davis, Optoelectr. Rev. **10**, 261 (2002)
85. A. Sakai, H. Sunakawa, A. Kimura, A. Usui, Appl. Phys. Lett. **76**, 442 (2000)
86. C.I.H. Ashby, C.C. Mitchell, J. Han, N.A. Missert, P.P. Provencio, D.M. Follstaedt, G.M. Peake, L. Griego, Appl. Phys. Lett. **77**, 3233 (2000)
87. S. Gradecak, P. Stadelmann, V. Wagner, M. Ilegems, Appl. Phys. Lett. **85**, 4648 (2004)
88. P. Gibart, Rep. Prog. Phys. **67**, 667 (2004)
89. K. Hiramatsu, K. Nishiyama, M. Onishi, H. Mizutani, M. Narukawa, A. Motogaito, H. Miyake, Y. Iyechika, T. Maeda, J. Cryst. Growth **221**, 316 (2000)
90. Z.L. Weber, M. Benamara, W. Swider, J. Washburn, J. Park, P.A. Grudowski, C.J. Eiting, R.D. Dupuis, MRS Internet J. Nitride Semicond. Res. **4S1**, 4.6 (1999)
91. P. Vennéguès, B. Beaumont, V. Bousquet, M. Vaille, P. Gibart, J. Appl. Phys. **87**, 4175 (2000)
92. J.P. Hirth, J. Lothe, Theory of Dislocations (Wiley, New York, 1982)
93. V. Bousquet, P. Vennéguès, B. Beaumont, M. Vaille, P. Gibart, Phys. Stat. Sol. B **216**, 691 (1999)

94. H.K. Cho, D.C. Kim, H.J. Lee, H.S. Cheong, C.H. Hong, Superlattice. Microst. **36**, 385 (2004)
95. H.S. Cheong, Y.K. Hong, C.H. Hong, Y.H. Choi, S.J. Leem, H.J. Lee, Phys. Stat. Sol. A **192**, 377 (2002)
96. X. Zhang, R.R. Li, P.D. Dapkus, D.H. Rich, Appl. Phys. Lett. **77**, 2213 (2000)
97. H.S. Cheong, M.K. Yoo, H.G. Kim, S.J. Bae, C.S. Kim, C.-H. Hong, J.H. Baek, H.J. Kim, Y.M. Yu, H.K. Cho, Phys. Stat. Sol. (b) **241**, 2763 (2004)
98. T. Wang, Y. Moreshima, N. Naoi, S. Sakai, J. Cryst. Growth **213**, 188 (2000)
99. Y. Chen, S.K. Hong, H.J. Ko, V. Kirshner, H. Wenisch, T. Yao, K. Inaba, Y. Segawa, Appl. Phys. Lett. **78**, 3352 (2001)
100. C.J. Tun, C.H. Kuo, Y.K. Fu, C.W. Kuo, C.J. Pan, G.C. Chi, Appl. Phys. Lett. **90**, 212109 (2007)
101. M.J. Kappers, R. Datta, R.A. Oliver, F.D.G. Rayment, M.E. Vickers, C.J. Humphreys, J. Cryst. Growth **300**, 70 (2007)
102. Y. Fu, Y.T. Moon, F. Yun, Ü. Özgür, J.Q. Xie, S. Dogan, H. Morkoç, C.K. Inoki, T.S. Kuan, L. Zhou, D.J. Smith, Appl. Phys. Lett. **86**, 043108 (2005)
103. M.A. Moram, Y. Zhang, M.J. Kappers, Z.H. Barber, C.J. Humphreys, Appl. Phys. Lett. **91**, 152101 (2007)
104. J.S. Ha, H.J. Lee, S.W. Lee, H.J. Lee, S.H. Lee, H. Goto, M.W. Cho, T. Yao, S.K. Hong, R. Toba, J.W. Lee, J.Y. Lee, Appl. Phys. Lett. **92**, 091906 (2008)
105. M. Benamara, Z. Liliental-Weber, S. Kellermann, W. Swider, J. Washburn, J. Mazur, E.D. Bourret-Courchesne, J. Cryst. Growth **218**, 447 (2000)

7

Optical Properties of GaN and ZnO

J.-H. Song

Abstract. A brief review on the optical properties of wurtzite ZnO and GaN is presented in this chapter with an emphasis on comparison between the materials. The properties of free excitons and impurity-bound excitons, such as their energetic positions and binding energies, are summarized. The localization energy and the ionization energy of the dominant impurities obtained by emission spectroscopy are also presented. Typical aspects of emissions from donor–acceptor pairs, free-to-bound transition, and deep level recombination are discussed. Several experimental characteristics of the relevant heterostructures, InGaN/GaN and MgZnO/ZnO, are also given below. Basic optical methods characterizing the effects of internal electric fields and carrier-localization are summarized. The unique properties of polarization sensitive emissions from nonpolar films are presented. Based on the valence band structures, the polarization selection rules can be obtained in simpler forms. Some recent reports will also be introduced stating that the anisotropic strain in nonpolar films plays an important role in deciding the polarization selectivity. The results of Raman spectroscopy are summarized in the end, with the emphasis on deciding the residual strain and the carrier concentration.

7.1 Introduction

ZnO and GaN are two of the most intensely studied semiconductors in optoelectronic research field. Due to the recent realization of commercial products such as high-brightness light emitting diodes (LED) and laser diodes (LD) [1,2], GaN and its related materials have attracted world-wide attentions from researchers in both academic and industrial fields in the past decade. More than 2,000 GaN-related papers were published in 2006 and 2007 (about 2,500 in 2006 and 2,300 in 2007, respectively, according to the Web of Science). The slight decrease in the number of publications in 2007 may imply the maturation in the field of GaN-related material system, although many technical and fundamental issues are still pretty much intact, such as the efficiency droop in high injection current, the nonpolar based devices, the identification of defects, and thermal management problems for general lighting devices. ZnO also enjoys its renaissance since mid-1990s, when the epitaxially grown films and heterostructures of high quality became available. While ZnO has many similar features as GaN in optical properties, such as its wide bandgap (3.437 eV in ZnO and 3.507 eV in GaN at low temperature), the valence band structure [3], and the presence of piezo- and spontaneous-electric fields [4, 5], it also has its unique properties most represented by its large exciton binding energy (60 meV in ZnO, 26 meV in GaN). Current ZnO production has already reached $\sim 10^5$ tons per year, largely for the rubber and concrete industry ([6] and references therein). Zinc is much more abundant and environmental-friendly than, e.g., Indium or Gallium. ZnO can be produced by various methods which include relatively cheaper techniques, like gas-transport, evaporation, or sputtering. As a potential alternative of GaN related materials, and a material for new devices such as excitonic emitter, nonlinear optical devices, cheap and environmental-friendly transparent electrodes, and nanostructure-based sensors, ZnO also attracts intense attention currently as indicated by the number of publications, about 3,200 in 2006 and 3,500 in 2007 respectively. Some basic parameters of ZnO and GaN are shown in Table 7.1.

Since both ZnO and GaN have a direct bandgap, the applications of these materials mainly include optical devices. As an obvious result, the optical properties of these materials have been extensively studied in the past decade. Several comprehensive reviews on this subject can be found in [6, 8–12]. In this chapter, the optical properties of ZnO and GaN will be reviewed with the emphasis on comparison between the materials. Some reported values of relevant optical parameters will be tabulated for the reader's sake. Without employing rigorous theoretical calculations, the underlying physics of optical properties will be presented in simpler forms.

Although the cubic phase is often observed in ZnO and GaN, and has its own importance [13–22], ZnO and GaN both are thermodynamically stable in the wurtzite phase, which is most common and actively investigated. For this reason, the optical properties of wurtzite structures will be specifically discussed. The emission properties of GaN and ZnO will be reviewed first, and then several experimental characteristics of the relevant heterostructures, InGaN/GaN and MgZnO/ZnO, will be followed. The optical properties of nonpolar films, with the selection rules of the transitions will be discussed in

Table 7.1. Some physical parameters of GaN and ZnO

	GaN	ZnO
Band gap at T < 10 (eV)	3.51	3.44
Lattice constant (Å)	$a = 3.189$, $c = 5.185$	$a = 3.249$, $c = 5.207$
Melting point (K)	1700	2250
Exciton binding energy (meV)	26	60
Ionicity parameter (ZC/ω_p)	0.84 [6]	0.95 [6]

terms of the valence band structures. Recent reports, which show that the modification of the selection rules by the anisotropic strain is substantial, will be given. The results of Raman spectroscopy are summarized in the end, with the emphasis on deciding the residual strain and the carrier concentration.

7.1.1 Basics

The fundamental equation that governs all the emissions (and absorptions) in solid is the Fermi's golden rule, given by [23],

$$W_\mathrm{m} = \frac{2\pi}{\hbar} \sum_{\text{final states}} \delta\left(E_\mathrm{f} - E_\mathrm{i} \pm \hbar\omega\right) |\langle f| H' |i\rangle|^2, \qquad (7.1)$$

where W_m is the transition rate from the initial state $|i\rangle$ to the final state $|f\rangle$, H' is the spatial part of the perturbing Hamiltonian, E_f and E_i are the energy of the final and initial states, and $\hbar\omega$ is the energy of the involved photon, respectively. The upper sign is for photon emission and the lower one is for absorption. The equation is the direct consequence of the first order time-dependent perturbation theory, where the photon, the electromagnetic radiation, is considered as a time-dependent harmonic perturbation. The delta function part represents the conservation of energy, and the transition strength and therefore the selection rule is determined by the matrix elements, $p_\mathrm{if} = |\langle f| H' |i\rangle|^2$. $|i\rangle$ and $|f\rangle$ are Bloch-type wavefunctions (the products of a plane wave part and a bonding orbital that is periodic) of electrons in the valence and conduction band. $H' = \frac{ie\hbar}{m_0} A \cdot \nabla$, where A is the plane wave vector potential [23]. Since the momentum of photon in semiconductors for most energies of interest ($\hbar\omega = 0.1$–$3.5\,\mathrm{eV}$) is very small compared to the electron momentum, that is, the wavelength of the electromagnetic wave is large enough compared to the lattice constant, the dipole approximation ($k \approx 0$) can be applied, simplifying the transition matrix to $p_\mathrm{if} \sim \langle u_\mathrm{v0}| \nabla_i |u_\mathrm{c0}\rangle$ where u_v0 and u_c0 represent the valence and conduction band central cell bonding orbitals, and ∇_i is the differential operator in the direction of the light polarization [24].

7.1.2 Valence Band Structure

The conduction band in semiconductors originates from the empty antibonding states of sp^3 hybrid orbitals and the valence band from the filled bonding sp^3 orbitals (Fig. 7.1). As a result, the electronic states in the conduction band at $k = 0$, or the Γ point in group theory representation, are s-type while the states in the valence band are p-type which are 6-fold degenerated (p_x, p_y and p_z, including spin). In zinc-blend semiconductors, the wavefunction of the valence band edge is the linear combination of p_x, p_y and p_z when ignoring the spin–orbit interaction. However, in wurtzite structure like ZnO and GaN, the degenerated valence bands are split even without spin–orbit interaction

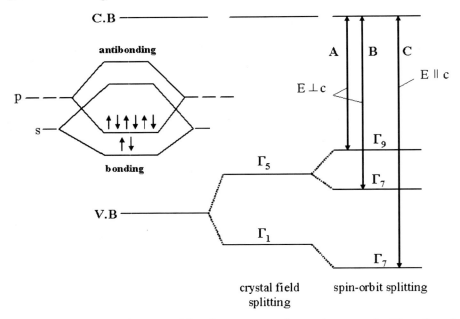

Fig. 7.1. Schematic diagram of band structures of wurtzite crystals. The allowed transitions for the respective polarizations of the electric field of the photon are indicated

due to the influence of the strong crystal–field interaction which is originated from the strong polarity in [0001] direction. Inclusion of spin gives a further splitting and shift due to the spin–orbit coupling, resulting in three twofold-degenerated valence bands which are usually denoted by A (Γ_9 or heavy hole), B (Γ_7 or light hole) and C (Γ_7, or crystal–field split-off) bands from higher to lower electronic energies. A schematic diagram of the band structure, ignoring strain effects, for wurtzite crystal is shown in Fig. 7.1.

This valence band ordering, A-Γ_9, B-Γ_7 and C-Γ_7, is well followed in most of wurtzite-type semiconductors such as ZnS, CdS, CdSe and GaN, while the ordering in ZnO has been controversial [6]. If z-axis is chosen to be parallel to the outstanding axis (c-axis), the wavefunctions of A and B bands can be represented mainly by the linear combination of p_x and p_y. The wavefunction for the C band is supposed to be originated from p_z, denoted by

$$|u_{v0}\rangle (A) \cong |X \pm iY\rangle, \quad |u_{v0}\rangle (B) \cong |X \pm iY\rangle, \quad |u_{v0}\rangle (C) \cong |Z\rangle. \quad (7.2)$$

Keeping in mind that p_x (p_y, p_z) is an odd function in x (y, z) direction, the transition matrix, p_{if}, involving A and B bands have nonzero elements when the light is polarized to E$\perp$c, while p_{if} involving C band has nonzero elements for E $\parallel$ c. Therefore, emissions from the conduction band to A and B bands are mainly polarized to E$\perp$c, and the C band is mainly polarized to E $\parallel$ c in unstrained GaN and ZnO.

Ignoring strain, the analytic equation for the band-edge energies of A, B and C bands are [25],

$$E(A:\Gamma_9) = E_0 + \Delta_1 + \Delta_2$$
$$E(B,C:\Gamma_7^{\pm}) = E_0 + \frac{\Delta_1 - \Delta_2 \pm \sqrt{(\Delta_1 - \Delta_2)^2 + 8\Delta_3^2}}{2}, \quad (7.3)$$

where Δ_1 is crystal field splitting (Δ_{cr}) and, Δ_2 and Δ_3 are spin–orbit splitting (Δ_{so}). Although anisotropy in Δ_2/Δ_3 has sometimes been reported, $\Delta_2 = \Delta_3 = \Delta_{so}/3$ is taken in most cases in GaN. Inclusion of strain effects, however, makes the analytic expression (7.3) much more complex with additional seven more parameters [26], while compressive (tensile) strain causes overall blue-shift (red-shift) in the bandgap by increasing interaction between the bonding orbitals of the nearest neighbors. The reported values for Δ_{cr} and Δ_{so} have been summarized by Vurgaftman et al. [27]. Although the values have some variations, the recommended values by the authors were $\Delta_{cr} \approx 10\,\text{meV}$ and $\Delta_{so} \approx 17\,\text{meV}$. These values are also similar to the recently reported values, $\Delta_{cr} = 9.2\,\text{meV}$ and $\Delta_{so} = 18.9\,\text{meV}$ by Misra et al. [28].

7.2 Emission Properties of GaN

7.2.1 Band Edge Emissions

Free Excitons in GaN

The envelope function approximation tells that an electron in the conduction band and a hole in the valence band can behave like a free particle with the effective mass in the corresponding band. The Coulomb attraction between the electron and the hole can make a quasi particle, in the same way that an electron and a proton form a hydrogen atom in free space. This quasi particle is called an exciton. Free exciton is an exciton of which the center-of-mass moves freely in the semiconductor. The binding of the electron and hole lowers the total energy of the crystal, so that free exciton emission line is observed at the energy lower than the bandgap by the exciton binding energy. The energy of an emitted photon from free exciton transitions becomes

$$\hbar\omega = E_g - \frac{E_{Xb}}{n^2}, \quad (7.4)$$

where E_g is the bandgap of the related bands, and E_{Xb} is the binding energy of the exciton in the ground state ($n = 1$). E_{Xb} is the energy necessary to dissociate the exciton. Table 7.2 shows the reported emission energies of free excitons at low temperature in strain-free GaN. The energetic position of A, B and C free excitons are approximately 3.478, 3.485 and 3.500 eV, respectively.

The exciton binding energy can be accurately determined at low temperature if one can observe both $n = 1$ and $n = 2$ transitions. The energetic

Table 7.2. Reported energetic positions of A, B, and C free excitons in strain-free GaN and ZnO at low temperature (T < 10K)

	GaN	ZnO
A exciton (eV)	3.4785 [35]	$A_L = 3.377l$, $A_T = 3.3757$ [91]
	3.4776 [36]	$A_L = 3.3773$, $A_T = 3.3756$ [92]
	3.4771 [30]	3.378 [93]
B exciton (eV)	3.4832 [35]	3.3898 [91]
	3.4827 [36]	3.3895 [92]
	3.4817 [30]	3.385 [93,94]
C exciton (eV)	3.499 [35]	3.419 [94]
	3.5015 [36]	3.425 [95]
	3.4986 [30]	3.4198 [96]

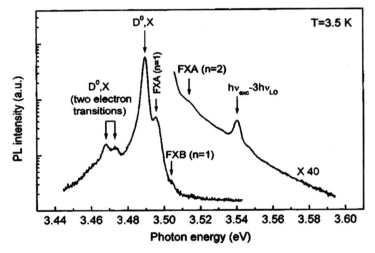

Fig. 7.2. 3.5 K exciton resonant PL spectrum of the GaN epilayer [29]. Reprinted with permission from [29]. Copyright (2002), American Institute of Physics. By observing the transitions from $n = 1$ and $n = 2$ states, one can accurately determine the exciton binding energy

distance between hydrogen-like ground state and the first excite state is given by, $\hbar\omega_{n=2} - \hbar\omega_{n=1} = \frac{3}{4}E_{Xb}$. The exciton binding energy of A free excitons (FX_A) was determined to be 25–26 meV by this method [29–32], while there were some earlier reports with slight smaller values ($E_{Xb} = 18$–25 meV). From Fig. 7.2, it can be seen that the energy separation between $n = 1$ state and $n = 2$ state of the A free exciton is 18.7 meV, which gives us the binding energy of the A free exciton 24.9 meV. The theoretically [33] and experimentally obtained binding energies are nearly equal for the three excitons. The reason for this is that the binding energy is mainly determined by the reduced mass of exciton, and the reduced mass is dominantly determined by electron

Table 7.3. Exciton binding energies of A, B, and C free excitons in GaN and ZnO. The estimated Bohr radiuses of the excitons are indicated

	GaN	ZnO
A exciton (meV)	26.3 [29], 24.8 [30]	59.7 [92]
B exciton (meV)	24.7 [30]	60.0 [92]
C exciton (meV)	26.8 [30]	60
Excitonic Bohr radius (nm)	3.0	1.8

mass because the hole mass is much larger. The estimated Bohr radius of the excitons is ∼3.0 nm in GaN. Some reported exciton binding energies of A, B, and C free excitons in GaN and ZnO are shown in Table 7.3.

Since excitonic transitions also follow the selection rule of their related valence band, A and B excitonic emissions are supposed to be mainly polarized to E⊥c, and C excitons are mainly polarized to E ∥ c. However, for GaN films grown along the c-axis (polar GaN), the measurement system can usually have only the α polarization (E⊥c and k ∥ c), which makes it difficult to detect the transitions with E ∥ c polarization unless the film is very thick. The polarization selection rule has been demonstrated in an 80 μm-thick c-GaN layer [34]. The photoluminescence (PL) spectra of the GaN layer for σ polarization (E⊥c and k⊥c) and π polarization (E ∥ c and k⊥c) were measured in order to check the selection rule. The intensities of the donor-bound excitons and acceptor-bound excitons were substantially lower in the π-polarized spectrum compared to the σ polarization because the bound A excitons are dipole forbidden, in the E ∥ c geometry. It should be noted that the polarization selectivity in emission and absorption is much more relevant in nonpolar films. Since the c-axis is perpendicular to the growth axis in nonpolar films, the surface normal emissions can be purely polarized to either E ∥ c or E⊥c. The polarization anisotropic emissions in nonpolar GaN and ZnO will be discussed in Sect. 7.4.

Bound Excitons in GaN

Excitons are easily localized near a neutral or ionized impurity by lowering the energy by the Coulomb interaction between the impurity and the electron and hole. The localization energy of an exciton bound to a donor or an acceptor is denoted by $E_{\rm BXb}$. The energy of an emitted photon from the ground state of the bound exciton becomes,

$$\hbar\omega = E_{\rm g} - E_{\rm Xb} - E_{\rm BXb}. \qquad (7.5)$$

In unintentionally doped GaN, an emission from an exciton bound to a neutral donor (D⁰X), is usually dominant at low temperature. D⁰X can be regarded as an analog of the neutral hydrogen molecule H_2 except for the different binding energy [37]. Figure 7.3 shows an example of low-temperature

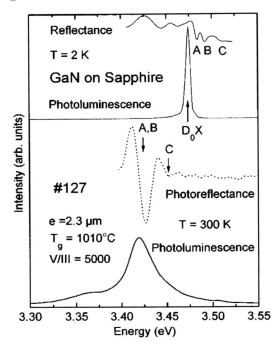

Fig. 7.3. 2 K (*top*) and room-temperature (*bottom*) spectroscopy on a typical GaN epilayer [26]. Reprinted with permission from [26]. Copyright (1996), American Institute of Physics. D^0X_A emission is dominating the low temperature PL spectrum

and room-temperature PL and reflectance spectra of a typical GaN [26]. D^0X_A line, A exciton bound to a neutral donor, with 6–7 meV localization energy is commonly observed in PL in GaN at low temperature, regardless of growth techniques and substrate choices. In strain-free GaN, the energy position of D^0X_A is at around 3.471 eV.

Figure 7.4 shows energetic positions of D^0X_A and FX_A as a function of layer thickness at 1.5 K for GaN layer deposited on sapphire and 6H-SiC substrates [38]. It should be noted that GaN on sapphire undergoes compressive strain, while GaN on SiC is under tensile strain. Under biaxial compression on sapphire substrates, the blue-shift of the FX_A exciton is about 17 meV, whereas under biaxial tension on 6H-SiC substrates the red-shift is 10 meV. One also notes that the localization energy of D^0X_A exciton increases from 6.2 to 7.2 meV as the film thickness increases [38]. The reported emission energies of D^0X_A at low temperature and its localization energy are presented in Table 7.4. The origin of the donor, which should be responsible for the as-grown n-type conductivity of GaN, is still controversial. Native defects as well as extrinsic impurities are believed to be responsible, while the nitrogen vacancies or the gallium interstitials are the prime candidates for intrinsic defects [41–44].

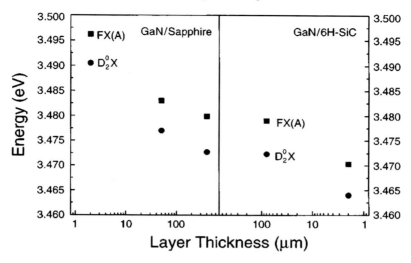

Fig. 7.4. Energetic positions of the neutral-donor-bound and free exciton as a function of layer thickness for GaN layer deposited on sapphire and 6H-SiC substrates [38]. Reprinted with permission from [38]. Copyright (1996), American Physical Society

Table 7.4. Energetic positions of commonly observed D^0X_A, A^0X_A, and TES of D^0X_A in GaN and ZnO. The localization energies of D^0X_A and A^0X_A are also indicated

	GaN	ZnO
Energy of dominant D^0X_A (eV)	3.4709 [30] 3.469 [39] 3.473 [40]	3.3605 [91] 3.359–3.363 [8]
Localization energy of D^0X_A	6.2 meV [30] 7.2 meV [39] 6 meV [40]	16.6 meV [92] 13–17 meV [8]
Energy of A^0X_A (eV)	3.466 [30, 51]	
Localization energy of (A^0X_A)	12 meV [51]	
TES of D^0X_A (eV)	3.44–3.46	3.32–3.34

In some cases, a sharp but weak peak (sometimes a series of peaks) are measured at around 3.44–3.46 eV [45, 46] as shown in Fig. 7.5. It is the region where one can expect the two-electron satellite (TES) recombination lines of the neutral donor bound excitons. During the recombination of an exciton bound to a neutral donor, the donor's final state can be the $n = 1$ (normal D^0X line) or the $n = 2$ (TES-line). Since the energy of the final state is larger, the emission energy of TES lines is smaller than D^0X line by the difference

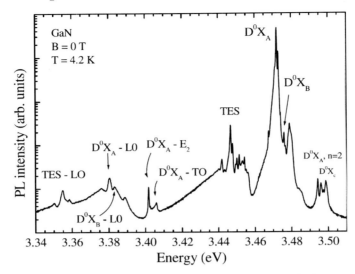

Fig. 7.5. Low-temperature PL spectrum of freestanding GaN [45]. Reprinted with permission from [45]. Copyright (1999), American Institute of Physics. The fine structures related to TES transitions of D^0X_A are clearly observable

in the ground- and excited-state energies of the neutral donor [45, 46]. The related donor binding energy, the energy needed to dissociate the electron from the neutral donor, can be determined by the energetic distance between the D^0X and its TES, which is 3/4 of the donor binding energy. The donor binding energy of the donor related to the dominating D^0X_A is reported to be about 30 meV in GaN [45].

An exciton can also bind to a neutral acceptor forming the complex of acceptor bound exciton (A^0X), which is often observed at about 3.466 eV in undoped strain-free GaN (Fig. 7.6) [30]. Although A^0X line is less common in undoped GaN than D^0X because GaN is as-grown n-type, A^0X emission can be dominant in Mg-dope GaN [47–50]. The localization energy of the Mg-related acceptor bound exciton, which can be deduced from the energetic distance from the free exciton line, has been reported to about 12 meV [51–53].

7.2.2 Defect-Related Emissions

Donor–Acceptor-Pair Transitions (DAPs) in GaN

GaN, more often in p-type GaN, can contain both donors and acceptors. Such materials are said to be compensated because some of the electrons from the donors can be captured (or compensated) by the acceptors. As a result, the sample can have both ionized donors (D^+) and acceptors (A^-). If the distance between D^+ and A^- is close enough, this donor–acceptor-pair can be an effective radiative recombination center after capturing photogenerated electron

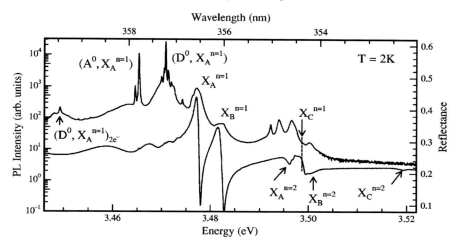

Fig. 7.6. Low-temperature PL and reflectance spectra of the homoepitaxial GaN layer in the band-edge region. The resolution is 0.12 meV [30]. Reprinted with permission from [30]. Copyright (1999), American Physical Society. The A^0X lines are observed at 3.466 eV

(in conduction band) and a hole (in valence band). This donor–acceptor-pair transition (DAP) can be represented by

$$D^0 + A^0 \rightarrow \hbar\omega + D^+ + A^-. \tag{7.6}$$

Since the energy of the final state is lowered by the Coulomb attraction between ionized donor and acceptor, the energy of DAP emission becomes

$$\hbar\omega = E_\mathrm{g} - E_\mathrm{D} - E_\mathrm{A} + \frac{e^2}{\varepsilon R}, \tag{7.7}$$

where E_D and E_A are the donor and acceptor binding energy and R is the distance between D^+ and A^- [37]. The donor (acceptor) binding energy is defined as the energy necessary to dissociate the electron (hole) from the donor (acceptor). Typical shallow DAP lines with phonon replica are usually observed at 3.26–3.27 eV at low temperature [11]. The energy separation between the phonon replicas, which is the energy of longitudinal optical (LO) phonon, is about 92 meV in GaN. Fine structures of DAP lines with higher emission energy, which means closer distance between D^+ and A^-, has been observed in Mg-doped p-type GaN [52]. Figure 7.7 shows the variation of the energy of the DAP transitions as a function of the reciprocal DAP separation in Mg-doped GaN [52].

The dependence of the DAP transition energies versus reciprocal DAP separation well follows (7.8). From the extrapolation of $R \rightarrow \infty$, one can determine the sum of the isolated donor and acceptor binding energies from (7.8), (296 meV).

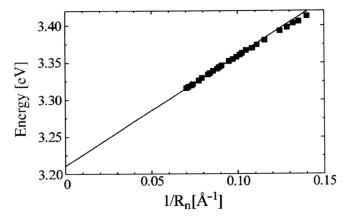

Fig. 7.7. Variation of the energy of the DAP transitions as a function of the reciprocal DAP separation in Mg-doped GaN [52]. Reprinted with permission from [52]. Copyright (1998), Wiley

Free-to-Bound Transitions (e–A^0) in GaN

Donors in GaN or ZnO have much smaller binding energy than acceptors. When the sample temperature is high enough to ionize the donors, or the concentration of defects is not high enough to form DAP, free electrons in the conduction band can radiatively recombine with holes in the neutral acceptor. This transition, involving a free carrier (an electron in this case) and a charge trapped in an impurity (a hole in this case), are called free-to-bound transition. The emission energy of this free-electron-to-bound-hole (e–A^0) is $E_g - E_A + \frac{1}{2}k_B T$ at a given temperature, where $1/2\ k_B T$ is mean thermal energy of (free) electrons in the conduction band [54]. Thus measurement of energetic position of e–A^0 can be a simple way to determine the acceptor binding energy. The e–A^0 transitions have been observed in GaN system for both the same acceptor as in DAP [54] and acceptor-like defects [55]. Figure 7.8 shows the onset of e–A^0 from the DAP line as the temperature increases [54]. From the peak position of e–A^0 at 3.282 eV and the average kinetic energy $kT/2$ of free electrons, 1.5 meV at 35 K, one can obtain $E_A = 220$ meV, which agrees well with the theoretical binding energy (224 meV) of the Ga-substitutional Mg acceptor [56].

If the Mg concentration is very high, typically above 10^{19} cm^{-3}, the PL is usually dominated by a broad band with its peak at about 2.9 eV [57–59]. One of the typical features of this blue emission is its redshifts as the excitation intensity decreases [60–63]. While the origin of the blue emission band is still controversial, the band is mostly assigned to transitions from a deep donor to the Mg$_{Ga}$ shallow acceptor [11,60,61].

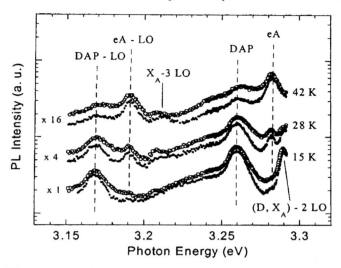

Fig. 7.8. PL spectrum of the shallow DAP band at different temperatures [54]. Reprinted with permission from [54]. Copyright (2001), American Institute of Physics. Onset of the free-to-bound transition is observed at elevated temperatures

7.2.3 Deep Level Emissions

In semiconductors, highly localized states can be formed with energies in the middle of the bandgap by a defect or complex of defects such as broken bonds, vacancy, strain associated with displacement of atoms, and other impurities. Although these states, which are called deep centers, are nonradiative in many cases, deep centers can act as radiative recombination centers, and emit photons in the energy far below the fundamental bandgap. The most famous deep level emission in GaN is the yellow luminescence (YL), which is measured at around 2.2 eV with the width of ∼400 meV. Typical energetic positions of DAP and deep level emissions in GaN and ZnO at low temperature PL are summarized in Table 7.5. The YL is observed in most of the unintentionally and intentionally doped n-type GaN samples regardless of growth conditions and substrate choices. One example of YL is shown in Fig. 7.9 [64]. In Fig. 7.9a, the YL emissions were suppressed as the growth temperature of the buffer layer increased. It implied that the YL emissions could be closely related to extended defects, and the elimination of the YL was actually the result of increased hexagonal crystallites of GaN grown on a buffer layer deposited at relatively high temperatures [64]. For laterally overgrown GaN, a much lower density of extended defects substantially suppresses the YL emission (Fig. 7.9b). The chemical origin of the YL has been one of the topics of heated discussion and extensive study for the past decade. While the $V_{Ga}-O_N$ complex is one of the main candidates, there is however still no universal consensus on the origin of YL [11].

Table 7.5. Typical energetic positions of DAP and deep level emissions in GaN and ZnO at low temperature. The donor binding energies were derived by measuring TES positions, and the acceptor binding energies were derived by measuring the position of e–A^0 transitions (see text)

	GaN	ZnO
DAP (eV)	3.26–3.27	~3.22 (undoped)
		3.235 (ZnO:N)
Donor binding energy (meV)	33	50–55
Acceptor binding energy (meV)	220	190 (undoped)
		180 (ZnO:N)
LO phonon energy (meV)	92	72
Deep Level emission	~2.2 eV (YL)	2.5 eV (GL)

7.2.4 Emission Properties of InGaN/GaN Quantum Wells

Since InGaN/GaN quantum well structures are widely used in GaN-based LEDs, the optical properties of InGaN and its quantum well have been a topic of interest. Among the various features for the optical process in InGaN-based structures, two aspects are considered to be the most outstanding: Carrier localization and strain-induced piezoelectric field. Experimental characteristics of these two mechanisms are summarized in this section.

Carrier Localization Effect in InGaN/GaN Quantum Wells

Theoretically indium is not soluble in GaN at the typical growth temperatures [65], which means that In in InGaN tends to segregate from GaN. The In segregation, or phase-separation, can easily form the In-rich region which has the local potential minima. This potential fluctuation was observed in many other alloys of III–V and II–VI semiconductors [66, 67]. The potential fluctuation in alloys was generally considered to be a negative effect because it can cause, for example, inhomogeneous broadenings, reduction in mobility, and well thickness fluctuations. On the other hand in InGaN-related system, the potential fluctuation is believed to make a positive effect. It has been widely accepted that the relatively high internal quantum efficiency of InGaN/GaN LEDs despite of high dislocation density, could come from capturing the carriers in localization centers in quantum dot-like In-rich region [1, 68–70]. It is likely that this carrier localization enhances the quantum efficiency by suppressing the probability that carriers will encounter nonradiative recombination centers which could be caused by, for example, dislocations [71].

In addition to the random localization by the In phase-separation, another mechanism of carrier localization was also proposed recently by Hitzel et al. [72]. The authors observed a reduction of InGaN/GaN quantum-well thickness in the vicinity of V-shaped defects, resulting in significantly larger bandgap than that of the regular quantum well. After analyzing both the transmission

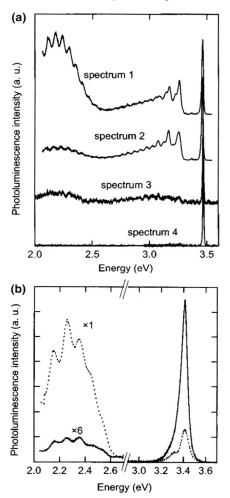

Fig. 7.9. (a) Low-temperature photoluminescence of undoped GaN with different growth temperatures of the buffer layer. Spectrum 1: grown at 495°C. Spectrum 2: at 555°C. Spectrum 3: 565°C. Spectrum 4: at 565°C. (b) Room temperature photoluminescence of laterally overgrown GaN (*solid lines*) on Si_3N_4 and GaN directly grown on the low-temperature GaN buffer layer (*dashed lines*) [64]. Reprinted with permission from [64]. Copyright (1999), American Institute of Physics

electron microscopy and near-field scanning optical microscopy images near the V-shaped defects, they proposed that this higher bandgap around every V-shaped pit can effectively shield the dislocation line-defect from mobile carriers localized in the thicker InGaN quantum well (Fig. 7.10). This effective screening would maintain a high radiative efficiency despite the presence of dislocation.

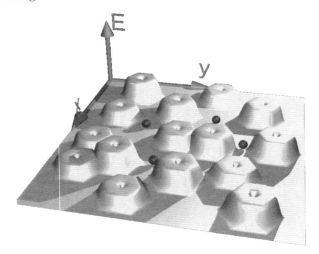

Fig. 7.10. Visualization of the energy landscape resulting from V-shaped hexagonal pits exhibiting increased band gap in the sidewalls. Carriers in the regular c-plane QW's (with lower band gap) in between have to overcome a large energy barrier to reach the defects and to recombine nonradiatively [72]. Reprinted with permission from [72]. Copyright (2005), American Physical Society

Several experimental methods have been performed to characterize the effects of carrier localization. The first method would be time-integrated temperature dependent PL. Figure 7.11 shows typical InGaN-related PL spectra for the InGaN/GaN quantum well in the temperature range from 10 to 300 K [73]. Instead of the steady red-shift of the bandgap, the energetic PL peak position shows an "S-shaped" (red–blue–red shift) temperature dependence for InGaN-related PL with increasing temperature. The blue-shift is caused by thermal activation of localized carriers from deeper localization centers to shallower localization centers.

In the conventional macroscopic PL, the inhomogeneously distributed localized centers are contributing to the emission spectra. In order to probe individual localization centers, temperature dependence of PL with reduced aperture (∼350 nm) was also performed [74]. The temperature and excitation power dependent PL spectra of the individual localization centers behaved much differently from their macroscopic spectra. It is likely that the complex interplay between the effect of localization and strong internal electric fields makes strong influence on every aspect in InGaN/GaN quantum well structures.

A second and a direct method to characterize the carrier localization effect is to measure the luminescence with high spatial resolution. Transmission electron microscopy is reported to be not a good choice to directly measure the indium-composition fluctuation due to high-energy electron-beam

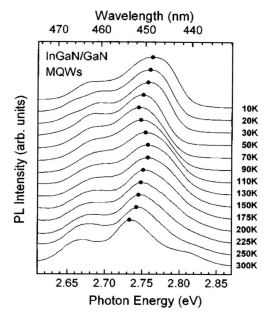

Fig. 7.11. Typical InGaN-related PL spectra for the InGaN/GaN MQWs in the temperature range from 10 to 300 K. The main emission peak shows an S-shaped shift with increasing temperature (*solid circles*) [73]. Reprinted with permission from [73]. Copyright (1998), American Institute of Physics

induced damage [75]. High spatial resolution cathodoluminescence (CL) [76] and near-field scanning optical microscopy (NSOM) [77] have been successfully performed to detect the localization effect.

Another typical feature of carrier localization can be obtained by measuring luminescence life time as a function of emission energy. A typical experimental result is shown in Fig. 7.12, where the PL lifetime increases with decreasing emission photon energy [78]. If the emission is composed of ensembles of localized states, carriers in higher energy states can either be transferred to the tail states or radiatively recombine at the site. The additional channel in the higher energy states, other than radiative recombination, would lead to a much faster PL life time of higher energy states.

Figure 7.12 shows the increase of the PL lifetime with decreasing emission photon energy, which can be characteristic of a localized electronic system. This faster decay time of higher energy states also can cause so-called ' time-lapse red-shift' that is the appearance of the red-shift with evolving time after pulsed excitation. Although the emission energy dependence emission life time is often referred to be an evidence of carrier localization and sometimes is used for comparison of relative magnitude of localization, care must be needed to interpret the result because the screening of the piezoelectric fields in the quantum well also can cause a similar effect [79].

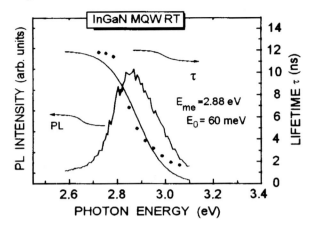

Fig. 7.12. PL spectra and lifetime of InGaN MQWs as a function of emission photon energy measured at RT [78]. Reprinted with permission from [78]. Copyright (1999), American Institute of Physics

Piezoelectric Field in the InGaN/GaN Quantum Wells

It is well known that GaN-related material have huge ($\sim$MV/cm) spontaneous polarization as well as strain-induced or piezoelectric polarization along the c-axis due to the symmetry of wurtzite structure [80–83]. The direction and magnitude of the total polarization fields depend on polarity (Ga-face or N-face), strain (compressive or tensile), well-width, and alloy composition. Direction of the piezoelectric field in compressive strained (Ga-face) InGaN/GaN quantum wells is toward the sapphire substrate [84]. The direction of total electric field is opposite to the direction of the total polarization field. Since InGaN quantum well layers are under compressive strain, the direction of the total electric field by the polarizations (some authors simply refer it to piezoelectric field) in conventional (Ga-face) InGaN/GaN LEDs grown by metal–organic chemical vapor deposition technique on c-sapphire is towards the sapphire substrate, which is opposite to the built-in electric field formed by p–n junction. A schematic band diagram of InGaN active region is shown in Fig. 7.13 [71].

The electric field produces a considerable red-shift of the confined energy level in the quantum well, known as the quantum confined Stark effect (QCSF). This macroscopic electric field in the quantum well plays a crucial role in device performance in many aspects. Most notably, the existence of a strong polarization field reduces the internal quantum efficiency. In quantum well structures, the transition matrix elements, p_{if}, mentioned in Sect. 7.1.1, become the product of overlap integral $\langle \Psi_{\mathrm{e}} | \Psi_{\mathrm{h}} \rangle$ and dipole transition matrix by applying the envelope function approximation [85], where Ψ_{e} and Ψ_{h} is the envelope wavefunctions of a electron and a hole in the quantum well, respectively. Due to the strong electric field that spatially separates electrons and

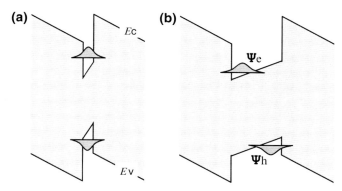

Fig. 7.13. Schematic band diagram of (**a**) thin and (**b**) thick InGaN/GaN active regions with polarization fields for Ga-face growth [71]. The electron and hole envelope wavefunctions are depicted

holes, the overlap integral becomes smaller, and therefore the radiative transition probability or transition rate (Γ_r) are reduced. As a result, the internal quantum efficiency ($\eta_{\text{int}} = \frac{\Gamma_r}{\Gamma_r + \Gamma_{nr}}$) is reduced. The reduction in the overlap integral is more severe in thicker wells (Fig. 7.13).

One of the most straightforward techniques to measure the direction and the magnitude of the piezoelectric field is reverse-biased photo- (or cathodo-) luminescence measurement. Figure 7.14 [86] shows the reported results of the reverse-bias cathodoluminescence (CL) spectra on their specific LEDs [86]. As the reverse bias increases, the luminescence peak position shows a distinct blue-shift due to the *decrease* of the total electric field in the quantum well. This is clear evidence that the direction of the piezoelectric field is opposite to the built-in electric field formed by the junction.

Figure 7.14 shows a series of room-temperature SEM-CL spectra of the InGaN MQW LED as a function of the reverse bias. The CL emission peak blueshifts 52 meV (11 nm) before it begins to redshift above −15 V. The circles indicate the peak positions of the CL spectra. The intensity of the luminescence decreases as the reverse bias increases due to the increase of the carrier escape at higher reverse bias [87]. As the reverse bias further increases up to the value that can fully compensate the piezoelectric field, the blue shift no longer proceeds and the peak position starts to redshift as the total internal electric field in the quantum well starts to *increase* in the opposite direction to the original piezoelectric field. By the obtained value of the reverse bias that fully compensates the piezoelectric field, one can obtain accurately the magnitude of piezoelectric field if the thickness of the active layers is known. In Fig. 7.14, the compensating voltage was estimated to be 14.5 V. With the width of the depletion layer of 140 nm, the piezoelectric field of ∼1 MV cm^{-1} are obtained in the InGaN/GaN quantum well grown on a SiC substrate [86]. A similar result was also reported by the PL measurement [87]. The estimated

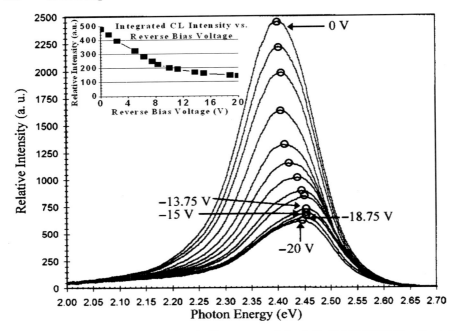

Fig. 7.14. Room-temperature SEM-CL spectra of the InGaN MQW LED collected at several reverse biases. The CL emission peak blueshifts 52 meV (11 nm) before it begins to redshift above 15 V. The circles indicate the peak positions of the CL spectra. The insert shows the integrated CL intensity as a function of reverse bias [86]. Reprinted with permission from [86]. Copyright (2005), American Institute of Physics

electric field in InGaN/GaN in [87], which was grown on a sapphire substrate, was ~2 MV cm^{-1}. One shortcoming of this technique is that the luminescence is not able to sustain the reverse bias until when the bias reaches the fully compensating condition due to the strong carrier-escape in many cases. It is still an issue to experimentally derive the magnitude of the internal electric field of any given sample structure.

Another profound effect of piezoelectric field in the quantum well is the strong well-thickness dependence of PL lifetime at low temperature. As discussed above, the existence of strong piezoelectric field in the quantum well significantly reduces the radiative transition rate. In general the PL lifetime (τ_{PL}) is the summation of radiative (τ_r) and nonradiative (τ_{nr}) component, given by [88]

$$\frac{1}{\tau_{PL}} = \frac{1}{\tau_r} + \frac{1}{\tau_{nr}}. \tag{7.8}$$

Because transitions at low temperature (T < 10 K) are almost purely radiative in InGaN/GaN quantum wells ($\Gamma_{nr} = 1/\tau_{nr} \approx 0$), PL lifetime at a low temperature can simply represent the radiative lifetime which is the

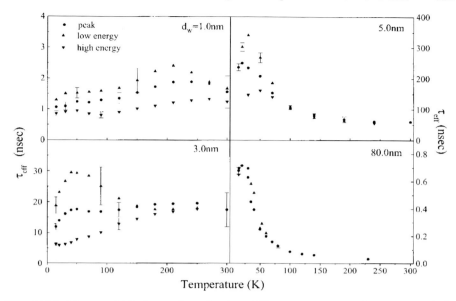

Fig. 7.15. Measured PL decay times for four different quantum well structures [89]. Reprinted with permission from [89]. Copyright (2000), American Physical Society. Low temperature τ_r $(1/\Gamma_r)$ drastically increases by two orders-of-magnitude as the quantum well thickness increases from 1 to 5 nm

reciprocal of the radiative transition rate $(1/\Gamma_r)$. In thicker wells, Γ_r would be smaller since the overlap integral is significantly smaller than in thinner wells. Figure 7.15 [89] shows low temperature τ_r $(1/\Gamma_r)$ drastically increases by two orders-of-magnitude as the quantum well thickness increases from 1 to 5 nm. The situation is more complex at room temperature because of the significant contribution of nonradiative recombination on the PL lifetime.

7.3 Emission Properties of ZnO

ZnO, like GaN, preferentially crystallizes in the wurtzite structure. ZnO reportedly has similar bandstructure as GaN [90]. The valence band ordering of A-Γ_9, B-Γ_7 and C-Γ_7 is well followed in GaN as in most of wurtzite-type semiconductors. However, the ordering in ZnO has been a subject of long-standing debate whether the ordering is the usual one, or it is A-Γ_7, B-Γ_9 and C-Γ_7 called inverted valence band ordering. The detailed summary of the debate can be found in [7] and [8]. In either ordering, A (lowest energy) and B (next higher energy) excitonic emissions are supposed to be mainly polarized to E⊥c, and C excitons (highest energy) are mainly polarized to E ∥ c. (See the Sects. 7.1.2 and 7.2.1) The reported values of Δ_{cr} are in the range of 30–40 meV [8,90,91] while Δ_{so} is 10–20 meV. ZnO has larger Δ_{cr} than in GaN mainly due to the stronger ionicity of ZnO.

7.3.1 Band Edge Emissions

Free Excitons in ZnO

Exciton transitions have been readily measured at around 3.377 eV in low temperature photoluminescence (PL) from high quality bulk ZnO [92,93]. Excitons with their oscillators in the plane normal to the c-axis (Γ_5 states) are called longitudinal excitons [7]. The transverse exciton (A_T) has a Γ_1 oscillator elongated parallel to the c-axis. As shown in Fig. 7.16 [92], the splitting of the transverse and longitudinal (A_L) excitons is measurable (1–2 meV) in high quality ZnO while A_T has the lower energy.

The reported ground state energies of A_L and A_L free excitons from [93] are 3.3773 and 3.3756 eV, respectively. While the energetic position of C exciton has some uncertainty, transitions of B and C free excitons are observed at around 3.389 and 3.420 eV, respectively [92–97]. Several reported the ground state energies of B and C free excitons are listed in Table 7.2. The energetic distance between A and B excitons (E_{AB}), and B and C exciton (E_{BC}) is larger in ZnO ($E_{AB} \approx 12$ meV, $E_{BC} \approx 34$ meV) than in GaN ($E_{AB} \approx 5$ meV, $E_{BC} \approx 16$ meV) mainly due to the larger Δ_{cr} in ZnO (7.3). In contrast to GaN, ZnO bulk crystals can readily be grown by a number of methods [98]. Polarization selection rules of A, B, and C excitons, therefore, can be measured by reflectance spectroscopy from a bulk ZnO. Figure 7.17 [93] shows reflection spectra for the orientation of E⊥c and E ∥ c, where the A

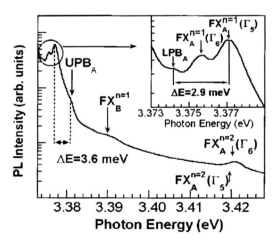

Fig. 7.16. Free excitonic fine structure region of the 10 K PL spectrum for the forming gas annealed ZnO substrate [92]. Reprinted with permission from [92]. Copyright (2004), American Physical Society

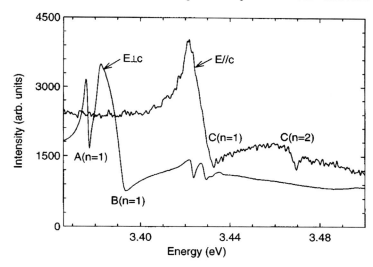

Fig. 7.17. Reflection spectra for the orientation E⊥c and E ∥ c. A, B, and C exciton transitions are depicted [93]. Reprinted with permission from [92]. Copyright (1999), American Physical Society

and B excitonic transitions are mainly polarized to E⊥c, and the C exciton is polarized to E ∥ c.

One of the outstanding properties of ZnO is its large exciton binding energy (∼60 meV). As the E_{Xb} is much larger than the room temperature thermal energy ($k_B T = 26$ meV), ZnO also attracts much attention as a promising material for excitonic devices [98–101] that can be operative at room temperature. (It should be noted that the excitonic effect on the *lasing* action at room-temperature from ZnO has been recently questioned [101].) As in GaN, the reported binding energies of A, B and C excitons in ZnO are nearly equal, which is also due to the fact that the reduced mass is dominantly determined by electron mass because the hole mass is much larger as discussed in the Sect. 7.2.1. One recent report of the binding energies of excitons can be found in [93]. By measuring $n = 2$ (and $n = 3$) energies of excitons, the authors reported the binding energy of A_T, A_L, and B excitons of 59.7, 60, and 57 meV, respectively [93] (See the Table 7.3).

Bound Excitons in ZnO

Like in GaN (Sect. 7.2.1), emissions from D^0X are dominant in low temperature PL spectra in ZnO. The strongest PL line from a D^0X_A is often observed at around 3.360 eV [92,103–105]. Figure 7.18 shows a low temperature PL spectrum from an unstrained ZnO with the dominating D^0X_A at 3.3605 eV [92]. The E_{BXb} of this D^0X_A is 16.6 meV with the energy of A_T as the reference. Several reports [9, 106, 107] support the assignment that the impurity

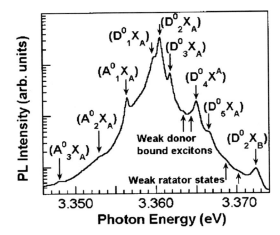

Fig. 7.18. Bound excitonic features of the 10 K PL spectrum for a ZnO bulk substrate [92]. Reprinted with permission from [92]. Copyright (2004), American Physical Society

involved in the D^0X_A at 3.3604 or 3.3608 eV is most probably Al, and the impurity for the D^0X_A at 3.3593 or 3.3598 eV is Ga. Many different D^0X_A lines have been observed in the energy range from 3.359 to 3.363 eV with the E_{BXb} of 13–17 meV ([8, 9] and references therein). The chemical origins of these D^0X_A lines are not yet clear. In general, the E_{BXb} of D^0X_A lines in ZnO are larger than in GaN (6–7 meV, Table 7.4), which is expected from its larger exciton binding energy. The emission at 3.367 eV is usually attributed to an exciton bound to an ionized donor [108]. The emissions often observed at 3.353–3.356 eV were earlier attributed to the Na or Li related acceptor bound exciton [109], however the exact nature of these emissions is still under debate [9].

TES lines of D^0X_A lines are also often observed in ZnO in the spectral region of 3.32–3.34 eV (Fig. 7.19). As discussed in Sect. 7.2.1, one can accurately determine the donor binding energy if TES line of the corresponding D^0X_A is observed. From the energetic difference between TES and D^0X_A, the donor binding energy of D^0X_A at 3.359–3.361 eV has been estimated to be 50–55 meV [8, 9, 92, 110]. As one can see in Table 7.4, the binding energy of the donor in dominating D^0X_A is larger in ZnO than in GaN (∼33 meV).

7.3.2 Defect-Related Emissions

Donor–Acceptor-Pair Transitions (DAPs) in ZnO

There have been many DAP lines reported in ZnO for different growth condition and ion-implantation [111]. Among them, a DAP line at 3.22 eV is often observed in undoped-ZnO [8, 9, 92, 110] as shown in Fig. 7.19. Observation of

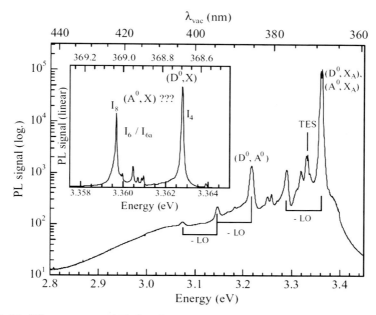

Fig. 7.19. PL spectrum of ZnO substrate material recorded at 4.2 K. The bound exciton lines at ~3.36 eV are followed by a two-electron replica ("TES") 30 meV low in energy. At 3.220 eV, the donor–acceptor pair transition (denoted by D^0, A^0) is observed [110]. Reprinted with permission from [110]. Copyright (2001), Elsevier

DAP emission in undoped-ZnO imply that acceptors, as well as donors, are unintentionally incorporated into the as-grown ZnO. The chemical identity of the residual acceptor, however, remains to be determined. Binding energy of the acceptor was reported to be 185 meV [9]. Another DAP line of interest is a line observed at ~3.235 eV in Nitrogen-doped samples [9, 112]. The DAP lines at around 3.235 eV was monitored before and after N-implantation in the low temperature PL spectra [112]. The DAP line at around 3.235 eV was enhanced by the N-implantation, which supported the assignment of N as a shallow acceptor in DAP luminescence. The authors obtained the binding energy of 177 meV [112], while Meyer et al. obtained 165 meV for the N-acceptor [9] (Table 7.5).

Free-to-Bound Transitions (e–A^0) in ZnO

As discussed in Sect. 7.2.2, DAP transition can be transformed to e–A^0 transition at elevated temperature. As shown in Fig. 7.20 [110], the intensity of the DAP transition at 3.220 eV decreases between 30 and 100 K, and the adjacent e–A^0 transition at 3.236 eV increases. From the measurement of energetic position of e–A^0 (Sect. 7.2.2), the acceptor binding energy is determined to be ~195 meV [110].

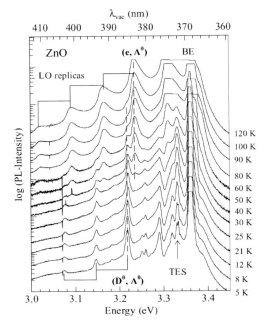

Fig. 7.20. Series of PL spectra of ZnO substrate material recorded at different temperatures. Between 30 and 100 K, the intensity of the DAP transition at 3.220 eV decreases, and the adjacent free-to-acceptor transition at 3.236 eV increases [110]. Reprinted with permission from [110]. Copyright (2001), Elsevier

Various p-type dopants, such as N, P, and As, are being tried actively to establish good p-type conductivity in ZnO. The details are summarized in [113]. It is notable that the acceptors of interest in ZnO generally have smaller binding energies than in GaN [113].

Structural defects sometimes can act as acceptors. It is possible for free electrons in a conduction band to recombine with a hole bound to the acceptors, which can be another type of e–A^0 transition. In ZnO, an emission at around 3.315 eV is often observed at low temperature. Schirra et al. recently reported [114] that the emission is originated from an e–A^0 transition. By analyzing TEM images and spatially resolved CL of an undoped-ZnO at different temperatures, the authors concluded that the acceptors were formed by a structural defect related to stacking faults. The emission is too strong to be assigned as a TES, which is as strong as the D^0X. Although the energetic position of the emission is coincident with the previously assigned A^0X in a p-ZnO with nitrogen dopants [115] and with other p-type dopants [116–118], it is unlikely that the emission is related to a shallow acceptor because the sample is undoped. Schirra et al estimated the acceptor binding energy to be 130 meV. As shown in Fig. 7.21 [114], the e–A^0 transition is only observed (bright) along specific lines while no D^0X and FX is observed along the lines.

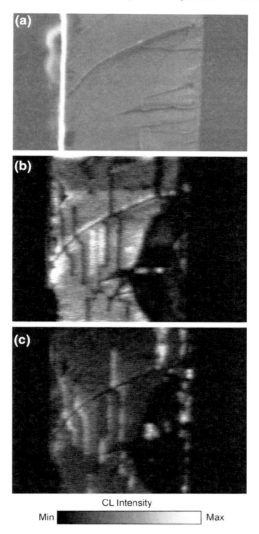

Fig. 7.21. Cross-section SEM micrograph of a 3 mm thick layer (*top*), monochromatic CL intensity mapping in the range of FX_A and D^0X exciton emission (3.349–3.379 eV) (*middle*), and intensity mapping of the (e, A^0) transition (3.296–3.325 eV) (*bottom*). The (e, A^0) transition is only observed along specific directions [114]. Reprinted with permission from [114]. Copyright (2007), Elsevier

Schirra et al. seriously questioned, the result described above, with the previous assignment of the ∼3.315 eV line to the acceptor bound exciton in p-ZnO. Similar emission line of e–A^0 was also observed in ZnO nanocrystals with typical temperature dependency of free-to-bound transitions [119].

7.3.3 Deep-Level Emissions

The research on the deep centers in ZnO has a decade-long history. Comprehensive reviews on this subject can be found in [6] and [8], and references therein. The most characteristic deep center in undoped-ZnO appeared at 2.5 eV, called green luminescence band (GL). Although the exact nature of the GL band still remains controversial, two of the main candidates are most widely accepted. The green emission with clear phononic fine structures is likely due to the Cu impurities. This 'Cu-green' band is one of the most commonly observed deep centers in II–VI compound. GL band of the smooth spectrum most likely comes from oxygen (or zinc) vacancy [8]. The deep center responsible for the red emission is believed to be the Fe impurity [7]. Li-doped ZnO emits luminescence at around 2.2 eV, which is in the yellow spectra region [120]

7.3.4 Emission Properties of MgZnO/ZnO Quantum Wells

The growth of quantum well structures of high quality is another major issue for realizing high efficiency devices with ZnO. MgZnO has been the choice of materials for UV spectral region. Since the bandgap energy of MgO ($E_g \sim 6.4$ eV) is larger than in ZnO, MgZnO layer is used for the barrier material. The research on MgZnO/ZnO quantum wells is not as mature as its InGaN/GaN counterpart. Much effort is being made to improve their quality and understand the electronic and physical properties.

ZnO has also large piezoelectric constants, as shown in Table 7.6. The piezoelectric constants are about ten times larger than in conventional III–V and II–VI, and comparable to those of GaN [121]. Therefore, MgZnO/ZnO quantum well structures are also expected to have large piezoelectric field in the well, as in InGaN/GaN system. However, results reported so far are not very consistent. Several results have been reported that the polarization-induced electric field is relatively small [122, 123]. Others report that the electric field is as large as in InGaN/GaN quantum wells [124–126].

Morhain et al. reported 0.9 MV cm^{-1} of internal electric field in Mg$_{0.22}$Zn$_{0.78}$O/ZnO quantum well structures by analyzing their cw and time-resolved PL data as a function of the well width [124]. In the wider well-width structures, emissions *below* the ZnO bandgap were observed, which can be a direct evidence of existence of strong electric field in the quantum well.

Table 7.6. Spontaneous and piezoelectric constants in GaN and ZnO [121]

	GaN	ZnO
Spontaneous polarization constant P^{eq} (C/m^2)	−0.029	−0.057
Piezoelectric constant e$_{33}$ (C/m^2)	0.73	0.89
Piezoelectric constant e$_{31}$ (C/m^2)	−0.49	−0.51

They could derive the relation, $\sim 4.1\,x$ MV cm^{-1} where x is the Mg composition. The obtained magnitude of the internal electric field is as large as in InGaN/GaN structures with similar In compositions. Zhang et al. recently reported a similar result, ~ 0.3 MV cm^{-1} with $x = 0.1$. [126]. Morhain et al. attributed the previous report of the lack of electric field to the too small barrier and well width [124]. On the other hand, Park and Ahn claimed that negligible electric field should be observed in low Mg composition ($x < 0.2$) and thin well width (L$_\text{w}$ < 4.6 nm), due to the cancellation of the sum of piezoelectric and spontaneous polarizations between the well and the barrier [123]. More works obviously are needed to be done to clarify the effect of the polarization field in MgZnO/ZnO.

Carrier localization effect is not supposed to exist in MgZnO/ZnO quantum well, because the well layer is no longer an alloy. But the conventional localization due to the well-thickness fluctuation has been recently reported [127, 128].

7.4 Emission Properties of Nonpolar GaN and ZnO

As discussed above, huge macroscopic static electric field exists in the quantum well structures of GaN and ZnO. In polar crystals grown on c-sapphire substrates, the polarization fields are along the growth axis, which reduces the internal quantum efficiency by reducing the overlap integral of electron and hole envelope wavefunctions (see the Sect. 7.2.4). In addition to that, the existence of the polarization fields and the related interface charges make the band structures complex leading to increase of leakage current and the operating voltage. Kim et al. recently claimed that the piezoelectric fields produce potential barriers in the InGaN/GaN quantum well region, so that the electronic energy of the conduction (and the valence band) would be *higher* in n-GaN region [129]. The potential barriers in the active region could be the source of increase of the leakage current and the operating voltage. Kim et al. also claimed that the polarization induced potential barriers in the quantum wells could be the main reason of, the *efficiency droop* or the lowering the LED efficiency as the injection current increases [129].

In order to avoid these drawbacks in device efficiency, the growth of nonpolar films has become one of the hottest issues in GaN and ZnO research fields (See the Chap. 5 in this book). Since the polar c-axis is perpendicular to the growth direction in nonpolar GaN or ZnO epitaxial layers, the quantum-confined Stark effect (QCSF) no longer exist. As of today, December 2008, the device efficiency of nonpolar GaN-based LEDs is still not as good as their polar counterparts, but it is improving rapidly. The external quantum efficiency (EQE) of 40% in 407 nm nonpolar InGaN/GaN LED was reported, with a minimal efficiency droop [130].

7.4.1 Polarized Emission

One of the profound characteristics of nonpolar film is the polarization anisotropy of the emission. Since the radiative recombination usually occurs in the lowest possible state due to the very efficient energy relaxation process, the emissions involving A (or A and B at elevated temperature) valence band are dominating in luminescence. The recombination of an electron in the conduction band with a hole in A (or B) valence band is allowed when the emitting photons are polarized to E⊥c by the selection rule discussed in the Sects. 7.1.2 and 7.2.1. However, for epitaxially-grown polar-GaN or ZnO, it is difficult to obtain the polarization selectivity since all the photons emitted to the surface normal direction satisfy the selection rule. On the other hand, the photons emitted to the surface normal direction from a nonpolar film is strongly polarized to the E⊥c direction, resulting in strong polarization anisotropy in emission. The polarization selectivity in nonpolar films can be a big advantage for the polarization sensitive devices, like backlight units for liquid–crystal display. The conventional way to obtain polarized light is to use an unpolarized light-source and a polarizer, which absorbs good portion of the light from the source. Therefore, an *intrinsic* polarized light source without any external polarizer would be advantageous in terms of cost, efficiency and large-scale integration. Figure 7.22 [131] shows polarization-dependent PL spectra from nonpolar and polar AlGaN/GaN quantum wells. The PL spectra of the nonpolar (m-plane) sample show strong polarization anisotropy, while the c-plane sample shows no change by polarization. The degree of polarization ρ, defined as $\rho = (I_\perp - I_\parallel)/(I_\perp + I_\parallel)$, where $I_\perp$ is the intensity of light with polarization perpendicular to the c-axis and $I_\parallel$ is the intensity of light with polarization parallel to the c-axis, is 90% in Fig. 7.22. $\rho = 96\%$ has been reported from an m-plane InGaN/GaN grown by molecular beam epitaxy in 10 K photoluminescence [132]. The polarization anisotropy in electroluminescence was also reported in with $\rho \sim 80\%$ [133]. The ρ generally decreases as either temperature or injection current increases.

The research on the optical properties of epitaxially grown nonpolar ZnO is still at its early stage. Polarized PL spectra from a-plane ZnO on r-plane sapphire substrate have been recently reported [134,135]. An example is included in Fig. 7.23 [135]. The PL spectrum is mainly polarized to E⊥c, which means the emissions have A-excitonic characters. In Fig. 7.23, the PL peak at 3.384 and 3.326 eV were assigned to be a D^0X_A, and a free-to-bound transition, respectively. It is notable that the emission lines were significantly blueshifted from in bulk ZnO by overall compressive strain in the epilayer. The selection rule in highly strained nonpolar films will be discussed in the Sect. 7.4.2. In Fig. 7.23, the ρ was estimated to be 93% in 12 K photoluminescence [135].

7.4.2 Strain Effects

In the heterostructres of GaN and ZnO, the layers are often under strong strain due to the mismatch of the lattice constants and thermal expansion

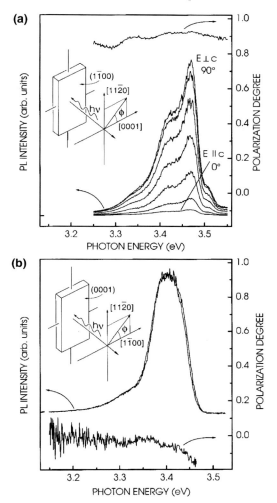

Fig. 7.22. Polarization-dependent PL spectra of (**a**) the m-plane and (**b**) the c-plane MQW at 5 K. The PL of the c-plane sample shows no change within measurement accuracy. The insets show a schematic sketch of the measurement geometry and sample orientation [131]. Reprinted with permission from [131]. Copyright (2000), American Institute of Physics

coefficients. It is well-known that biaxial strain can significantly change the bandstructures and their related selection rules. The analytic expression of energetic positions of three valence bands including the strain effects can be found in [26], which has seven more parameters (A parameters) in addition to (7.3). The biaxial strain not only shifts further the band-edge energies of conduction and valence bands, but changes the oscillator strength of the respective transition as well. The situation becomes even more complex in

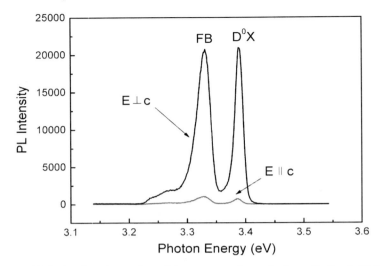

Fig. 7.23. Polarization-dependent PL spectra of an A-plane ZnO at 12 K [135]. Reprinted with permission from [135]. Copyright (2008), Korean Physical Society

nonpolar GaN and ZnO, since the biaxial strain itself is strongly anisotropic, which is due to the different lattice mismatches in c-axis and a- or m-axis. [136–141]. Ghosh et al. reported that the in-plane anisotropic compressive strain beaks the degeneracy in the $x-y$ plane of the wurtzite crystal in m-plane GaN [140]. The original $|X \pm iY\rangle$-like valence band states of unstrained wurtzite GaN are broken into $|X\rangle$-like and $|Y\rangle$-like ones. (The growth axis is parallel to y-direction and c $\parallel$ z in this case, Sect. 1). The energy of $|X\rangle$-like state shifts to higher energy (smaller bandgap), while the $|Y\rangle$-like state shifts down below the $|Z\rangle$-like state, resulting in $|X\rangle$-like, $|Z\rangle$-like, $|Y\rangle$-like ordering. Instead of observing A and B excitonic transitions in an unstrained case for E⊥c (for example in Fig. 7.17), only a single resonance at lower energy for E⊥c was observed in strained m-plane GaN as shown in Fig. 7.24 [140]. A similar result was also reported for A-plane ZnO where it showed only a single feature in the respective polarized photoreflectance (PR) spectrum [138]. Higher strain or different orientation of the substrate, however, can change the selection rule further by mixing two or three of the $|X\rangle$-like, $|Y\rangle$-like, $|Z\rangle$-like states [140–142]. The nomenclature of E_1, E_2 and E_3 (or T_1, T_2, and T_3) excitons, instead of A, B and C excitons, is sometimes used in this case, because the crystal symmetry of anisotropically strained film is different from that of strain-free film.

Polarization selectivity in PR and PL spectra from highly strained a-plane ZnO are shown in Fig. 7.25. The E_1 exciton (lowest transition energy) was mainly polarized to E⊥c and weakly polarized to E $\parallel$ c under strong biaxial compressive strain in the 100 nm-thick film. The E_2 exciton (next-higher

7 Optical Properties of GaN and ZnO 343

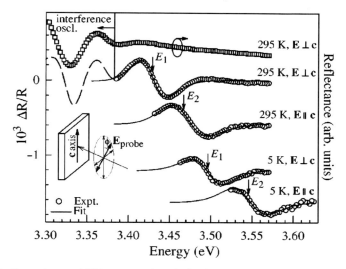

Fig. 7.24. Experimental PR spectra (*circles*) of the M-plane GaN film for different polarizations of the probe beam relative to the c-axis of GaN [140]. Reprinted with permission from [140]. Copyright (2002), American Physical Society

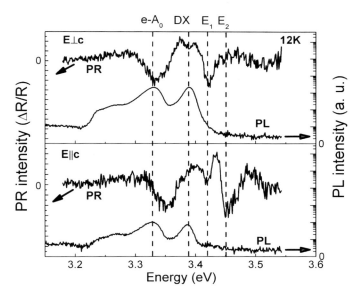

Fig. 7.25. Polarization dependent PR and PL spectra at 12 K. The upper and lower panel represent E⊥c and E ∥ c polarization, respectively [142]. Reprinted with permission from [142]. Copyright (2008), American Institute of Physics

energy transition) was exclusively polarized to E ∥ c. It is notable that the luminescence from strained nonpolar films is still mostly polarized to E⊥c because the $|X\rangle$-like state has the smallest band-gap in all the cases reported so far.

7.5 Raman Scattering Properties of GaN and ZnO

Raman scattering is an inelastic light scattering by phonons in materials. Raman spectroscopy has been one of the powerful techniques for characterizing lattice and electronic properties of semiconductors. When a monochromatic light (a laser) is incident on the material, the light can be scattered inelastically by phonons. As a result, photons with a slightly less energy than the incident light are scattered out with emission of a phonon (Stokes's lines). The scattered light can gain energies if phonon-absorption is involved (anti Stoke's lines). In Raman spectroscopy, the intensity of the scattered light as a function of the energy difference from the incident light is measured. This Raman shift represents the energy of the vibrational mode involved. The experimental geometry can be specified by four vectors: k_i and k_s (the direction of the incident and scattered photons, respectively) and e_i and e_s (the polarizations of the incident and scattered photons, respectively). These four vectors define the scattering configurations usually represented as k_i (e_i, e_s) k_s. In this section, the experimental characteristics of Raman scattering in GaN and ZnO will be briefly summarized with the specific focus on determination of the strain and the carrier concentration.

7.5.1 Strain Effects

Raman scattering properties of GaN and ZnO have been studied extensively for decades. A recent review on GaN can be found in [143]. In the wurtize GaN structure, the unit cell consists of two Ga–N atoms. Since the unit cell has four atoms, 12 phonon modes are possible in principle. Among them, four modes (A_1, E_1, and two E_2 modes) are Raman-active. A_1 and E_1 mode represent the vibration of both Ga and N atoms parallel and perpendicular to the c-axis, respectively. Each of the nearest-neighbored atom is vibrating in the opposite direction (optical mode) [143]. E_2-low (high) mode represents the vibration of two Ga (N) atoms in the unit cell, whose displacement is perpendicular to the c-axis. The A_1 and E_1 each has longitudinal-optical (LO) and transverse-optical (TO) modes. Therefore six Raman-active modes are observable. The Raman selection rule can be obtained by calculating the intensity of the scattered radiation from the time-averaged power radiated by the incident light-induced polarization into unit solid angle. The allowed modes in given configuration in wurtzite structures are shown in Table 7.7. In the backscattering geometry from the c-plane, the higher-frequency E_2 (often called E_2-high) and A_1(LO) modes are allowed, while E_2-high gives the

Table 7.7. The allowed modes in given configuration in wurtzite structures [143]

Mode	Configuration
A_1 (TO)	x (y, y) − x, x (z, z) − x
A_1 (LO)	z (x, x) − z
E_1 (TO)	x(y, z) − x, x(y, z)y
E_1 (LO)	x (y, z) y
E_2	z(y, y) − z, z(x, y) − z, x(y, y) − x, x(y, y) z

Table 7.8. The reported room-temperature frequencies (cm^{-1}) of the six Raman-active modes in unstrained wurtzite GaN and ZnO

	GaN [143]	ZnO [144]
E_2–low	144	100
A_1(TO)	533	380
E_1(TO)	561	410
E_2-high	569	440
A_1 (LO)	735	577
E_1(LO)	743	590

strongest signal, which is common in wurtzite compounds. The reported frequencies of the six Raman-active mode in unstrained GaN and ZnO is shown in Table 7.8.

The Raman frequencies are supposed to be sensitive to the strain and the lattice constant since the frequencies represent eigenfrequencies of the vibrational mode. The residual strain in wurtzite GaN epitaxial layers can be easily measured by looking at the shift of the E_2-high mode. Figure 7.26 shows an example where the spectra of E_2-high modes are observed in wurtzite c-GaN grown on SiC and sapphire substrates [146]. It should be noted that c-GaN on SiC is under tensile strain, and c-GaN on sapphire is under compressive strain due to the difference in the thermal expansion coefficient.

The strain free bulk sample has the E_2-high peak at 566.2 cm^{-1}. While the GaN on SiC sample has the E_2-high resonance at a smaller frequency, the GaN on sapphire sample has a larger E_2-high frequency than in strain-free bulk GaN. In fact, the Raman scattering measurement is believed to be one of the best methods to determine the residual strain in GaN epitaxial films. For example, the gradual relaxation of compressive strain in a very thick (220 μm) GaN layer grown on sapphire has been demonstrated by Siegle et al. [147]. By measuring the shift of E_2-mode, the authors showed that the residual strain is not completely relaxed until the distance from the interface reached 100 μm (Fig. 7.27). Thus, conventional LED structures grown on sapphire, which are typically several μm thick, undergo a significant residual compressive strain.

Strain evaluation by Raman spectroscopy also has been reported in ZnO system [148]. By monitoring the relative peak position of E_2-high mode

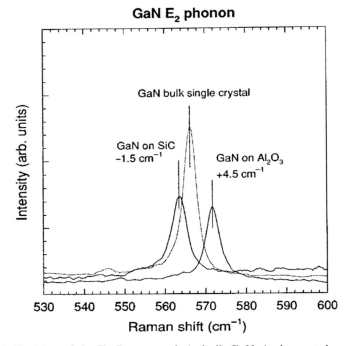

Fig. 7.26. Position of the E_2 Ramanmode in bulk GaN single crystals and in films grown on SiC and on sapphire. Homoepitaxially grown MBE films and bulk GaN exhibit a position of the E_2 Raman mode at 566.2 cm^{-1} (stress-free standard). The shifts of the line position are caused by a tensile stress in the film grown on SiC and compressive one if grown on sapphire [146]. Reprinted with permission from [146]. Copyright (1996), American Physical Society

($\sim$440 cm^{-1}) from an unstrained film, the authors obtained the sign and the magnitude of the biaxial strain in epitaxially grown ZnO. As is the case with GaN, E_2-high mode shifted to smaller (larger) frequency side when the film is under tensile (compressive) strain.

7.5.2 Carrier Concentration

A_1(LO) and E_2-high modes are readily observable in polar-GaN and ZnO, while E_2-high modes shift more sensitively by the biaxial strain. This is the reason why E_2-high modes are generally monitored for evaluating the strain. Insensitivity of A_1(LO) to strain can be an advantage when evaluating carrier concentrations.

Carrier concentration can also be determined by Raman spectroscopy by measuring the shift of so called LO-phonon–plasmon coupled mode. Plasmons, the collective excitation of free carriers, are one of the elementary excitations in solids. As the plasmon resonance frequency is in the order of the LO phonon

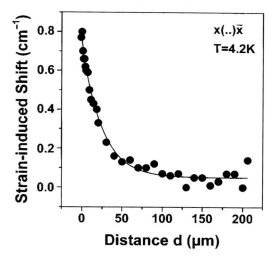

Fig. 7.27. Strain-induced shift of the E_2 Raman mode as a function of distance d to the substrate interface. The data were taken at 4.2 K. The solid line represents an exponential fit to the data points [147]. Reprinted with permission from [147]. Copyright (1997), American Institute of Physics

frequency and is quite sensitive to the free-carrier concentration of the material, the frequency can be tuned to match the LO phonon frequency by adjusting the carrier concentrations. When the energy of the two resonances becomes closer by changing the concentration, the interaction between the two becomes stronger. If the energy of the two resonances becomes identical, a plasmon can excite a phonon and the excited phonon can excite the plasmon again. In this case, there is no way to separate these two resonances, resulting in forming another quasi particle – LO-phonon–plasmon coupled-mode. This is a universal behavior that can happen in any two excitations in solids, which is often called quantum-mechanical-level-anticrossing. As the plasmon frequency is moving closer to the LO phonon frequency, the LO phonon frequency is pushed away and splits into two modes by forming quasi-particle of the plasmon and phonon. Figure 7.28 shows an example of observing the LO-phonon–plasmon-coupled mode in a n-type GaN [149]. As the plasmon mode moves toward the LO phonon mode, the position of LPP$^+$ mode is pushed to higher-frequency-side, and a new mode, denoted by LPP$^-$, appears in the lower frequency range. By measuring the frequency position of these two modes, one can calculate the carrier concentration. This can be an alternative way to determine the carrier concentration when a direct probe, such as Hall measurement, is not possible.

It is worth mentioning another interesting aspect of Raman scattering in p-type GaN films. Since the Raman scattering is sensitive to the specific bonding, the local modes of Mg–N bonding can be observed [150, 151]. It is well

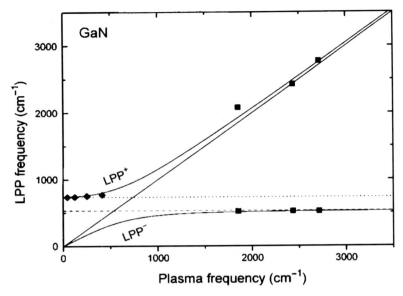

Fig. 7.28. Measurement of LO-phonon–plasmon-coupled modes in a n-type GaN. As the plasmon mode moves toward the LO phonon mode, the position of LPP$^+$ mode is pushed to higher-frequency-side, and a new mode, denoted by LPP$^-$, appears in the lower frequency range [149]. Reprinted with permission from [149]. Copyright (1995), American Institute of Physics

known that Mg doped GaN layers need to be thermally annealed to obtain p-type conductivity [152]. Although the exact nature of the Mg activation process is still under debate, the formation of Mg–H complex is commonly believed to be the main origin of the passivation [153]. Harima et al. observed local modes related to Mg–N and Mg–H bondings in Mg-doped GaN [150]. The Mg–N related mode became stronger after activation, whereas the Mg–H related mode disappeared. By comparing the local modes as a function of the annealing condition, the signatures of optimal activating conditions could be found.

The realization of a highly conductive p-type ZnO is one of the major problems for optoelectronic application. Therefore, p-doped ZnO films are attracting attention for characterizing their electronic and vibration properties using Raman spectroscopy. A study on the LO-phonon–plasmon-coupled mode in N–In codoped p-ZnO films was recently reported [154]. The local modes in phosphorus-doped ZnO were studied [155]. The modes observed at 364 cm^{-1} and 478 cm^{-1} were attributed to the vibration of Zn–P and P–O bondings. The Raman study of N$^+$-implanted ZnO was also recently reported [156]. The peak observed at 275 cm^{-1} was assigned to the local mode related to Zn–N bonding.

References

1. S. Nakamura, G. Fasol, *The Blue Laser Diode* (Springer, Berlin, 1997)
2. S. Nakamura, in *High Brightness Light Emitting Diodes*, ed. by S.B. Stringfellow, M.G. Craford (Academic, New York, 1997), pp. 391
3. B. Gil, Phys. Rev. B **64**, 201310(R) (2001)
4. A.R. Hutson, Phys. Rev. Lett. **4**, 505 (1960)
5. F. Bernardini, V. Fiorentini, Phys. Stat. Sol. B **216**, 391 (1999)
6. C. Klingshirn, Phys. Stat. Sol. B **244**, 3027 (2007)
7. P. Lawaetz, Phys. Rev. B **5**, 4039 (1972)
8. U. Ozgur, Ya, I. Alivov, C. Liu, A. Teke, M.A. Reshchikov, S. Dogan, V. Avrutin, S.J. Cho, H. Morkoc, J. Appl. Phys. **98**, 041301 (2005)
9. B.K. Meyer, H. Alves, D.M. Hofmann, W. Kriegseis, D. Forster, F. Bertram, J. Christen, A. Hoffmann, M. Straßburg, M. Dworzak, U. Haboeck, A.V. Rodina, Phys. Stat. Sol. B **241**, 231 (2004)
10. S.C. Jain, J. Narayan, R. Van Overstraeten, J. Appl. Phys. **87**, 965 (2000)
11. M.A. Reshchikova, H. Morkoç, J. Appl. Phys. **97**, 061301 (2005)
12. M.O. Manasreh, H.X. Jiang, in *III-Nitride Semiconductors, Optical properties I*, ed. by M.O. Manasreh (Taylor, New York, 2002), pp. 1
13. S. Desgreniers, Phys. Rev. B **58**, 14102 (1998)
14. A.B.M. Almamun Ashrafi, A. Ueta, A. Avramescu, H. Kumano, I. Suemune, Y.-W. Ok, T.-Y. Seong, Appl. Phys. Lett. **76**, 550 (2000)
15. G.M. Dalpian, Y. Yan, S.H. Wei, Appl. Phys. Lett. **89**, 011907 (2006)
16. J. Menniger, U. Jahn, O. Brandt, H. Yang, K. Ploog, Phys. Rev. B, **53**, 1881 (1996)
17. U. Strauss, H. Tews, H. Riechert, R. Averbeck, M. Schienle, B. Jobst, D. Volm, T. Streibl, B.K. Meyer, W.W. Rühle, Semicond. Sci. Technol. **12**, 637 (1997)
18. A.V. Andrianov, D.E. Lacklison, J.W. Orton, T.S. Cheng, C.T. Foxon, K.P. O'Donnell, J.F.H. Nicholls, Semicond. Sci. Technol. **12**, 59 (1997)
19. D.J. As, F. Schmilgus, C. Wang, B. Schöttker, D. Schikora, K. Lischka, Appl. Phys. Lett. **70**, 1311 (1997)
20. D. Xu, H. Yang, J.B. Li, D.G. Zhao, S.F. Li, S.M. Zhuang, R.H. Wu, Y. Chen, G.H. Li, Appl. Phys. Lett. **76**, 3025 (2000)
21. J. Wu, H. Yaguchi, K. Onabe, R. Ito, Y. Shiraki, Appl. Phys. Lett. **71**, 2067 (1997)
22. X. Liu, A.R. Goni, K. Syassen, H. Siegle, C. Thomsen, B. Schöttker, D.J. As, D. Schikora, J. Appl. Phys. **86**, 929 (1999)
23. Jasprit Singh, *Semiconductor Optoelectronics* (McGraw-Hill, New York, 1995), pp. 170–232
24. R. Shankar, *Principles of Quantum Mechanics* (Springer, New York, 1994)
25. J.J. Hopfield, J. Phys. Chem. Solids **15**, 97 (1960)
26. M. Tchounkeu, O. Briot, B. Gil, J.P. Alexis, R.L. Aulombard, J. Appl. Phys. **80**, 5352 (1996)
27. I. Vurgaftman, J.R. Meyer, J. Appl. Phys. **94**, 3675 (2003)
28. P. Misra, U. Behn, O. Brandt, H.T. Grahn, B. Imer, S. Nakamura, S.P. DenBaars, James S. Speck, Appl. Phys. Lett. **88**, 161920 (2006)
29. S.J. Xu, W. Liu, M.F. Li, Appl. Phys. Lett. **81**, 2959 (2002)
30. K. Kornitzer, T. Ebner, K. Thonke, R. Sauer, C. Kirchner, V. Schwegler, M. Kamp, M. Leszczynski, I. Grzegory, S. Porowski, Phys. Rev. B **60**, 1471 (1999)

31. R. Stepniewski, K.P. Korona, A. Wysmołek, J.M. Baranowski, K. Pakuła, M. Potemski, G. Martinez, I. Grzegory, S. Porowski, Phys. Rev. B **56**, 15151 (1997)
32. A.V. Rodina, M. Dietrich, A. Göldner, L. Eckey, A. Hoffmann, Al. L. Efros, M. Rosen, B.K. Meyer, Phys. Rev. B **64**, 115204 (2001)
33. G.D. Chen, M. Smith, J.Y. Lin, H.X. Jiang, S.-H. Wei, M.A. Khan, C.J. Sun, Appl. Phys. Lett. **68**, 2784 (1996)
34. P.P. Paskov, T. Paskova, P.O. Holtz, B. Monemar, Phys. Status Solidi B **228**, 467 (2001)
35. M. Mayer, A. Pelzmann, M. Kamp, K.J. Ebeling, H. Teisseyre, G. Nowak, M. Leszczynski, I. Grzegory, S. Porowski, G. Karczewski, Jpn. J. Appl. Phys. Part 2 **36**, L1634 (1997)
36. K.P. Korona, A. Wysmołek, K. Pakuła, R. Stepniewski, J.M. Baranowski, I. Grzegory, B. Łucznik, M. Wro'blewski, S. Porowski, Appl. Phys. Lett. **69**, 788 (1996)
37. P.Y. Yu, M. Cardona, *Fundamentals of Semiconductors* (Springer, Berlin, 1996)
38. D. Volm, K. Oettinger, T. Streibl, D. Kovalev, M. Ben-Chorin, J. Diener, B.K. Meyer, J. Majewski, L. Eckey, A. Hoffmann, H. Amano, I. Akasaki, K. Hiramatsu, T. Detchprohm, Phys. Rev. B **53**, 16543 (1996)
39. Z.X. Liu, K.P. Korona, K. Syassen, J. Kuhl, K. Paku, J.M. Baranowski, I. Grzegory, S. Porowski, Solid State Commun. **108**, 433 (1998)
40. Z.X. Liu, S. Pau, K. Syassen, J. Kuhl, W. Kim H. Morkoc, M.A. Khan, C.J. Sun, Phys. Rev. B **58**, 6696 (1998)
41. P. Boguslawski, E. Briggs, T.A. White, M.G. Wensell, J. Bernholc, in *Diamond, SiC and Nitride Wide Bandgap Semiconductors*, ed. by C.H. Carter Jr., G. Gildenblat, S. Nakamura, R.J. Nemanichi, MRS Symposia Proceedings No. 339 (Materials Research Society, Pittsburgh, 1994), p. 693
42. P. Boguslawski, E. Briggs, J. Bernholc, Phys. Rev. B **51**,17255 (1995)
43. J. Neugebauer, C.G. Van de Walle, Phys. Rev. B **50**, 8067 (1994)
44. T.L. Tansley, R.J. Eagan, Phys. Rev. B **45**, 10 942 (1992)
45. B.J. Skromme, J. Jayapalan, R.P. Vaudo, V.M. Phanse, Appl. Phys. Lett. **73**, 2358 (1999)
46. A. Wysmolek, M. Potemski, R. Stepniewski, J.M. Baranowski, D.C. Look, S.K. Lee, J.Y. Han, Phys. Stat. Sol. B **235**, 36 (2003)
47. B. Monemar, J. Phys. Condens. Matter **13**, 7011 (2001)
48. B.J. Skromme, G.L. Martinez, Mater. Res. Soc. Symp. Proc. **5S1**, W9.8 (2000)
49. M. Leroux, B. Beaumont, N. Grandjean, P. Lorenzini, S. Haffouz, P. Vennegues, J. Massies, P. Gibart, Mater. Sci. Eng. B **50**, 97 (1997)
50. M. Leroux, N. Grandjean, B. Beaumont, G. Nataf, F. Semond, J. Massies, P. Gibart, J. Appl. Phys. **86**, 3721 (1999)
51. R. Stepniewski, A.Wysmołek, M. Potemski, K. Pakuła, J.M. Baranowski, I. Grzegory, S. Porowski, G. Martinez, P. Wyder, Phys. Rev. Lett. **91**, 226404 (2003)
52. R. Stepniewski, A. Wysmolek, M. Potemski, J. Lusakowski, K. Korona, K. Pakula, J.M. Baranowski, G. Martinez, P. Wyder, I. Grzegory, S. Porowski, Phys. Status Solidi B **210**, 373 (1998)
53. R. Stepniewski, A. Wysmolek, Acta Phys. Pol. A **90**, 681 (1996)
54. M.A. Reshchikov, D. Huang, F. Yun, L. He, H. Morkoç, D.C. Reynolds, S.S. Park, K.Y. Lee, Appl. Phys. Lett. **79**, 3779 (2001)

55. S.J. Rhee, S. Kim, E.E. Reuter, S.G. Bishop, R.J. Molnar, Appl. Phys. Lett. **73**, 2636 (1998)
56. H. Wang, A.-B. Chen, Phys. Rev. B **63**, 125212 (2001)
57. U. Kaufmann, M. Kunzer, M. Maier, H. Obloh, A. Ramakrishnan, B. Santic, P. Schlotter, Appl. Phys. Lett. **72**, 1326 (1998)
58. S. Hess, R.A. Taylor, J.F. Ryan, N.J. Cain, V. Roberts, J. Roberts, Phys. Status Solidi B **210**, 465 (1998)
59. L. Eckey, U. von Gfug, J. Holst, A. Hoffmann, A. Kaschner, H. Siegle, C. Thomsen, B. Schineller, K. Heime, M. Heuken, O. Schön, R. Beccard, J. Appl. Phys. **84**, 5828 (1998)
60. M.A. Reshchikov, G.-C. Yi, B.W. Wessels, Phys. Rev. B **59**, 13176 (1999)
61. H. Obloh, K.H. Bachem, U. Kaufmann, M. Kunzer, M. Maier, A. Ramakrishnan, P. Schlotter, J. Cryst. Growth **195**, 270 (1998)
62. Y. Koide, D.E. Walker Jr., B.D. White, L.J. Brillson, M. Murakami, S. Kamiyama, H. Amano, I. Akasaki, J. Appl. Phys. **92**, 3657 (2002)
63. L.S. Wang, W.K. Fong, C. Surya, K.W. Cheah, W.H. Zheng, Z.G. Wang, Solid State Electron. **45**, 1153 (2001)
64. G. Li, S.J. Chua, S.J. Xu, W. Wang, P. Li, B. Beaumont, P. Gibart, Appl. Phys. Lett. **74**, 2821 (1999)
65. I. Ho, G.G. Stringfellow, Appl. Phys. Lett. **69**, 2701 (1996)
66. J.H. Song, E.D. Sim, Y.S. Joh, Y.G. Kim, K.S. Baek, S.K. Chang, Solid State Commun. **28**, 413 (2003)
67. H. Wang, M. Jiang, D.G. Steel, Phys. Rev. Lett. **65**, 1255 (1990)
68. S.F. Chichibu, A.C. Abare, M.S. Minsky, S. Keller, S.B. Fleischer, J.E. Bowers, E. Hu, U.K. Mishra, L.A. Coldren, S.P. DenBaars, T. Sota, Appl. Phys. Lett. **73**, 2006 (1998)
69. Y. Narukawa, Y. Kawakami, M. Funato, S. Fujita, S. Fujita, S. Nakamura, Appl. Phys. Lett. **70**, 981 (1997)
70. Y. Narukawa, Y. Kawakami, Sz. Fujita, Sh. Fujita, S. Nakamura, Phys. Rev. B **55**, R1938 (1997)
71. E.F. Schubert, *Light-Emitting Diodes*, 2nd edn. (Cambridge Press, Cambridge, 2006)
72. F. Hitzel, G. Klewer, S. Lahmann, U. Rossow, A. Hangleiter, Phys. Rev. Lett. **95**, 127402 (2005)
73. Y.H. Cho, G.H. Gainer, A.J. Fischer, J.J. Song, S. Keller, U.K. Mishra, S.P. DenBaars, Appl. Phys. Lett. **73**, 1370 (1998)
74. H. Schomig, S. Halm, A. Forchel, G. Bacher, J. Off, F. Scholz, Phys. Rev. Lett. **92**, 106802 (2004)
75. T.M. Smeeton, M.J. Kappers, J.S. Barnard, M.E. Vickers, C.J. Humphreys, Appl. Phys. Lett. **83**, 5419 (2003)
76. S.F. Chichibu, K. Wada, J. Mullhauser, O. Brandt, K.H. Ploog, T. Mizutani, A. Setoguchi, R. Nakai, M. Sugiyama, H. Nakanishi, K. Korii, T. Deguchi, T. Sota, S. Nakamura, Appl. Phys. Lett. **76**, 1671 (2000)
77. F. Hitzel, G. Klewer, S. Lahmann, U. Rossow, A. Hangleiter, Phys. Rev. B **72**, 081309(R) (2005)
78. S.F. Chichibu, H. Marchand, M.S. Minsky, S. Keller, P.T. Fini, J.P. Ibbetson, S.B. Fleischer, J.S. Speck, J.E. Bowers, E. Hu, U.K. Mishra, S.P. DenBaars, T. Deguchi, T. Sota, S. Nakamura, Appl. Phys. Lett. **74**, 1460 (1999)
79. J.K. Son, T. Sakong, S.N. Lee, H.S. Paek, H. Ryu, K.H. Ha, O. Nam, Y. Park, J.S. Hwang, Y.H. Cho, Appl. Phys. Lett. **90**, 051918 (2007)

80. T. Gessmann, Y.-L. Li, E.L. Waldron, J.W. Graff, E.F. Schubert, J.K. Sheu, Appl. Phys. Lett. **80**, 986 (2002)
81. F. Bernardini, V. Fiorentini, Phys. Rev. B **58**, 15292 (1998)
82. R.D. King-Smith, D. Vanderbilt, Phys. Rev. B **47**, 1651 (1993)
83. R. Resta, Rev. Mod. Phys. **66**, 899 (1994)
84. F. Bernardini, V. Fiorentini, Appl. Surf. Sci. **166**, 23 (2000)
85. G. Bastard, *Wave Mechanics Applied to Semiconductor Heterostructures* (Halsted, New York, 1988)
86. K.L. Bunker, R. Garcia, P.E. Russell, Appl. Phys. Lett. **86**, 082108 (2005)
87. Y.D. Jho, J.S. Yahng, E. Oh, D.S. Kim, Appl. Phys. Lett. **79**, 1130 (2001)
88. J.I. Pankove, *Optical Processes in Semiconductors* (Dover, New York, 1971)
89. E. Berkowicz, D. Gershoni, G. Bahir, E. Lakin, D. Shilo, E. Zolotoyabko, A.C. Abare, S. Pl. Denbaars, L.A. Goldren, Phys. Rev. B **61**, 10994 (2000)
90. Bernard Gil, Phys. Rev. B **64**, 201310(R) (2001)
91. A. Mang, K. Reimann, St. Rubenacke, Solid State Commun. **94**, 251 (1995)
92. A. Teke, Ü. Özgür, S. Doğan, X. Gu, H. Morkoç, B. Nemeth, J. Nause, H.O. Everitt, Phys. Rev. B **70**, 195207 (2004)
93. D.C. Reynolds, D.C. Look, B. Jogai, C.W. Litton, G. Cantwell, W.C. Harsch, Phys. Rev. B **60**, 2340 (1999)
94. D.W. Hamby, D.A. Lucca, M.J. Klopfstein, G. Cantwell, J. Appl. Phys. **93**, 3214 (2003)
95. D.G. Thomas, J. Phys. Chem. Solids **15**, 86 (1960)
96. S.F. Chichibu, A. Tsukazaki, M. Kawasaki, K. Tamura, Y. Segawa, T. Sota, H. Koinuma, Appl. Phys. Lett. **80**, 2860 (2002)
97. J. Lagois, Phys. Rev. B **16**, 1699 (1977); Phys. Rev. B **23**, 5511 (1981)
98. D.C. Look, Mater. Sci. Eng. B **80**, 381 (2001)
99. D.M. Bagnall, Y.F. Chen, Z. Zhu, T. Yao, S. Koyama, M.Y. Shen, T. Goto, Appl. Phys. Lett. **70**, 2230 (1997)
100. Z.K. Tang, G.K.L. Wong, P. Yu, M. Kawasaki, A. Ohotomo, H. Koinuma, Y. Segawa, Appl. Phys. Lett. **72**, 3270 (1998)
101. D.M. Bagnall, Y.F. Chen, Z. Zhu, T. Yao, M.Y. Shen, T. Goto, Appl. Phys. Lett. **73**, 1038 (1998)
102. C. Klingshirn, R. Hauschild, J. Fallert, H. Kalt, Phys. Rev. B. **75**, 115203 (2007)
103. D.C. Reynolds, D.C. Look, B. Jogai, C.W. Litton, T.C. Collins, W. Harsch, G. Cantwell, Phys. Rev. B **57**, 12151 (1998)
104. H.J. Ko, Y.F. Chen, S.K. Hong, H. Wenisch, T. Yao, Appl. Phys. Lett. **77**, 3761 (2000)
105. D.C. Reynolds, D.C. Look, B. Jogai, T.C. Collins, Appl. Phys. Lett. **79**, 3794 (2001)
106. M. Schilling, R. Helbig, G. Pensl, J. Luminesc. **33**, 201 (1985)
107. H.J. Ko, Y.F. Chen, S.K. Hong, H. Wenisch, T. Yao, Appl. Phys. Lett. **77**, 3761 (2000)
108. D.C. Reynolds, C.W. Litton, T.C. Collins, Phys. Rev. **140**, A1726 (1965)
109. A. Kobayashi, O.F. Sankey, J.D. Dow, Phys. Rev. B **28**, 946 (1983)
110. K. Thonke, Th. Gruber, N. Teofilov, R. Schönfelder, A. Waag, R. Sauer, Physica B **308–310**, 945 (2001)
111. D.C. Reynolds, C.W. Litton, T.C. Collins, J.E. Hoelscher, J. Nause, Appl. Phys. Lett. **88**, 141919 (2006)

112. G. Xiong, K.B. Ucer, R.T. Williams, J. Lee, D. Bhattacharyya, J. Metson, P. Evans, J. Appl. Phys. **97**, 043528 (2005)
113. D.C. Look, B. Claflin, Phys. Stat. Sol. B **241**, 624 (2004)
114. M. Schirra, R. Schneider, A. Reiser, G.M. Prinz, M. Feneberg, J. Biskupek, U. Kaiser, C.E. Krill, R. Sauer, K. Thonke, Physica B **401–402**, 362 (2007)
115. D.C. Look, D.C. Reynolds, C.W. Litton, R.L. Jones, D.B. Eason, G. Cantwell, Appl. Phys. Lett. **81**, 1830 (2002)
116. J.D. Ye, S.L. Gu, F. Li, S.M. Zhu, R. Zhang, Y. Shi, Y.D. Zheng, X.W. Sun, G.Q. Lo, D.L. Kwong, Appl. Phys. Lett. **90**, 152108 (2007)
117. F.X. Xiu, Z. Yang, L.J. Mandalapu, D.T. Zhao, J.L. Liu, Appl. Phys. Lett. **87**, 252102 (2005)
118. H.S. Kang, G.H. Kim, D.L. Kim, H.W. Chang, B.D. Ahn, S.Y. Lee, Appl. Phys. Lett. **89**, 181103 (2006)
119. D.H. Chi, L.T. Binh, N.T. Binh, L.D. Khanh, N.N. Long, Appl. Surf. Sci. **252**, 2770 (2006)
120. O.F. Schirmer, D. Zwingel, Solid State Commun. **8**, 1559 (1970)
121. F. Bernardini, V. Fiorentini, D. Vanderbilt, Phys. Rev. B **56**, R10024 (1997)
122. H.D. Sun, T. Makino, N.T. Tuan, Y. Segawa, M. Kawasaki, A. Ohtomo, K. Tamura, H. Koinuma, Appl. Phys. Lett. **78**, 2464 (2001)
123. S.H. Park, D. Ahn, Appl. Phys. Lett. **87**, 253509 (2005)
124. C. Morhain, T. Bretagnon, P. Lefebvre, X. Tang, P. Valvin, T. Guillet, B. Gil, T. Taliercio, M. Teisseire-Doninelli, B. Vinter, C. Deparis, Phys. Rev. B **72**, 241305(R) (2005)
125. T. Makino, K. Tamura, C.H. Chia, Y. Segawa, M. Kawasaki, A. Ohtomo, H. Koinuma, Appl. Phys. Lett. **81**, 2355 (2002)
126. B.P. Zhang, B.L. Liu, J.Z. Yu, Q.M. Wang, C.Y. Liu, Y.C. Liu, Y. Segawa, Appl. Phys. Lett. **90**, 132113 (2007)
127. M. Al-Suleiman, A. El-Shaer, A. Bakin, H.-H. Wehmann, A. Waag, Appl. Phys. Lett. **91**, 081911 (2007)
128. D Zhao, B. Lia, Chunxia Wu, Y. Lu, D. Shen, J. Zhang, X. Fan, J. Lumin. **119**, 304 (2006)
129. M.H. Kim, M.F. Schubert, Q. Dai, J.K. Kim, E.F. Schuberta, Yongjo Park, Appl. Phys. Lett. **91**, 183507 (2007)
130. M.C. Schmidt, K.C. Kim, H. Sato, N. Fellows, H. Masui, S. Nakamura, S.P. DenBaars, J.S. Speck, Jpn. J. Appl. Phys. **46**, L126 (2007)
131. B. Rau, P. Waltereit, O. Brandt, M. Ramsteiner, K.H. Ploog, J. Puls, F. Henneberger, Appl. Phys. Lett. **77**, 3343 (2000)
132. Y.J. Sun, O. Brandt, M. Ramsteiner, H.T. Grahn, K.H. Ploog, Appl. Phys. Lett. **82**, 3850 (2003)
133. N.F. Gardner, J.C. Kim, J.J. Wierer, Y.C. Shen, M.R. Krames, Appl. Phys. Lett. **86**, 111101 (2005)
134. T. Koida, S.F. Chichibu, A. Uedono, T. Sota, A. Tsukazaki, M. Kawasaki, Appl. Phys. Lett. **84**, 1079 (2004)
135. Y.S. Nam, S.W. Lee, K.S. Baek, S.K. Chang, J.W. Ryu, J.H. Song, S.K. Han, S.K. Hong, T. Yao, J. Korean Phys. Soc. **53**, 288 (2008)
136. P. Misra, U. Behn, O. Brandt, H.T. Grahn, B. Imer, S. Nakamura, S.P. DenBaars, J.S. Speck, Appl. Phys. Lett. **88**, 161920 (2006)
137. S. Ghosh, P. Misra, H.T. Grahn, B. Imer, S. Nakamura, S.P. DenBaars, J.S. Speck, J. Appl. Phys. **98**, 026105 (2005)

138. T. Koida, S.F. Chichibu, A. Uedono, T. Sota, A. Tsukazaki, M. Kawasaki, Appl. Phys. Lett. **84**, 1079 (2004)
139. B. Gil, A. Alemu, Phys. Rev. B **56**, 12446 (1997)
140. S. Ghosh, P. Waltereit, O. Brandt, H.T. Grahn, K.H. Ploog, Phys. Rev. B **65**, 075202 (2002)
141. T. Koyama, T. Onuma, H. Masui, A. Chakraborty, B.A. Haskell, S. Keller, U.K. Mishra, J.S. Speck, S. Nakamura, S.P. DenBarrs, T. Sota, S.F. Chichibu, Appl. Phys. Lett. **89**, 091906 (2006)
142. Y.S. Nam, S.W. Lee, K.S. Baek, S.K. Chang, Ja.H. Song, Ju.H. Song, S.K. Han, S.K. Hong, T. Yao, Appl. Phys. Lett. **92**, 201907 (2008)
143. H. Harima, J. Frandon, F. Demangeot, M.A. Renucci, in *III-Nitride Semiconductors, Optical properties I*, ed. by M.O. Manasreh (Taylor, New York, 2002), pp. 283–377
144. T. Deguchi, D. Ichiryu, K. Sekiguchi, T. Sota, R. Matsuo, T. Azuhata, M. Yamaguchi, T. Yagi, S. Chichibu, S. Nakamura, J. Appl. Phys. **86**, 1860 (1999)
145. Landolt-Börnstein, in *New Series, Group III*, Vol. 17B, 22, 41B, ed. by U. Rössler (Springer, Heidelberg, Berlin, 1999)
146. C. Kisielowski, J. Kruger, S. Ruvimov, T. Suski, J.W. Ager III, E. Jones, Z.L. Weber, M. Rubin, E.R. Weber, M.D. Bremser, R.F. Davis, Phys. Rev B **54**, 17745 (1996)
147. H. Siegle, A. Hoffmann, L. Eckey, C. Thomsen, J. Christen, F. Bertram, D. Schmidt, D. Rudloff, K. Hiramatsu, Appl. Phys. Lett. **71**, 2490 (1997)
148. Th. Gruber, G.M. Prinz, C. Kirchner, R. Kling, F. Reuss, W. Limmer, A. Waag, J. Appl. Phys. **96**, 289 (2004)
149. P. Perlin, J. Camassel, W. Knap, T. Taliercio, J.C. Chervin, T. Suski, I. Grzegory, S. Porowski, Appl. Phys. Lett. **67**, 2524 (1995)
150. H. Harima, T. Inoue, S. Nakashima, M. Ishida, M. Taneya, Appl. Phys. Lett. **75**. 1383, (1999)
151. W. Gotz, N.M. Johnson, D.P. Bour, M.D. McCluskey, E.E. Haller, Appl. Phys. Lett. **69**, 3725 (1996)
152. S. Nakamura, N. Iwasa, M. Senoh, T. Mukai, Jpn. J. Appl. Phys. **31**, 1258 (1992)
153. W. Gotz, N.M. Johnson, D.P. Bour, M.D. McCluskey, E.E. Haller, Appl. Phys. Lett. **69**, 3725 (1996)
154. J.F. Kong, H. Chen, H.B. Ye, W.Z. Shen, J.L. Zhao, X.M. Li, Appl. Phys. Lett. **90**, 041907 (2007)
155. J.D. Ye, S.L. Gu, S.M. Zhu, S.M. Liu, Y.D. Zheng, R. Zhang, Y. Shi, Q. Chen, H.Q. Yu, Y.D. Ye, Appl. Phys. Lett. **88**, 101905 (2006)
156. J.B. Wang, H.M. Zhong, Z.F. Li, W. Lua, Appl. Phys. Lett. **88**, 101913 (2006)

8

Electrical Properties of GaN and ZnO

D.-C. Oh

Abstract. Conductivity control is one of the important issues in GaN and ZnO because various types of native defects that hamper high conductivity or high resistivity are easily formed in the materials during growth. These defects are generated as a result of crystal imperfections, which have an influence on the material properties and device performance by introducing shallow or deep levels into the bandgap. In this chapter, first, ohmic contacts to GaN and ZnO are described. In the cases of GaN and ZnO, fortunately, just by placing metals on them, ohmic behavior, rather than rectifying behavior, appears because of the large densities of residual electrons existing in these materials. Second, Schottky contacts to GaN and ZnO are described. The Schottky contacts are not easily formed, differently from the ohmic contacts, even though the Schottky barrier is theoretically formed at the interface, because conducting layers present at the surface or donors and dislocations present in the bulk cause a leakage current. Third, electrical properties of GaN and ZnO are described. GaN and ZnO films are generally grown on Al_2O_3 substrates with the large lattice misfits, which induce various native defects that hamper the transport of free electrons. Their electron transport mechanisms are characterized by a combination of Hall measurements and theoretical calculation of the scattering mechanisms. The defects form deep levels in the bandgap. Finally, the deep levels of GaN and ZnO are described, which are strongly influenced by the growth conditions.

8.1 Introduction

GaN and ZnO are open and attractive playgrounds for physicists and engineers towards the implementation of optoelectronic devices in the blue and ultraviolet (UV) region. In the last two decades, the enormous development in single-crystal-growth technology by the epitaxy technique has resulted in the realization of blue and UV light-emitting devices, laser diodes, and photodetectors.

In the development of GaN and ZnO, the two materials have suffered from the basic problem of the lack of appropriate substrates, in addition to the troublesome problems of excessively high melting points and high vapor pressures. Fortunately, the right breakthroughs at the right periods changed their fates. Sophisticated low-temperature buffer layer growth made it possible to obtain high-quality films. Thus, the current epitaxy of GaN and ZnO almost means heteroepitaxy on Al_2O_3 substrates. However, even though the heteroepitaxy could succeed in getting high-quality films, this does not mean that the large density of defects induced by lattice mismatch is cleared out [1, 2]. In this chapter, the defects are emphasized and the problems induced by the defects are discussed. This chapter is organized in four parts. In the first part, ohmic contacts to GaN and ZnO are described. In the second part, Schottkty contacts to these materials are described. In the third part, electrical properties of GaN and ZnO are described. Finally, deep levels of these materials are described.

8.2 Ohmic Contacts to GaN and ZnO

Semiconductor devices are composed of several layers having their own roles. This means that the semiconductor devices have interfaces as many as the number of the layers. Ohmic contact is the first layer and the first interface that the supplied electrons meet with, when biased to the device, because its purpose is to allow the supplied electrons to flow into and out of the semiconductor material. Ohmic contact is characterized by the contact resistance at the metal/semiconductor interface. In the cases of GaN and ZnO, fortunately, just by placing metals on semiconductor materials ohmic behaviors, rather than rectifying ones, appear because of large densities of residual electrons in the materials. Nevertheless, more stable ohmic contacts over time and temperature are required to realize a long-life operation of optical and electrical devices.

8.2.1 Principle of Ohmic Contact

Generally, most of the metals used as electrodes in semiconductor devices form the Schottky barriers on GaN and ZnO because the workfunctions of GaN and ZnO are smaller than those of the metals. Therefore, the barriers should be sufficiently narrow at or near the bottom of the conduction bands to allow thermally excited electrons to be tunneled directly. The contact resistance between the two materials is dependent on doping concentration of semiconductors, engineering of interface structures, and unintentionally formed interface states. Figure 8.1 shows the energy band diagrams in the metal/semiconductor interface before and after junction formation [3].

In estimating the contact resistance, it is important to differentiate it from the measured resistance, which can be derived by a simple trick, named the transfer length method (TLM) [4]. Figure 8.2 shows a schematic diagram of the TLM, which can be mathematically described by

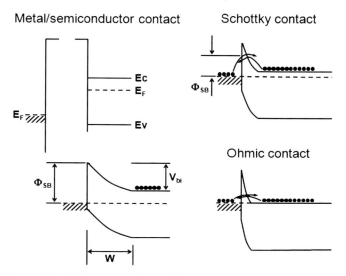

Fig. 8.1. Energy band diagrams of metal/semiconductor contacts according to the Schottky model [3]

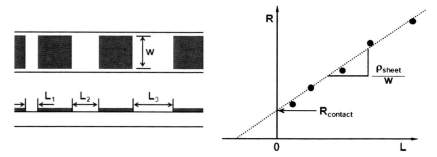

Fig. 8.2. Schematic explanation of TLM [4]. (*Left*) Basic pattern used to determine contact resistance. Contact metals are separated by a constant spacing or increasing ones. (*Right*) Plot of measured resistance as a function of contact spacing

$$R_{\text{measured}} = 2R_{\text{contact}} + R_{\text{semiconductor}}$$
$$= 2R_{\text{contact}} + \frac{\rho_{\text{sheet}}}{w}L. \quad (8.1)$$

Contact metals are deposited on the surface of semiconductor materials at a distance. The measured resistance (R_{measured}) consists of the two contact resistances ($2R_{\text{contact}}$) and the semiconductor resistance ($R_{\text{semiconductor}}$). Here, the latter is determined by the sheet resistivity (ρ_{sheet}), which is dependent on the contact width (w) and the contact spacing (L), while the former is determined by the contact resistance, which is constant. Hence, a plot of the measured resistance as a function of the spacing yields a straight line. The slope of this line gives information on the sheet resistivity and the intercept gives information on the contact resistance.

8.2.2 Ohmic Contacts to GaN

In the formation of GaN ohmic contact four parameters are of importance: annealing, surface treatment, doping concentration, and surface polarity. Annealing forms an intermediate compound at the interface or generates donor-type defects on the surface, which helps electron transport by tunneling. Surface treatment produces mechanical damage to the surface, which induces the donor-type defects. Doping concentration has a direct relationship with the Schottky barrier narrowing. Also, different polar surfaces result in different contact characteristics.

Figure 8.3 shows the effect of annealing on Al/n-GaN and Au/n-GaN contacts, where the n-type GaN layers were grown by molecular-beam epitaxy (MBE) [5]. It is found that the as-deposited Al/n-GaN contact has a linear current–voltage (I–V) characteristic, while the as-deposited Au/n-GaN contact has a nonlinear I–V characteristic. However, when these samples are subjected to the annealing process at a reduced atmosphere for 10 min at 575°C, it is shown that their I–V characteristics are changed exactly: the characteristic of the Au/n-GaN contact becomes linear, while that of the Al/n-GaN contact loses its linearity. This indicates that thermal treatment can induce opposite results depending on the materials. In the Au contact, Au appears to be diffused into the GaN surface upon the annealing, which improves the ohmic characteristic, while in the Al contact the annealing might have resulted in the formation of a thin insulating AlN interlayer, which deteriorated the ohmic characteristic. However, the relationships of these ohmic characteristics and physical mechanisms remain controversial.

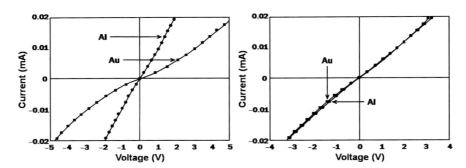

Fig. 8.3. Effect of annealing on Al/n-GaN and Au/n-GaN contacts, where n-type GaN layers were grown by MBE. (*Left*) I–V curves for the as-deposited contacts. (*Right*) I–V curves after annealing at a reduced atmosphere for 10 min at 575°C. Reprinted with permission from [5]. Copyright (1993), American Institute of Physics

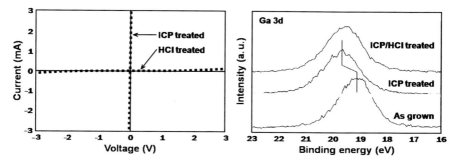

Fig. 8.4. Effect of HCl and Cl_2 ICP treatments on n-type GaN layers grown by MOCVD. (*Left*) I–V curves for Au/Ti/n-GaN contacts. (*Right*) XPS spectra of (*left*) Ga 3d and (*right*) O 1s for n-type GaN layers. Reprinted with permission from [6]. Copyright (2001), American Institute of Physics

Figure 8.4 shows the effect of surface treatment on Al/Ti/n-GaN contacts by Cl_2 inductively coupled plasma (ICP), where the n-type GaN layers were grown by metal-organic chemical-vapor deposition (MOCVD) [6]. It is found that the surface treatment of the GaN layer changes nonlinear I–V characteristics to linear ones. Then, the specific contact resistivity was improved from a completely rectifying behavior up to 9.4×10^{-6} $\Omega\,cm^2$. Here, it should be noted that the surface treatments using the ICP process and HCl solution resulted in a shift of the binding energy of Ga–N bonds toward higher energy by ~0.5 eV, which means that the Fermi level at the surface moves up near the conduction band edge. It is attributed to the fact that the surface treatment generates donor-type defects such as N vacancies, resulting in a decrease of the effective Schottky barrier height for electron transport.

Figure 8.5 shows the effect of doping concentration on Ag/Ti/n-GaN contacts, where the n-type GaN layers were grown by MOCVD [7]. It is found that increasing doping concentration in the GaN layers improves the linearity of the I–V curves. And, it is shown that the specific contact resistivity also decreases with the doping concentration: the contact with a large electron concentration of 1.7×10^{19} cm^{-3} shows a low specific contact resistivity of 6.5×10^{-5} $\Omega\,cm^2$, while the contact with a small electron concentration of 1.5×10^{17} cm^{-3} shows a high specific contact resistivity of 2.2×10^{-2} $\Omega\,cm^2$. As a result, even though thermal annealing is not followed, good ohmic characteristics are observed in contacts with electron concentration higher than 1×10^{18} cm^{-3}. It is attributed to the fact that increasing the doping concentration induces the narrowing of the Schottky barrier, which decreases the specific contact resistivity by increasing tunneling current.

On the other hand, GaN has surface polarity, because it has lack of inversion symmetry due to its wurzite structure. In the wurzite structure, dipole moments are not cancelled on the surface (Ga-polar face and N-polar face) along the c-axis, which induces the spontaneous surface polarization.

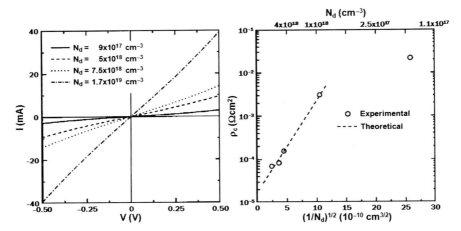

Fig. 8.5. Effect of doping concentration on Ag/Ti/n-GaN contacts, where n-type GaN layers were grown by MOCVD. (*Left*) *I–V* curves for Ag/Ti/n-GaN contacts. (*Right*) Relationship of specific contact resistivity and doping concentration. Reprinted with permission from [7]. Copyright (1996), American Institute of Physics

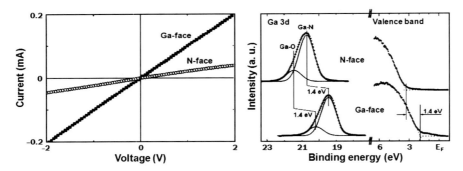

Fig. 8.6. Polarity dependence of Au/Ni/Al/Ti contacts on Ga- and N-faced GaN layers, where GaN layers were grown by MOCVD and contacts were annealed at 700°C. (*Left*) *I–V* curves for Au/Ni/Al/Ti contacts. (*Right*) XPS spectra of (*left*) Ga 3d and (*right*) valence bands for GaN layers. Reprinted with permission from [8]. Copyright (2002), American Institute of Physics

Figure 8.6 shows the polarity dependence of Au/Ni/Al/Ti contacts on the Ga- and the N-faced GaN layers, where the GaN layers were grown by MOCVD (these samples were annealed at 700°C after contact fabrication) [8]. It is found that the Ga face shows better *I–V* characteristics than the N face. Then, the specific contact resistivities of Ga and N faces were determined to be 8.3×10^{-4} and 7.0×10^{-2} $\Omega\,\text{cm}^2$, respectively. This indicates that the Ga face is favorable for ohmic contact formation. Here, it should be noted that the binding energy of the Ga–N bonds for the N face is shifted to higher value by 1.4 eV, as shown in Figure 8.6, compared to the Ga face. This means that the

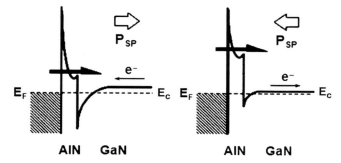

Fig. 8.7. Schematic energy band diagrams for Au/Ni/Al/Ti ohmic contacts on Ga-faced (*left*) and N-faced (*right*) GaN. Reprinted with permission from [8]. Copyright (2002), American Institute of Physics

Fermi level at the N face moves toward the conduction band, compared to the Ga face (the result is consistent with the valence-band spectrum). Figure 8.7 shows the schematic band diagrams for the two faces [8]. It is expected that in the Ga-polar face, the polarization direction by the Schottky barrier is the same as that of the crystal polarity, which enhances the downward band bending of the Shottky barrier, while in the N-polar face, the polarization directions are opposite, which reduces the band bending. Moreover, it was reported that the annealing of those samples results in thin AlN layers at the interfaces. Thus, it was assumed that the AlN interlayer generates a degenerated GaN region in the AlN/GaN interface, which enhances the downward band bending of the Shottky barrier in the case of Ga face and helps the electron tunneling.

Up to now, the ohmic contacts to GaN were discussed with respect to annealing, surface treatment, doping concentration, and surface polarity. The annealing process generally follows the ohmic contact fabrication. However, the thermal process can deteriorate their characteristics, which depend on the materials. Surface treatment using plasma is helpful for the ohmic contact formation, which is ascribed to the fact that the plasma treatment generates donor-type point defects on the surface. The doping concentration directly influences the characteristics of the ohmic contacts. A higher carrier concentration is favorable for the contact formation.

On the other hand, various metal systems such as Au, Al, Ti, V, NiAl, and In have been tried for the fabrication of stable ohmic contacts. In surface treatments, chemical treatments using simple etching solutions such as BOE (buffered oxide etchant), CH_3CSNH_2/NH_4OH, and $H_2SO_4:H_2PO_4:H_2O$ + $HCl:H_2O$ are also helpful for the contact fabrication. Laser irradiation of the GaN surface is also reported to reduce the contact resistivity, which is ascribed to the increase of donor-type defects by mechanical damage. In Table 8.1, characteristics of the reported ohmic contacts of GaN are summarized.

Table 8.1. Summary of characteristics of the GaN ohmic contacts

Metal	Growth	Material	Electron con. (cm^{-3})	Annealing	Surface treatment	Contact resistivity ($\Omega\,cm^2$)	Ref.
Au	MBE	GaN	3×10^{18}	As grown	–	Nonlinear	[5]
Al				Air, 575°C, 10 min	–	Linear	
				As grown	–	Linear	
				Air, 575°C, 10 min	–	More resistive	
Al/Ti	MOCVD	GaN:Si	1×10^{18}		HCl	Nonlinear	[6]
					Cl_2 ICP	9.4×10^{-6}	
Al/Ti	MOCVD	GaN:Si	3×10^{18}		As grown	Nonlinear	[9]
					CH_3CSNH_2/NH_4OH	4.8×10^{-4}	
				N_2, 700°C, 1 min	CH_3CSNH_2/NH_4OH	3.1×10^{-6}	
Al/Ti	MOCVD	GaN:Si	8×10^{16}	As grown	–	Nonlinear	[10]
				Laser irradiated	–	1.7×10^{-6}	
Ag/Ti	MOCVD	GaN:Si	2×10^{17}		BOE solution	2.2×10^{-2}	[7]
			2×10^{18}		BOE solution	4.0×10^{-3}	
			2×10^{19}		BOE solution	6.5×10^{-5}	
Al/Ti	HVPE	GaN:Si	2×10^{17}	N_2, > 600°C, 30 sec	BOE solution	2.0×10^{-5} (Ga face)	[11]
						Nonlinear (N face)	
Au/Ni/Al/Ti	MOCVD	GaN	3×10^{16}	Air (?), 700°C, 1 min		8.3×10^{-4} (Ga face)	[8]
						7.0×10^{-2} (N face)	
V	MOCVD	GaN:Si	2×10^{18}	As grown	BOE solution	Nonlinear	[12]
				N_2, 850°C, 2 min	BOE solution	2.3×10^{-4}	
Au/Ti/V				As grown	BOE solution	Nonlinear	
				N_2, 850°C, 2 min	BOE solution	4.0×10^{-6}	
NiAl	MOCVD	GaN:Si	3×10^{17}	As grown	$H_2SO_4{:}H_3PO_4{:}H_2O$ + $HCl{:}H_2O$	Nonlinear	[13]
				Ar, 850°C, 5 min	$H_2SO_4{:}H_3PO_4{:}H_2O$ + $HCl{:}H_2O$	9.4×10^{-6}	

8.2.3 Ohmic Contacts to ZnO

It should be noted that ZnO and GaN have the same crystal structure of wurzite, where the structure is composed of two atoms with opposite ionicity, Ga^+/N^- and Zn^{2+}/O^{2-} and the constituent ions are located in first neighboring positions in the periodic table. However, the ohmic contact is not yet an issue in ZnO-based device technologies because the ohmic behaviors appear just by placing metals on ZnO films. It is attributed to the fact that the ZnO films severely suffer from large densities of residual electrons in general due to native defects and external impurities, which enhance the electron transport by tunneling at the interface.

Figure 8.8 shows the effect of annealing on Au/Ti/n-ZnO contacts, where the n-type ZnO layers were grown by radio frequency (rf) sputtering [14]. It was found that the annealing process on the ZnO layer improves the I–V characteristic, but its strongly affected by annealing temperature. The specific contact resistivity of $2 \times 10^{-4}\,\Omega\,cm^2$ was obtained when annealed at 300°C for 1 min in N_2 atmosphere. However, after annealing at temperatures over 300°C, the contact was degraded. The improvement of the contact is ascribed to the increase of electron concentration that is induced by the oxygen vacancies formed by the out-diffusion of oxygen atoms to the ZnO surface, while the degradation of the contact is expected to be related to the disruption of the interface.

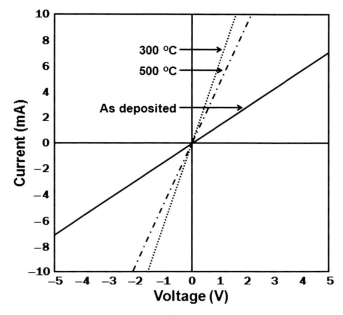

Fig. 8.8. Effect of annealing on Au/Ti/n-ZnO contacts, where n-type ZnO layers were grown by sputtering. Reprinted with permission from [14]. Copyright (2000), American Institute of Physics

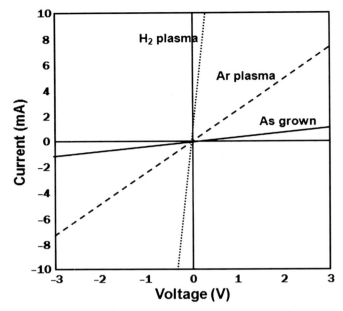

Fig. 8.9. Effect of Ar plasma and H_2 ICP treatments on n-type ZnO layers grown by sputtering. I–V curves for Au/Ti/n-ZnO contacts. Reprinted with permission from [15]. Copyright (2001), American Institute of Physics

Figure 8.9 shows the effect of surface treatments on Au/Ti/n-ZnO contacts by Ar and H_2 ICP, where the n-type ZnO layers were grown by rf sputtering [15]. It is found that the plasma treatments on the ZnO layers significantly increases the linearity of the I–V curves, especially for the H_2 plasma. Then, the specific contact resistivities decreased from 7.3×10^{-3} to 4.3×10^{-5} and 5.0×10^{-4} $\Omega\,cm^2$ by using the H_2 and the Ar plasma treatments, respectively. The former is attributed to the fact that hydrogens are acted as carriers and they passivated the deep-level defects in the surface region of the ZnO layer. The latter is attributed to the oxygen vacancies formed on the ZnO surface by Argon ion bombardment, which are acted as shallow donors. Hydrogen and oxygen vacancies are noted as native donors, which will be described in Sect. 8.4.3.

Figure 8.10 shows the effect of doping concentration of Au/Pt/Al/Ti/n-ZnO:P contacts as a function of post-annealing temperature, where the n-type ZnO:P layers were grown by pulsed-laser deposition (PLD) [16]. It is shown that the background electron concentration in the ZnO layers was suppressed by increasing the annealing temperature. And, it is found that the specific contact resistivity increases with the decrease of the electron concentration. Phosphorous is noted as p-type dopant. Therefore, the reduction of the background electron concentration by the annealing process was attributed to the carrier compensation by the activation of the incorporated

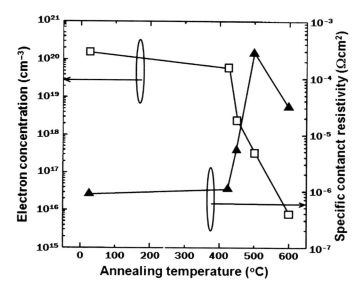

Fig. 8.10. Dependence of doping concentration for Au/Pt/Al/Ti/n-ZnO:P contacts, where n-type ZnO:P layers were grown by PLD. The plot shows the relationship of electron concentration and specific contact resistivity as a function of annealing temperature. Reprinted with permission from [16]. Copyright (2004), American Institute of Physics

phosphorous atoms. This means that the increase of the specific contact resistivity is due to the Schottky barrier broadening by the decreased electron concentration.

Up to now the contacts to ZnO were discussed in the aspects of annealing, surface treatment, and doping concentration. The basic principles and approaches of the ZnO ohmic contacts are almost the same as those for GaN, because the two materials have similar structural and electrical properties. For the fabrication of stable ohmic contacts to ZnO, Ti, Al, In, and Ru have been tried. The annealing process below a critical temperature after the contact fabrication improves the characteristics, which is ascribed to the generation of oxygen vacancies on the surface the Ar and H_2 plasma treatments are helpful for the ohmic contact formation, which is ascribed to the generation of donor-type defects of oxygen vacancies and hydrogen interstitials on the surface. The annealing of P-doped ZnO films results in the increase of the specific contact resistivity, which is ascribed to the Schottky barrier broadening by the suppression of background electron concentration.

On the other hand, in the ZnO ohmic contact formation, surface treatments using chemical solutions are not tried well, differently from GaN, which can be attributed to the facts that ZnO is easily etched out by acid etchants and the control of etching is difficult. Moreover, surface polarity almost does not have any influence on the ZnO ohmic contacts because the

Table 8.2. Summary of characteristics of the ZnO ohmic contacts

Metal	Growth	Material	Electron con. (cm^{-3})	Annealing	Surface treatment	Contact resistivity (Ωcm^2)	Ref.
Au/Ti	rf sputter	ZnO:Al	2×10^{17}	As grown	HCl solution	2.0×10^{-2}	[14]
				N$_2$, 300°C, 1 min	HCl solution	2.0×10^{-4}	
				N$_2$, 500°C, 1 min	HCl solution	1.0×10^{-3}	
Au/Ti	rf sputter	ZnO:Al	7×10^{17}	–	As grown	7.3×10^{-3}	[15]
				–	Ar ICP	5.0×10^{-4}	
				–	H$_2$ ICP	4.3×10^{-5}	
Au/Pt/Al/Ti	PLD	ZnO:P	1×10^{20}	As grown	–	8.0×10^{-7}	[16]
			8×10^{15}	O$_2$, 600°C, 60 min	–	3.0×10^{-5}	
Al	rf sputter	ZnO:Al	2×10^{18}	As grown	–	8.0×10^{-4}	[17]
Pt/Al				As grown	–	1.2×10^{-5}	
Ru	rf sputter	ZnO:Al	3×10^{18}	As grown	–	2.1×10^{-3}	[18]
				N$_2$, 700°C, 1 min	–	3.2×10^{-5}	
				N$_2$, 700°C, 10 min	–	Degraded	

heteroepitaxially grown ZnO films have a high background electron concentration, which cancels the effect of carrier depletion and accumulation by surface polarity. In Table 8.2, characteristics of the reported ohmic contacts of ZnO are summarized.

8.3 Schottky Contacts to GaN and ZnO

Schottky contact plays a role opposite to that of ohmic contact, because it partially blocks the flow of supplied electrons. The rectifying contact is not easily formed, differently from the ohmic contact, even though the Schottky barrier can be theoretically simply formed in the metal/semiconductor interface. It is attributed to the fact that the conducting layer present at the surface or the donors and the dislocations present in the bulk cause a leakage current. The Schottky contact is characterized by evaluating the Schottky diode based on the metal/semiconductor contact. In GaN- and ZnO-based device technologies, high-quality Schottky contacts are an issue, because they play crucial roles in the development of applications such as high-electron-mobility transistors (HEMTs) and UV detectors. They are also necessary for capacitance measurements for the studies of deep levels.

8.3.1 Principle of Schottky Contact

The principle of Schottky contact can be started with the same point as ohmic contact. However, the transport mechanism that allows electrons to

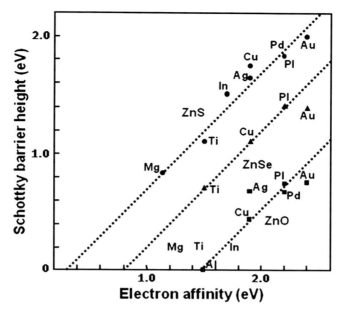

Fig. 8.11. Relationship of Schottky barrier height and electron affinity for metal contacts on ZnSe, ZnS, and ZnO [19]

pass the Schottky barrier is different. In the ohmic contact, electron transport is determined by the field emission that allows electrons to tunnel into the Schottky barrier, while in the Schottky contact it is determined by the thermionic emission that allows electrons to jump over the Schottky barrier [3]. Theoretically, the Schottky barrier is determined by the difference of the metal workfunction and the semiconductor electron affinity, as shown in Fig. 8.11 [19]. This means that the Schottky barrier height is independent of interface structures and states, and surface and bulk conductivities. However, the actually obtained Schottky barrier height, that is, the effective Schottky barrier height, is different from the theoretical value because various factors such as interface states, residual electrons, and edge and screw dislocations generate leakage current paths.

To evaluate the Schottky contact is to estimate the Schottky diode characteristics using I–V measurements. This is derived by theoretical calculation and fitting using the diode equation, which is expressed by [19]

$$I = I_\mathrm{o} \exp\left(\frac{q(V-IR)}{nkT}\right),$$
$$I_\mathrm{o} = A^* A T^2 \exp\left(-\frac{\Phi_\mathrm{SB}}{kT}\right) \quad (8.2)$$

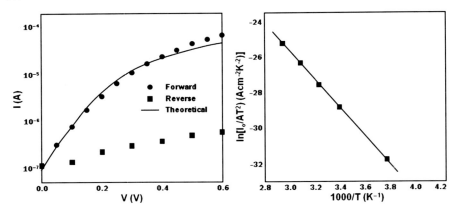

Fig. 8.12. Examples of I–V analyses using the diode equation for a Schottky contact. Reprinted with permission from [20]. Copyright (2005), American Institute of Physics

where I is the current, I_o is the reverse saturation current, q is the unit charge of an electron, V is the applied voltage, R is the series resistance, n is the ideality factor, k is the Boltzman constant, T is the temperature, A^* is the Richardson constant ($=32$ A cm^{-2} K^{-2}), A is the contact area, and Φ_{SB} is the Schottky barrier height. Figure 8.12 shows an example of the I–V analysis [20]. The ideality factor is estimated from a theoretical fitting to the forward bias current. The Schottky barrier height is calculated from the reverse-bias saturation current, which can also be derived from a plot of reverse saturation current as a function of the inverse of temperature. On the other hand, the estimated parameters should be cross-checked by capacitance–voltage (C–V) measurements [20].

Figure 8.13 shows the C–V analysis [20]. First, it should be noted that measured built-in potential varies with measurement frequency when series resistance due to trap centers are dominant. Therefore, the real built-in potential is estimated by using the two equations below (8.3 and 8.4):

$$\left(\frac{1}{C^2}\right) = \frac{2}{q\varepsilon A^2 N_{\mathrm{d}}}(V - V_{\mathrm{bi}}^{\mathrm{meas}}), \tag{8.3}$$

$$V_{\mathrm{bi}}^{\mathrm{meas}} = V_{\mathrm{bi}}^{\mathrm{real}} + 2R^2\omega^2\left(\frac{\mathrm{d}C^{-2}}{\mathrm{d}V}\right)^{-1}, \tag{8.4}$$

where C is the capacitance, ε is the dielectric constant, N_{d} is the donor density, $V_{\mathrm{bi}}^{\mathrm{meas}}$ is the measured built-in potential, $V_{\mathrm{bi}}^{\mathrm{real}}$ is the real built-in potential, R is the series resistance, and ω is the measurement frequency. Second, the Schottky barrier height is determined by the built-in potential and the donor density, which is expressed by

$$\Phi_{\mathrm{SB}} = qV_{\mathrm{bi}}^{\mathrm{real}} + kT\ln\left(\frac{N_{\mathrm{c}}}{N_{\mathrm{d}}}\right), \tag{8.5}$$

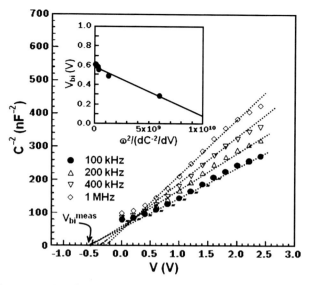

Fig. 8.13. An example of C–V analysis for a Schottky contact. Reprinted with permission from [20]. Copyright (2005), American Institute of Physics

where N_c is the effective density of states in the conduction band. By inserting the real built-in potential obtained from (8.3) and (8.4) into (8.5), the Schottky barrier can be estimated. The results estimated from C–V measurements should be consistent with those from I–V measurements.

8.3.2 Schottky Contacts to GaN

In the formation of GaN Schottky contact, two points are to be emphasized: current leakage and surface polarity. The current leakage is present as a major technological drawback in GaN-based optoelectronic devices, which is due to structural defects rather than residual electrons at the surface or in the bulk. It can be reduced by inserting current blocking layers into the devices. The surface polarity in the Schottky contacts is more important than the cases for the GaN ohmic contacts because the Schottky contacts are generally formed on the films or bulks of low electron concentration. On the other hand, the background electron concentration has a direct relationship with the Schottky contact characteristics, which will be introduced in Sect. 8.3.3.

First of all, the most representative current leakage mechanisms in GaN-based Schottky contacts are shown in Fig. 8.14 [21, 22]. The first one is the field-emission tunneling through the metal/GaN interface, which appears at large reverse-bias voltages. It is dominant at low temperatures. The second one is the trap-assisted tunneling, which is associated with the dislocation-related leakage current path. It is dominant above approximately 275 K. The third one is the Schottky barrier thinning, which assumes the unintentional

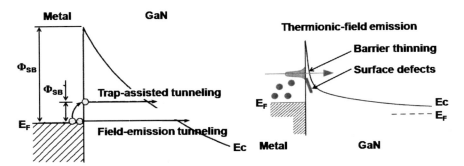

Fig. 8.14. Schematic diagrams for leakage current mechanisms of Schottky contacts. Reprinted with permission from [21, 22]. Copyright (2004), American Institute of Physics

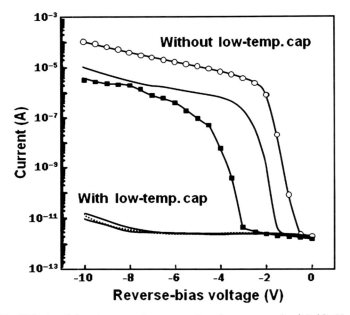

Fig. 8.15. Effect of low-temperature capping layers on Au/Ni/GaN contacts, where GaN layers were grown by MOCVD. I–V curves in the reverse-biased region. Reprinted with permission from [23]. Copyright (2005), American Institute of Physics

introduction of a high density of donor defects near the surface. It helps the electron tunneling through the Schottky barrier in both forward and reverse directions by the thermionic field emission or the field emission mechanisms depending on temperature.

The leakage currents are effectively blocked by inserting the current blocking layers into the Schottky contacts. Figure 8.15 shows the effect of LT GaN

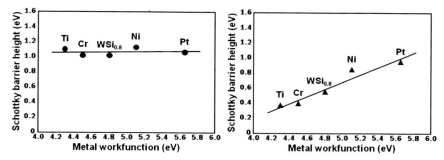

Fig. 8.16. Effect of low-temperature capping layers on Schottky contacts on n-type GaN layers grown by MOCVD. The plots show Schottky barrier heights as a function of workfunction for metal contacts with (*left*) and without (*right*) low-temperature caps. Reprinted with permission from [24]. Copyright (2006), American Institute of Physics

capping layers on the Au/Ni/GaN contacts, where the GaN layers were grown by MOCVD [23]. It is found that reverse saturation current is significantly improved, which is reduced from 1×10^{-6} to 1×10^{-11} A, when the LT cap is introduced. Also, it was observed from atomic-force microscopy images that the surface pits of threading dislocation termination, exposed nominally in the conventional structures, were mostly not exposed in the new structures with the LT caps. This indicates that the LT caps can block most of the leakage paths, which are related to the threading dislocations. On the other hand, this LT GaN capping layer influences the Schottky barrier height. Figure 8.16 shows the effect of the LT GaN capping layers on the Schottky barrier heights for the metal contacts of $Wsi_{0.8}$, Cr, Ti, Pt, and Ni on n-type GaN layers, where the n-type GaN layers were grown by MOCVD [24]. It is found that in the cases of conventional structures the Schottky barrier heights are dependent on their metal workfunctions, ranging from 0.3 to 0.9 eV, while in the cases of new structures with the LT caps they are higher than those of the conventional structures and the values are in the range of 1.02–1.13 eV, independent of the metal workfunctions. Therefore, it is expected that the LT capping layers may be causing the Fermi-level pinning effect at the metal/semiconductor interfaces. It was confirmed by resistivity measurements that the LT caps were electrically highly resistive.

The electrical conductivity on the surface structure is also influenced by the surface polarity of GaN. Thus, in the Schottky contact that requires the relatively low electron concentration the surface polarity becomes important. Figure 8.17 shows the polarity dependence for Pt contacts on the Ga- and the N-faced GaN layers grown by MBE [25]. It is found that the Ga-face shows better rectifying characteristics than the N-face. It should be noted that the ohmic contacts are fabricated on relatively highly doped GaN films, but the Schottky contacts are relatively lowly doped GaN films. It is attributed to

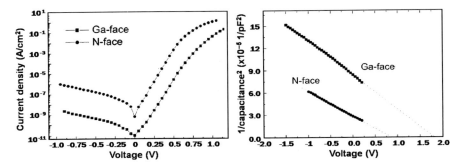

Fig. 8.17. Polarity dependence of I–V (*left*) and C–V (*right*) characteristics for Pt/GaN contacts on Ga- and N-faced GaN layers, grown by MOCVD. Reprinted with permission from [25]. Copyright (2000), American Institute of Physics

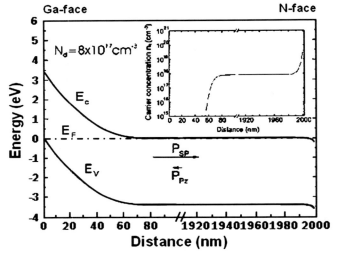

Fig. 8.18. Energy band diagrams simulated taking into account the spontaneous polarization for n-type GaN. Reprinted with permission from [25]. Copyright (2000), American Institute of Physics

the fact that in the ohmic contact doping control is necessary for enhancing the electron transport by the field emission, while in the Schottky contact doping control is necessary for reducing the field emission. Therefore, in the case of the Schottky contact the bound charges at the surface due to the spontaneous polarization are not completely compensated by free electrons, which induce the enhancement/lowering of the electron affinity at the N-/Ga-faced surface. This will influence the Schottky barrier height.

Figure 8.18 shows the simulated conduction and valence band profiles including the spontaneous polarization [25]. The strong increase of the conduction band energy at the Ga-face is caused by the lack of free holes to compensate the negative polarization charges, while the slight decrease at the N-face arose from the nearly complete compensation of the positive sheet charges by the free electrons. Theoretically, the band bending at the Ga-face results in the reduction of electron affinity by about 3.4 eV. Of course, this value is too ideal because the several effects such as structural defects, oxidants, and adsorbates influencing the Fermi level are present at the actual surface. Here, it should be noted that the binding energy of Ga–N bonds in the Ga-face was just 1.4 eV larger than that of the N-face as mentioned in Fig. 8.6.

Up to now, the Schottky contacts to GaN were discussed with regard to current leakage and surface polarity. The leakage currents in the Schottky contacts are ascribed to dislocations, point defects, and a high density of native donors. These can be effectively blocked by inserting the highly resistive LT capping layers into the the Schottky contacts. The surface polarity is important in the Schottky contact formation. The Ga-polar face is more favorable. It is ascribed to the fact that on the Ga-faced surface free electrons are depleted and on the N-faced surface the free electrons are accumulated. On the other hand, various metals such as Pt, Ni, Ti, Cr, $WSi_{0.8}$, and Pd have been tried for the fabrication of high-quality Schottky contacts. Exposure to air was also reported to be good for enhancing the Schottky barrier height, which was attributed to the generation of a native oxide layer. However, post-annealing is not followed in the Schottky contact formation, differently from the ohmic contacts, because it can reduce the effective Schottky barrier height by forming intermediate compounds or donor-type defects in the interface. In Table 8.3, characteristics of the reported Schottky contacts of GaN are summarized.

8.3.3 Schottky Contacts to ZnO

In the ZnO Schottky contact formation, three points are to be emphasized: surface treatment, background electron concentration and surface polarity. The surfaces of ZnO films and bulks are exposed to various intrinsic defects and extrinsic contaminants. The surface treatment eliminates conductive states, which induce the current leakage on the surface rather than electron-trapping centers, resulting in the Fermi-level pinning. The background electron concentration in ZnO has a direct relationship with the effective Shottky barrier height. The surface polarity is also important, because the Schottky contacts are generally formed on the films or bulks with relatively low electron concentration.

Figure 8.19 shows the surface treatment effect on Au/n-ZnO contacts, where the n-type ZnO was the as-supplied ZnO bulks, produced by Eagle-Picher Inc. [31]. It was found that the Au contacts on the as-supplied ZnO bulks have the leakage current of $\sim\mu A$, while those on the plasma-cleaned samples have the very small leakage current of ~ 10 nA. On the other hand, it is shown that OH^- bonding states exist on the surface of as-supplied samples and they are removed by the plasma treatments using

Table 8.3. Summary of characteristics of the GaN Schottky contacts

Metal	Growth	Material	Electron con. (cm^{-3})	Surface treatment	Reverse-bias current (A)	Schottky barrier height (eV)	Ideality factor (Dimensionless)	Ref.
Au/Pt	MOCVD (Homoepitaxy)	GaN	3×10^{16}	–	3×10^{-8} at 100 V	1.37	1.4	[26]
Pt	MBE	GaN	5×10^{17}	As grown	1×10^{-1} at 0.5 V	0.50	–	[27]
				O_2, oxidation, 700°C	1×10^{-6} at 0.5 V	0.80	–	
Pt	MBE	GaN	8×10^{17}	HF vapor	3×10^{-9} A cm^{-2} at 1 V	1.1 (Ga face)	–	[25]
					1×10^{-8} A cm^{-2} at 1 V	0.9 (N face)	–	
Au/Ni	MBE	GaN	–	As grown	5×10^{-1} A cm^{-2} at 5 V	–	–	[28]
				KOH solution	1×10^{-9} A cm^{-2} at 5 V	–	–	
Ni	MBE	GaN	5×10^{16}	As grown	2 A cm^{-2} at 20 V	0.80	–	[29]
				Anodizing by NaOH solution	1×10^{-3} A cm^{-2} at 20 V	0.86	–	
Ti				As grown	5×10^{-2} at 5 V	0.36	–	[24]
Cr				As grown	9×10^{-3} at 5 V	0.39	–	
$WSi_{0.8}$	MOCVD	GaN	–	As grown	9×10^{-6} at 5 V	0.55	–	
Ni				As grown	1×10^{-8} at 5 V	0.85	–	
Pt				As grown	5×10^{-12} at 5 V	0.95	–	
Pd	MBE	GaN:Si	–	In situ	1×10^{-2} A cm^{-2} at 1 V	0.75	1.6	[30]
				Ex situ	1×10^{-5} A cm^{-2} at 1 V	0.90	1.9	

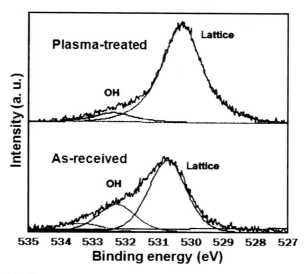

Fig. 8.19. XPS O 1s core level spectra for as-received and plasma-treated surfaces of ZnO (000–1) bulks. Reprinted with permission from [31]. Copyright (2003), American Institute of Physics

20% O_2/80% He gas mixture. These results mean that the hydrate (OH^-) adsorbed on the ZnO surface increases the surface conductivity. Here, it should be mentioned that the ZnO surface can be easily exposed to various extrinsic contaminants such as hydrogen and water vapor because of their abundance in nature. In addition, it was found that the leakage current, when the ZnO bulks are exposed to unignited plasma gases, significantly reduces until 20 pA. It was attributed to the fact that the adsorbed O species on the ZnO surface act as the electron acceptors, which lower the surface conductivity. On the contrary, the surface conduction on the ZnO surface was more severe when exposed to an H_2 atmosphere. Figure 8.20 shows the effect of N_2 atmosphere on Pt/n-ZnO contacts, where the n-type ZnO was the as-supplied ZnO bulks, produced by Cermet Inc. [32]. It was found that the I–V curves change from a rectifying behavior to the ohmic characteristics with the injection of the H_2 gas into the N_2 atmosphere. It is attributed to the fact that the H_2 gas decomposes on Pt metallization and diffuses rapidly through the underlying Pt/ZnO interface. (Here, it should be mentioned that hydrogen is predicted to be a shallow donor from the density functional theory.)

Figure 8.21 shows the relationship of background electron concentration and Schottky contact characteristics of Au/ZnO:N contacts, where the ZnO:N layers were grown by MBE [20]. It is found that the I–V characteristic is strongly dependent on the growth temperature and the surface polarity of ZnO:N layers. Here it should be noted that the N-doping was carried out for the control of conductivity in ZnO layers, where it was controlled by the growth temperature and the surface polarity [20]. As a result, it is shown that

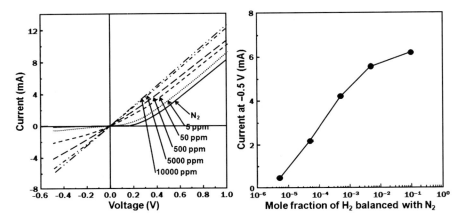

Fig. 8.20. Effect of N_2 atmosphere on Pt contacts on ZnO bulks. I–V curves in pure N_2 (*left*) and current at -0.5 V with the injection of H_2 gas in N_2 ambient (*right*). Reprinted with permission from [32]. Copyright (2004), American Institute of Physics

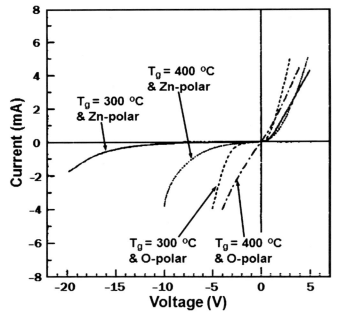

Fig. 8.21. Effect of N incorporation on Au/ZnO:N contacts, where ZnO: layers were grown by MBE. I–V curves for Zn-polar and $T_g = 400°$C; Zn-polar and $T_g = 300°$C; O-polar and $T_g = 400°$C; and O-polar and $T_g = 300°$C. Reprinted with permission from [20]. Copyright (2005), American Institute of Physics

in the cases of ZnO:N layers with the same polar direction, ZnO:N layers grown at 300°C exhibit larger breakdown voltages and smaller leakage currents in the reverse bias region than the samples grown at 400°C, while in the cases of ZnO:N layers grown at the same growth temperature, the Zn-polar ZnO:N layers exhibit larger breakdown voltages and smaller leakage currents than the O-polar samples. Therefore, it can be concluded that the Zn-polar ZnO:N layer grown at 300°C exhibits the largest breakdown voltage and the smallest leakage current. Here, the Schottky barrier height and the ideality factor were estimated to be 0.66 eV and 1.8, respectively. Figure 8.22 and Table 8.4 show the electrical properties of the ZnO:N layers and the Au/ZnO:N contacts, respectively [20]. These indicate that (1) the ZnO:N layers with lower growth temperature and Zn-polarity have larger N incorporation than those with higher growth temperature and O-polarity; (2) the resistivities of ZnO:N layers are proportional to the incorporated N concentrations; (3) the ZnO:N layers with higher resistivity show better Schottky characteristics. Consequently, it can be expected that the higher resistivity is favorable for the Schottky contact

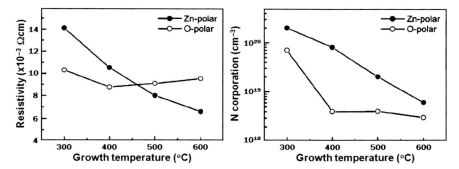

Fig. 8.22. Resistivity and incorporated N concentrations for Zn-polar ZnO:N layers and O-polar ZnO:N layers grown at 300–600°C. Reprinted with permission from [20]. Copyright (2005), American Institute of Physics

Table 8.4. Schottky characteristics for Au contacts on ZnO:N layers grown by MBE [20]. Schottky barrier heights (Φ_{SB}) and ideality factors (n) for Au Schottky contacts on ZnO:N layers, and resistivities (ρ) and incorporated N concentrations (N) of ZnO:N layers

T_g (°C)	Polar dir.	Φ_{SB} (eV)	n (-)	ρ (Ωcm)	N (cm^{-3})
400	O-polar	–	–	8.8×10^{-3}	4×10^{18}
400	Zn-polar	0.55	3.2	1.1×10^{-2}	8×10^{19}
300	O-polar	0.37	3.5	1.0×10^{-2}	7×10^{19}
300	Zn-polar	0.66	1.8	1.4×10^{-2}	2×10^{20}

fabrication, which is attributed to the suppression of the background electron concentration by the N incorporation.

Here, the N incorporation mechanism should be considered. N incorporation into ZnO layers can be approached in two ways. First, the N incorporation is strongly influenced by the growth temperature. Nitrogen is well known as a very efficient p-type dopant in ZnSe. Qiu et al. reported that the N incorporation into ZnSe increases at lower growth temperatures, which is attributed to the increased sticking coefficient [38]. This explains well the fact that the N incorporation into ZnO layers increases at lower growth temperatures. Second, the polarity can influence the N incorporation. The polarity effect in the case of impurity incorporation has also been reported for GaN:Mg. However, the origin is not yet fully understood. Oh et al. has proposed a simple model of the N-incorporation mechanism, as shown in Fig. 8.23 [20]. In the case of a Zn-polar ZnO layer, the surface is composed of Zn atoms with one dangling bond and three occupied Zn–O bonds. The Zn atoms, forming electropositive centers, tend to bond with the electronegative N atoms. These Zn–N bonds compete with the Zn–O bonds during the growth of ZnO:N and affect the N incorporation. For the case of an O-polar ZnO layer, the surface is composed of O atoms with one dangling bond and three occupied O–Zn bonds. It is relatively difficult for the N atoms to bond with the O atoms since the difference in electronegativity between N and O is small. Consequently, it is expected that the Zn-polar direction is favorable for the N incorporation in the growth of ZnO:N layers.

Figure 8.24 shows the surface polarity effect for Ag contacts on Zn-/O-polar faces of ZnO bulks, produced by Tokyo Denpa inc. [35]. It is found that the Zn face shows better rectifying characteristics than the O face. As already mentioned in the case of GaN, it is expected that the conduction band energy increases in the Zn-face owing to the depletion of free electrons, while it decreases in the O-face owing to the accumulation of free electrons. This band bending will help the Schottky barrier formation for the Zn-face and hamper it for the O-face of ZnO.

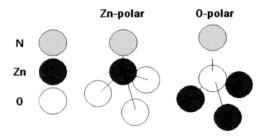

Fig. 8.23. Schematic diagram for the N-incorporation model for Zn-polar/O-polar directions [20]

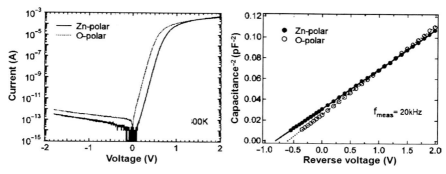

Fig. 8.24. Polarity dependence of I–V (*left*) and C–V (*right*) characteristics for silver oxide Schottky contacts on Zn-polar and O-polar ZnO bulks. Reprinted with permission from [35]. Copyright (2007), American Institute of Physics

Up to now, the Schottky contacts to ZnO were discussed with respect to surface treatment, background electron concentration, and surface polarity. Surface treatment has strong influence on the Schottky contact characteristics, which depends on the surrounding gas. The O_2/He plasma treatment is helpful for the Schottky contact formation, which is ascribed to the removal of adsorbed hydrate (OH^-) on the ZnO surface. The exposure of the ZnO surface to O_2 gas is also favorable for the Schottky contact formation, which is ascribed to the adsorption of O species that acts as an electron acceptor. However, the exposure to H_2 gas breaks the Schottky contact, which is ascribed to the diffusion of hydrogen atom that is noted as a donor. The background electron concentration should be as low as possible for the high-quality Schottky contact. It is attributed to the fact that the carrier concentration shows a direct relationship with the Schottky barrier height, i.e., the latter is inversely proportional to the carrier concentration. the N doping into the ZnO films is one method for reducing their electron concentrations. The surface polarity should be considered in the Schottky contact formation. In the ZnO Schottky contact, the Zn-face is favorable, which could be ascribed to the depletion of free electrons.

On the other hand, various metals such as Au, Ag, Pt, and Pd have been used for the fabrication of high-quality Schottky contacts. For the surface treatments, simple solutions such as the organic solvent, H_2O_2, and organic solvent + toluene + DMSO are also helpful in the contact formation. On the contrary, one should be careful with the surface treatments using an acid such as HCl and HNO_3 because they deteriorate the Schottky contact characteristics. In Table 8.5, characteristics of the reported Schottky contacts of ZnO are summarized.

Table 8.5. Summary of characteristics of the ZnO Schottky contacts

Metal	Growth	Material	Electron con. (cm^{-3})	Surface treatment	Reverse-biased current (A)	Schottky barrier height (eV)	Ideality factor (Dimensionless)	Ref.
Au	Bulk (Eagle-Picher Inc.)	ZnO	5×10^{16}	As grown	5×10^{-6} at 4V	–	>2	[31]
				20%O_2/80%He ICP	4×10^{-8} at 4V	0.67	1.9	
				Ignited plasma gas added	2×10^{-11} at 4V	0.60	1.0	
Au	Bulk (Eagle-Picher Inc.)	ZnO	9×10^{16}	Organic solvent	9×10^{-8} at 1V	–	1.8	[33]
				HCl solution	2×10^{-6} at 1V	–	1.6	
				HNO_3 solution	2×10^{-6} at 1V	–	1.8	
Ag				HCl solution	3×10^{-6} at 1V	–	1.6	
				HNO_3 solution	2×10^{-6} at 1V	–	1.8	
Ag	MOCVD	ZnO	1×10^{17}	(112–0) R plane	1×10^{-10} at 1V	0.90	1.3	[34]
Ag (Oxide)	Bulk (Tokyo Denpa Inc.)	ZnO	–	–	–	1.1 (Zn face)	1.1 (Zn face)	[35]
						1.0 (O face)	1.1 (O face)	
Pt	Bulk (MAHK Inc.)	ZnO	2×10^{17}	Organic solvent	5×10^{-2} at 5V	–	–	[36]
				H_2O_2 solution	7×10^{-8} at 5V	0.90	–	

8 Electrical Properties of GaN and ZnO

Metal	Deposition	Material	Doping	Treatment		Value 1	Value 2	Ref
Pt	PLD	ZnO:P	8×10^{15}	N_2, 300°C, 1 min (Pre)	—	0.61	1.7	[32]
				N_2, 300°C, 1 min (post)	—	0.42	4.3	
Pd	PLD	ZnO	—	Organic solvent	—	0.63	1.7	[37]
				Organic solvent + toluene + DMSO	—	0.68	1.4	
	Bulk (Eagle Picher Inc.)			N_2O plasma	—	0.60	2.0	
				Organic solvent	—	0.74	2.0	
				Organic solvent + toluene + DMSO	—	0.70	1.8	
				N_2O plasma	—	0.60	1.4	

8.4 Electrical Properties

Conductivity control is one of the most important issues in wide-bandgap compound semiconductors, because various types of defects are easily formed in the materials during growth and hamper high conductivity or high resistivity. These native defects, unintentionally generated as a result of crystal imperfections, supply extra electrons and trap free electrons. Actually, most literature commonly report that epitaxially grown GaN and ZnO films exert n-type conductivity, i.e., high electron concentration and low electron mobility. This indicates that the native defects have n-type conductivity, resulting in the degradation of electron transport. One of the possible origins for the n-type conductivity is attributed to the fact that GaN and ZnO films are generally grown on Al_2O_3 substrates with the large lattice misfits of 14–16%, which induce point defects and dislocations.

8.4.1 Electron Transport Mechanism

Electron transport mechanism can be theoretically predicted by considering various scattering terms. The scattering mechanisms are classified into lattice scattering and defect scattering [39]. In wide-bandgap compound semiconductors, the lattice scattering mechanisms by crystal lattices are composed of optical-polar phonon, piezoelectric potential, and deformation potential terms. The lattice vibration of an optical mode induces a dipole moment by the interaction of ionic charges, which is the source of optical-polar phonon scattering ($\mu_{optical}$). The lattice vibration of an acoustic mode induces the second potential by atomic displacement, which is the source of piezoelectric potential scattering (μ_{piezo}). The lattice vibration of an acoustic mode also causes the change of lattice spacing, and this induces a change of the bandgap, which is the source of deformation potential scattering (μ_{deform}). The lattice scatterings, which are not related to impurities and defects, are proportional to temperature. On the other hand, the defect scattering mechanisms are dominated by ionized impurities. Ionized impurities scatter electrons through their screened coulomb potential, which is the source of ionized impurity scattering ($\mu_{ionized}$). Here, it should be noted that among the five scattering mechanisms only ionized impurity scattering is dependent on the crystal qualities of the samples, while the others are dependent on the characteristics of the materials themselves.

On the other hand, these scattering mechanisms are simultaneously involved in the electron transport of semiconductor materials. According to the Matthiessen's rule, the total electron mobility (μ_{total}) is expressed by [39, 40]

$$\frac{1}{\mu_{\text{total}}} = \frac{1}{\mu_{\text{optical}}} + \frac{1}{\mu_{\text{piezo}}} + \frac{1}{\mu_{\text{deform}}} + \frac{1}{\mu_{\text{ionized}}}. \tag{8.6}$$

However, in the electron transport mechanisms of GaN and ZnO another scattering term should be considered because they are heteroepitaxially grown on

Table 8.6. Physical parameters of GaN and ZnO [41, 42]

Parameter	GaN	ZnO
Electron effective mass, m_n^* (kg)	$0.22 \times m_o$	$0.318 \times m_o$
Longitudinal elastic constant, c_1 (dyne cm^{-2})	2.65×10^{12}	2.10×10^{12}
Deformation potential, E_1 (eV)	8.54	3.8
Piezoelectric coefficient, P (dimensionless)	0.118	0.21
Debye temperature, T_o (K)	1044	847
High frequency dielectric constant, ε_∞ (F cm^{-1})	$5.47 \times \varepsilon_o$	$3.72 \times \varepsilon_o$
High frequency dielectric constant, ε_o (F cm^{-1})	$10 \times \varepsilon_o$	$8.12 \times \varepsilon_o$

Al$_2$O$_3$ substrates with a large lattice misfit, which means that the grown films suffer from structural defects, especially dislocations. The dislocations can have the negative charges at the cores, which is the source of dislocation scattering ($\mu_{\text{dislocation}}$), expressed by

$$\mu_{\text{dislocation}} = \frac{30(2\pi)^{1/2}\varepsilon^2 a^2 (kT)^{3/2}}{N_{\text{dis}} q^3 f^2 \lambda_D m^{1/2}}, \tag{8.7}$$

where ε is the dielectric constant, a is the distance between the cores, N_{dis} is the dislocation density, f is the electron occupation probability of the cores, λ_D is the Debye screening length, and m is the effective mass of electron. The charged dislocation not only hamper the electron transport, but also can form another current path, because they are generally localised at the heterointerface. Look et al. have proposed a simple way to differentiate the electrical properties of bulk and interface from the experimentally measured values by applying the two-layer model [40].

$$\mu_{\text{total}} = \frac{n_b \mu_b^2 + n_i \mu_i^2}{n_b \mu_b + n_i \mu_i}, \tag{8.8}$$

where n_b is the electron concentration in the bulk, n_i is the electron concentration in the interface, μ_b is the electron mobility in the bulk, and μ_i is the electron mobility in the interface.

Table 8.6 summarizes the material parameters of GaN and ZnO necessary for the theoretical calculation of electron mobility [41, 42].

8.4.2 Electrical Properties of GaN

The difficulty of conductivity control in GaN originates from the actual problem that GaN films are heteroepitaxially grown on Al$_2$O$_3$ substrates with a lattice misfit of 16% and a thermal mismatch of 25%. Therefore, the grown films on the substrates have large densities of threading dislocations and non-stoichiometric point defects, which are known to traverse vertically from the interface to the surface. Especially, the highly dislocated GaN/Al$_2$O$_3$ heterointerface becomes the source of detrimental line and point defects that

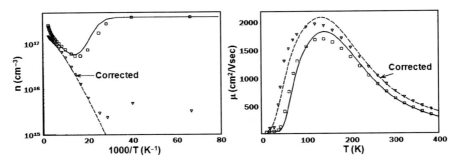

Fig. 8.25. Influence of interface states on Hall data. *Solid lines* and *open squares* are uncorrected data and *dashed lines* and inverse *open triangle* are corrected ones. Reprinted with permission from [40]. Copyright (1997), American Institute of Physics

deteriorate the electrical properties of the subsequently grown GaN film, because those defects are not electrically neutral but active. The line defects are effectively reduced by inserting the LT buffer between the GaN/Al$_2$O$_3$ interface, which has a role in suppressing the dislocations generated in the interface by forming dislocation loops. The point defects are effectively suppressed by manipulating the molar ratio of group V and group III sources, that is, the V/III ratio, because group III and V atoms have high vapor pressures at the growth temperature.

Figure 8.25 represents the typical temperature dependence of electron concentration and electron mobility obtained by Hall measurements from heteroepitaxially grown GaN thick films on Al$_2$O$_3$ substrates by hydride vapor-phase epitaxy (HVPE), where thin sputtered ZnO layers were inserted as buffers between the GaN films and the Al$_2$O$_3$ substrates but were etched out during the growth and were no longer present after the growth was completed [40]. It is found that the electron concentration increases with the decrease of temperature in the temperature region below 50 K. This anomalous phenomenon is not explained by the conventional electron transport mechanism which states that free electrons lose their activation energies as temperature decreases and are thermally quenched and captured on ionized donors below 100 K. Gotz et al. reported that the electrical characterization of GaN films grown on Al$_2$O$_3$ substrates is influenced by the heterointerface: a sheet of electrons of the density of $\sim 10^{15}$ cm^{-2} was estimated to be confined at the GaN/Al$_2$O$_3$ interface in terms of Hall measurements, while the net doping density in the film obtained through C-V measurements was about 5×10^{16} cm^{-3}, which corresponded to the sheet concentration of about 7×10^{13} cm^{-2} [43]. Podor reported that plastic deformation strongly influences the electrical properties of single-crystalline materials [44]. If the dislocations generated by plastic deformation have an edge component, they introduce acceptor centers along the dislocation lines which capture electrons.

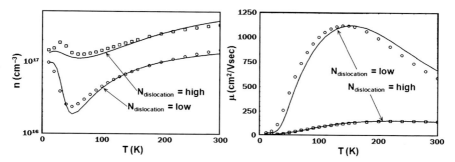

Fig. 8.26. Influence of dislocation scattering on electrical properties of GaN layers grown on Al_2O_3 substrates by MOCVD. *Solid lines* are the theoretical results and *open circles* and *squares* are the experimental ones. Reprinted with permission from [45]. Copyright (1999), American Physical Society

Thus, the dislocation lines become negatively charged, and space charges are formed around it. The resulting potential field hampers the electron transport and reduces the electron mobility. Look et al. developed the theory of the charged-dislocation scattering within the framework of the Boltzmann transport equation and applied it to the temperature-dependent Hall measurement interpretation using the two-layer model, i.e., one is bulk layer and the other is interface, to separate the properties of the bulk and the interface. In Fig. 8.25, two kinds of data are shown: one is the measured, and the other is the theoretically corrected. It can be found that the theoretically calculated results are well fitted into the experimentally obtained ones. On the other hand, it is seen that the real electron concentration and electron mobility in the GaN film are a little overestimated and underestimated, respectively. Figure 8.26 compares the electrical properties for two types of GaN layers with different dislocation densities, i.e., one has lower dislocation density and the other has higher dislocation density [45]. It was seen that the GaN layer of higher dislocation density has larger electron concentration and lower electron mobility, and the increasing trend of electron concentration at the low tempered region is weaker in the GaN layer of lower dislocation density. Therefore, it is expected that the threading dislocations generated in the GaN/Al_2O_3 interface are electrically active, which hamper the electron transport in the GaN film and induce the distortion of the measured electrical properties. Hence, the dislocation densities were controlled by manipulating the LT buffers. The effect of LT buffers on the electrical properties will be introduced in Chap. 9 on the electrical properties of ZnO.

Figure 8.27 shows the influence of molar ratio of group V and group III sources on the electrical properties of GaN thick films grown on Al_2O_3 substrates by HVPE [46]. It is found that the electron concentration decreases as the V/III ratio increases, while the electrical resistivity and the electron

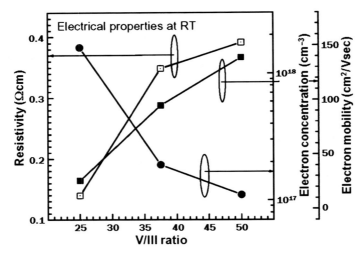

Fig. 8.27. Impact of V/III ratio on the electrical properties for GaN thick films grown on Al_2O_3 substrates by HVPE. The plot shows resistivity, electron concentration, and electron mobility as a function of the V/III ratio. Reprinted with permission from [46]. Copyright (2007), American Institute of Physics

mobility increase simultaneously with the increase of the V/III ratio. Here, it looks worthwhile to refer to the case of ZnSe films briefly, which had suffered from similar problems in conductivity control [47–49]. In the ZnSe films doped with impurities such as Cl, Ga, or Al, their electrical properties are strongly dependent on doping: when the doping concentration decreases, the electron concentration decreases, while the electrical resistivity and the electron mobility increase simultaneously. This indicates that increasing the V/III ratio in growing the GaN films suppresses electron-feeding sources by reducing the donor-type defects not by generating the electron-trapping centers. On the other hand, it was found that the full width at half-maximum of (0002) X-ray rocking curves decreases with the V/III ratio, which means that the higher V/III ratio growth condition is helpful in reducing the crystalline point and line defects in the GaN films.

Figure 8.28 shows a representative temperature dependence of electron mobility and its theoretical fitting results for the GaN films [46]. It can be found that the defect scattering terms are more influential on the electron transport of GaN films than the lattice scattering terms. Also, (1) in the low-temperature region below 150 K, the ionized-impurity scattering and the dislocation scattering are dominant; and (2) near room temperature above 150 K, the ionized-impurity scattering and the local-defect scattering become dominant. Here, the local-defect scattering considers the local defects such as point defects or local strains and it is empirically expressed in the form of $C/T^{1.5}$, where the proportionality constant C expresses the degree of scattering. Moreover, it

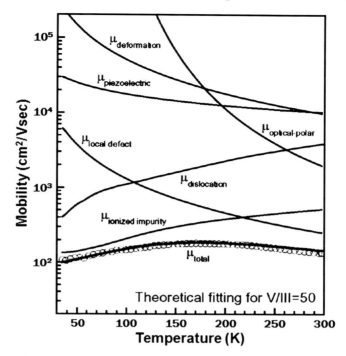

Fig. 8.28. Temperature-dependent curve of electron mobility in the range of 30–300 K for a thick GaN film grown in the V/III ratio 50 and its theoretical fitting by using Matthiesen's rule. Reprinted with permission from [46]. Copyright (2007), American Institute of Physics

was found from the theoretical fittings that the influence of the defect scattering factors of ionized impurities, charged dislocations, and local defects in the GaN films decreases with the increase of the V/III ratio. These results mean that the generation of donor-type defects that supply free electrons in the GaN films is more suppressed by the higher V/III ratio, which induces the lower background electron concentration and the higher electron mobility. However, this is controversial because it seems that the electrical properties are dependent on the growth conditions themselves.

Look et al. reported that the electrical resistivity in GaN layers grown by MBE increases with increasing the N flux, while both the electron concentration and the electron mobility decrease [50]. They suggested that increasing the N flux induces point defects in GaN lattices, which form deep levels in the bandgap to reduce the free electrons generated by native defects. Saarinen et al. also reported similar results in GaN layers grown by MOCVD, and suggested that the decreased electron concentration at higher V/III ratios is due to the carrier compensation by Ga vacancies [51, 52]. As a result, it is clear

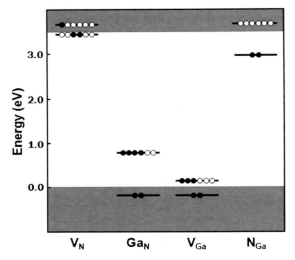

Fig. 8.29. Energy levels and electron occupancies of neutral native defects in GaN. Reprinted with permission from [54]. Copyright (1989), American Physical Society

that the V/III ratio strongly influences the electrical properties of GaN films, but it should be remembered that the growth conditions also affect it.

At this point, it should be asked why undoped GaN films commonly exert n-type conductivity. It is believed for long that nitrogen vacancy (V_N) is the nonstoichiometric point defect that induces the n-type conductivity. Maruska et al. reported that the undoped GaN bulks grown by VPE have very high inherent electron concentrations, above 10^{19} cm^{-3}, and proposed that the origin is probably related to V_N [53]. This assignment is consistent with the electronic structures of nonstoichiometric point defects theoretically calculated by using the tight-binding model by Jenkins and Dow [54]. Figure 8.29 shows the energy levels and electron occupancies of neutral native defects in GaN [54]. It should be noted that the V_N is the shallowest donor, while the gallium vacancy (V_{Ga}) is the shallowest acceptor. However, Neugebauer et al. pointed out the limited accuracy in the theoretical result by the tight-binding model, though it could predict energy levels [55]. They calculated the formation energies of the native defects by using the total energy calculations from the first principles based on the density functional theory. Figure 8.30 shows the defect formation energies as a function of Fermi level for all the native defects in GaN under nitrogen-rich conditions [55]. It is found that the V_N dominates the conductivity of GaN only under p-type conduction, while under n-type conduction the V_{Ga} dominates. Most of all, it should be noted that the formation energy of the V_N in n-type conduction is very high, of the order of 4 eV, which means that the V_N is difficult to be found in appreciable concentrations in thermodynamic equilibrium.

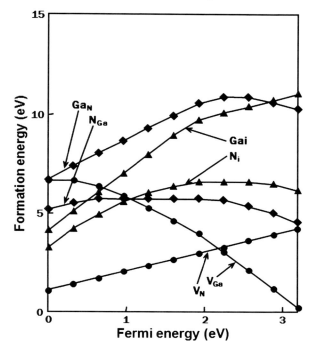

Fig. 8.30. Defect formation energies as a function of the Fermi level for all native defects in GaN under N-rich conditions ($F_F = 0$ corresponds to the top of the valence band). Reprinted with permission from [55]. Copyright (1994), American Physical Society

On the other hand, Wright et al. applied the density functional theory to charged dislocations in GaN [56]. Figure 8.31 shows the formation energy vs. Fermi level for four representative dislocation structures of full core and open core structures (stoichiometric defects) and V_N and V_{Ga} types (nonstoichiometric defects) [56]. A structure having the V_{Ga} at the dislocation core was predicted to be most stable in the n-type GaN grown under nitrogen-rich condition, while the core structures without vacancies were stable in the p-type GaN. In the n-type GaN grown under gallium-rich condition, a variety of structures could be present simultaneously, while in the p-type GaN a structure having the V_N at the dislocation core was predicted to be the most stable.

Recently, it has been reported that extraneously incorporated oxygen induces n-type conductivity in GaN. Seifert et al. insisted that the assumption that V_N is responsible for the n-conductivity shown in the undoped GaN bulks grown by HVPE is questionable [57]. Instead, they postulated that substitutionally incorporated oxygen atoms into nitrogen sites could be the cause. This proposition was supported by extensive work by Chung et al. [58]. Figure 8.32 shows the electron concentration in GaN layers grown by MOCVD, in which

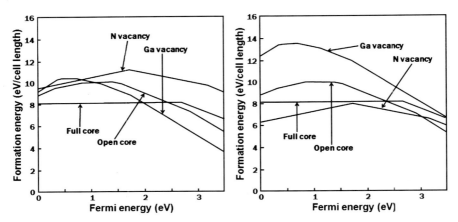

Fig. 8.31. Plots of the formation energy vs. Fermi level under (*left*) N-rich and (*right*) Ga-rich conditions. Reprinted with permission from [56]. Copyright (1998), American Institute of Physics

it is significantly dependent on the growth conditions of oxygen gettering and growth temperature [58]. First, it is noted that the electron concentrations of the GaN layers become low by using gallium eutectic bubblers, though they have very high background electron concentrations, much above $10^{19}\,\mathrm{cm}^{-3}$, and the electron concentration increases up to around $10^{20}\,\mathrm{cm}^{-3}$ with the intentional injection of water vapor. Second, it is found that the electron concentration decreases with the increase of growth temperature, which is inconsistent with the assumption regarding the V_N as the origin because the V_N is expected to increase with the increase of growth temperature from thermodynamic considerations. Moreover, it was found that the oxygen concentration detected by electron-probe microanalysis (EPMA) decrease with the increase of growth temperature. These observations imply that the increased electron concentration is related to the incorporated oxygen.

Up to now, the electrical properties of undoped GaN films and bulks were discussed in the lights of electron transport and native conductivity. The electrical properties of GaN are significantly influenced by its crystal quality, such as the densities of dislocations and point defects, because the GaN films are heteroepitaxially grown, in general, on Al_2O_3 substrates with large lattice misfit and thermal mismatch. This hampers the electron transport, and the observed behavior could be explained by considering the dislocation scattering and the local defect scattering in the conventional scattering mechanism. On the other hand, the undoped GaN films and bulks commonly have n-type conductivity. Its origin is still controversial. Up to now, the V_N as the non-stoichiometric point defect and the oxygen as the extrinsically incorporated impurity are considered as the most probable origins of the n-type conductivity of the undoped GaN.

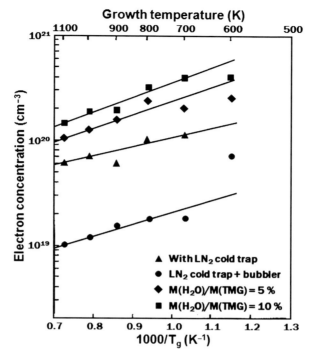

Fig. 8.32. Temperature dependence of the electron concentration for GaN layers grown at different growth temperatures. The data indicate the results of the layers grown with different gettering techniques and with different amounts of intentional water injection during growth. Reprinted with permission from [58]. Copyright (1992), American Institute of Physics

8.4.3 Electrical Properties of ZnO

The essential problem of conductivity control shown in GaN also appears in heteroepitaxially grown ZnO films on Al_2O_3 substrates. It is attributed to the large lattice mismatches that exist between ZnO and Al_2O_3: the lattice misfit of ∼18% and the thermal mismatch of ∼57%. This induces the electrically active line defects and point defects, which hamper the electron transport in ZnO films. Thus, the scattering mechanisms that consider those defects should be considered in the interpretation of the electron transport mechanism of ZnO films. On the other hand LT buffers are known to be the most effective in improving the electrical properties of the heteroepitaxially grown ZnO films by reducing the dislocation density, which is indispensable for the growth of high-quality ZnO films with low electron concentrations and high electron mobilities.

Figure 8.33 shows the reproducibility of obtained electron mobility as a function of electron concentration for ZnO layers grown on Al_2O_3 substrates

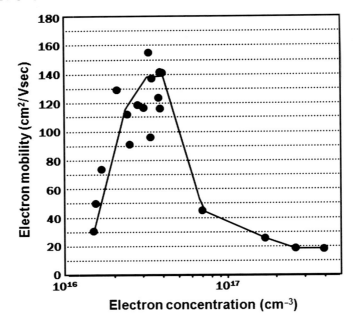

Fig. 8.33. Influence of growth buffer on the electrical properties of ZnO layers grown on Al$_2$O$_3$ by PLD. The plot summarizes the experimentally obtained electron mobilities for various samples as a function of electron concentration. Reprinted with permission from [59]. Copyright (2003), American Institute of Physics

by PLD [59]. It is found that the ZnO layers have high electron mobilities of 115–155 cm^2 V^{-1} s^{-1} in a narrow region of electron concentration ranged in 2–5×10^{16} cm^{-3}. It is found that the electron mobility decreases when the electron concentration deviates from this range. Above all, it was shown that the anomalous electrical properties are strongly dependent on the growth conditions of the buffer layers that influence the crystal qualities of the subsequently grown ZnO layers. This means that the electrical properties of the ZnO layers do not follow the conventional electron transport mechanism, which is generally determined by the ionized-impurity scattering and the optical polar phonon scattering.

Figure 8.34 represents a theoretical approach to the electron transport of ZnO layers grown on Al$_2$O$_3$ substrates by MBE, where LT ZnO and MgO buffers were inserted between the ZnO layers and the Al$_2$O$_3$ substrates [60]. In the theoretical calculation of the scattering mechanisms, first, the ionized-impurity scattering should be considered. It was found in the depth profile of secondary-ion mass spectroscopy (SIMS) that metallic impurities such as Al, Si, and K are detected in the ZnO layers, which are attributed to the quartz discharge tube used in the plasma cell of the oxygen source. However, their concentrations were estimated to be relatively low: [Al] = 2×10^{16} cm^{-3},

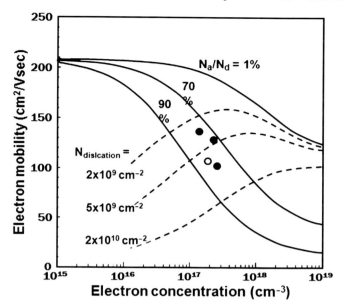

Fig. 8.34. Influence of dislocation scattering on ZnO layers with different dislocation densities, grown on Al_2O_3 substrates by MBE. Solid lines are the theoretically calculated electron mobilities without dislocation scattering, dashed lines are the theoretically calculated with dislocation scattering, and circles are the experimentally obtained values. Reprinted with permission from [60]. Copyright (2002), The Institute of Pure and Applied Physics

$[Si] = 4 \times 10^{15}\,cm^{-3}$, $[Na] = 2 \times 10^{14}\,cm^{-3}$, and $[K] = 2 \times 10^{13}\,cm^{-3}$. Even in the fully ionized case (100% ionization) of extraneously incorporated all donor (Al) and acceptor (Na and K) impurities, the compensation rate (N_a/N_d) was just 1%. It is shown that the calculated values for the compensation of 1% are much higher than the experimental ones. On the contrary, it should be noted that the calculated result is well fitted into the experimental results when dislocation scattering is included in the calculation: the compensation ratio is fitted to 70–90%. These mean that the decrease of electron mobility in the lower electron concentration region is due to dislocation scattering, which is more severe in the case with a higher dislocation density.

Undoped ZnO intrinsically exerts n-type conductivity with very high electron densities of 10^{16}–$10^{18}\,cm^{-3}$. Up to now, though various propositions and experimental results have been reported, the origin for the n-type conductivity is still controversial. One of the proposed possible origins is intrinsic defects such as vacancies and interstitials caused by a deviation from stoichiometry. Kohan et al. theoretically calculated the formation energies of native defects in ZnO and their electronic structures that determine the quantity of the defects by using the first principles and the plane-wave pseudopotential technique [61,62]. In this theory, the formation energy (E^{f}) of a point defect in a charged point defect in ZnO is given by

$$E^{\mathrm{f}} = E^{\mathrm{tot}}(n_{\mathrm{Zn}}, n_{\mathrm{O}}) - n_{\mathrm{Zn}}\mu_{\mathrm{Zn}} - n_{\mathrm{O}}\mu_{\mathrm{O}} + qE_{\mathrm{F}}, \qquad (8.9)$$

where $E^{\mathrm{tot}}(n_{\mathrm{Zn}}, n_{\mathrm{O}})$ is the total energy of the crystal system containing the number of Zn atoms n_{Zn} and the number of O atoms n_{O}, μ_{Zn} and μ_{O} are the chemical potential of a Zn atom and an O atom, q is the charge of the defect (including its sign), and E_{F} is the Fermi energy. Then the concentration (c) of point defects in the crystal system depends on their formation energy, and it is expressed by

$$c = N_{\mathrm{sites}} \exp\left(-\frac{E^{\mathrm{f}}}{kT}\right), \qquad (8.10)$$

where N_{sites} is the concentration of atomic sites in the crystal system where the point-type defects and the impurities can be formed. Thus, a low formation energy implies a high equilibrium concentration of the defect, while a high formation energy means that the defects are unlikely to be formed. Figure 8.35 shows the formation energy as a function of Fermi energy for oxygen and zinc vacancies, interstitials, and antisites in ZnO, where 0 in the Fermi level is the valence-band maximum and 3.4 is the conduction-band minimum [63]. It is found that zinc vacancy (V_{Zn}) and oxygen vacancy (V_{O}) are most likely to be formed in n-type ZnO, in which the densities of the vacancies depend on the partial pressure under growth conditions.

In particular, under Zn-rich growth conditions the V_{O} has lower formation energy than the zinc interstitial (Zn_{i}), while the V_{Zn} should dominate in an O-rich growth condition. This means that the V_{O} originates the n-type conductivity in ZnO. However, Look et al. suggested that the Zn_{i} rather than

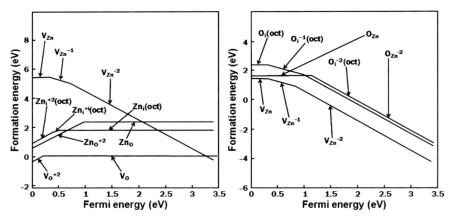

Fig. 8.35. Calculated defect formation energy as a function of Fermi level at (*Left*) high Zn partial pressure and (*right*) low Zn partial pressure. Reprinted with permission from [61,62]. Copyright (2000), American Physical Society

the V_O is the dominant donor in ZnO [63]. They found a shallow donor with the ionization energy of 30 meV in ZnO bulks intentionally damaged by high electron irradiation. They identified the shallow donor as the Zn_i, because the production rate of the donors is much higher for the Zn-face ZnO than the O-face ZnO. When the ZnO bulks are irradiated by high-energy and high-density electrons, various types of nonstoichiometric defects can be generated: in the Zn-face, the displacement of Zn atoms is relatively easy, which means the Zn_i and the V_{Zn} are abundant; in the O-face, the displacement of O atoms is relatively easy, which means the O_i and the V_O are abundant.

On the other hand, it was recently reported that extraneously incorporated hydrogen induces n-type conductivity in ZnO. Hydrogen is amphoteric, i.e., occurs as H^+ in the case of p-type conduction and H^- in case of n-type conduction. However, using calculations from the first principles calculation based on the density functional theory, it was reported that hydrogen in ZnO is always positive, i.e., always acts as a donor, and the formation energy of H^+ is low enough to allow a large solubility of hydrogen atoms in n-type ZnO [64]. Also, it was reported that the formation energy of H^+ becomes even less in p-type conduction. Compensation by the hydrogen donors is therefore an important concern when the acceptor doping of ZnO is attempted. This theoretical result is supported by the results of Chen et al. [65]. Figure 8.36 shows the electrical properties of ZnO:H layers vs. the injected amount of H_2 in the ZnO:H layers, where the ZnO:H layers were deposited by rf-magnetron sputtering [65]. It was found that the electrical resistivity is

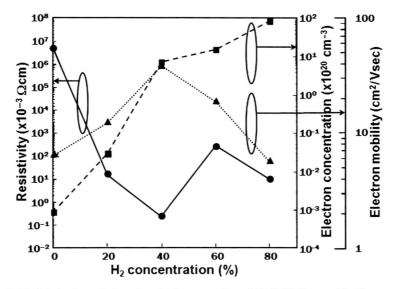

Fig. 8.36. Variation of the electrical properties of ZnO:H films with H_2 composition. Reprinted with permission from [65]. Copyright (2004), American Institute of Physics

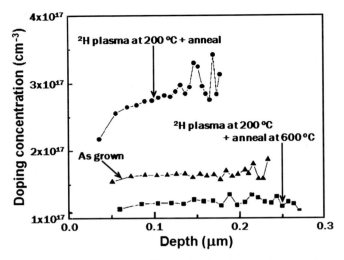

Fig. 8.37. Donor concentration profiles in ZnO before and after plasma exposure and after subsequent annealing. Reprinted with permission from [66]. Copyright (2003), American Institute of Physics

significantly reduced, and the electron concentration is increased by the addition of H_2 gas in the surrounding Ar gas during rf sputtering. As a result, the electrical resistivity decreases from 5×10^3 to 1×10^{-4} Ω cm and the electron concentration increases from 1×10^{17} to 1×10^{22} cm^{-3}. This means that the incorporated hydrogens in ZnO crystals induced from the hydrogen radicals in plasma increase the n-type conductivity.

Figure 8.37 shows the donor concentration profile in ZnO bulks with different treatments [66]. It is found that H_2 plasma treatment causes an increase in donor concentration, while subsequent annealing causes a decrease in donor concentration below an initial value of the as-prepared ZnO bulks. The increased donor concentration by the H_2 plasma treatment is explained by the fact that the incorporated hydrogen atom induces a donor state in the bandgap of ZnO, thereby increasing the free electron concentration. However, an alternative explanation is also possible: the H_2 plasma treatment induces the hydrogen passivation of compensating acceptor impurities present in the as-grown ZO bulks. Above all, it should be noted that the subsequent annealing reduces the carrier density below the initial value, which may indicate that the as-grown ZnO bulks already contained the hydrogen atoms that came from the growth process, which is eliminated in the annealing process. This implies that hydrogen appears to be a shallow donor, but it is not necessarily the dominant shallow donor.

Up to now, the electrical properties of undoped ZnO films and bulks were discussed in the aspect of electron transport and native defects. The electrical properties of ZnO are significantly influenced by its crystal quality, such as the densities of dislocations and point defects, because ZnO films are heteroepitaxially grown, in general, on Al_2O_3 substrates with large lattice misfit and thermal mismatch. This hampers the electron transport, and the observed behaviors can be explained by considering scattering from the dislocations and the local defects in the conventional scattering mechanism. On the other hand, undoped ZnO films and bulks commonly have n-type conductivity. Its origin is still controversial. Up to now, the V_O and the Zn_i as the nonstoichiometric point defect and the hydrogen as the extrinsically incorporated impurity are considered as the most probable origins of the n-type conductivity in the undoped ZnO.

8.5 Deep Levels of GaN and ZnO

GaN and ZnO suffer from various types of unintentionally generated defects. Those defects are generated as a result of crystal imperfections and have an influence on the material properties and device performance by introducing deep levels into the bandgap. These deep levels may compensate free carriers and limit the maximum attainable carrier concentration below the level required for device fabrication. Furthermore, the compensated centers hamper the transport of electrons through various scattering mechanisms, resulting in lowering the obtainable electron mobility. In Sect. 8.4.3, electrical properties of GaN and ZnO, especially the various defects and impurities that affect the conduction of GaN and ZnO, were introduced. However, those defects and impurities were evaluated just by the synthetic suspicion from indirect measurements and theoretical approaches, not by accurate characterizations. In the next section, the deep levels generated by the defects will be identified by quantifying them through direct characterizations.

8.5.1 Characterization of Deep Levels

In detecting deep levels in semiconductors, deep-level transient spectroscopy (DLTS) has been widely used. This technique utilizes the capacitance change induced by the trapping of the majority carriers in the depletion region [67,68]. If trap centers exist in a semiconductor, the total space charge in the depletion region of a Schottky junction diode is a sum of the ionized dopants and the trap centers (deep levels), i.e., $N_d \rightarrow N_d + N_T - n_T \rightarrow N_d + N_T(1-\exp(-e_n t))$ in an n-type semiconductor, where N_T is the total density of the deep level, n_T is the density of the electrons partially trapped in the deep level, and e_n is the emission rate of the electrons freely released from the deep level. When a cycle of bias pulse is applied to the Schottky junction diode, the carriers captured in the trap centers are gradually emitted into the conduction band (or the valence band) and the capacitance $C(t)$ exhibits a transient process, which is described by

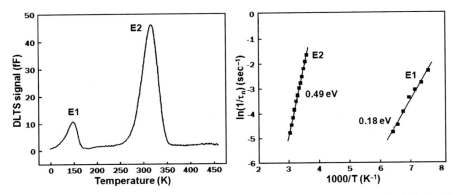

Fig. 8.38. Representative DLTS spectra for n-type GaN layers grown by MOCVD. Reprinted with permission from [69, 70]. Copyright (1994), American Institute of Physics

$$C(t)^2 = B\left[N_\mathrm{d} + N_\mathrm{T}\left\{1 - \exp(-e_n t)\right\}\right]$$
$$B = \frac{q\varepsilon A^2}{2(V - V_\mathrm{bi}^\mathrm{meas})}. \tag{8.11}$$

Here, the rate at which the capacitance changes during a cycle of bias pulse strongly depends on temperature. In the capacitance-transient (C–t) curves obtained at low and high temperatures, there is no difference between the two sampling times due to very slow and very fast transients, respectively. However, the C–t curve obtained at a temperature through a time window produces a maximum output. Such a plot is called a DLTS spectrum, as shown in Fig. 8.38 [69, 70]. The essential feature of DLTS is the implementation of a rate window (or a time window) for monitoring the capacitance transient as a function of temperature. The DLTS signal $S(t_1, t_2)$ for a given time window (t_1 and t_2) is given by

$$S(t_1, t_2) = C(t_1)^2 - C(t_2)^2. \tag{8.12}$$

Substituting (8.11) into (8.12) gives

$$S(t_1, t_2) = -BN_\mathrm{T}\left\{\exp\left(-\frac{t_1}{\tau}\right) - \exp\left(-\frac{t_2}{\tau}\right)\right\}, \tag{8.13}$$

where τ is the time constant, defined as $\tau = 1/e_n$.

On the other hand, in the study of deep levels in high-resistivity materials, admittance spectroscopy (AS) is well suited, because the capture rate of free carriers is very small in these cases [71–73]. The essential feature of AS comes from the capacitance change due to the frequency response of deep levels during a temperature sweep. Therefore, the large bias voltage pulse used to fill the deep levels in the DLTS measurement is not required in the AS technique. In a Schottky junction diode, the total measurement capacitance is composed of the high-frequency junction capacitance (that the trap centers cannot respond to the applied ac signal) and the additional capacitance (that trap centers can respond to the applied ac signal), which are given by

$$C_\text{T} = \frac{e_n^2}{e_n^2 + \omega^2}\left(\frac{N_T}{n}\right)C_\text{O}, \tag{8.14}$$

where C_O is the capacitance due to the shallow levels and C_T is the capacitance due to the deep levels. Hence, the total capacitance becomes $C = C_\text{O} + C_\text{T}$. The additional capacitance is determined by the emission rate of the trap centers, by assuming a weak temperature dependence of n and C_O. Consequently, at the temperature at which the emission rate becomes the same as the frequency of the applied ac signal, the trap centers respond to the applied ac signal and the total measurement capacitance begins to increase abruptly (if $e_n \ll \omega$, $C = C_\text{O}$, and if $e_n \gg \omega$, $C = C_\text{O} + (N_\text{T}/n)C_\text{O}$).

8.5.2 Deep Levels in GaN

Thermal stability of defects in GaN is a topic of concern in the reliability of many types of devices including high-power and high-temperature devices. If the defects that act as traps, recombination or generation centers, and scattering centers change their characteristics during the operation of the device, they can have a detrimental impact on the reliability of the device. In this section, various kinds of deep levels in GaN are described.

Figure 8.38 shows the representative DLTS spectrum for n-type GaN layers grown by MOCVD, where the electron concentration is $2 \times 10^{17}\,\text{cm}^{-3}$ [69, 70]. It is shown that two peaks, labeled E1 and E2, exist at 150 and 320 K. The obtained DLTS signals arise from the difference of the measured capacitance in the depletion region formed by the n-type Schottky contact, which means the electron emissions from the two electronic deep levels into the conduction band. The thermal activation energies of the two electron trap centers were estimated to be 0.18 eV for E1 and 0.49 eV for E2, respectively, from the Arrhenius plots for the peak shifts with changing sampling times. The deep level densities were estimated to be $7 \times 10^{13}\,\text{cm}^{-3}$ for E1 and $6 \times 10^{14}\,\text{cm}^{-3}$ for E2 from the DLTS peak signals.

Figure 8.39 shows the representative DLTS plot for GaN layers grown by MOCVD, where TMGa (trimethyl gallium) or TEGa (triethyl gallium) was used as gallium source [74]. Here, the electron concentrations of samples grown using the TMGa and the TEGa were 2×10^{16} and $2 \times 10^{17}\,\text{cm}^{-3}$, respectively. It is found that the two samples exhibit clearly different features in the DLTS spectrum in the temperature range: three deep levels, labeled E1′, E2′, and E3′, are detected from the film grown with the TMGa, while just the level corresponding to E3′ in the TMGa sample is seen in the film grown with TEGa. The activation energies of E1′, E2′, and E3′ were estimated to be 0.14, 0.49, and 1.44 (1.63) eV, respectively. First, it is thought that the E2′ is very similar to the E2 of the n-type GaN layer,

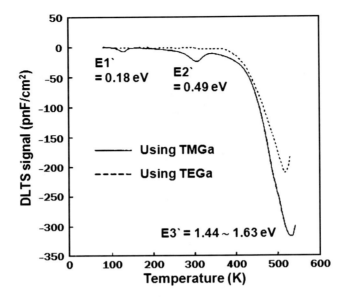

Fig. 8.39. Influence of metal organic sources on deep levels of GaN layers grown by MOCVD. DLTS spectra for TMGa and TEGa sources. Reprinted with permission from [74]. Copyright (1995), American Institute of Physics

which was discussed in Fig. 3.38, because both emission peaks were found at a similar temperature of 300 K and their activation energies were also the same, 0.49 eV. Second, it was suspected that the origins of E1' and E2' are due to the carbon atoms and the hydrogen atoms in the ethyl and methyl radicals. In the thermal reaction in MOCVD, TEGa pyrolyzes by eliminating β-hydride with formation of ethylene without the production of reactive carbon-containing species, and reduces carbon incorporation. TMGa pyrolyzes by producing highly reactive CH_3 radicals, which leads to carbon contamination. However, substitutional carbon has been theoretically predicted to be just a shallow level impurity in GaN. On the other hand, in the breaking of radicals during the thermal reaction of MOCVD, hydrogen atom can be incorporated into the films, especially for the TMGa source composed of CH_3 rather than the TEGa source composed of CH_2. Hydrogen atoms have been identified as a source of the deep levels in GaN.

Figure 8.40 shows the representative DLTS spectra for n-type GaN layers grown by HVPE, where the electron concentration is in the range $1-3 \times 10^{17}$ cm^{-3} [75]. Here, the GaN layers, grown on Al_2O_3 substrates, cover thicknesses from 5 to 68 μm. It is found that various deep levels, labeled A, A_1, A_x, B, C, and D exist between 100 and 400 K. In the thinnest film ($t = 50$ μm), all the mentioned deep levels A, A_1, A_x, B, C, and D are seen and their signals are considerably large, while in the thickest film ($t = 68$ μm) just two levels

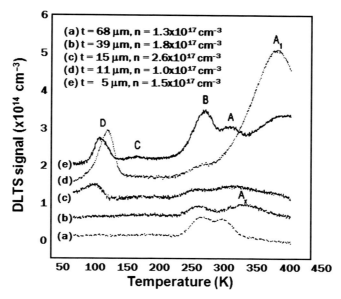

Fig. 8.40. Typical DLTS spectra for HVPE GaN samples with different thicknesses. Reprinted with permission from [75]. Copyright (2001), American Institute of Physics

A and B are seen and their signals are much weaker. This means that the number of deep levels and the density of deep levels seem to be dependent on the film thickness. Their thermal activation energies were estimated to be 0.67 eV for A, 0.89 eV for A_1, 0.72 eV for A_x, 0.61 eV for B, 0.41 eV for C, and 0.23 eV for D. Figure 8.41 shows the cross-sectional transmission electron microscopy (TEM) images for the nominal 68-µm-thick sample. It is found that dislocation density increases from 1×10^8 cm^{-2} in the top region ($t = 55$ µm) to 3×10^8 cm^{-2} in the middle region ($t = 20$ µm), and then to 2×10^9 and 1×10^{10} cm^{-2} in the region near the interface ($t = 3$ and 1 µm, respectively). Therefore, from Figs. 8.40 and 8.41, it can be concluded that the formation of the deep levels in GaN layers is strongly related to the presence of the threading dislocations. Among the deep levels, A_1 and B are dominant in thin GaN layers grown by HVPE. However, it was revealed that a reverse relationship exists between the A_1 and the B: that is, a strong increase of B is accompanied by a significant reduction of A1. The deep level B is believed to correspond to the previously introduced E2 level because their thermal activation energies are similar and their emission peak positions are almost the same.

The origin of B (i.e., E2) has been controversial for a long time. Earlier studies suggest that it could be due to chemical impurities such as C or Mg. Another study shows that E2 could be effectively suppressed by isoelectronic In doping and it is suspected of being an antisite point defect N_{Ga}. On the

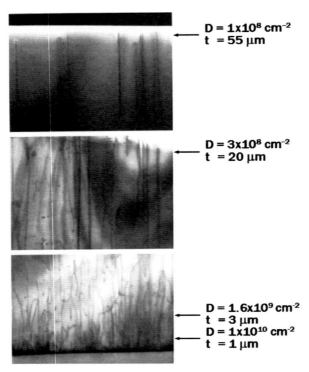

Fig. 8.41. Cross-sectional TEM images of a thick HVPE GaN sample. Reprinted with permission from [75]. Copyright (2001), American Institute of Physics

other hand, the peaks of A, A_1, and A_x have often been reported from electron-irradiated samples. The level A_1 is tentatively assigned to a N_i-related defect. But, the others are unclear. The peak D (0.17–0.23 eV) also showed a correlation with the dislocation density. Theoretically, threading dislocations in GaN have been predicted to have V_{Ga} or V_N defects in their core structures, depending on doping and growth stoichiometry. Formation of the V_{Ga} structure is favored under N-rich conditions and is most stable in n-type material, while formation of the V_N structure is favored under Ga-rich conditions. It is expected that the level D could possibly be a defect complex involving V_{Ga} as an electron trap, since the V_{Ga} is easily available in n-type GaN with a high dislocation density.

Figure 8.42 shows the representative DLTS spectra for n-type GaN layers grown by MBE, where the electron concentration is in the range of 6×10^{16} cm^{-3} [76]. It is found that five peaks, labeled by $E1''$, $E2''$, $E3''$, $E4''$, and $E5''$, exist between 100 and 500 K. Their thermal activation energies were estimated to be 0.23 eV for $E1''$, 0.58 eV for $E2''$, 0.66 eV for $E3''$, 0.96 eV for $E4''$, and 0.24 eV for $E5''$. Among these levels, the peaks $E1''$, $E2''$, and

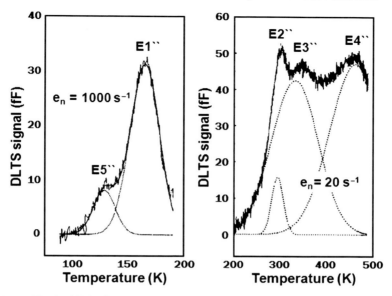

Fig. 8.42. Typical DLTS spectra for MBE GaN samples. Reprinted with permission from [76]. Copyright (1998), American Institute of Physics

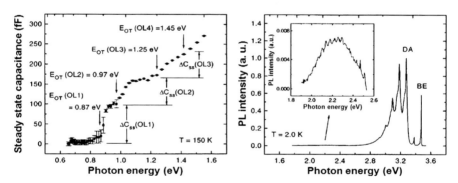

Fig. 8.43. PHCAP spectra and PL spectra for GaN layers grown by MOCVD. Reprinted with permission from [69, 70]. Copyright (1994), American Institute of Physics

E3″ are interpreted as the same deep levels in the n-type GaN layers grown by HVPE and MOCVD, previously described in Figs. 8.39 and 8.40. However, the levels E4″ and E5″ did not correspond to any previously reported defect levels. The exact origin of these deep levels remains as an open question.

On the other hand, Fig. 8.43 shows the representative photocapacitance (PHCAP) spectrum for GaN layers grown by MOCVD [69,70]. It is found that four steps exist at the energy levels of 0.87, 0.97, 1.25, and 1.45 eV. PHCAP employs the photoionization of deep levels in the depletion region. In PHCAP, the carriers trapped in the deep levels are excited to the conduction band or the valence band, which induces an increase or decrease of the capacitance when the illuminated optical energies equal to the thermal activation energies of deep levels. However, when the photon energy is smaller than half the bandgap ($E < E_g/2$), the photoexcitation is just related to the conduction band, which results in the appearance of a step in the PHCAP spectra as shown in Fig. 8.43. Therefore, the four levels in Fig. 8.43 correspond to the position at 0.87–1.45 eV above the valence band maximum. Therefore, it is expected that the photogenerated deep levels may participate in the deep green-yellow emission at 2.2 eV observed in photoluminescence (PL) spectra.

Up to now, the deep levels in GaN were discussed with respect to their electrical properties. In Table 8.7, the various deep levels detected in GaN materials are summarized. The structural point defects N_{Ga}, N_i, and V_{Ga} have been suspected as the most probable origins for the deep levels in GaN. However, the deep level densities are relatively small compared to the background electron concentration.

Table 8.7. Summary of the deep levels in GaN films

Process	Electron con. (cm^{-3})	Energy level (eV)	Trap density (cm^{-3})	Meas.	Ref.
MOCVD	2×10^{17}	0.18 (E1)	7×10^{13}	DLTS	[69,70]
		0.49 (E2)	6×10^{14}		
MOCVD	2×10^{17}	0.87	–	PHCAP	[69,70]
		0.97	–		
		1.25	–		
		1.45	–		
MOCVD	2×10^{16} (TMGa)	0.14 (E1′)	2×10^{14}	DLTS	[74]
		0.49 (E2′)	3×10^{14}		
		1.44 (E3′)	3×10^{15}		
	2×10^{17} (TEGa)	1.63	5×10^{15}		
MBE	6×10^{16}	0.23 (E1″)	8×10^{14}	DLTS	[76]
		0.58 (E2″)	1×10^{15}		
		0.66 (E3″)	4×10^{15}		
		0.96 (E4″)	8×10^{15}		
		0.24 (E5′)	2×10^{14}		
HVPE	$\sim 10^{17}$	0.67 (A)	$\sim 10^{14}$	DLTS	[75]
		0.61 (B)			
		0.23 (D)			
		0.89 (A$_1$)	$\sim 10^{15}$		
MOCVD (Electron irradiated)	2×10^{16}	0.62 (B)	–	DLTS	[77]
		0.18 (E)			

8.5.3 Deep Levels in ZnO

The first study of deep levels in ZnO started from ZnO ceramics in order to find out the origin of the rectifying behavior of ZnO ceramic varistors. Actually, the ZnO ceramics synthesized by sintering have the ohmic behavior. However, they showed the excellent rectifying behavior by the incorporation of small amounts of additives such as Bi_2O_3, CoO, MnO, and Sb_2O_3 in ZnO matrixes. On the other hand, later studies on deep levels have focused on single-crystalline ZnO bulks and films in order to identify the native defects that hamper the conductivity control. Actually, they have relatively poor electrical properties of the high background electron concentration and the low electron mobility due to the native defects. In the current point many types of deep levels in ZnO materials are detected and identified although most of those researches are performed on the ZnO bulks.

Figure 8.44 shows the representative DLTS spectrum for ZnO ceramics synthesized by sintering [78]. It is seen that two peaks, labeled by L1 and L2, exist between 100 and 200 K. The negative DLTS signals mean that the L1 and the L2 are the majority electron-trap centers. Their thermal activation energies are estimated to be 0.24 eV for L1 and 0.33 eV for L2. The L1 has been commonly observed in ZnO ceramics in the range 0.12–0.24 eV below the conduction band minimum, but its origin is unclear. The L2 is situated at around 0.23–0.33 eV below the conduction band minimum, and also has been commonly observed in ZnO ceramics. It has been proposed that the L2 is the oxygen vacancy (V_o) or the zinc interstitial (Zn_i). However, it has been controversial for long time. From the theoretical calculations based on the local density approximation, it appears that the V_o model, and not the Zn_i model, may be correct in the explanation of the L2. Therefore, the L2 is ascribed to V_o. On the other hand, another deep level labeled L3 is sometimes reported in the range 0.20–0.32 eV below the conduction band minimum, which is detected

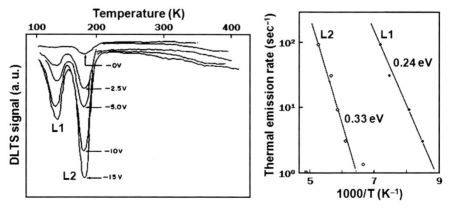

Fig. 8.44. Representative DLTS spectra of ZnO bulks produced by sintering process. Reprinted with permission from [78]. Copyright (1980), The Institute of Pure and Applied Physics

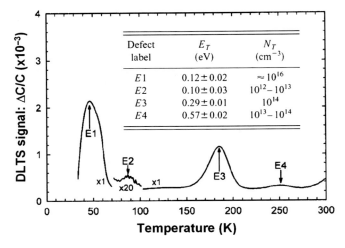

Fig. 8.45. Representative DLTS spectra of ZnO bulks produced by VPT. Reprinted with permission from [79]. Copyright (2002), American Institute of Physics

only in polycrystalline ZnO bulks. Therefore, the L3 is expected to be linked to the multiple, closely spaced point defects.

Figure 8.45 shows the representative DLTS spectrum for ZnO bulks grown by vapor-phase transport (VPT), where the background electron concentration is 5×10^{16} cm^{-3} [79]. It is shown that four peaks labeled by E1, E2, E3, and E4 exist at 50, 90, 190, and 250 K, respectively. Their thermal activation energies were estimated to be 0.12 (E1), 0.10 (E2), 0.29 (E3), and 0.57 eV (E4). Among these deep levels, the E1 and E3 exhibit much larger emission intensities than the E2 and E4, which means that the E1 and E3 are the relatively dominant electron-trap centers in the single crystalline ZnO bulks. The E2 is regarded as a minor deep level because its emission signal is much small and its density is estimated to be below 10^{13} cm^{-3}. The E4 is suggested as having its origin not from a simple point defect but from an extended defect because its apparent capture cross-section is unrealistically large at 2×10^{-12} cm^2. On the other hand, in the cases of the E1 and the E3, the relatively shallow level E1 has a deep level density of $\sim 10^{16}$ cm^{-3}, which corresponds to the background electron concentration. However, in the temperature dependent Hall measurement, two shallow levels at 31 and 61 meV were also found from the same materials, which are not close to 0.12 eV. A possible explanation is that E1 is a defect with a large temperature-activated capture cross-section. If the barrier for this process is about 59 meV, the true level of E1 would correspond to 61 meV. The E3 is believed to correspond to the structural point defect L2 detected in the polycrystalline ZnO bulks in Fig. 8.44 because their emission peaks are located at similar positions and their thermal activation energies are almost the same.

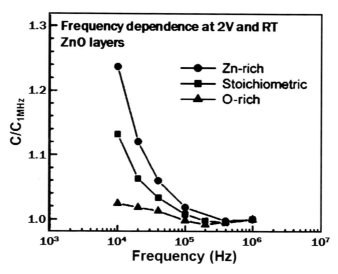

Fig. 8.46. Capacitance vs.frequency curves for ZnO layers, measured at 2 V and room temperature. Reprinted with permission from [80]. Copyright (2005), American Institute of Physics

Figure 8.46 shows the frequency dependence of capacitance for ZnO layers grown by MBE at different Zn/O flux conditions: Zn-rich, stoichiometric, and O-rich [80]. It is found that the capacitance of Zn-rich and stoichiometric ZnO layers is strongly dependent on the measurement frequency, while the capacitance of an O-rich ZnO layer slightly varies with the measurement frequency. The observed variation of the capacitance is ascribed to the dispersion effect that occurs when the deep levels are unable to follow the high-frequency voltage modulation: the low-frequency capacitance will be influenced by the deep levels, while the high-frequency capacitance will not be influenced by the deep levels. Therefore, it is expected that the ZnO layers grown under the Zn-rich and stoichiometric flux conditions suffer from larger densities of deep levels than the ZnO layer grown under the O-rich flux condition.

Figure 8.47 compares the AS spectra for these Zn-rich, stoichiometric, and O-rich ZnO layers [80]. It is shown that the three samples commonly have two capacitance steps (ET1 and ET2) at similar positions, while the Zn-rich ZnO layer exhibits another capacitance step (ET3) at 110 K. In AS, the thermal activation energy can be derived from plotting step shift against measurement frequency vs. the inverse of peak temperature ($\ln(T_{\text{step}}^2/\omega)$ vs. $1/T_{\text{step}}$). The thermal activation energies of ET1 and ET2 were estimated to be 0.033–0.046 eV and 0.12–0.15 eV, respectively. The thermal activation energy of ET3 was estimated to be 0.065 eV. On the other hand, the deep level density is determined by $N_T = n \cdot (C_{\text{low}}/C_{\text{1MHz}} - 1)$, where N_T is the deep level density, n is the free electron concentration, C_{low} is the capacitance at low frequency,

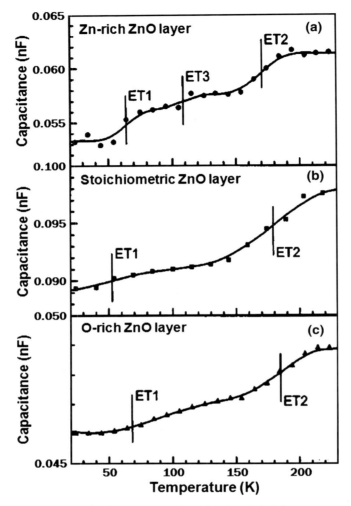

Fig. 8.47. Impact of VI/II flux ratio on deep levels of ZnO layers grown by MBE. AS spectra of (**a**) Zn-rich, (**b**) stoichiometric, and (**c**) O-rich ZnO layers. Reprinted with permission from [80]. Copyright (2005), American Institute of Physics

and $C_{1\mathrm{MHz}}$ is the capacitance at 1 MHz. The deep level densities of the O-rich and stoichiometric ZnO layers were estimated to be 2×10^{15} and $8 \times 10^{15}\,\mathrm{cm^{-3}}$, respectively. The deep level density of the Zn-rich ZnO layer was estimated to be $3 \times 10^{16}\,\mathrm{cm^{-3}}$. The energy levels and densities are summarized in Table 8.8. It is found that the deep levels ET1 and ET2 are commonly observed in all the ZnO layers, while the deep level ET3 is observed only in the Zn-rich ZnO layer. Above all, it should be noted that the deep level density of ET2 is larger

Table 8.8. Thermal activation energies and trap densities for three trap centers (ET1, ET2, and ET3) found in ZnO layers grown under Zn-rich, stoichiometric, and O-rich flux conditions [80]

	ET1		ET2		ET3	
	ET1 (eV)	N_{ET1} (cm^{-3})	E_{ET2} (eV)	N_{ET2} (cm^{-3})	E_{ET3} (eV)	N_{ET3} (cm^{-3})
Zn-rich	0.033	3×10^{15}	0.14	2×10^{16}	0.065	3×10^{15}
Stoichiometric	0.033	2×10^{15}	0.15	6×10^{15}	–	–
O-rich	0.046	4×10^{14}	0.12	2×10^{15}	–	–

than that of ET1 or ET3 in all the cases and it increases with increasing the Zn flux. This means that the large dispersion effects observed in the Zn-rich and the stoichiometric ZnO layers are ascribed to the high density of the deep level ET2.

Here, it is important to compare the deep levels ET1, ET2, and ET3 found in ZnO films with the trap centers reported in ZnO bulks. Shohata et al. reported the presence of a deep level (L1) situated at 0.12–0.24 eV below the conduction band minimum using DLTS in ZnO ceramic varistors synthesized by sintering. Auret et al. observed L1 in ZnO bulks grown by VPE [79]. Moreover, Cordaro et al. observed L1 by AS measurements, in which a capacitance step was observed at 185 K [81]. These results suggest that the deep level ET2 found in ZnO layers is the trap center that corresponds to the L1 reported in bulk ZnO. Sukkar et al. reported that the second ionization energy of Zn_i is estimated to be 0.2 eV [82]. Therefore, the Zn_i is suggested as the native defect that forms the L1. On the other hand, Roth et al. reported that the thermal activation energy of a shallow donor level in ZnO layers, formed by oxidation method, can be estimated to be 0.03–0.06 eV [83]. Ziegler et al. observed the similar donor levels of 0.03–0.05 eV from ZnO layers prepared by CVD and found that the value is varied with growth temperature [84]. Natsume et al. suggested that the widely ranged donor levels of 0.02–0.2 eV are ascribed to different types of Zn_i because metallic zinc atoms observed by X-ray photoelectron spectroscopy (XPS) could be interpreted as interstitial zinc atoms [85,86]. Hagenark et al. also suggested that the shallow donor level of 0.043–0.045 eV is due to singly ionized Zn_i [87]. All these observed energy levels are very close to those of ET1 and ET3. Consequently, it is suggested that the relatively shallow deep levels ET1 and ET3 are the trap centers that commonly exist in the ZnO bulks and the ZnO layers.

Up to now, the deep levels in ZnO were discussed in the aspect of their electrical properties. The structural point defects Zn_i and V_o have been suspected as the most possible origin for the deep levels in ZnO. However, the deep-level densities are relatively small compared to the background electron concentration. In Table 8.9, the various deep levels detected in ZnO materials, fabricated by MBE, VPE, flux method, and sintering, are summarized.

Table 8.9. Summary of the deep levels in ZnO ceramics, single-crystalline films, and bulks

Process	Electron con. (cm^{-3})	Energy level (eV)	Trap density (cm^{-3})	Meas.	Ref.
Sintering (Ceramics)	–	0.12–0.24 (L1) 0.24–0.33 (L2)	– –	DLTS	[78]
Sintering (Ceramics)	5×10^{17} 6×10^{16}	0.16 (L1) 0.23 (L2) 0.32 (L3)	8×10^{15} 9×10^{15} 3×10^{14}	DLTS	[88]
Sintering (Ceramics)	–	0.17 0.33	– –	AS	[81]
Sintering (Ceramics)	$0.8–2 \times 10^{17}$	0.17 (L1) 0.26 (L2) 0.20–0.30 (L3)	1×10^{14} 5×10^{13} 9×10^{14}	DLTS	[89]
Sintering (Ceramics)	–	0.138 (L1) 0.23 (L2)	8×10^{15} 1×10^{16}	DLTS	[90]
Hydrothermal (Single crystal)	$2–8 \times 10^{14}$	0.30	–	AS	[91]
Flux method (Single crystal)	$0.2–2 \times 10^{18}$	0.30–0.34	$0.3–19 \times 10^{17}$	AS	[92]
VPE (Single crystal)	$4–6 \times 10^{16}$	0.12 (E1) 0.10 (E2) 0.29 (E3) 0.57 (E4)	1×10^{16} $10^{12}–10^{13}$ 1×10^{14} $10^{13}–10^{14}$	DLTS	[79]
Flux method (Single crystal)	–	0.27	–	ICTS	[93]
Flux method (Cu-doped)	–	0.27	–	AS	[94]
Flux method (Ag-doped)	–	0.27	–	AS	[95]
VPE (Proton bombardment)	$4–6 \times 10^{16}$	0.54 0.78	$\sim 10^{14}$ $\sim 10^{14}$	DLTS	[96]
VPE (Proton-bombardment)	9×10^{16}	0.55 0.75 0.90	$\sim 10^{14}$ $\sim 10^{14}$ $\sim 10^{14}$	DLTS	[97]
MBE (Single crystal)	3×10^{17}	0.033–0.046 (ET1) 0.12–0.15 (ET2) 0.065 (ET3)	2×10^{15} 6×10^{15} –	AS	[80]

References

1. M.A. Reshchikov, H. Morkoc, J. Appl. Phys. **97**, 061301 (2005)
2. U. Ozgur, Y.A. Alivov, C. Liu, A. Teke, M.A. Reshchikov, S. Dogan, V. Avrutin, S.J. Cho, H. Morkoc, J. Appl. Phys. **98**, 041301 (2005)
3. D.K. Schroder, *Semiconductor Material and Device Characterization* (Wiley, Singapore, 1990), p. 99
4. R. Williams, *Modern GaAs Processing Methods* (Artech House, 1990), p. 211
5. J.S. Foresi, T.D. Moustakas, Appl. Phys. Lett. **62**, 2859 (1993)
6. H.W. Jang, C.M. Jeon, J.K. Kim, J.L. Lee, Appl. Phys. Lett. **78**, 2015 (2001)
7. J.D. Guo, C.I. Lin, M.S. Feng, F.M. Pan, G.C. Chi, C.T. Lee, Appl. Phys. Lett. **68**, 235 (1996)
8. H.W. Jang, J.H. Lee, J.L. Lee, Appl. Phys. Lett. **80**, 3955 (2002)
9. J.O. Song, S.J. Park, T.Y. Seong, Appl. Phys. Lett. **80**, 3129 (2002)
10. H.W. Jang, J.K. Kim, J.L. Lee, J. Schroeder, T. Sands, Appl. Phys. Lett. **82**, 580 (2003)
11. J.S. Kwak, K.Y. Lee, J.Y. Han, J. Cho, S. Chae, O.H. Nam, Y. Park, Appl. Phys. Lett. **79**, 3254 (2001)
12. J.O. Song, S.H. Kim, J.S. Kwak, T.Y. Seong, Appl. Phys. Lett. **83**, 1154 (2003)
13. D.B. Ingerly, Y. Chen, R.S. William, T. Takeuchi, Y.A. Chang, Appl. Phys. Lett. **77**, 382 (2000)
14. H.K. Kim, S.H. Han, T.Y. Song, W.K. Choi, Appl. Phys. Lett. **77**, 1647 (2000)
15. J.M. Lee, K.K. Kim, S.J. Park, W.K. Choi, Appl. Phys. Lett. **78**, 3842 (2001)
16. K. Ip, Y.W. Heo, K.H. Baik, D.P. Nortons, S.J. Pearton, F. Ren, Appl. Phys. Lett. **84**, 544 (2004)
17. H.K. Kim, K.K. Kim, S.J. Park, T.Y. Seong, I. Adesida, J. Appl. Phys. **94**, 4225 (2003)
18. H.K. Kim, K.K. Kim, S.J. Park, T.Y. Seong, Y.S. Yoon, Jpn. J. Appl. Phys. **41**, L546 (2002)
19. D.K. Schroder, *Semiconductor Material and Device Characterization* (Wiley, Singapore, 1990), p. 130
20. D.C. Oh, J.J. Kim, H. Makino, T. Hanada, M.W. Cho, T. Yao, H.J. Ko, Appl. Phys. Lett. **86**, 042110 (2005)
21. E.J. Miller, E.T. Yu, P. Waltereit, J.S. Speck, Appl. Phys. Lett. **26**, 535 (2004)
22. T. Hashizume, J. Kotani, H. Hasegawa, Appl. Phys. Lett. **84**, 4884 (2004)
23. J.K. Sheu, M.L. Lee, W.C. Lai, Appl. Phys. Lett. **86**, 052103 (2005)
24. M.L. Lee, J.K. Sheu, S.W. Lin, Appl. Phys. Lett. **88**, 032103 (2006)
25. U. Karrer, O. Ambacher, M. Stutzmann, Appl. Phys. Lett. **77**, 2012 (2000)
26. H. Lu, R. Zhang, X. Xiu, Z. Xie, Y. Zheng, Z. Li, Appl. Phys. Lett. **91**, 172113 (2007)
27. O. Weidemann, E. Monroy, E. Hahn, M. Stutzmann, M. Eickhoff, Appl. Phys. Lett. **86**, 083507 (2005)
28. J. Spradlin, S. Dogan, M. Mikkelson, D. Huang, L. He, D. Johnstone, H. Morkoc, R.J. Molnar, Appl. Phys. Lett. **82**, 3556 (2003)
29. E.J. Miller, D.M. Schaadt, E.T. Yu, P. Waltereit, C. Poblenz, J.S. Speck, Appl. Phys. Lett. **82**, 1293 (2003)
30. O. Weidemann, M. Hermann, G. Steinhoff, H. Wingbrant, A.L. Spetz, M. Stutzmann, M. Eickhoff, Appl. Phys. Lett. **83**, 773 (2003)
31. B.J. Coppa, R.F. Davis, R.J. Nemanich, Appl. Phys. Lett. **82**, 400 (2003)

32. S.K. Kim, B.S. Kang, F. Ren, K. Ip, Y.W. Heo, D.P. Norton, S.J. Pearton, Appl. Phys. Lett. **84**, 1698 (2004)
33. A.Y. Polyakov, N.B. Smirnov, E.A. Kozhukhova, V.I. Vdovin, K. Ip, Y.W. Heo, D.P. Norton, S.J. Pearton, Appl. Phys. Lett. **83**, 1575 (2003).
34. H. Sheng, S. Muthukumar, N.W. Emanetoglu, Y. Lu, Appl. Phys. Lett. **80**, 2132 (2002).
35. M.W. Allen, P. Miller, R.J. Reeves, S.M. Durbin, Appl. Phys. Lett. **90**, 062104 (2007)
36. K. Ip, Y.W. Heo, K.H. Baik, D.P. Norton, S.J. Pearton, S. Kim, J.R. LaRoche, F. Ren, Appl. Phys. Lett. **84**, 2835 (2004)
37. H. von Wenckstern, E.M. Kaidashev, M. Lorenz, H. Hochmuth, G. Biehne, J. Lenzner, V. Gottschalch, R. Pickenhain, M. Grundmann, Appl. Phys. Lett. **84**, 79 (2004)
38. J. Qiu, J.M. Depuydt, H. Cheng, M.A. Haase, Appl. Phys. Lett. **59**, 2992 (1991)
39. D.C. Look, *Electrical Characterization of GaAs Materials and Devices* (Wiley, Singapore, 1989), p. 74
40. D.C. Look, R.J. Molnar, Appl. Phys. Lett. **70**, 3377 (1997)
41. S. Adachi, *Handbook on Physical Properties of Semiconductors*, Vol 2. III–V Compound Semiconductors (Kluwer, USA, 2004), p. 233
42. S. Adachi, *Handbook on Physical Properties of Semiconductors*, Vol 3. II–VI Compound Semiconductors (Kluwer, USA, 2004), p. 65
43. W. Gotz, J. Walker, L.T. Romano, N.M. Johnson, R.J. Molnar, Mater. Res. Soc. Symp. Proc. **449**, 525 (1997)
44. B. Podor, Phys. Stat. Sol. **16**, K169 (1966)
45. D.C. Look, J.R. Sizelove, Phys. Rev. Lett. **82**, 1237 (1999)
46. D.C. Oh, S.W. Lee, H. Goto, S.H. Park, I.H. Im, T. Hanada, M.W. Cho, T. Yao, Appl. Phys. Lett. **91**, 132112 (2007)
47. T. Ninna, T. Minato, K. Yoneda, Jpn. J. Appl. Phys. **21**, L387 (1982)
48. K. Ohkawa, T. Mitsuyu, O. Yamazaki, J. Appl. Phys. **62**, 3216 (1987)
49. T. Yao, M. Ogura, S. Matsuoka, T. Morishita, Appl. Phys. Lett. **43**, 499 (1983)
50. D.C. Look, D.C. Reynolds, W. Kim, O. Aktas, A. Botchkarev, A. Salvador, H. Morkoc, J. Appl. Phys. **80**, 2960 (1996)
51. K. Saarinen, T. Laine, S. Kuisma, J. Nissila, P. Hautojarvi, L. Dobrzynski, J.M. Baranowski, K. Pakula, R. Stepniewski, M. Wojdak, A. Wysmolek, T. Suzuki, M. Leszczynski, I. Grzegory, S. Porowski, Phys. Rev. Lett. **79**, 3030 (1997)
52. K. Saarinen, P. Seppala, J. Oila, P. Hautojarvi, C. Corbel, O. Briot, R.L. Aulombard, Appl. Phys. Lett. **73**, 3253 (1998)
53. H.P. Maruska, J.J. Tietjen, Appl. Phys. Lett. **15**, 327 (1969)
54. D.W. Jenkins, J.D. Dow, Phys. Rev. B **39**, 3317 (1989)
55. J. Neugebauer, C.G. Van de Walle, Phys. Rev. B **50**, 8067 (1994)
56. A.F. Wright, U. Grossner, Appl. Phys. Lett. **73**, 2751 (1998)
57. W. Seifer, R. Franzheld, E. Butter, H. Sobotta, V. Riede, Cryst. Res. Technol. **18**, 383 (1983)
58. B.C. Chung, M. Gershenzon, J. Appl. Phys. **72**, 651 (1992)
59. E.M. Kaidashev, M. Lorenz, H. von Wenckstern, A. Rahm, H.C. Semmelhack, K.H. Han, G. Benndorf, C. Bundesmann, H. Hochmuth, M. Grundmann, Appl. Phys. Lett. **84**, 3901 (2003)
60. K. Miyamoto, M. Sano, H. Kato, T. Yao, Jpn. J. Appl. Phys. **41**, L1203 (2002)

61. A.F. Kohan, G. Ceder, D. Morgan, C.G. Van de Walle, Phys. Rev. B **61**, 15019 (2000)
62. C. G. Van de Walle, Physica B **308–310**, 899 (2001)
63. D.C. Look, J.W. Hemsky, J.R. Sizelove, Phys. Rev. B **22**, 2552 (1999)
64. C.G. Van de Walle, Phys. Rev. B **85**, 1012 (2000)
65. L.Y. Chen, W.H. Chen, J.J. Wang, F.C.N. Hong, Y.K. Su, Appl. Phys. Lett. **85**, 5628 (2004)
66. K. Ip, M.E. Overberg, Y.W. Heo, D.P. Norton, S.J. Pearton, C.E. Stutz, B. Luo, F. Ren, D.C. Look, J.M. Zavada, Appl. Phys. Lett. **82**, 385 (2003)
67. D.V. Lang, J. Appl. Phys. **45**, 3023 (1974)
68. D.K. Schroder, *Semiconductor Material and Device Characterization* (Wiley, Singapore, 1990), p. 297
69. W. Gotz, N.M. Johnson, H. Amano, I. Akasaki, Appl. Phys. Lett. **65**, 463 (1994)
70. W. Gotz, N.M. Johnson, R.A. Street, H. Amano, I. Akasaki, Appl. Phys. Lett. **66**, 1340 (1995)
71. D.L. Losee, J. Appl. Phys. **46**, 2204 (1975)
72. G. Vincent, D. Bois, P. Pinard, J. Appl. Phys. **46**, 5173 (1975)
73. D.L. Polla, C.E. Jones, J. Appl. Phys. **51**, 6233 (1980)
74. W.I. Lee, T.C. Huang, J.D. Guo, M.S. Feng, Appl. Phys. Lett. **67**, 1721 (1995)
75. Z.Q. Fang, D.C. Look, J. Jasinski, M. Benamara, Z. Liliental-Weber, R.J. Molnar, Appl. Phys. Lett. **78**, 322 (2001)
76. C.D. Wang, L.S. Yu, S.S. Lau, E.T. Yu, W. Kim, A.E. botchkarev, H. Morkoc, Appl. Phys. Lett. **72**, 1211 (1998)
77. Z.Q. Fang, J.W. Hemsky, D.C. Look, M.P. Mack, Appl. Phys. Lett. **72**, 448 (1998)
78. N. Shohata, T. Matsumura, T. Ohno, Jpn, J. Appl. Phys. **19**, 1793 (1980)
79. F.D. Auret, S.A. Goodman, M.J. Legodi, W.E. Meyer, Appl. Phys. Lett. **80**, 1340 (2002)
80. D.C. Oh, T. Suzuki, J.J. Kim, H. Makino, T. Hanada, M.W. Cho, T. Yao, Appl. Phys. Lett. **86**, 032909 (2005)
81. J.F. Cordaro, Y. Shim, J.E. May, J. Appl. Phys. **60**, 4186 (1986)
82. M.H. Sukkar, H.L. Tuller, *Advances in Ceramics*, Vol. 7, ed. by M.F. Yan, A.H. Heuer (American Ceramic Society, Columbus, OH, 1983), pp. 71–90
83. A.P. Roth, D.F. Williams, J. Appl. Phys. **52**, 6685 (1981)
84. E. Ziegler, A. Heinrich, H. Oppermann, G. Stoever, Phys. Stat. Sol. A **66**, 635 (1981)
85. Y. Natsume, H. Sakata, T. Hirayama, Phys. Stat. Sol. A **148**, 485 (1995)
86. Y. Natsume, H. Sakata, T. Hirayama, H. Yanagida, J. Appl. Phys. **72**, 4203 (1992)
87. K.I. Hagemark, L.C. Chacka, J. Solid Stat. Chem. **15**, 261 (1975)
88. A. Nitayama, H. Sakaki, T. Ikoma, Jpn. J. Appl. Phys. L743 (1980)
89. A. Rohatgi, S.K. Pang, T.K. Gupta, W.D. Straub, J. Appl. Phys. **63**, 5375 (1988)
90. W.I. Lee, R.L. Young, W.K. Chen, Jpn. J. Appl. Phys. **35**, L1158 (1996)
91. J.C. Simpson, J.F. Cordaro, J. Appl. Phys. **63**, 1781 (1988)
92. Y. Kanai, Jpn. J. Appl. Phys. **29**, 1426 (1990)
93. N. Ohashi, J. Tanaka, T. Ohgaki, H. Haneda, N. Ozawa, T. Tsurumi, J. Mater. Res. **17**, 1529 (2002)
94. Y. Kanai, Jpn. J. Appl. Phys. **30**, 703 (1991)

95. Y. Kanai, Jpn. J. Appl. Phys. **30**, 2021 (1991)
96. F.D. Auret, S.A. Goodman, M. Hayes, M.J. Legodi, H.A. van Laarhoven, D.C. Look, Appl. Phys. Lett. **79**, 3074 (2001)
97. A.Y. Polyakov, N.B. Smirnov, A.V. Govorkov, E.A. Kozhukhova, V.I. Vdovin, K. Ip, M.E. Overberg, Y.W. Heo, D.P. Norton, S.J. Pearton, J.M. Zavada, V.A. Dravin, J. Appl. Phys. **94**, 2895 (2003)

9
GaN and ZnO Light Emitters

J.-S. Ha

Abstract. In the recent several decades, there are huge concerns in solid-state light emitters based on semiconductor compound materials, which emit light of ultraviolet to red light. Light-emitting diode (LED) fabrication technology for this application is now relatively mature. Currently, the lifetime of blue or green light-emitter are apparently determined mostly by light-induced degradation of a packaging unit encapsulating the LED. New renaissance is taking place in research societies and industries of LEDs because of the straight possibilities and needs for LED-based solid-state lighting in human life instead of the conventional ones employing incandescent, halogen, fluorescent lightings etc. Among various semiconductor compounds applicable to LEDs, GaN and ZnO are regarded as promising materials for solid-state lighting because ultraviolet- or blue-light emitters, which are applicable to white-light LEDs, based on these materials are possible. In this chapter, current technologies and researches on GaN- and ZnO-based LEDs are described. A special emphasis is given to the efficiency of the LEDs in the review of the GaN-based LEDs, while current status technologies in LED applications of ZnO-based materials have been reviewed.

9.1 Introduction

Since the exhaustion of fossil energy resource and protection of our environments are one of the most critical and hot issues for mankind, efficient and environmentally friendly use of energy are becoming more and more important. Energy consumption for lighting has reached about 20% of the total electrical energy consumption in the world. This is one of the reasons for the accelerated researches on solid-state lighting, which is expected to give us new ways for saving energy, increasing the efficiency of energy use, and protecting and healing the environments in a long-term time scale. Therefore, LEDs for solid-state lighting will open huge industrial markets and business chances in making monitary benefits.

III-nitride semiconductors including GaN, AlN, InN, and related alloys give us wide and versatile bandgap engineering because InN, GaN, and AlN have bandgaps of $0.70 \sim 0.78\,\text{eV}$ $3.35 \sim 3.51\,\text{eV}$, and $6.10 \sim 6.15\,\text{eV}$, respectively [1,2], at room temperature (RT), which can cover light emitters ranging from visible to invisible lights. This feature is quite different from other III–V material systems like As-based and P-based ones. On the other hand, II-oxide semiconductors including ZnO, MgO, CdO, and related alloys can offer bandgap engineering spanning from 2.4 to 8.2 eV considering the bandgap energies of $2.2 \sim 2.68$ [3], 3.37 [4], and 7.8 eV [5] for CdO, ZnO, and MgO, respectively, at RT.

In 1972, the first blue LED using III-nitride materials (GaN) was fabricated by Pankove et al. [6] with the metal-i–n structure. However, because of the difficulty in getting p-type conductivity in GaN-based materials, the success of GaN-based LED with a homo p–n junction had been delayed. In the late 1980s, Amano et al. [7,8] developed the AlN buffer layer growth [7] and low-energy electron beam irradiation (LEEBI) techniques to obtain conductive p-type GaN [8] and reported the first GaN-based blue LED with a homo p–n junction [8]. After that, Nakamura et al. reported double-heterostructure (DH) blue LED by employing an n-type InGaN layer [9]. On the basis of these initiative works on GaN-based blue LEDs, lots of breakthroughs and advances in technologies have been achieved. Since the report of the really low efficiency of GaN LED by Amano [10], the bright LEDs which show the high *internal quantum efficiency* of $60\% \sim 70\%$ ($I = 20\,\text{mA}$) are available now and continuous progress is being reported. As the devices with higher power capabilities have become available, new applications of LEDs are appearing continuously.

In the case of ZnO-based LED researches, lot of advances have been reported since the reports on the lasing from the single-crystalline ZnO films [11,12]. Similar to the GaN, obtaining the reliable p-type conductivity applicable to ZnO-based LED fabrications was a big obstacle. Recent progresses in epitaxy techniques make this possible to get p-type conductivity and to fabricate ZnO-based LEDs [13–15].

In this chapter, the current technologies and researches on GaN- and ZnO-based LEDs are described. A special emphasis is given to the efficiency of the LEDs in the review of the GaN-based LEDs, while current status technologies in the LED applications of ZnO-based materials have been reviewed. Various topics from the LEDs' history to today's technological edge are touched. Following sections (9.2, 9.3, and 9.4) address the recent research progresses and results on GaN-based LEDs and ZnO-based LEDs.

9.2 Light-Emitting Diodes Basic

In order to get better understanding on the LEDs, it is important to know the physics governing the light emissions from semiconductors. Figure 9.2 schematically shows the p–n junction. When p-type and n-type semiconductors are fully contacting each other, the electrons will transfer freely between the layers and form the thermal equilibrium status. This is called as

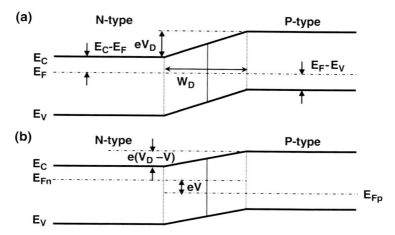

Fig. 9.1. (a) p–n junction in zero bias; (b) p–n junction in forward bias

"p–n junction." At this time, the electrons of the n-region flow to the p-region, while the holes of the p-region flow to the n-region resulting in the formation of a depletion layer, where the carriers do not exist in the middle of the p–n junction. Also, there occurs the relative voltage difference between n- and p-type regions and we call this diffusion potential (Fig. 9.1a) as V_D, which is expressed in (9.1)

$$V_D = \frac{kT}{e} \ln \frac{N_A N_D}{n_i^2}. \tag{9.1}$$

If a positive voltage is applied to the p-region with respect to the n-region, the potential barrier is reduced. Figure 9.1b shows a p–n junction with an applied voltage V. The junction is no longer in the thermal equilibrium state. The applied bias voltage lowers the potential barrier so that the majority carrier electrons from the n-region are injected across the junction into the p-region. Therefore, the current flow increases. These electrons will finally emit photons by recombination with holes in the LED.

If we assume that the individual electron and the hole currents were a continuous function and a constant through the space-charged region, the total current is the sum of the electron and the hole currents and is a constant through the entire junction. Total current density in the p–n junction can be written as

$$J = J_p + J_n = \left[\frac{eD_p p_{n0}}{L_p} + \frac{eD_n p_{p0}}{L_n}\right]\left[\exp\left(\frac{eV}{kT}\right) - 1\right], \tag{9.2}$$

where J is electronic current density (A cm^{-2}), J_n and J_p are electron and hole electric current density (A cm^{-2}), respectively, L_n and L_p are minority carrier electron and hole diffusion length (cm), respectively, e is electronic charge, D_n and D_p are minority carrier electron and minority carrier hole diffusion coefficient (cm^2 sec^{-1}), respectively, and p$_{n0}$ and n$_{p0}$ are thermal equilibrium minority carrier electron and minority carrier hole concentration (cm^{-3}), respectively.

Equation (9.2) shows the ideal current–voltage (I–V) relationship of a p–n junction. If J_s is defined as

$$J_s = \left[\frac{eD_p p_{n0}}{L_p} + \frac{eD_n p_{p0}}{L_n}\right], \tag{9.3}$$

(9.2) will be expressed as

$$J = J_s \left[\exp\left(\frac{eV}{kT}\right) - 1\right]. \tag{9.4}$$

Equation (9.4) is well known as the ideal-diode equation, which gives a good description of the I–V characteristics of the p–n junction over a wide range of currents and voltages. Equation (9.4) plotted as a function of forward-bias voltage V is shown in Fig. 9.2. If the voltage V becomes negative by a few kT/e volts, then the reverse-bias current density becomes independent of the reverse-bias voltage. The parameter J_s is then referred to as the reverse-saturation current density. The I–V characteristics of the p–n junction diode are obviously not bilateral.

However, there exists main difference between an ideal diode and a real p–n junction diode. That is the value of non-zero forward voltage drop V_F. Therefore, William Bradford Shockley, who is famous as one of the inventors of the transistor, suggested (9.5), the Shockley equation of

$$i_D = i_0 \left[\exp\left(\frac{V_D}{nV_T}\right) - 1\right], \tag{9.5}$$

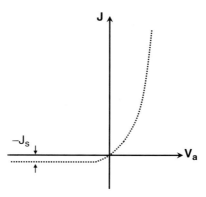

Fig. 9.2. Ideal I–V characteristics of a p–n junction diode

where i_D is diode current, i_0 is reverse saturation current (minority carrier diffusion), n is diode ideality factor, V_T is thermal voltage ($V_T = kT/q = 25.7$ mV at 25°C), k is Boltzmann constant ($k = 1.38 \times 10^{-23}$ J/K), T is the absolute temperature, and q is electron charge ($q = 1.6 \times 10^{-19}$ C).

When the V_D does not very small value, (9.5) is usually derived to following equation

$$i_D = i_0 \left[\exp\left(\frac{V_D}{nV_T}\right) \right]. \quad (9.6)$$

Here, n is an ideality factor of the diode. For a perfect diode, the value of n is 1 and for real diodes, this typical ideality factors show 1.1 ~ 1.5. However, in the cases of the GaN-based heterojunction diodes the value of ideality factor were represented to higher value than 2 and also, it become 6.8 for InGaN/AlGaN diodes DH blue LEDs [16]. Shah et al. reproted that the diodes fabricated from a bulk GaN p-n junction and a p–n junction and a p–n junction structure with a p-type AlGaN/GaN superlattice display ideality factors of 6.9 and 4.0, respectively [17] (Fig. 9.3.). Practically, the characteristics of real diodes are governed by various other factors including series resistance, shunt, and sub-threshold turn on [18].

9.3 Light-Emitting Diodes Based on GaN

9.3.1 Issues for High Internal Quantum Efficiency

When positive bias is applied to the LEDs, the electrons are injected into the p-type semiconductor and the holes are injected into the n-type material. The excess minority carriers diffuse into the neutral semiconductor regions, where they recombine with the majority carriers. If this recombination process is a direct band-to-band process, photons are emitted. The wavelength of emission (in μm) is expressed as

$$\lambda = \frac{hc}{E_g} = \frac{1.24}{E_g}, \quad (9.7)$$

where E_g is band gap energy (eV), h is Plank's constant, and c is the speed of light.

When electrons and holes recombine, the "recombination" energy (the sum of potential and kinetic energies) must be returned to the system. When the energy is given off as a photon, it is called as a "radiative transition" and when the energy is given to the lattice it is called as a "nonradiative transition." For the generation of light in LED devices, electrons and holes should be recombined in a radiative way; if not, the energy is absorbed in the device and is a loss. Therefore, the high internal efficiency, most of all, the recombination must occur by the radiative transition.

Internal quantum efficiency (η_{int}) is an important part of optical property of the LED. This is the value of the "number of photons emitted from active region per second" divided by the "number of electrons injected into LED

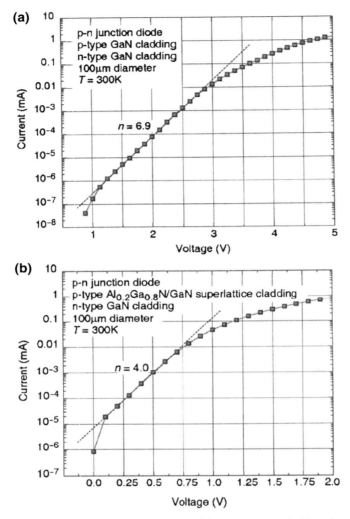

Fig. 9.3. I-V characteristics of p-n junction (a) with p-type GaN and n-type GaN cladding layers; (b) with p-type Al$_{0.2}$Ga$_{0.8}$N/GaN superlattice and n-type GaN cladding layer [17]. Copyright (2003), American Institute of Physics

per second" [19]. This describes a function of the injection efficiency and a percentage of radiative recombination events, compared with the total number of recombination events as expressed by (9.8),

$$\eta_{\text{ext}} = \frac{\text{\# of photons emitted from active region per second}}{\text{\# of electrons injected into LED per second}} = \frac{P_{\text{int}}/(h\nu)}{I/e}, \tag{9.8}$$

where P_{int} is optical power emitted from the active region and I is injection current.

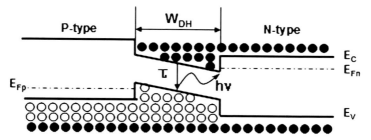

Fig. 9.4. Schematic diagram of heterojunction under forward bias

There is a report that a peak internal quantum efficiency of ∼60% at an operating wavelength of 460 nm has been achieved from the state-of-the-art blue LEDs [20]. In order to achieve the high internal quantum efficiency, various approaches are given. First, the formation of double-heterostructured active region is considered as an important factor. A DH is formed when two semiconductor materials, with one having a less energy gap, are joined together. This consists of the active region, in which the electrons and holes recombine and two confinement layers are cladding the active region. The semiconductor with the larger energy gap is used for the cladding layer while the one with smaller energy gap semiconductor forms the "filling" in this DH semiconductor sandwich. Figure 9.4 shows the schematic diagram of heterojunction under the forward bias. When a current is applied to the ends of the cladding-confinement-cladding structure, electrons and holes are confined to the active region with a smaller energy gap by the energy barrier higher than the thermal energy of carriers. Since the report of Nakamura et al. [21], which demonstrated candela-class high-brightness InGaN/AlGaN DH blue LEDs; all the high brightness LEDs are substantially adapting this double-heterostructured active region. Besides, the thickness of DH is also regarded to play an important role [22]. If the active region is thick, the carriers diffuse like that at the homojunction, so the heterostructure does not show any effect of carrier confinement. However, if the active region is too thin, the active region is apt to be overflown when the current injection is high. Therefore, the control of the thickness of the DH is one of the key factors for the high internal quantum efficiency.

Moderate doping is also regarded as an important factor affecting the efficiency of DH LEDs. The carrier lifetime is known to be dependent on the concentration of majority carrier concentration. In the low excitation regime, a lifetime of the radiative carrier decreases with increasing the free carrier concentration, that is influenced by the doping concentration. Therefore, the radiative efficiency increases. However, with too high concentration, the highly concentrated intentional dopants lead to an increased concentration of native defects that can be acted as nonradiative transition sources.

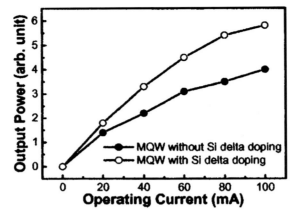

Fig. 9.5. Power–current (L–I) characteristic of a 385-nm UV LEDs with and without Si delta-doping. Reprinted with permission from [24]. Copyright (2005). American Institute of Physics

In addition, various doping techniques are investigated for the high internal quantum efficiency. In InGaN/GaN multiquantum-well (MQW) structures for ultraviolet (UV) LEDs, owing to the poor confinement of carriers by small band offset between the InGaN-well layer and the GaN barrier [23], it is necessary to improve the internal quantum efficiency of UV LEDs by increasing the carrier injection into the quantum-well layer and carrier confinement in the InGaN/GaN MQW structures. Kwon et al. reported that the PL intensity of a MQW and the output power of the UV LED were increased when a Si delta-doped barrier layer was employed [24]. The light-output power of UV LED with delta-doping method is 40% higher than that of the UV LED without the delta doping at an input current of 100 mA as shown in Fig. 9.5. This enhancement is thought to be due to the efficient injection of electrons from the Si delta-doped GaN barrier layer into the InGaN quantum-well layer and an increase in the hole confinement in the valence-band region.

For the high internal quantum efficiency, electron blocking layer (EBL) was developed. The EBL is a thin layer and a material with a wider bandgap than the material of the active region used for it. This layer is inserted between the active layer and the p-type layer of the device, the purpose of which is to prevent the injected electrons from overflowing the active region. Therefore, it forces the main recombinations to take place in the GaN n-type layer. Consequently, carrier recombinations are less affected by the nonradiative defects existing in the p-type layer. In the case of GaN-based LEDs, the EBLs are typically a 30–60 nm $Al_xGa_{1-x}N$ layers with the aluminum concentration of 0.2 [25, 26]. Dussaigne et al. inserted a 10 nm-thick Mg-doped AlGaN EBL between the n- and p-type GaN regions [25]. The EL spectrum of this p–n junction emitted at 368 nm with a full width at half-maximum (FWHM) of 19 nm, which was twice smaller than that of the reference junction without

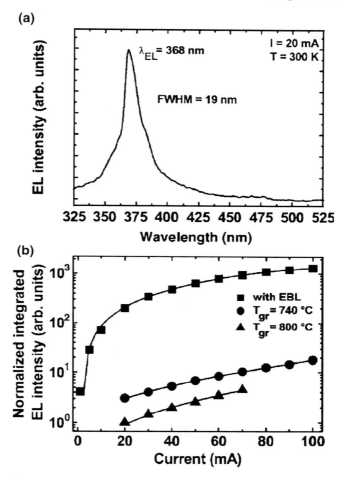

Fig. 9.6. (a) EL spectrum of p–n junction with EBL at 20 mA; (b) Integrated EL intensity as a function of the current for GaN homojunctions with a p-GaN layer grown at 800°C, 740°C, and 740°C with an $Al_{0.1}Ga_{0.9}N$ EBL. Reprinted with permission from [25]. Copyright (2008), American Institute of Physics

the EBL. The integrated EL intensities at 20 mA and at 100 mA were 200 times larger than those of the reference junction without the EBL (Fig. 9.6).

For the high efficient LEDs, high quality InGaN-based QWs structure and heterointerfaces between GaNs and InGaNs are of great importance. Because the optimum growth temperature of the GaN barrier layer is substantially higher than that of the InGaN well layer, during the ramping and barrier growth the InGaN well can be exposed to a higher temperature resulting in reevaporation of indium and interface degradation [27]. This problem becomes more severe when intending to increase the indium concentration in the InGaN well for the fabrication of GaN based LEDs with a longer wavelength. Therefore, some research groups introduced a thin well protection layer (WPL) into the MQW [27, 28]. Ju et al. [29] investigated the effects of the WPL on the emission properties of GaN based InGaN/GaN green MQWs. In order to increase the emission wavelength by preventing the volatilization of InGaN well, they grew a thin GaN WPL subsequently on each well layer at the InGaN growth temperature before ramping-up the temperature to the GaN barrier growth temperature. It was found that the WPL directly influenced the indium content and optical properties of the MQW. They also tried to embed a superlattice electron reservoir layer (ERL) composed of ten pairs of InGaN/GaN between the MQW and the n-GaN enhance the quantum efficiency by increasing the electron capture rate. The electroluminescence (EL) intensity from this device with the ERL was up to three times higher than that of the diode without the ERL. These results imply that the carrier capture by the MQW is significantly improved by the additional superlattice ERL, which consequently leads to the enhancement of the quantum efficiency.

Preventing the possibility of nonradiative recombination is also crucial for the internal quantum efficiency. Electron-hole pair can recombine at point defects, impurities, dislocations, and even defected surface such as mesa etched face. These are called to "Deep level carrier traps" existing at energies within the bandgap of semiconductor and lead to reduction of the light-emission efficiency. The InGaN active layer has a large number of threading dislocations, from 1×10^8 to 1×10^{12} cm^{-2}, originated from the interface between the GaN and sapphire substrate due to the large lattice mismatch of 15% and threaded into the InGaN layer. However, it was reported that III–V nitride-based LEDs are less sensitive to dislocations than conventional III–V semiconductors. Even though there is report that the GaN based materials show the less sensitivity to dislocations [30], it is evident that the low concentration of defects in the GaN crystalline gives a better performance because they could be the cause of nonradiative recombination sources. Therefore, the efforts to acquire good crystallinity of GaN are still proceeded. By the report of the epitaxial laterally overgrown GaN (ELOG) by Usui et al. [31], the dislocation density was reduced to $\sim 10^6$ cm^{-2}. With employing the ELOG technique, Mukai [32] reported that the UV GaN LED on ELOG has a much higher (approximately twofold) output power than that on sapphire. He explained that this is because of the less number of dislocations in a GaN LED on ELOG (Fig. 9.7). Therefore, for the high internal quantum efficiency, nonradiative recombination traps should be removed as possible as.

There are some reports that even the internal quantum efficiency shows a high value at low current density, but it decreases with increasing the drive current [33–37]. This is commonly referred to as efficiency droop. This phenomenon is regarded as a severe limitation for high power devices that operate at high current densities. For the explanation of this phenomenon, many researches are undertaking. Mukai et al. [33] explained the efficiency droop by the escape of more carriers from localized indium-rich states in the QW at

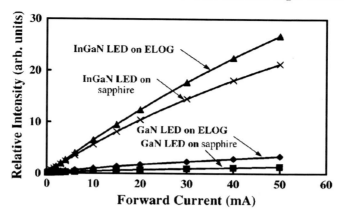

Fig. 9.7. Relative output power of UV InGaN and GaN LEDs as a function of forward current. Reprinted with permission from [32]. Copyright (2002), IEEE

high carrier density. Figure 9.8 shows the output power and external quantum efficiency of blue and green InGaN QW structure LEDs as a function of the forward current. At a current of about 0.1 mA, the carriers recombined radiatively from the localized energy state, therefore the efficiency of the LEDs was high at the low current. However, at a current of 20 mA, some carriers overflowed from the localized energy states and reached to the nonradiative recombination centers formed by a large number of dislocations.

Rozhansky et al. suggested that the efficiency droop is comes from a decrease in the carrier injection efficiency because of insufficient EBL effect caused by the large piezoelectric field [34]. As a solution for this problem, they proposed devices with p-type active region which shall increase the hole injection efficiency. Also, polarization fields were reported to enhance the leakage of injected electrons into the p-type GaN cladding layer, causing the efficiency droop. Kim et al. [35] and Schubert et al. [36] suggested quaternary AlInGaN barriers in place of the conventional GaN barriers. There is a positive sheet charge which can attract the electrons at the interface between the GaN barrier and AlGaN EBL in conventional LED. As a result, the conduction band of EBL can be pulled down, which reduces the effective barrier height for electrons. Figure 9.9 shows the difference of light-output power between the conventional MQW LEDs and the GaInN/AlGaInN LEDs as a function of forward current density up to 300 A cm^{-2} [36].

In addition, Auger recombination was reported to be a limiting factor for the quantum efficiency of LEDs at high current densities. Shen et al. investigated the cause of the drop in internal quantum efficiency with increasing excitation. Relatively thick InGaN layers, capped by GaN on GaN (0001)/sapphire template with various threading dislocation densities, were grown with different alloying composition and thickness of the layers [37]. According to the wavelength-selective steady-state photoluminescence (PL) technique for

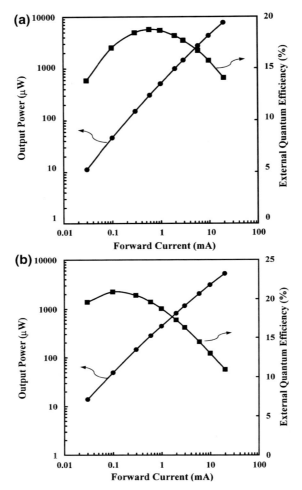

Fig. 9.8. Output power and the external quantum efficiency of (a) blue and (b) green InGaN QW structure LEDs as a function of the forward current. Reprinted with permission from [33]. Copyright (1999), The Institute of Pure and Applied Physics

quantifying the Auger recombination coefficient, they determined that the Auger coefficient of $In_xGa_{1-x}N$ (0001) in which x varies from 9% to 15% is in the range from 1.4×10^{-30} cm^6 s^{-1} to 2.0×10^{-30} cm^6 s^{-1}, and this value is large enough to decrease the internal quantum efficiency of InGaN/GaN QW LEDs.

As another approach for the improvement of internal quantum efficiency, the nonpolar GaN-based LEDs are introduced [38–47]. Because of the non-centrosymmetric nature of wurtzite GaN grown in the [0001] direction, a strong built-in electrostatic field could be formed by the spontaneous and

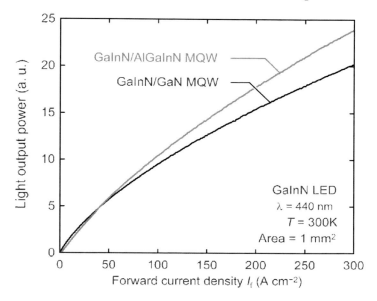

Fig. 9.9. Light-output power for GaInN/AlGaInN MQW LEDs and reference GaInN/GaN MQW LEDs as a function of forward current density. Reprinted with permission from [36]. Copyright (2008), American Institute of Physics

piezoelectric polarization. The manifestation of the built-in electric field includes a red shift of the PL peak because of the quantum-confined Stark effect (QCSE) and the reduction of the peak intensity as a result of the reduced spatial overlap of the electron and hole wave functions [41, 42]. The way to overcome this problem is to fabricate the GaN based LED having its growth axis different from the [0001] axis. In fact, it was reported that the heterostructures grown along directions perpendicular to the c-axis, such as [1$\bar{1}$00] (m-plane) [38, 39, 47] or [11$\bar{2}$0] (a-plane) [40, 43–46] were free of the internal electrostatic fields along the growth direction. Kuokstis et al. [39] compared the RT PL properties of GaN/AlGaN MQW grown with m-plane and those grown with c-plane. As shown in Fig. 9.10, the energy of the peak of the PL spectrum were blue-shifted with increasing excitation intensity for the c-plane oriented MQWs, while the peak positions for the m-plane MQWs remained unchanged. This blue shift agrees with calculations based on the built-in electric field. Theoretical calculations and comparison with the PL data confirm that the excitation-induced blue shift of the PL line is because of the screening of the built-in electric field by photoinjected carriers. Also, it was observed that PL intensity was much weaker for the c-plane MQWs compared to the m-plane MQWs at low excitation levels. These results indicate the existence of much stronger built-in electric fields in the c-plane quantum structures in comparison with the m-plane MQWs. Similarly, Ng reported the results about nonpolar a-plane GaN [40]. Figure 9.11 shows a

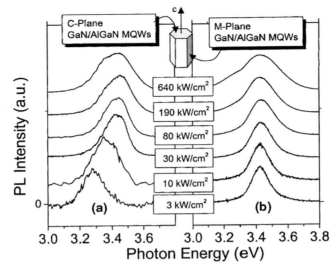

Fig. 9.10. RT PL spectra for C-plane (**a**) and M-plane (**b**) GaN/AlGaN MQWs as a function of ArF excimer laser excitation power density. Reprinted with permission from [39]. Copyright (2002), American Institute of Physics

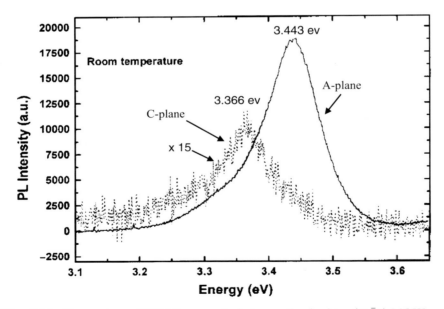

Fig. 9.11. Comparison of RT PL property between the A-plane (11$\bar{2}$0) MQW and c-plane (0001) MQW. Reprinted with permission from [40]. Copyright (2002), American Institute of Physics

comparison of the PL properties at RT between a-plane ($11\bar{2}0$) GaN/AlGaN MQWs on r-plane ($10\bar{1}2$) Al$_2$O$_3$ substrates and c-plane GaN/AlGaN MQWs on c-plane (0001) Al$_2$O$_3$ substrates [40]. Here, the peak intensity for the c-plane MQW was lower by about a factor of 30 than that of the a-plane MQW. This could be attributed to the effect of the built-in electric field causing a spatial separation of the electron and hole wave functions and reducing the radiative transition probability. Craven et al. [46] reported that a-plane ($11\bar{2}0$) GaN/AlGaN MQW structure grown on r-plane ($1\bar{1}02$) Al$_2$O$_3$ substrates by metalorganic chemical vapor deposition (MOCVD). They demonstrated that the nonpolar MQW did not experience the QCSE using RT PL while c-plane MQWs grown exhibited characteristics affected by QCSE.

For the devices, Chen et al. [45] fabricated the nonpolar UV LED using a-plane GaN/AlGaN MQWs over r-plane sapphire. Figure 9.12a shows the schematic illustration of the nonpolar a-plane GaN/AlGaN LED structure. All the layers were grown on r-plane sapphire substrate using low-pressure MOCVD at 76 Torr. The I–V characteristic of the a-plane MQW LED structure is shown in the inset of Fig. 9.12b. The device turned on at about 4 V, and showed a series resistance of about 65 Ω. Figure 9.12b shows that the peak position at 363 nm does not shift with the pulsed current, which establishes a clear absence of the polarization fields for the quantum well band-edge emission.

Chakraborty et al. [47] reported the nonpolar m-plane InGaN/GaN MQW LEDs. The nonpolar LED structure was grown by MOCVD on 250-μm-thick free-standing m-plane GaN substrates. This LED device consisted of a 2.2 μm Si-doped n-type GaN based layer with an electron concentration of 2×10^{18} cm^{-3}, followed by the active region, which consisted of a five period MQW stack with 16 nm Si-doped GaN barriers and 4 nm In$_{0.17}$Ga$_{0.83}$N quantum wells. A 16 nm undoped GaN barrier separated the active region from the 0.3 μm Mg-doped p-type GaN layer with a hole concentration of 6×10^{17} cm^{-3}. A 40 nm p$^+$-GaN layer was used to cap the structure for improved electrical contact to the p-GaN layer (Fig. 9.13a). The minimal blue shift at high drive currents shown in the EL spectra of the devices under DC drive current (Fig. 9.13b) indicates the absence of polarization-induced electric fields in the M-plane quantum wells.

In addition, Koida et al. [45] applied the lateral epitaxial overgrowth (LEO) method to the nonpolar growth. They showed improvements in the optical properties of ($11\bar{2}0$) A-plane AlGaN/GaN MQWs. The improved internal quantum efficiency was thought to be due to the tremendous reduction in the densities of nonradiative defects and the bound states by the use of GaN templates prepared by LEO.

9.3.2 Issues for High External Quantum Efficiency

Another important parameter of the LEDs is the external quantum efficiency [48]. It is the fraction of generated photons that are actually emitted from the semiconductor and defined as

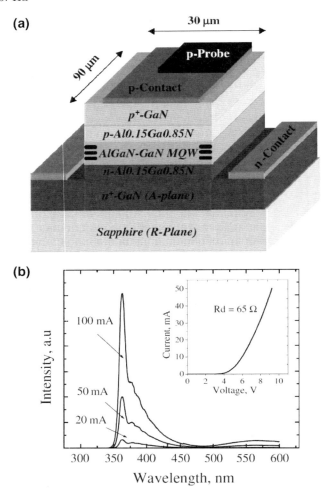

Fig. 9.12. (a) Schematic illustration of nonpolar a-plane UV LED on R-plane sapphire and (b) LED emission spectra as a function of pulsed pump current. Inset shows I–V curve for a $30 \times 90\,\mu\text{m}^2$ device. Reprinted with permission from [45]. Copyright (2003), The Institute of Pure and Applied Physics

$$\eta_{\text{ext}} = \frac{\#\ \text{of photons emitted into free space per second}}{\#\ \text{of electrons injected into LED per second}} = \frac{P/(h\upsilon)}{I/e}, \quad (9.9)$$

where P is the optical power emitted into free space. If the extraction efficiency is defined as

$$\eta_{\text{extraction}} = \frac{\#\ \text{of photons emitted into free space per second}}{\#\ \text{of photons emitted from active region per second}} = \frac{P/(h\upsilon)}{P_{\text{int}}/h\upsilon}, \quad (9.10)$$

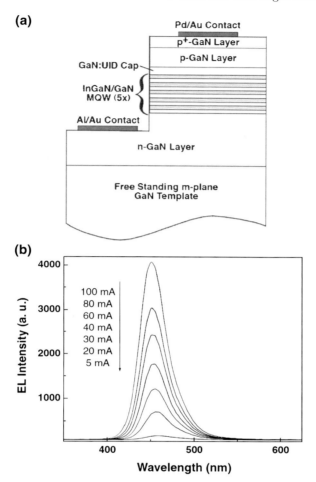

Fig. 9.13. (a) Schematic illustration of nonpolar m-plane InGaN/GaN LED on free-standing m-plane GaN template and (b) EL spectra as a function of the drive current. Reprinted with permission from [47]. Copyright (2005), The Institute of Pure and Applied Physics

the external quantum efficiency is expressed as

$$\eta_{\text{ext}} = \eta_{\text{int}} \eta_{\text{extraction}}. \quad (9.11)$$

The external quantum efficiency is normally a much smaller number than that of the internal quantum efficiency. Once a photon has been produced in the semiconductor, three loss mechanisms that the photons may encounter: photon absorption within the semiconductor, Fresnel Loss, and Snell's law.

If the substrate absorbs the generated light, photons can be emitted into any direction. Since the emitted photon energy must satisfy the condition

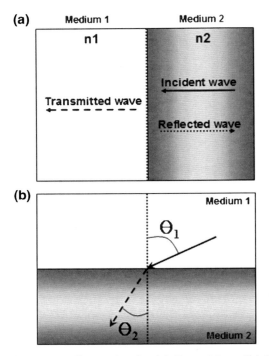

Fig. 9.14. Schematic illustration for (**a**) Fresnel loss; (**b**) Snell's law

$h\nu \geq E_g$, the emitted photons can be reabsorbed within the semiconductor material. The majority of photons will actually be emitted away from the surface and reabsorbed in the semiconductor. Also, when the photons collide with metallic contacts, they could be absorbed by metals.

Photons must be emitted from the semiconductor into air. Thus, the photons must be transmitted across a dielectric interface. Figure 9.14a shows the incident, reflected, and transmitted waves. If the index of refraction for air is n_1 and the index of refraction for semiconductor is n_2, the reflection coefficient (Γ) is defined by

$$\Gamma = \left(\frac{n_2 - n_1}{n_2 + n_1}\right)^2. \tag{9.12}$$

This effect is called *Fresnel loss* [49]. The reflection coefficient Γ is the fraction of incident photons that are reflected back into the semiconductor. For example, if the index of refraction for GaN $n_2 = 2.45$ and for air $n_1 = 1$, the reflection coefficient will be 0.42. The reflection coefficient $\Gamma = 0.42$ means that 42% of the photons incident from the GaN on the GaN/air interface are reflected back into the LED. These reflected photons can be absorbed in the semiconductor again.

When the photons are incident on the interface between two media that have refractive indices n_1 and n_2, it will be transmitted and reflected. Refraction and transmission angles are governed by the Snell's law. Equation (9.13) shows the Snell's law which states that the ratio of the sines of the angles of incidence and refraction is equivalent to the ratio of velocities in the two media, or equivalent to the reverse ratio of the indices of refraction (Fig. 9.14b).

$$\frac{\sin\theta_1}{\sin\theta_2} = \frac{v_1}{v_2} = \frac{n_2}{n_1}, \quad (9.13)$$

where θ_1 is incident angle, θ_2 is refracted angle, v_1 is incident velocity of wave, v_2 is refracted velocity of wave, n_1 is refractive index of medium 1, and n_2 is refractive index of medium 2.

In conventional LEDs, the external efficiency is limited by the total internal refraction at the semiconductor–air interface because of the different refractive indices between the semiconductors. Thus, it is difficult for the internal light to escape from the semiconductor into air. In order to improve the external quantum efficiency, various researches are proceeded employing materials such as transparent conductive oxide (TCO) for high refractive index p-contact, surface roughening of p-contact layer, and photonic crystal. Conventionally, GaN-based LEDs were fabricated by using semitransparent p-type electrodes such as Ni/Au, Pt, and Pd because these materials have large work functions for the ohmic contact with the p-type GaN. However, these materials highly absorb the light resulting in low light extraction efficiency. To improve the light extraction efficiency, the transparent conducting oxides, such as indium tin oxide (ITO) [50,51] and Cu-doped indium oxide [52], have been developed as the ohmic metals for the p-GaN. When the ITO is used for this purpose, it also increases the light escape angle, which results in much light output from the GaN epi layer. As shown in Fig. 9.15, the escape angle increased from 24° to 29° by forming the air/ITO/GaN structure compared to the case for the structure of air/GaN because the ITO has a high refractive index of 2.06. Kim et al. [50] and Margalith et al. [51] also reported that the ITO metals are effective in increasing the light output of LEDs because of their higher transmittance characteristic.

However, in case of directly employing ITO as a p-type GaN contact layer, a significantly large operation voltage in the range of $\sim 10^{-1}\Omega$ cm^2 is required [50]. Thus, it is also necessary to reduce the p-contact resistance. Sheu et al. [53] and Kuo et al. [54] introduced an n$^+$-InGaN/GaN short period superlattice (SPS) tunneling contact layer on the top of p-GaN layer to reduce the p-contact resistance. By employing this SPS structure, it becomes possible to achieve a good ohmic contact through tunneling when the n$^+$ (InGaN/GaN)–p(GaN) junction was properly reverse-biased.

Another method to increase the light extraction efficiency in LEDs is introducing the surface roughening or surface texturing. It has been reported that roughening the LED surface is very effective in increasing the light extraction efficiency of nitride-based LEDs [55–59] near to $\sim$30%. However, since these

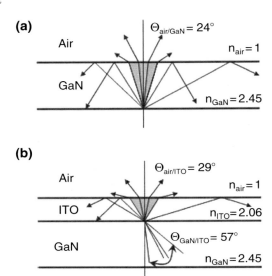

Fig. 9.15. Escape path of light (**a**) in air/GaN; (**b**) in air/ITO/GaN

methods generally need precise photographic techniques and dry *etch*ing, the surface roughening techniques have some problems, including surface damages induced by plasma ions, destruction of a large part of the junction, and reductions in light generation [60]. Consequently, it is not desirable to do direct *etch*ing of the p-GaN layer. To avoid damaging of the thin p-GaN layer, transparent contact layers such as the ITO layer on p-GaN layer, have been used as the roughened layer instead of the p-GaN layer. Horng et al. reported that the surface-textured ITO increased the output power of LED by about 28% at 20 mA compared to the LED with the planar surface [61, 62]. Figure 9.16a shows the schematic illustration of the GaN LED structure with the textured ITO surface [62]. The output powers for the LEDs with and without the textured ITO surface are shown in Fig. 9.16b [62].

In order to increase much more the light extraction efficiency, photonic crystals (PCs) have been employed [63–68]. The PCs are periodic dielectric material arrays that can be characterized by photonic bandgaps (PBGs) or electromagnetic stop bands [69, 70]. Progress has been achieved in fabricating the photonic crystal LEDs (PC–LEDs) with emissions ranging from the ultraviolet to the infrared. For the fabrication of photonic crystal pattern, the e-beam lithography method is used in general. By using the e-beam lithography, various kinds of patterns as small as a few tens of nanometer can be formed. Boroditsky et al. described a thin-slab LED design, which uses a highly efficient coherent external scattering of trapped light by employing the two-dimensional (2D) photonic crystal formed through

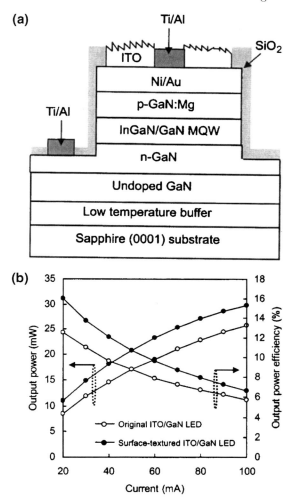

Fig. 9.16. (a) Schematic diagram of GaN LED structure with a textured ITO Surface; (b) Output power and external quantum efficiency as functions of injection current. Reprinted with permission from [62]. Copyright (2005), American Institute of Physics

the e-beam lithography. Here, the light-generating region was an unpatterned heterostructure surrounded by the light-extraction region. Consequently, a sixfold enhancement of PL was observed compared to the unpatterned thin-film LED and this corresponded to improvement by 70% in the external quantum efficiency [63]. However, it should be pointed out that the e-beam lithography technique has a very low throughput and hard to be applied for the mass production of LEDs.

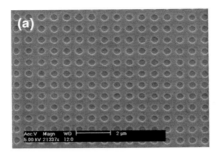
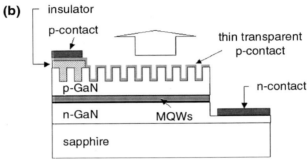

Fig. 9.17. (a) SEM image for a PCLED device surface. The square-lattice air-hole array pattern was generated by the holographic double-exposure method. The lattice period of this specific example is 700 nm. (b) Schematic illustration of the PC–LED. Reprinted with permission from [65]. Copyright (2005), American Institute of Physics

Kim et al. reported an enhancement in the light output from the GaN-based LEDs by 2D square-lattice air-hole array patterns with a period that varied from 300 to 700 nm using laser-holography method [65]. Differently from the e-beam lithography technique, the holographic method can make patterns over a large area with relatively high throughput. Figure 9.17 shows the SEM image and schematic illustration of PC LED by laser holographic method [65]. Figure 9.18 shows the images of the devices during LED operation taken at the same injection current of 1 mA [65]. Here, we can see that compared with the reference device (Fig. 9.18a) the PC LED with a period of 500 nm (Fig. 9.18c) has the strongly enhanced light output.

On the other hand, nano-imprint lithography (NIL) method is also used to fabricate nanosized patterns for the PCs [67, 68]. The NIL technique has the capabilities of high-resolution patterning and high-throughput patterning even with low cost. Byeon et al. reported the fabrication process of PCs in a p-GaN layer by using the NIL accomplished with the inductively coupled plasma (ICP) etching process. As shown in Fig. 9.19, the p-GaN layer was well patterned by this technique and it was confirmed that the PL intensity from the p-GaN patterned LED was significantly enhanced as compared to the unpatterned LED [67].

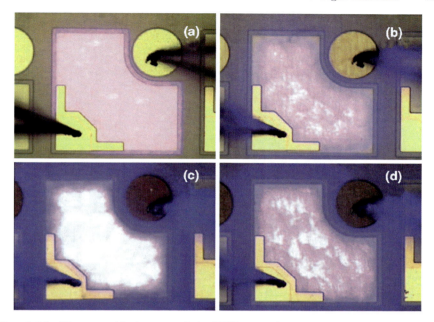

Fig. 9.18. Photographic images of the LEDs at a low injection current of 1 mA, (**a**) reference planar device; (**b**) 300 nm; (**c**) 500 nm, and (**d**) 700 nm. Reprinted with permission from [65]. Copyright (2005), American Institute of Physics

9.3.3 Packaging

The fabricated LED chips are mounted in a package that mostly consists of two electrical leads, a transparent optical window for the escape of light to escape, and a thermal path for heat dissipation. Typical package for low-power devices is shown in Fig. 9.20. The active device is die-bonded or soldered to the bottom of a cup-like depression ("reflector cup") with one of the lead wires (usually the cathode lead). A bond wire connects the LED top contact to the other lead wire (usually the anode lead). This LED package is frequently referred to as the "5 mm" or the "T1-3/4" package [71]. This package consists of the encapsulant with a hemisphere shape to maintain the angle of incidence to the encapsulant–air interface normal.

The standard 5 mm LED package was originally designed to be used in indicator applications. Therefore, it is not suitable to get sufficient heat transfer from the LED chip, which is necessary to keep the LED cool during the operation. Thus, new packages for high-power LEDs have been designed for LEDs of illumination applications. The main idea in the packages for LEDs of illumination applications is to make the heat-removing path as large as in surface and as short as in length. For these purposes, the flip-chip technology has been adapted. The flip-chip package is based on the technology employing the "epi-side down" different from the conventional packaging technique with the

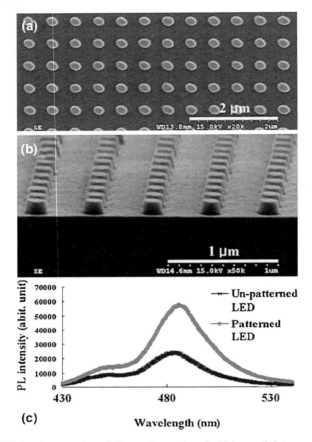

Fig. 9.19. SEM micrographs of the patterned p-GaN layer (a) top and (b) tilted views, (c) PL spectrum of unpatterned and patterned LED samples at room temperature. Reprinted with permission from [67]. Copyright (2007), American Institute of Physics

"epi-side up." This technology has been widely used in the fields of Si-based semiconductor fabrication because it offers a number of advantages such as a fast signal exchange through its short contact distance and a thermal management through its wide solder bump bonding area. In the applications to optical devices, the flip-chip packaging provides the other advantages. Firstly it reduces obscuration of light by removing the contacts from the light-emitting surface. In addition, the contact for cathode electrode in flip-chip packaging acts as high-optical-reflecting medium. The thick and opaque metallic contacts make it possible that the devices are operatable at a current density two times more than that of the conventional LEDs. Also, since the Si wafer is used for the substrate, electrostatic discharge (ESD) protective circuitry can be easily embedded in the substrate. Moreover, owing to the direct heat-sunk system,

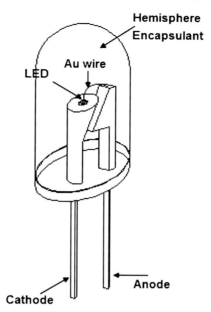

Fig. 9.20. Schematic illustration of LED indicator

it has a strong point in the view of the thermal management. The "Luxeon K2 lamps," as shown in Fig. 9.21, is regarded as a typical LED packaging using the flip-chip technology showing impressive performances including the highest operating junction temperature 185°C, the high-drive current of 1,500 mA, and the low thermal resistance of 9°C/W [72].

The thermal resistance of LED becomes one of the important factors in packaging. Because the high-power LEDs are used in the applications such as automotive headlights, interior lighting, *etc.* that need extremely high brightness or high output, if the heat management is not seriously considered, the lifetime of LED will be significantly decreased. Therefore, in the packaging, the junction temperature representing a temperature at the p–n junction should be maintained cool during the operation. To maintain satisfactory junction temperature as the reliable operating condition, the thermal resistance should be controlled to remove the heat well from active region through all of the materials in the package. The thermal resistance (R_{th}) means the temperature (°C) change per unit power (W). Therefore, as the thermal resistance increases the junction temperature will increase as shown in (9.14)

$$\frac{dQ_{heat}}{dt} = -\kappa A \frac{dT}{dx} = \frac{\Delta T}{R_{th}}, \qquad (9.14)$$

where t is time, and Q is heat, A is area, T is temperature, and k is thermal conductivity.

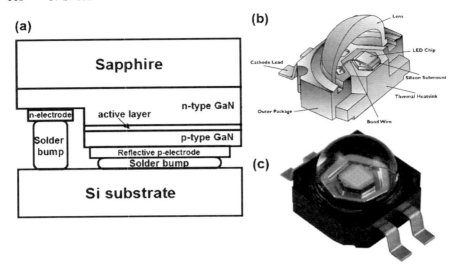

Fig. 9.21. (a) Schematic diagrams of flip-chip technologies; (b) Cross section through high-power package using this technology (Lumileds); (c) Photograph of Philips Lumileds LUXEON® K2. Reprinted with permission from [72]. Copyright (2007), Philips Lumileds Lighting Company

In addition, since the sapphire substrate is electrically insulating, the ESD-induced electrical failure becomes one of the main reliability concerns in optoelectronic devices employing the sapphire substrates. To solve this problem, the LED was electrically connected to the Si Zener diodes through the flip-chip process [73]. By using this technique, significant reduction of the potential damage of GaN LEDs is possible since the ESD-induced pulse current will flow through the underneath Si Zener diodes. Another possible way to solve this problem is to build an internal GaN Schottky diode inside the LED chip [74]. If the GaN Schottky diode is electrically in parallel with the GaN LED itself, ESD-induced pulse current will then flow through the internal GaN Schottky diode without making damages on the LED.

9.3.4 Vertical Light-Emitting Diode

The most essential factor for the high-power LED is thought to be the capability of high current injection to the LED device. For the high current injection, many researches have been conducted with the goal of removing sapphire substrates, which have poor thermal and electrical conductivities. The poor thermal and electrical conductivities cause an increase in the junction temperature in the case of the high current injection, which results in a decrease of a luminous efficacy by 5% for every increase of 10°C. Furthermore, it can cause destruction of the LED chips by ESD [75]. Therefore, the replacement of sapphire substrate by the one with good thermal and electrical conductivities will improve both the heat extraction from the active region of GaN-based devices and the ESD problem.

On the other hand, if we can fabricate a vertical current-flow device with a top n-contact and a bottom p-contact after removing the sapphire substrate, the performance at the LED chips will be improved.

In order to improve the LED performance by considering these factors, the fabrication of vertical-type free-standing GaN LED without the sapphire substrate have been developed by using the laser lift-off (LLO) process [76–79], which separates the GaN LED from the sapphire substrate by using irradiation of high energy laser to the interface between sapphire and GaN LED structure. The irradiation of high-energy laser causes decomposition of GaN to liquid gallium and gaseous nitrogen resulting in the separation of GaN LED structure from the sapphire substrate.

However, it has been reported that the LLO process induces several detrimental things including generation of dislocations in the GaN layer [80], damages by laser [81], and change of the crystal structure of GaN by the abrupt heat generation at the GaN/sapphire interface [82]. Nevertheless, the high-performance vertical LEDs based on the LLO technique have been reported. Tran et al. reported the characteristics of GaN-based vertical LEDs on metal alloyed substrate by using the LLO process [83]. The vertical LEDs showed a very good I–V behavior with low serial dynamic resistance of 0.7 V and low operated voltage of 3.2 V at 350 mA. High-current operation up to 3 A in continuous light-emitting mode was demonstrated without any performance deterioration. The high thermal conductivity of metal alloyed substrate exhibits excellent heat dissipation capability. A light-output efficiency of 70 lumens/W or better was achieved in the single chip or the multiple chips package. By coupling good reliability and mass production ability, the vertical LEDs are very suitable for general lighting application. Figure 9.22 shows the schematic illustrations of conventional GaN LED and vertical-type LED. In the case of the conventional GaN LED on sapphire (Fig. 9.20a), the p- and n-electrodes are fabricated on the same side because of the sapphire substrate. Therefore, about 20–30% of emission area is sacrificed to form the n-electrode. In addition, the current transport from an anode to a cathode is lateral along the n-GaN layer. Therefore, the current crowding effect may be occurring which results in high serial dynamic resistance. Table 9.1 summarizes typical features of conventional-type and vertical-type LEDs.

Figure 9.23a shows the I–V characteristics for the vertical LEDs on metal alloyed Substrate (VLEDMS), and the conventional LEDs on sapphire [83]. The voltages at 350 mA were about 3.4 V for the conventional LEDs on sapphire and 3.2 V for the VLEDMS. From the slope of I–V curve, the dynamic serial resistances were 0.7 Ω for the VLEDMS and 1.1 Ω for the conventional LEDs on sapphire. The higher dynamic serial resistances of the conventional LEDs on sapphire were due to the lateral current path and the current crowded effect on the bottom of n-electrode as mentioned before. The lower operation

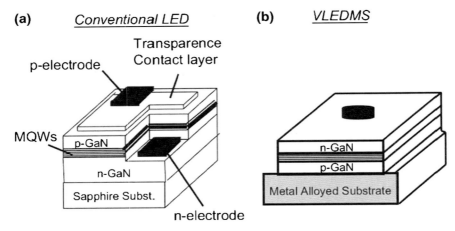

Fig. 9.22. Schematic diagram and comparison of (**a**) GaN LEDs on sapphire substrate and (**b**) GaN vertical light-emitting diodes on metal alloyed substrate (VLEDMS). Reprinted with permission from [83]. Copyright (2007), Elsevier

Table 9.1. Typical features of conventional-type and vertical-type LEDs

	Conventional LED	Vertical LED
Light-emission Area	Loss of Mesa etched area	Emission from whole chip surface Increase of no. of chip per wafer
ESD/Voltage	Local current concentration Low reliability	Uniform current spreading Improved ESD
Wire bonding	2 Bonding	1 Bonding
Substrate	Sapphire substrate – Thermal, electrical insulator	Metal substrate – High thermal and electrical conductivities – Sapphire reusable

voltage and the lower dynamic serial resistance of the VLEDMS provided the better light-output efficiency and operation performance comparing with the conventional GaN LEDs on sapphire substrate. Figure 9.23b shows the comparison of the light-output power–current (L–I) characteristics for the VLEDMS and the conventional LED on sapphire with the same chip size. The light-output power form the conventional LED peaked at around 1,000 mA and then dropped down fastly over 1,000 mA. The poor heat dissipation of the sapphire was mostly the main reason for the degradation. In contrast, the light-output power from the VLEDMS was increased almost linearly up to the current of 3,000 mA.

On the other hand, differently from the LLO technique, Ha et al. reported the fabrication of vertical-type GaN-based LED by employing the so-called chemical lift-off (CLO) process [84]. The CLO process detaches the GaN LEDs from the sapphire substrate by wet chemical etching solution. Here, the CrN

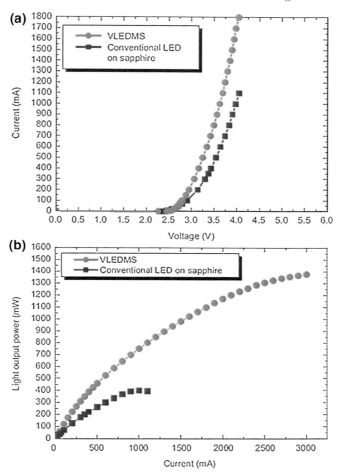

Fig. 9.23. (**a**) I–V characteristics for the vertical light-emitting diodes on metal alloyed substrate (VLEDMS) and the conventional LEDs on sapphire; (**b**) Comparison of the L–I characteristics for the vertical light-emitting diodes on metal alloyed substrate and the conventional LEDs on sapphire. Reprinted with permission from [83]. Copyright (2007), Elsevier

metallic buffer layer in between the sapphire substrate and the GaN layer was selectively *etch*ed. The CrN metallic layer worked as a buffer layer to grow high-quality GaN layers and simultaneously as a sacrificial *etch*ing layer that was *etch*ed out as shown in Fig. 9.24 [84]. The GaN-based vertical LED employing the CLO technique showed the equivalent I–V performance with the low series resistance of 0.65 Ω and the low operated voltage of 3.11 V at 350 mA to those of the LEDs by laser lift-off process [83]. In fact, the LED

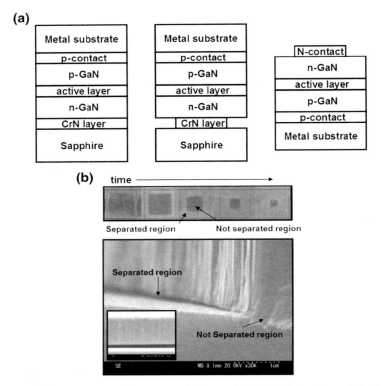

Fig. 9.24. (a) Schematic diagrams of Vertical LED fabrication; (b) Optical-microscope image of GaN top surface. The CrN layers shown as *dark area* was in the procedure of chemical *etching*. SEM cross-sectional image of chemical lifted-off region. Reprinted with permission from [84]. Copyright (2008), IEEE

operated at a much higher injection forward current of 1,118 mA at 3.70 V. Figure 9.25 shows the I–V characteristics from the GaN-based vertical LED employing the CLO technique [84].

9.4 Light-Emitting Diodes Based on ZnO

9.4.1 Hybrid LED

Because of the difficulty in getting a reliable p-type ZnO layer, the hybrid LED structure, that is, employing a p-type layer based on material different from ZnO has been reported. For this application, several kinds of p-type materials, including p-type TCO [85, 86], p-GaN [87, 88], p-AlGaN [89, 90], and p-SiC [91, 92], were adopted.

In 2000, Ohta et al. attempted to fabricate the UV−ZnO LED by using the p-$SrCu_2O_2$ as a p-type conductor [85]. Among the various TCO

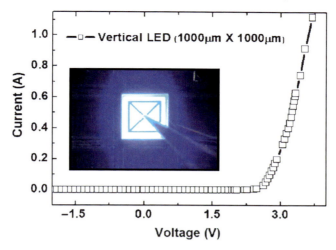

Fig. 9.25. *I–V* characteristics of the GaN-based vertical LED employing the CLO technique. Reprinted with permission from [84]. Copyright (2008), IEEE

candidates, $SrCu_2O_2$ was selected because it could be deposited at low temperature (350°C) and hence could minimize the chemical reactions at the $SrCu_2O_2$/ZnO interface. On the Y_2O_3-stabilized ZrO_2 (YSZ) single crystal (111) substrate, they deposited the ITO n-type electrode, n-ZnO layer and p-type $SrCu_2O_2$ layer by pulsed laser deposition (PLD) method sequentially. In order to get the p-type conductivity, K^+ ions were doped into Sr^{2+} sites in $SrCu_2O_2$. This device showed the emission band at 382 nm which could be assigned to the transition associated with electron-hole plasma in ZnO. The turn-on voltage was determined to be 3 V and the external efficiency was estimated to be less than 0.001%. Such low device performance was thought to be originated from the large lattice misfit existing at the heterointerface between ZnO and p-type material.

Since ZnO and GaN are materials similar in many of their physical properties [93] such as crystal structure and in-plane lattice parameter, p-type GaN layer had been employed for the substitute for p-type ZnO. Alivov et al. reported the growth, fabrication, and device characterization of heterojunction LEDs based on the n-ZnO/p-GaN material system [87]. The layer structure was achieved by first growing a Mg-doped GaN film of thickness 1 μm on Al_2O_3(0001) by molecular-beam epitaxy (MBE), then by growing a Ga-doped ZnO film of thickness 1 μm by chemical vapor deposition (CVD) on the p-GaN layer. From the *I–V* curve, rectifying diode characteristics was shown to be with the ∼3 V of forward bias. The EL spectra of the forward biased LED consist of a broad emission band with the maximum wavelength at about 430 nm, which lies in the violet region. From the correlation of this wavelength with CL spectra (Fig. 9.26), the emission was inferred that it was due to the radiative recombination in the GaN.

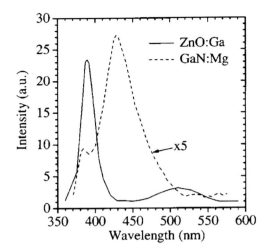

Fig. 9.26. RT CL spectra of (a) ZnO:Ga and (b) GaN:Mg films. Reprinted with permission from [87]. Copyright (2003), American Institute of Physics

Alivov et al. also proposed the n-ZnO/p-AlGaN structure as shown in Fig. 9.27a [88]. They fabricated the n-ZnO/p-AlGaN heterojunction LEDs, where the p-AlGaN layer was grown by hydride vapor phase epitaxy and the n-ZnO layer were grown by CVD on 6H-SiC substrate. Much better rectifying I–V characteristics than the former p-GaN case with the threshold voltage of $\sim$3.2 V and the low reverse leakage current of $\sim 10^{-7}$ A were observed at RT as shown in Fig. 9.27b. Under the forward bias, the n-ZnO/p-AlGaN device produced UV EL with a peak emission near 389 nm wavelength. By comparing this wavelength with the CL spectra in Fig. 9.27c, it could be concluded that the EL emission had emerged from the ZnO region contrary to the former case of n-type ZnO/p-type GaN LEDs, and the predominant device current is the hole injection from p-type AlGaN into the n-type ZnO region of the heterojunction. These differences between the LEDs with the p-GaN and the p-AlGaN could be explained by band structure. They suggested that the electron injection from the n-ZnO into the p-GaN should be more energetically favorable than the hole injection from the p-GaN into n-ZnO, while the hole injection from the p-AlGaN into the n-ZnO would be more favorable than the electron injection from the n-ZnO into the p-AlGaN.

Rogers et al. fabricated a n-ZnO/p-GaN:Mg heterostructure on c-Al$_2$O$_3$ substrate using PLD for the ZnO and MOCVD for the GaN:Mg [89]. Figure 9.28 shows the schematic illustration of the LED structure and I–V characteristics from n-ZnO/p-GaN:Mg heterojunction LED. The LEDs showed I–V characteristics confirming a rectifying diode behavior. The RT PL showed an intense main peak at 375 nm and RT EL measurement showed the EL peak at about 375 nm. From the correlation between the wavelength maxima for the EL and PL, it was suggested that the recombinations occurred in the

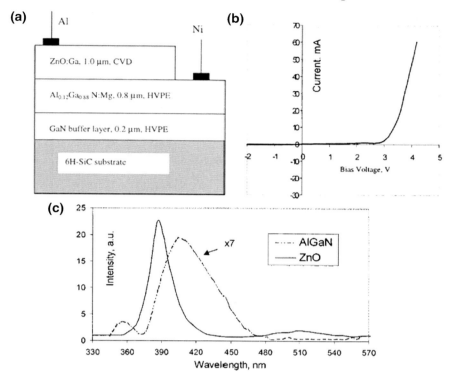

Fig. 9.27. (a) Schematic diagram of the n-ZnO/p-AlGaN heterojunction LED structure; (b) RT I–V characteristics of the-ZnO/p-AlGaN structure; (c) RT CL spectra of ZnO and AlGaN layers. Reprinted with permission from [88]. Copyright (2003), American Institute of Physics

ZnO layer through the significant hole injection from the p-type GaN:Mg into the n-type ZnO. This, in turn, indicates that it is possible to obtain the hole injection from the GaN:Mg into the intrinsically doped ZnO.

On the other hand, differently from other researches [87–89] which have employed Ga-doped ZnO film as an electron injection layer, Yuen et al. reported the low-temperature ($\sim$150°C) fabrication of n-ZnO:Al/p-SiC(4H) heterojunction LEDs as schematically shown in Fig. 9.29a [91]. The employment of n-type ZnO:Al film has some advantages in view of the fact that the low cost fabrication on n-doped ZnO films at a low substrate temperature ($<$150°C) can be achieved by using Al as the dopant [94]. The achieved resistivity ($<8 \times 10^{-4}$ Ω cm) and carrier concentration ($>10^{21}$ cm^{-3}) from the n-ZnO:Al films are as good as those of the n-ZnO:Ga films [94, 95]. This n-ZnO:Al/p-SiC(4H) heterojunction LEDs showed the diode-like rectifying I–V characteristics as shown in Fig. 9.29b [91].

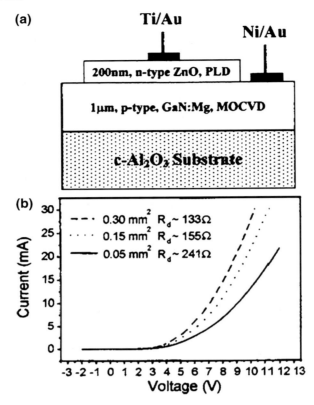

Fig. 9.28. (a) Cross-sectional schematic of the n-ZnO/p-GaN:Mg heterojunction LED mesa structure; (b) RT I–V characteristics for the n-ZnO/p-GaN:Mg mesa structures. Differential resistance R_d for each mesa area is also indicated on the graph. Reprinted with permission from [89]. Copyright (2006), American Institute of Physics

ZnO-based heterostructures are potential candidates for applications in ultraviolet–visible range, over the GaN-based material system [96]. Full-color EL from ZnO-based heterojunction diodes was reported by Nakamura et al. [92]. They demonstrated the red, the green, and the blue EL by using the ZnO-based heterojunction diodes consisting of n-ZnO/n-$Mg_yZn_{1-y}O$/$Zn_{1-x}Cd_xO$/p-SiC layers, which was grown by remote-plasma-enhanced MOCVD. Here, the emission color could be controlled by changing the cadmium content in the emission layer. Figure 9.30a shows a schematic diagram of the ZnO-based heterojunction LED. The normalized EL spectra of RGB diodes at RT are shown in Fig. 9.30b [92].

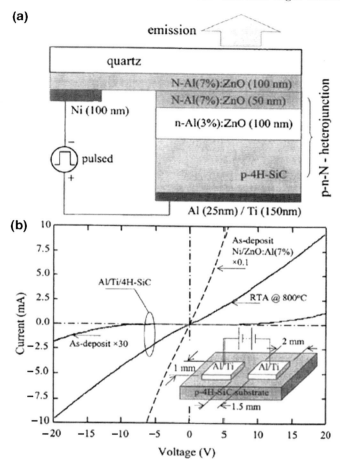

Fig. 9.29. (a) Schematic diagram of the n-ZnO:Al/p-SiC(4H) heterojunction LED structure; (b) Solid lines: I–V characteristics of the Al/Ti metallization on p-SiC before and after rapid thermal annealing at 800°C for 5 min. Dashed line: I–V characteristics of the Ni metallization on n-ZnO:Al (7%). Reprinted with permission from [91]. Copyright (2005), American Institute of Physics

9.4.2 ZnO LEDs with Homo p–n Junction

The fabrication of homo p–n junction ZnO LEDs have been reported by several researches [13–15, 97]. In 2000, Aoki et al. fabricated the ZnO diode by irradiation of excimer laser on the Zn_3P_2 layer to form a p-type ZnO layer on an n-type ZnO substrate [13]. After the growth of single-crystal ZnO wafers by hydrothermal method, a 35-nm-thick zinc phosphide (Zn_3P_2) film was deposited on the ZnO wafer as a phosphorous (P) source by evaporation method. After that, the KrF excimer laser with a wavelength of 248 nm, a pulse width of

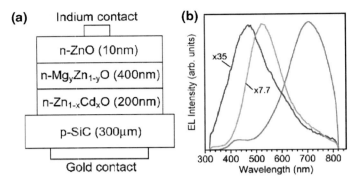

Fig. 9.30. (a) Cross-sectional view shows the structure of heterojunction grown on p-SiC substrate; (b) Normalized electroluminescence spectra of RGB diodes at RT. Reprinted with permission from [92]. Copyright (2007), American Institute of Physics

20 ns, and a power density of 150 mJ cm^{-2} was irradiated for the p-type doping in the reacting chamber containing either nitrogen or oxygen at a pressure of 4 atm. In this structure, I–V characteristics showed the diode characteristics, and white–violet EL was observed at 110 K. The spectrum showed a peak at about 370–380 nm, attributed to the band-edge emission, and a broad peak in the region of 400–500 nm because of the defect states.

Guo et al. fabricated the homojunction LED by depositing a relatively high resistivity p(i)-ZnO layer on an n-ZnO single-crystal wafer using the technique of N$_2$O plasma-enhanced PLD [14]. This ZnO LED showed nonlinear and rectifying I–V characteristics and current injection emission with bluish white light output.

Tsukazaki et al. reported that a near-band-edge bluish EL band centered at around 440 nm was observed from their ZnO p–i–n homojunction diodes, in which a semitransparent electrode was deposited on the p-type ZnO top layer [15]. The ZnO p–i–n homojunction structure was grown on lattice-matched (0001) ScAlMgO$_4$ (SCAM) substrate by laser MBE in 1×10^{-6} Torr of oxygen. Figure 9.31 shows optical microscope images of the operating devices and illustration of the structure. The p-type ZnO layer was obtained by n-doping. For nitrogen doping, a radio frequency radical source was used with a power of 350 W and a nitrogen flow of 5×10^{-6} Torr as a background equivalent pressure. After the growth of 100 nm ZnO buffer layer at 650°C, the layer was maintained in situ at 1,000°C in 1 mTorr of oxygen for 1 h. All successive layers were grown in a layer-by-layer growth mode as verified by persistent intensity oscillation of reflection high-energy electron diffraction. The n-type ZnO doped with Ga was deposited to the thickness of 400 nm. After the growth of undoped ZnO layer on the n-ZnO:Ga layer, the p-type ZnO was grown by repeated temperature modulation technique. A key to realize the p-type ZnO was to maintain both a high nitrogen concentration

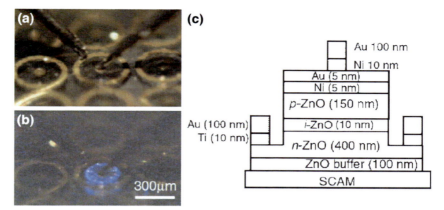

Fig. 9.31. Optical-microscope images of the device taken (**a**) under illumination; (**b**) in the dark with feeding a direct forward bias current of 5 mA; (**c**) Schematic cross-sectional view of a ZnO p–i–n homojunction diode. Semitransparent Au(5 nm)/Ni(nm) electrode is used for the contact to p-type ZnO layer. Reprinted with permission from [15]. Copyright (2005), The Institute of Pure and Applied Physics

($\sim 10^{20}$ cm^{-3}) and a low-defect density. In order to satisfy the former necessity of high nitrogen concentration, low temperature growth is preferred, while the latter necessity of low-defect density requires the high temperature crystal growth. By repeating the growth of an 15 nm-thick ZnO:N at the low temperature followed by the high-temperature annealing and subsequent growth of an 1-nm-thick ZnO at the high temperature, the p-type layers supposed to have the hole concentrations higher than 10^{16} cm^{-3} were obtained with reproducibility. The EL spectra of this LED in Fig. 9.32 shows three independent peaks at 395, 420, and 500 nm which satisfy the Bragg's multiple reflection conditions of $2d \sin \theta = m(\lambda/n)$, where d of 620 nm is the total ZnO layer thickness in this LED, θ is $\pi/2$ for the vertical reflection, m is a positive integer, and λ/n is the wavelength in the material. Therefore, these three maxima are considered as multiple internal reflection fringes of the EL at 440 nm, which is presumably because of the donor–acceptor pair recombination. In this LED, the broad yellow luminescence band was eliminated and hence the color of the emission was more bluish. The LEDs showed a good rectification property with a threshold voltage of 5.4 V.

Ryu et al. reported the ZnO-based UV LEDs employing the BeZnO/ZnO active layer between the n-type and the p-type ZnO and Be$_{0.3}$Zn$_{0.7}$O layers. The active layer was composed of seven quantum wells (QWs), in which the undoped Be$_{0.2}$Zn$_{0.8}$O and the ZnO form barrier and well layers, respectively, as shown in Fig. 9.33a [97]. The thickness of each BeZnO barrier layer was about 7 nm, and each ZnO well layer was about 4 nm. The p-type ZnO and BeZnO layers were formed with Arsenic as the acceptor dopant, while the

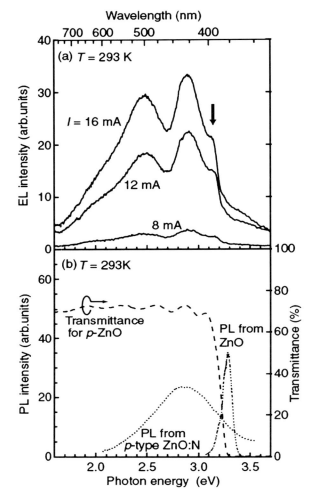

Fig. 9.32. (a) RT EL spectra taken from the top of the device feeding directly forward-bias current of 8, 12, and 16 mA. An arrow indicates the energy of sharp cut-off corresponding to the absorption edge of p-type ZnO top layer; (b) RT PL of undoped and p-type ZnO layers (*dotted lines*) and transmission spectrum (broken line) of p-type ZnO layer. Thickness of the undoped and p-type films are 1 mm and 500 nm, respectively. Reprinted with permission from [15]. Copyright (2005), The Institute of Pure and Applied Physics

n-type ZnO and BeZnO layer were formed with gallium as the dopant. These layers were deposited by the hybrid beam deposition (HBD) [98]. The LEDs showed two dominant EL peaks located in the ultraviolet spectral region between 358 and 369 nm, and a broad peak at 550 nm as shown in Fig. 9.33b. The

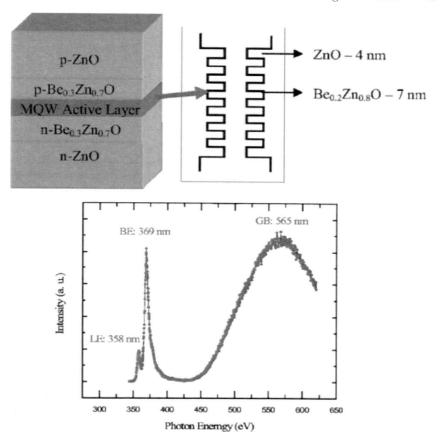

Fig. 9.33. (a) Schematic illustration of the structure of the ZnO-based UV LED devices that employ a BeZnO/ZnO active layer comprised of MQWs; (b) PL spectrum measured at 5 K for a ten QW superlattice structure deposited on ZnO and comprised of $Be_{0.2}Zn_{0.8}O$ and ZnO layers. Reprinted with permission from [97]. Copyright (2006), American Institute of Physics

peak around 358 nm was attributed to the localized exciton (LE) emissions from the QW region and the one at 369 nm to impurity-bound exciton emissions in the ZnO region, respectively. The broad peak at 565 nm was from deep-leveled impurities or defects. For device characterization, ohmic contacts were formed on each of p-type and n-type device surface layers using Ni/Au and Ti/Au. The I–V measurement demonstrated the p–n junction characteristics at the low reverse-bias current.

References

1. S.-H. Wei, X. Nie, I.G. Batyrev, S.B. Zhang, Phys. Rev. B **67**, 165209 (2003)
2. P. Carrier, S.-H. Wei, J. Appl. Phys. **97**, 033707 (2005)
3. Z. Zhao, D.L. Morel, C.S. Ferekides, Thin Solid Films **413**, 203 (2002)
4. C. Kligshirn, Phys. Status Solidi. B **71**, 547 (1975)
5. D.M. Roesler, W.C. Welker, Phys. Rev. **154**, 861 (1967)
6. J.I. Pankove, *Optical Processes in Semiconductors* (Courier Dover Publications, 1975)
7. H. Amano, N. Sawaki, I. Akasaki, Y. Toyoda, Appl. Phys. Lett. **48**, 353 (1986)
8. H. Amano, M. Kito, K. Hiramatsu, and I. Akasaki, Jpn. J. Appl. Phys. **28**, L2112 (1989)
9. S. Nakamura, M. Senoh, T. Mukai, Jpn. J. Appl. Phys. **32**, L8 (1993)
10. H. Amano, M. Kitoh, K. Hiramatsu, I. Akasaki, Proceedings of 16th International Symposium On Gallium Arsenide and Related Compounds, Karuizawa, 1989
11. D.M. Bagnall, Y.F. Chen, Z. Zhu, T. Yao, S. Koyama, M.Y. Shen, T. Goto, Appl. Phys. Lett. **70**, 2230 (1997)
12. Z.K. Tang, G.K.L. Wong, P. Yu, M. Kawasaki, A. Ohtomo, H. Koinuma, Y. Segawa, Appl. Phys. Lett. **72**, 25 (1998)
13. T. Aoki, Y. Hatanaka, D.C. Look, Appl. Phys. Lett. **76**, 3257 (2000)
14. X.-L. Guo, J.-H. Choi, H. Tabata, T. Kawai, Jpn. J. Appl. Phys. **40**, L177 (2001)
15. A. Tsukazaki, M. Kubota, A. Ohtomo, T. Onuma, K. Ohtani, H. Ohno, S.F. Chichibu, M. Kawasaki, Jpn. J. Appl. Phys. **44**, L643 (2005)
16. H.C. Casey, Jr., J. Muth, S. Krishnankutty, and J.M. Zavada, Appl. Phys. Lett. **68**, 2867 (1996)
17. J.M. Shah, Y.-L. Li, Th. Gressmann, and E.F. Schubert, J. Appl. Phys. **94**, 2627 (2003)
18. E.F. Schubert, Light-Emitting Diodes 2nd Edition (Cambridge University Press, 2006), p. 65~p. 67
19. E.F. Schubert, *Light-Emitting Diodes*, 2nd edn. (Cambridge University Press, Oxford, 2006), p. 86
20. M.G. Craford, Fourth China International Forum on Solid-State Lighting, Shanghai, China, 2007
21. S. Nakamura, T. Mukai, M. Senoh, Appl. Phys. Lett. **64**, 1687 (1994)
22. T. Mukai, D. Morita, S. Nakamura, J. Cryst. Growth **189/190**, 778 (1998)
23. H.X. Wang, H.D. Li, Y.B. Lee, H. Sato, K. Yamashita, T. Sugahara, S. Sakai, J. Cryst. Growth **264**, 48 (2004)
24. M.-K. Kwon, I.-K. Park, S.-H. Baek, J.-Y. Kim, S.-J. Park, J. Appl. Phys. **97**, 106109 (2005)
25. A. Dussaigne, B. Damilano, J. Brault, J. Massies, E. Feltin, N. Grandjean, J. Appl. Phys. **103**, 013110 (2008)
26. S. Grzanka, G. Franssen, G. Targowski, K. Krowicki, T. Suski, R. Czernecki, P. Perlin, M. Leszczyński, Appl. Phys. Lett. **90**, 103507 (2007)
27. S. Kim, K. Lee, K. Park, C.S. Kim, J. Cryst. Growth **247**, 62 (2003)
28. Y. Wang, X.J. Pei, Z.G. Xing, L.W. Guo, H.Q. Jia, H. Chen, J.M. Zhou, J. Appl. Phys. **101**, 033509 (2007)
29. J.-W. Ju, E.-S. Kang, H.-S. Kim, L.-W. Jang, H.-K. Ahn, J.-W. Jeon, I.-H. Lee, J.H. Baek, J. Appl. Phys. **102**, 053519 (2007)
30. S.D. Lester, F.A. Ponce, M.G. Craford, D.A. Steigerwald, Appl. Phys. Lett. **66**, 1249 (1995)

31. A. Usui, H. Sunakawa, A. Sakai, A.A. Yamaguchi, Jpn. J. Appl. Phys. **36**, L899 (1997)
32. T. Mukai, IEEE J. Select Top Quantum Electron **8**, 264 (2002)
33. T. Mukai, M. Yamada, S. Nakamura, Jpn. J. Appl. Phys. **38**, 3976 (1999)
34. V. Rozhansky, D.A. Zakheim, Phys. Stat. Sol. A **204**, 227 (2007)
35. M.H. Kim, M.F. Schubert, Q. Dai, J.K. Kim, E.F. Schubert, J. Piprek, Y. Park, Appl. Phys. Lett. **91**, 183507 (2007)
36. M.F. Schubert, J. Xu, J.K. Kim, E.F. Schubert, M.H. Kim, S. Yoon, S.M. Lee, C. Sone, T. Sakong, Y. Park, Appl. Phys. Lett. **93**, 041102 (2008)
37. Y.C. Shen, G.O. Müller, S. Watanabe, N.F. Gardner, A. Munkholm, M.R. Krames, Appl. Phys. Lett. **91**, 141101 (2007)
38. Y.J. Sun, O. Brandt, M. Ramsteiner, H.T. Grahn, K.H. Ploog, Appl. Phys. Lett. **82**, 3850 (2003)
39. E. Kuokstis, C.Q. Chen, M.E. Gaevski, W.H. Sun, J.W. Yang, G. Simin, M. Asif Khan, H.P. Maruska, D.W. Hill, M.C. Chou, J.J. Gallagher, B. Chai, Appl. Phys. Lett. **81**, 4130 (2002)
40. H.M. Ng, Appl. Phys. Lett. **80**, 4369 (2002)
41. T. Takeuchi, H. Amano, I. Akasaki, Jpn. J. Appl. Phys. **39**, 413 (2000)
42. T. Wang, J. Bai, S. Sakai, J.K. Ho, Appl. Phys. Lett. **78**, 2617 (2001)
43. T. Koida, S.F. Chichibu, T. Sota, M.D. Craven, B.A. Haskell, J.S. Speck, S.P. DenBaars, Appl. Phys. Lett. **84**, 3768 (2004)
44. M.D. Craven, S.H. Lim, F. Wu, J.S. Speck, S.P. DenBaars, Appl. Phys. Lett. **81**, 469 (2002)
45. C. Chen, V. Adivarahan, J. Yang, M. Shatalov, E. Kuokstis, M. Asif Khan, Jpn. J. Appl. Phys. **42**, L1039 (2003)
46. M.D. Craven, P. Waltereit, J.S. Speck, S.P. DenBaars, Appl. Phys. Lett. **84**, 496 (2004)
47. A. Chakraborty, B.A. Haskell, S. Keller, J.S. Speck, S.P. DenBaars, S. Nakamura, U.K. Mishura, Jpn. J. Appl. Phys. **44**, L173 (2005)
48. E.F. Schubert, *Light-Emitting Diodes*, 2nd edn. (Cambridge University Press, Oxford, 2006), pp. 86–87
49. D.A. Neamen, *Semiconductor Physics and Devices Basic Principles* (Richard D. Irwin, Illinois, USA, 1992) p. 644
50. D.W. Kim, Y.J. Sung, J.W. Park, G.Y. Yeom, Thin Solid Films **398**, 87 (2001)
51. T. Margalith, O. Buchinsky, D.A. Cohen, A.C. Abare, M. Hansen, S.P. DenBaars, L.A. Coldren, Appl. Phys. Lett. **74**, 3930 (1999)
52. J.-O. Song, J.S. Kwak, Y. Park, T.-Y. Seong, Appl. Phys. Lett. **86**, 213505 (2005)
53. J.K. Sheu, J.M. Tsai, S.C. Shei, W.C. Lai, T.C. Wen, C.H. Kou, Y.K. Su, S.J. Chang, G.C. Chi, IEEE Electron. Dev. Lett. **22**, 460 (2001)
54. C.H. Kuo, S.J. Chang, Y.K. Su, R.W. Chuang, C.S. Chang, L.W. Wu, W.C. Lai, J.F. Chen, J.K. Sheu, H.M. Lo, J.M. Tsai, Mater. Sci. Eng. B **106**, 69 (2004)
55. I. Schnitzer, E. Yablonovitch, C. Caneau, T.J. Gmitter, A. Scherer, Appl. Phys. Lett. **63**, 2174 (1993)
56. R. Windisch, B. Dutta, M. Kuijk, A. Knobloch, S. Meinlschmidt, S. Schoberth, P. Kiesel, G. Borghs, G.H. Dohler, P. Heremans, IEEE Trans. Electron Devices **47**, 1492 (2000)
57. R. Windisch, P. Heremans, A. Knobloch, P. Kiesel, G.H. Dohler, B. Dutta, G. Borghs, Appl. Phys. Lett. **74**, 2256 (1999)

58. T. Fujii, Y. Gao, R. Sharma, E.L. Hu, S.P. DenBaars, S. Nakamura, Appl. Phys. Lett. **84**, 855 (2004)
59. J.K. Sheu, C.M. Tsai, M.L. Lee, S.C. Shei, W.C. Lai, Appl. Phys. Lett. **88**, 113505 (2006)
60. S.M. Pan, R.C. Tu, Y.M. Fan, R.C. Yeh, J.T. Hsu, IEEE Photon. Technol. Lett. **15**, 649 (2003)
61. R.-H. Horng, D.-S. Wuu, Y.-C. Lien, W.-H. Lan, Appl. Phys. Lett. **79**, 2925 (2001)
62. R.-H. Horng, C.C. Yang, J.Y. Wu, S.H. Huang, C.E. Lee, D.S. Wuu, Appl. Phys. Lett. **86**, 221101 (2005)
63. M. Boroditsky, T.F. Krauss, R. Coccioli, R. Vrijen, R. Bhat, E. Yablonovitch, Appl. Phys. Lett. **75**, 1036 (1999)
64. T.N. Oder, K.H. Kim, J.Y. Lin, H.X. Jiang, Appl. Phys. Lett. **84**, 466 (2004)
65. D.-H. Kim, C.-O. Cho, Y.-G. Roh, H. Jeon, Y.S. Park, J. Cho, J.S. Im, C. Sone, Y. Park, W.J. Choi, Q. Park, Appl. Phys. Lett. **87**, 203508 (2005)
66. H.-G. Hong, S.-S. Kim, D.-Y. Kim, T. Lee, J.-O. Song, J.H. Cho, C. Sone, Y. Park, T.-Y. Seong, Appl. Phys. Lett. **88**, 103505 (2006)
67. K.-J. Byeon, S.-Y. Hwang, H. Lee, Appl. Phys. Lett. **91**, 091106 (2007)
68. J.-Y. Kim, M.-K. Kwon, K.-S. Lee, S.-J. Park, S.H. Kim, K.-D. Lee, Appl. Phys. Lett. **91**, 181109 (2007)
69. E. Yablonovitch, Phys. Rev. Lett. **58**, 2059 (1987)
70. S. John, Phys. Rev. Lett. **58**, 2486 (1987)
71. E.F. Schubert, *Light-Emitting Diodes*, 2nd edn. (Cambridge University Press, Oxford, 2006), p. 192
72. Philips Lumileds Lighting Company, Philips Application Brief AB29 (2007) http://www.lumileds.com/pdfs/AB29.pdf
73. T. Inoue, Japanese patent No. 3257455 (7 Dec 2001) (in Japanese)
74. S.J. Chang, C.H. Chen, Y.K. Su, J.K. Sheu, W.C. Lai, J.M. Tsai, C.H. Liu, S.C. Chen, IEEE Electron Dev. Lett. **24**, 129 (2001)
75. R. Saffa, LED J. **Jan./Feb.**, 4 (2007)
76. M.K. Kelly, O. Ambacher, R. Dimitrov, R. Handschuh, M. Stutzmann, Phys. Stat. Sol. A **159**, R1 (1997)
77. W.S. Wong, T. Sands, N.W. Cheung, Appl. Phys. Lett. **72**, 599 (1998)
78. W.S. Wong, T. Sands, N.W. Cheung, M. Kneissl, D.P. Bour, P. Mei, L.T. Romano, N.M. Johnson, Appl. Phys. Lett. **75**, 1360 (1999)
79. W.S. Wong, T. Sands, N.W. Cheung, M. Kneissl, D.P. Bour, P. Mei, L.T. Romano, N.M. Johnson, Appl. Phys. Lett. **77**, 2822 (2000)
80. Y.S. Wu, J.-H. Cheng, W.C. Peng, H. Ouyang, Appl. Phys. Lett. **90**, 251110 (2007)
81. W.H. Chen, X.N. Kang, X.D. Hu, R. Lee, Y.J. Wang, T.J. Yu, Z.J. Yang, G.Y. Zhang, L. Shan, K.X. Liu, X.D. Shan, L.P. You, D.P. Yu, Appl. Phys. Lett. **91**, 121114 (2007)
82. H.P. Ho, K.C. Lo, G.G. Siu, C. Surya, K.F. Li, K.W. Cheah, Mater Chem Phys **81**, 99 (2003)
83. C.A. Tran, C.-F. Chu, C.-C. Cheng, W.-H. Liu, J.-Y. Chu, H.-C. Cheng, F.-H. Fan, J.-K. Yen, T. Doan, J. Cryst. Growth **298**, 722 (2007)
84. J.-S. Ha, S.W. Lee, H.-J. Lee, H.-J. Lee, S.H. Lee, H. Goto, T. Kato, K. Fujii, M.W. Cho, T. Yao, IEEE Photon. Technol. Lett. **20**, 175 (2008)
85. H. Ohta, K.-I. Kawamura, M. Orita, M. Hirano, N. Sarukura, H. Hosono, Appl. Phys. Lett. **77**, 475 (2000)

86. H. Ohta, M. Orita, M. Hirano, H. Hosono, J. Appl. Phys. **89**, 5720 (2001)
87. Y.I. Alivov, J.E. Van Nostrand, D.C. Look, M.V. Chukichev, B.M. Ataev, Appl. Phys. Lett. **83**, 2943 (2003)
88. Y.I. Alivov, E.V. Kalinina, A.E. Cherenkov, D.C. Look, B.M. Ataev, A.K. Omaev, M.V. Chukichev, D.M. Bagnall, Appl. Phys. Lett. **83**, 4719 (2003)
89. D.J. Rogers, F. Hosseini Teherani, A. Yasan, K. Minder, P. Kung, M. Razeghi, Appl. Phys. Lett. **88**, 141918 (2006)
90. A. Osinsky, J.W. Dong, M.Z. Kauser, B. Hertog, A.M. Dabiran, P.P. Chow, S.J. Pearton, O. Lopatiuk, L. Chernyak, Appl. Phys. Lett. **85**, 4272 (2004)
91. C. Yuen, S.F. Yu, S.P. Lau, Rusli, T.P. Chen, Appl. Phys. Lett. **86**, 241111 (2005)
92. A. Nakamura, T. Ohashi, K. Yamamoto, J. Ishihara, T. Aoki, J. Temmyo, H. Gotoh, Appl. Phys. Lett. **90**, 093512 (2007)
93. Y. Chen, D.M. Bagnal, H.-J. Koh, K.-T. Park, K. Hiraga, Z. Zhu, T. Yao, J. Appl. Phys. **84**, 3912 (1998)
94. H.W. Lee, S.P. Lau, Y.G. Wang, K.Y. Tse, H.H. Hng, B.K. Tay, J. Cryst. Growth **268**, 596 (2004)
95. T. Makino, Y. Segawa, S. Yoshida, A. Tsukazaki, A. Ohtomo, M. Kawasaki, Appl. Phys. Lett. **85**, 759 (2004)
96. S. Nakamura, M. Senoh, N. Iwasa, S. Nagahama, Appl. Phys. Lett. **67**, 1868 (1995)
97. Y. Ryu, T.-S. Lee, J.A. Lubguban, H.W. White, B.-J. Kim, Y.-S. Park, C.-J. Youn, Appl. Phys. Lett. **88**, 241108 (2006)
98. Y. Ryu, T.-S. Lee, H.W. White, J. Cryst. Growth **261**, 502 (2004)

10
ZnO and GaN Nanostructures and their Applications

S.H. Lee

Abstract. The current researches on GaN- and ZnO-based nanostructures are reviewed with the emphasis on fundamental growth kinetics, characteristics, and applications. The nanostructured materials have unique properties and the devices employing the nanostructures show superior performances when compared with the conventional devices without the nanostructures. Control of shapes and sizes of the nanostructures has been possible by changing the synthetic methods or by processing. ZnO- and GaN-based nanostructures can successfully be used for fabrications of various sensors, transistors as well as for photonic device applications including light emitting diodes.

10.1 Introduction

Nanoscience or nanotechnology is currently quite familiar, in its terminology and even in its concept, to people. It is being closely connected and applied to many other different research areas. Much of the interest in nanoscience or nanotechnology has arisen from exploring new phenomena and overcoming current technological limitations in electronic devices caused by their structural features with nanometer-scale sized components and materials inside. Besides, the unique properties of nanomaterials have recently opened interdisciplinary research fields. Nanomaterials are being used as key components for many devices. A versatile chemical and physical property of semiconducting nanostructures depending on their dimensionality, size, and surface area makes the nanomaterials greatly potential building blocks for electronic, photonic, and sensing device fabrications.

There has been considerable interest especially in ZnO and GaN since this wide band gap material has been used in conventional planar structures to fabricate UV–blue LEDs and lasers, as well as in a range of other high-performance electronic devices. ZnO and GaN nanostructures include quantum dot, nanowire, and complex shapes with dimensions from zero to three. In this chapter, we will summarize the recent research about the quasi-one dimensional (1D) nanostructure of ZnO and GaN as an ideal system for investigating the electrical transport, optical properties and mechanical properties on size and dimensionality.

10.2 Control of ZnO and GaN Nanostructures

10.2.1 Synthetic Methods

The synthesis of quasi-1D ZnO and GaN nanostructures including nanowire, nanobelt, and nanotube have been achieved by various techniques such as hydrothermal reaction, chemical vapor deposition (CVD), pulsed laser deposition (PLD), and molecular beam epitaxy (MBE) via vapor–liquid–solid (VLS), vapor–solid (VS) or solution–liquid–solid (SLS) processes.

In the VLS process, metal clusters with sizes of a few tens or hundreds nanometer are used as a catalyst for reaction with gas-phased source material. It is usually so-called the catalytic growth. By elevating the temperature, catalysts efficiently provide the nucleation site for nanostructure by formation of alloy at the eutectic temperature and supersaturation. The growth of the 1D nanostructure proceeds to the nucleation site through a continuous feeding of source materials. Figure 10.1 shows typical ZnO and GaN nanowires on Si substrates by metal-catalyzed CVD. As a source for growth of the ZnO nanowires, ZnO powders were thermally decomposed by carbothermal reduction at 900°C under a mixture gas flow of Ar and O_2 (5%). Here, Au particles formed on the substrate by annealing the deposited thin Au film, were used as the catalyst. The ZnO nanowires showed a hexagonal cross-section with typical diameters in the range of 68–140 nm and lengths up to several micrometers (Fig. 10.1a). The GaN nanowires were prepared using ammonia gas and gallium metal as the N and Ga sources, respectively. The GaN nanowires prepared at 910°C using the nickel catalyst showed diameters ranging from 20 to 80 nm and lengths from 1 to 2 μm (Fig. 10.1b). The diameter and the length of nanowires could be controlled by growth conditions such as temperature and time.

Control of the growth direction and orientation of the nanowires by the crystallographical relationship between the semiconductor nanostructure and the substrate is considered as one of the characteristics of the semiconductor nanostructure. It is very helpful for integration and property modulation of devices based on the nanowires. The crystal structure of ZnO (and GaN) are hexagonal structure (space group P6$_3$mc) with lattice parameters a = 0.3250 (a = 0.3189) and c = 0.5207 (c = 0.5185) nm. Figure 10.2 shows the ZnO nanowires grown on various substrates. Vertically oriented ZnO nanowires could be obtained on the ZnO substrate with homostructure and on the AlN and GaN substrates with heterostructures, where the lattice misfits are 4.5 and 1.8%, respectively. In the case of Al_2O_3, the lattice mismatch between ZnO and Al_2O_3 is changed depending on its planes.

Fig. 10.1. SEM images of (a) ZnO and (b) GaN nanowires

The c-plane of Al_2O_3 and the c-plane of ZnO have the lattice misfit of about 18%, while a-plane Al_2O_3 and c-plane of ZnO have the misfit less than 2.7%. The growth direction (c-axis) of ZnO nanowires grown on c-Al_2O_3 substrate has revealed the relative angle of about 52° with the Al_2O_3 (0001) plane. Baxter et al. reported that this 51.8° results from the epitaxial relationship of $[0002]_{ZnO} \| [10\bar{1}4]_{sapphire}$ and $[10\bar{1}0]_{ZnO} \| [1\bar{2}10]_{sapphire}$, where the d-spacings of 2.61 Å for ZnO (0002) and 2.55 Å for Al_2O_3 ($10\bar{1}4$) make the lattice mismatch of only 2.2% [1]. Meanwhile, vertically oriented ZnO nanowires can be grown epitaxially on the ($11\bar{2}0$) plane of Al_2O_3. The large lattice misfit of about 40% between ZnO and Si (100) substrate leads to random growth directions of the ZnO nanowires.

The epitaxial relationships with substrates also make it possible to control the growth direction of nanostructures. Figure 10.3 shows GaN nanowires grown on (100) γ-$LiAlO_2$ and (111) MgO substrates [2]. In the case of γ-$LiAlO_2$ with lattice constant a = 0.517 and c = 0.628, the oxygen sublattice in the (100) plane with twofold symmetry matches well with the twofold symmetry

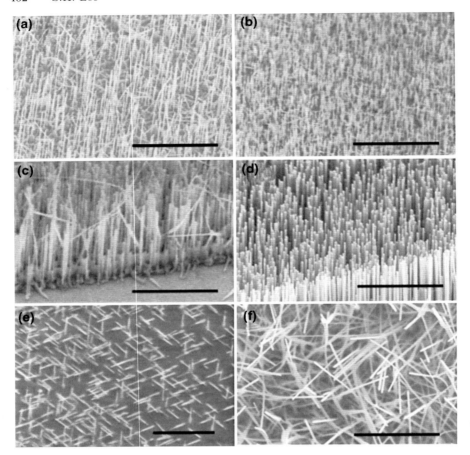

Fig. 10.2. Tilting view SEM images of ZnO nanowires on (**a**) (0001) ZnO, (**b**) (0001) AlN, (**c**) (0001) GaN, (**d**) a-plane (11$\bar{2}$0) Al$_2$O$_3$, (**e**) c-plane (0001) Al$_2$O$_3$, and (**f**) (100) Si substrates. The scale bar is 5 μm

of the (1000) plane of wurtzite GaN. In contrast, the (111) plane of MgO with a threefold symmetry have an interatomic distance of 0.298 nm for atoms in the (111) plane which matches well with the threefold symmetry of the (0001) plane of GaN with the lattice constant of a = 0.319 nm. The high resolution transmission electron microscopy (TEM) images of inset in Fig. 10.3 reveal the growth direction of GaN nanowires. The GaN nanowires on (100) γ-LiAlO$_2$ grew along the [1 $\bar{1}$ 00] direction with a triangular cross-section and the nanowires were enclosed by the side facets of (11 $\bar{2}$ 2), ($\bar{1}\bar{1}$ 2 2) and (0001) planes. On the other hand, the GaN nanowires with a hexagonal cross-section were vertically grown on (111) MgO substrate with the <0001> growth direction.

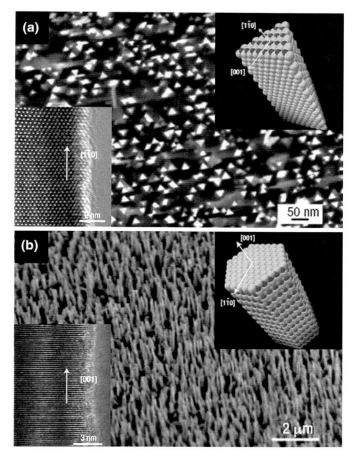

Fig. 10.3. SEM images of GaN nanowires grown on (**a**) (100) γ-LiAlO$_2$, (**b**) (111) MgO substrate. Inset shows the TEM images and structural models of each GaN nanowire with triangular and hexagonal cross-sections. Reprinted with permission from [2]. Copyright (2004), Macmillan

Many researchers successfully synthesized the nanostructures without using metallic catalyst under the suitable growth conditions. Anisotropic growth behavior due to the different surface energies of the facets of the crystal plays a pivotal role in the formation of 1D-nanostructures. The area of crystal facets usually depends on the surface energy under the thermodynamic equilibrium conditions: the facets with higher surface energy have smaller area while with lower energy have larger area. In the ZnO nanostructure, the polarity effect on the morphology of nanostructure was observed. Figure 10.4 shows the results of TEM observations of the ring-type nanobelts grown without catalyst by thermal evaporation of ZnO [3]. Bright and dark field TEM images in

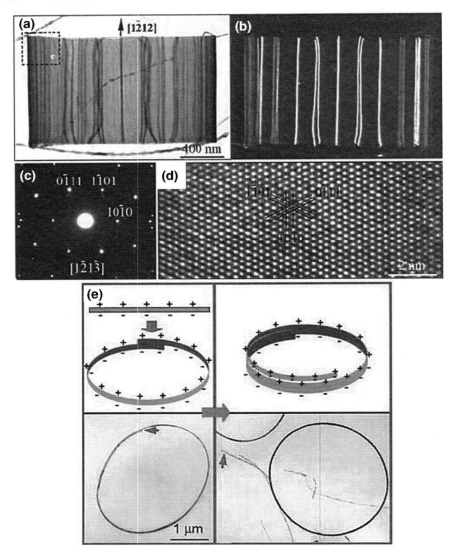

Fig. 10.4. (a) Bright-field and (b) dark-field TEM images, (c) electron diffraction pattern, and (d) HRTEM image observed at the central symmetric line in (a) of the ZnO nanoring. (e) Growth process of the single-crystal nanoring via self-coiling of a polar nanobelt. Reprinted with permission from [3]. Copyright (2004), AAAS

Fig. 10.4a and b reveals the complete ring. The different contrast line in ZnO ring was caused by the bending of the nanobelt. The single crystallinity and the growth direction of ZnO ring could be confirmed by the electron diffraction pattern and the high resolution TEM (HRTEM) image in Fig. 10.4c, d.

The plane of ring is (1 $\bar{2}$ 12) and the tangential direction is [10 $\bar{1}$ 0]. Formation mechanism of the ring shape was believed to be the results of stacking along the c-axis with the polar surfaces with negative (O-polar) and positive charges (Zn-polar) as shown in Fig. 10.4e. More complex nanostructures such as tetrapods, mesoporous structure have also been formed by the self-assembly or due to the structural defects in ZnO.

Beside the thermal growth methods, solution-based or biomimetic methods have been studied to decrease the growth temperature [4–7]. Usually, ZnO nanostructures have been synthesized by using zinc organic complex such as zinc nitrate hexahydrate ($Zn(NO_3)_2 \cdot 6H_2O$) or zinc acetate dihydrate ($Zn(CH_3COO)_2 \cdot 2H_2O$) at growth temperature below 100°C in base environment. These methods have a merit that various substrates including indium tin oxide (ITO) and fluorine doped tin oxide (FTO) coated glass are usable, where such substrates can be directly utilized for device fabrications.

10.2.2 Processing and Assembly

The integration and spatial alignment of 1D nanostructures are one of the most important factors for fabrication of devices. Basically, there are two available routes. One is to directly control the site, the size and the growth direction of the nanowire during the growth process, as mentioned in the Sect. 10.2.1. Usually the growth site of nanowires is specified by a nucleation site during the growth. For the superior control of the site of nanowires, various lithographic processes using photo, electron beam, dip-pen, nanoimprint, nanosphere and employment of microfluidic channels have been reported. Figure 10.5 displays the arrays of ZnO nanowires via various lithographic techniques. Spatially well separated ZnO nanowires were fabricated by using the patterned Au dots arrays with the several tens nanometer range on 6H-SiC substrate, where the arrays were formed by using the e-beam lithography (EBL) (Fig. 10.5a) [8]. The EBL technique has a merit of elaborated controllability for size and pitch of the catalyst, but requires too high costs due to the complex process and the long writing time.

For being cost-effective, simple, and applicable to large area-scale lithographic techniques, patterning methods by using the nanospheres and the polymers were reported [9–11]. Usually the submicrometer-scale polystyrene and silica spheres have been used as shadow masks. These spheres form hexagonal close-packed structures on the substrate by the self-organization. After the deposition of the catalyst for overall areas, the spheres are removed by chemical or plasma etching. Consequently, the periodical catalyst pattern could be obtained from the open areas between each sphere. It is the so-called nanosphere lithography (NSL). The pitch and the size of the catalyst pattern depend on the diameter of spheres. The ZnO nanowire arrays grown on Al_2O_3 substrate via NSL are shown in Fig. 10.5b. The nanowires consisted of a hexagonal lattice.

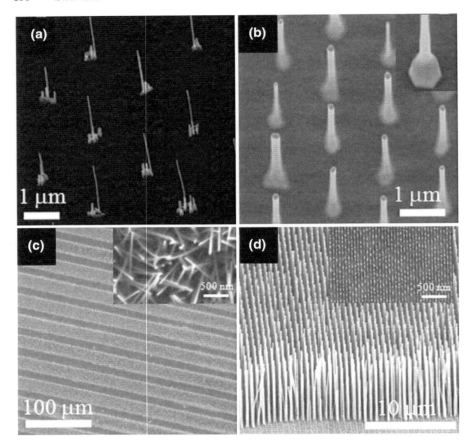

Fig. 10.5. SEM images of the ZnO nano and microarray using (**a**) electron beam [Reprinted with permission from [8]. Copyright (2004), American Chemical Society], (**b**) nanosphere [Reprinted with permission from [11]. Copyright (2006), American Chemical Society], (**c**) microfluidic channels, and (**d**) ZnO PPI template via holographic lithography

As another method, flexible polymer molders have been used as a tool for stamping and injecting into microfluidic channels of dispersed catalyst in solutions. Besides the metallic catalyst for the nucleation of nanowire, ZnO nanoparticles formed by the evaporation of zinc acetate (hydrate) solution were used as a seed [12]. The seed was patterned by the solution flowing through the poly(dimethylsiloxane) (PDMS) microfluidic channels [13]. Line-patterned ZnO nanowire arrays using the microfluidic channels are shown in Fig. 10.5c. The PDMS molder has the advantages of reuse after cleaning and applicability over a large area although the shape and the resolution of the patterns are limited by the initial template.

In the lithographic techniques by using light, holographic lithography (HGL) by multibeam interference of plane waves has been suggested in order to overcome the physical limitation of conventional photolithography [14–16]. HGL is a reliable, fast, and inexpensive lithographic tool for the fabrication of large area-scale periodical patterns. Figure 10.4d shows the ZnO nanowire arrays on the periodically polar-inverted (PPI) ZnO template which was fabricated using MBE via HGL. The polarity of ZnO was controlled by the thickness of the MgO buffer layer [16]. The vertically grown nanowires were spatially separated with a pitch of 650 nm on the ZnO polarity of the PPI template. The SEM observation reveals that the ZnO nanowires were formed only on the Zn-polar regions. It is probably due to the difference of the reactivity and thermal stability between the Zn- and the O-polar ZnO.

The other assembly route is to utilize suitable techniques to manipulate the as-grown nanowires as a building block. Several methods have been attempted for the assembly of nanowires including electric or magnetic fields assisted alignment, microfluidic alignment, and Langmuir–Blodgett compression. Figure 10.6 shows the schematic diagrams of nanowires alignment using microfluidic channels and Langmuir–Blodgett compression. Parallel and crossed arrays of nanowires can be achieved by passing suspensions of nanowires through fluidic channels which consist of a PDMS mold and a flat substrate (Fig. 10.6a) [17]. Meanwhile, Langmuir–Blodgett technique is suitable to fabricate high density nanowire arrays. Monolayer-surface functionalized nanowires on the water–air are uniaxially compressed into closed packed domain and the compressed layer is transferred to a planar substrate covered with nanowires (Fig. 10.6b) [18]. During the compression, the nanowires are aligned in the manner that their longitudinal axis is located parallel to the barrier irreversibly. Lieber group successfully fabricated the various nanowire arrays using the above two assembly methods and applied the arrays to the electronic and photonic devices based on GaN or other semiconductor nanowires [17, 18].

The alignment techniques of GaN and ZnO nanowires using electric and magnetic field between two or more electrodes have been studied. Figure 10.7a shows the GaN nanowires aligned on two electrodes (Ti/Al/Ti) by dielectrophoretic (DEP) force (1 kH, 20 V peak–peak ac signal) [19]. DEP is one of the electrokinetic phenomena to manipulate materials through the interaction of the induced dipoles in between materials and its surrounding medium under the nonuniform electric field. DEP technique has recently been expected to be employable in manipulating bio-material including cell, bacteria, viruses, and molecules as well as nanostructures. Here, the grown nanowires are detached from the substrate and suspended in a solvent such as water, ethanol, acetone, etc. Then, the dispersed nanowires in the solution are dropped on the two electrodes and the AC voltage is applied to the electrodes. Consequently, the nanowires move towards or away from the electrodes due to the differences in the induced dipoles between the nanowires and the solvent.

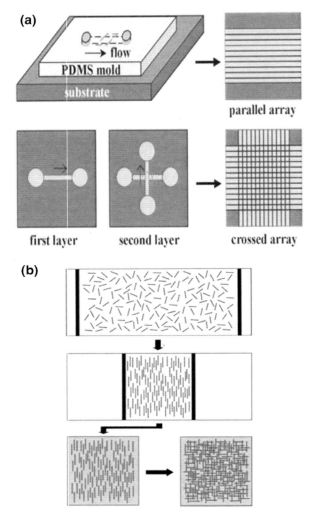

Fig. 10.6. Schematic diagram of typical nanowires alignment using (**a**) microfluidic channels [Reprinted with permission from [17]. Copyright (2001), AAAS.] and (**b**) Langmuir–Blodgett compression [Reprinted with permission from [18]. Copyright (2003), American Chemical Society]

Figure 10.7b shows the aligned ZnO nanowires on the two Ni electrodes patterns with different distances [20]. In order to provide the magnetic field to the nonmagnetic nanowires, the Ni layer was deposited on the top of the ZnO nanowires after the growth of the nanowires on the substrate and then dispersed on the Ni electrodes with a magnetic field (1.6 T). The density of nanowires is controllable by the distance between the electrode patterns, which result in the change of magnetic force.

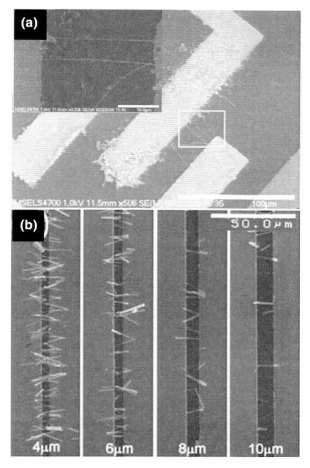

Fig. 10.7. GaN nanowires aligned on electrode using (**a**) DEP [Reprinted with permission from [19]. Copyright (2006), American Institute of Physics.] and (**b**) magnetic field [Reprinted with permission from [20]. Copyright (2007), American Institute of Physics]

Beside, new interesting methods such as the optical trapping by using exerting forces via a highly focused laser beam [21, 22] and the blown bubble film process of nanowire-polymer suspension [23] have been developed.

10.3 ZnO and GaN Nanostructures for Photonic Device Applications

ZnO and GaN have received much attention and developed as the materials for photonic devices due to their direct wide bandgaps. Optical properties and diverse geometries of ZnO and GaN nanostructures with variation of size, dimensionality, shape, and periodicity are interesting for strong-coupling, enhanced spontaneous emissions, and lasing. In addition, physical phenomena in nanometer-sized crystals including enhancement of optical gain, low threshold, and nondirectional laser emission are also highly interesting, which pursues the feasibility of their application to laser devices. In this chapter the application of nanostructures as photonic devices by utilizing the optical and morphological properties of quasi-1D ZnO and GaN nanostructure will be presented.

10.3.1 Optical Cavity and Lasing

Recent advances in the growth method of nano and micro-sized semiconductors have made it possible to create structures in various, [24, 25]. The excited photons are efficiently confined and amplified within small volumes due to the resonant recirculation by reflections at naturally formed crystal facets. Characteristics of lasing by the photonic confinement are determined by waveguide effects and optical cavity.

Theoretical Background

A similar sized nanostructure with the wavelength of light makes it possible to produce the photonic confinement and it determines the characteristic of lasing emissions. This effect in nanowire can approximately be estimated by considering the simple reflection law and waveguide theory. Considering the reflection at the facets of the nanostructure by the Fresnel law, in the case of ZnO and Air (refractive index, n is about 2.3 and 1, respectively), the reflectivity of 16% can be expected at the plane with normal to incident light as shown in Fig. 10.8a. With the assumption that the nanowire is a cylindrical dielectric waveguide, the number of modes (N) in the nanowire can be estimated from in the step index fiber as expressed by the following equation [26].

$$N \approx \left(\frac{4r}{\lambda_0}\right)^2 (n_{\text{ZnO}}^2 - n_{\text{Air}}^2), \tag{10.1}$$

where r is the radius of nanowire, λ_0 is the wavelength in vacuum, n_{ZnO} and n_{air} are the index of refraction of ZnO and air at λ_0, respectively.

Figure 10.8b shows the plot of total number of modes in the cylindrical ZnO nanowire as a function of the radius at $\lambda_0 = 390$ nm. From the calculations, it has been known that only a single mode exists from the nanowires with diameters in the range from 94 to 130 nm, while several different modes can occur in the ZnO nanowires with a diameter larger than 130 nm.

The mode cut-off diameter can be obtained from the characteristic equation [27]

$$\frac{\omega r}{c} = \frac{x_{lm}}{n}, \tag{10.2}$$

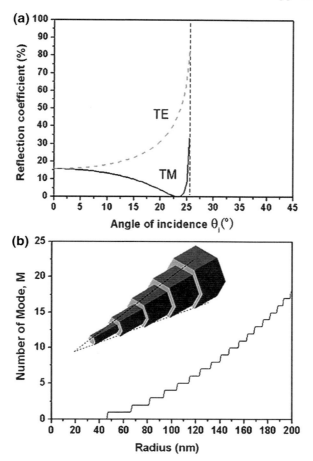

Fig. 10.8. (a) The reflectance as a function of incident angle between ZnO and air and (b) total number of modes vs radius of nanowires with assumption that the ZnO nanowire has a cylindrical shape

where ω is the frequency, c is the vacuum speed of light, r is the radius of the nanowire, and x_{lm} is m-th zero of Bessel function $J_l(x)$ (l is azimuthal index). Here ω obeys the relation of $\omega/c = k = 2\pi/\lambda_0$, where k is wave vector, λ_0 is the wavelength in vacuum. The cut-off diameter of transverse electric (TE_{0m}) and transverse magnetic (TM_{0m}) modes at $\lambda_0 = 390$ nm (at $n = 2.3$) are 129 nm for TE_{01} and TM_{01}, 296 nm for TE_{02} and TM_{02}, respectively.

Johnson et al. compared the electromagnetic field intensity within the nanowires using the fractional mode power (η) of cylindrical waveguide given by [28, 29]

$$\eta = 1 - \left(2.405\exp\left[-\frac{1}{V}\right]\right)^2 V^{-3}, \qquad (10.3)$$

where $V = kr(n_{ZnO}^2 - n_{Air}^2)^{1/2}$ and $k = 2\pi/\lambda$. The field intensity above 90% was retained in the nanowire for the lowest order guide mode for nanowire with $r > 100$ nm. If the diameter was decreased to 50 nm, however, <25% of the field intensity was present. Consequently, it indicates that the photon confinement, the optical loss to the surroundings, and the lasing mode are strongly related with the diameter of the nanostructures.

Optical cavity is one of the important factors for lasing in optical materials and devices. It confines and stores the light at resonance frequencies determined by its configuration. The light is repeatedly reflected or circulated within the cavity. For the planar-mirror resonator (usually called Fabry–Perot cavity) mentioned in Fig. 10.9a, a radiation propagating perpendicular to a pair of facets forms the standing wave in the cavity. The standing waves are formed whenever the cavity contains an integral number of half-wavelengths. For the cavity a with length of L, this condition is expressed by

$$m\frac{\lambda}{2n} = L, m = 1, 2, \ldots \qquad (10.4)$$

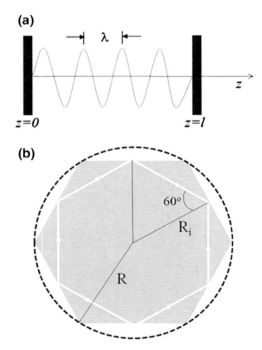

Fig. 10.9. Schematic of (a) Fabry–Perot and (b) whispering-gallery mode (WGM) in quasi-1D hexagonal nanostructure

Where λ is the wavelength and n is the index of refraction at λ.

When m is an integer and is the wavelength of the radiation in the semiconductor, the mode (it is called axial mode or longitudinal mode) spacing is given by [30]

$$\Delta\lambda = \frac{\lambda_0^2}{2L\left(1 - \lambda_0 \frac{\mathrm{d}n}{\mathrm{d}\lambda}\right)}, \quad (10.5)$$

where λ_0 is the wavelength of one of the modes, $\Delta\lambda$ is a mode spacing. The refractive index of ZnO according to wavelengths was determined by using the first-order Sellmeier equation, $n(\lambda)^2 = A + B\lambda^2/(\lambda^2 - C)$, where A, B and C are experimental fitting parameters [31].

Another representative optical cavity is a ring cavity within a microsphere or microdisk. The light may be confined in cavity by repeated reflections from the interior boundary. Such optical modes are known as whispering gallery mode (WGM). Figure 10.9b displays schematically the hexagonal cavity, i.e., a dielectric hexagonal surrounded by air. The theoretical study and the observation of ZnO nanostructure as the hexagonal dielectric cavity have been reported [32–34]. Considering the refractive index (n) of ZnO and the diameter of hexagonal cavity (R_i), the total phase shift of the wave along its path has to be an integer multiple of 2π for enforcing the interference. Here, the constructive interference condition is given by [32]

$$6R_\mathrm{i} = \frac{hc}{nE}\left[N + \frac{6}{\pi}\arctan\left(\beta\sqrt{3n^2 - 4}\right)\right] \quad (10.6)$$

where R_i is the radius of incircle ($R_\mathrm{i} = \sqrt{3}R/2$), h is the Plank's constant, c is the vacuum speed of light, N is the resonance interference order, β is n^{-1} for transverse magnetic (TM) and n for transverse electric (TE) polarization.

In the optical cavity, optical loss can arise due to an imperfection of the mirrors: (a) a partial transmittance is often intentionally used in a laser device to permit that the laser light in the cavity is escaped through it or (b) the finite size of mirror causes that a fraction of the light is to be leaked around them and thereby to be lost. Another factor generating the optical loss in cavity are absorption and scattering in a medium. The round-trip power attenuation factor associated with these effects is $\exp(2\alpha L)$, where α is the loss coefficient of the medium associated with absorption and scattering. Taking the mirror losses and other cavity losses into account, the threshold gain (g_th) can be expressed in the form of

$$g_\mathrm{th} = \alpha + \frac{1}{2L}\ln\left(\frac{1}{R_1 R_2}\right), \quad (10.7)$$

where α is the loss in the cavity, L is the cavity length, and $R_{1,2}$ is the end facet reflection [35, 36].

It is important to realize that the modes within the laser gain curve in order to produce one or two axial modes only. An important parameter is the linewidth of the individual modes. This is determined by the phase fluctuation, which is caused by a noise originated from the spontaneous emission process. In semiconductor lasers, the spectral linewidth of the individual mode is related to the material parameters. The Fabry–Perot cavity, like any other resonant cavity, can be characterized by the Q-factor, which is defined as

$$Q = \frac{\lambda}{\Delta\omega}, \qquad (10.8)$$

where $\Delta\omega$ is the linewidth of peak at the wavelength of λ.

More detail theoretical investigation on waveguide and lasing properties was performed by the analysis of electromagnetic field for the cylindrical waveguide mode using the finite-difference time-domain (FDTD) method [37–39].

Experimental Results and Discussions

The room temperature laser emission from ZnO and GaN film was firstly reported in the 1990s [40, 41]. After a few years later, the lasing from ZnO and GaN nanowires was observed at room temperature [42, 43]. Lasing behavior from ZnO and GaN nanowire was investigated using optical pumping. With increasing the excitation power, a spontaneous emission with linear characteristics under low density exciton regime transits to stimulated emission due to the increase of the exciton density.

Typical laser emission from single ZnO nanowire is shown in Fig. 10.10. The sample of single ZnO nanowire was prepared through the dispersion of nanowires suspended solution on the substrate and the single nanowire was confirmed, the morphology using optical microscopy under UV illumination and SEM observation (Fig. 10.10a, b). The diameter and the length of nanowire are about 240 nm and 19 µm, respectively. Optical pumping was carried out using micro photoluminescence (PL) system with the frequency-tripled output (355 nm) of Nd:YAG laser operating with a repetition rate of 10 Hz and 6 ns pulse-width. The emitted light from the sample was collected by an objective lens and the PL image was monitored by a CCD camera (Sony XC-EU50) having near UV sensitivity with a peak around 369 nm. The scattered light from the exciting laser light was excluded by a sharp cutoff absorption filter. Figure 10.10d shows the PL spectra as a function of excitation power. At a low excitation power density, broad UV emission was observed around at 380 nm. The lasing threshold of $30 \sim 70\,\mathrm{kW\,cm^{-2}}$ was typically obtained from the plot of PL intensity in terms of excitation power density. This value is quite lower than the previously reported values from lasing in particles or films [40, 44]. It is attributed to the an efficient photon confinement due to the high quality and good facets of the ZnO nanowire. Near the threshold, several sharp features were detected at low energy than the initial emission due to the exciton–exciton scattering process. As increasing the excitation power density, the intensity of these peaks became strong and additional lasing mode appeared

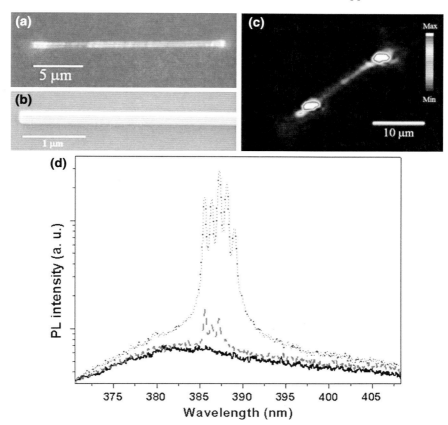

Fig. 10.10. (a) Optical microscope image obtained under the UV illumination and (b) SEM image of the single ZnO nanowire dispersed on the substrate (c) Laser emission and (d) excitation power dependent PL spectra recorded from the ZnO nanowire by optical pumping (solid: 0.075, dash: 0.09, and dot: 0.3 MW cm^{-2})

at the lower energy side as a result of bandgap renormalization. With the assumption that this nanowire is served as Fabry–Perot optical cavity like its hexagonal nanocylinder counterparts, the longitudinal modes spacing of 0.865 nm was obtained from the difference between peak to peak. The cavity length of the nanowire is estimated by using (10.5). The calculated value of the cavity length is about 20 μm and is almost the same with the length (19 μm) of the nanowire observed in the laser emission image of Fig. 10.10c. It could be considered that the formation of longitudinal Fabry–Perot cavity by the reflection at the end two facets of the nanowire occurred rather than the whispering gallery mode (WGM) in lateral planes. The Fabry–Perot cavity, like any other resonant cavity, can be characterized by the Q-factor. Analysis of the laser linewidth gives us the cavity Q factor values of between 902 and 1109 for most of peaks used for calculation.

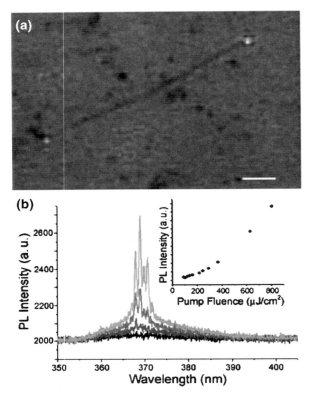

Fig. 10.11. (a) Far-field image and (b) power dependence spectra from a single GaN nanowire. Inset shows PL intensity as terms of pumping power. Reprinted with permission from [43]. Copyright (2005), American Chemical Society

Figure 10.11 shows the laser emission from the single GaN nanowire with a length of ∼40 μm and a diameter of ∼300 nm under optical pumping using UV pulses (with wavelength in the tunable range from 290 to 400 nm, 100 ∼ 200 fs duration, average power of 5 ∼ 10 mW, with a repetition rate of 1 kHz) from the frequency-quadrupled output of an optical parameter amplifier (OPA) and a regeneratively amplified Ti:sapphire oscillator [43]. The tendency of excitation power dependent PL spectra are similar to that of the ZnO nanowire. The localized laser emission was observed at the end of GaN nanowire due to the strong waveguide effect and the Fabry–Perot cavity as shown in Fig. 10.8a. Johnson et al. reported that the lasing threshold of ∼500 nJ cm^{-2} for the GaN nanowire was higher than that of the single ZnO nanowire [28]. They suggested that an electron-hole plasma (EHP) is a major lasing mechanism for the GaN nanowire due to the weak excitonic binding energy in GaN (∼26 meV) and enhanced Coulombic screening at high carrier concentration.

The wide angular emission due to the small diameter of nanowires was theoretically estimated by the electromagnetic modes [45]. It was experimentally demonstrated by the observation of interference and diffraction due to coherent laser emission at the end facets of individual ZnO nanowires dispersed onto SiO_2/Si substrate by using CCD with Nd:YLF laser (349 nm, 10 ns pulse, a repetition rate of 2–5 kHz) [46]. The detail observation of lasing was performed using CCD image as shown in Fig. 10.12a–c. The uniform and broad light is expanded from 3.2 µm-long ZnO nanowire below the lasing threshold. Above the lasing threshold, the light interference due to Fabry–Perot mode was clearly observed as shown in Fig. 10.12d, e. The corresponding emission spectra are shown in Fig. 10.12f–i. The broad peak under the lasing threshold changed to a sharp peak with mode spacing on increasing the excitation intensity.

The study on optical cavity dependent laser emission was investigated by using a length-controlled ZnO nanoribbon fabricated by focused ion beam (FIB) etching, and a shape-modified GaN nanowire by micromanipulator. The example of the etched ZnO nanoribbon using the FIB etching to control the length is shown in Fig. 10.13a [47]. From the obtained length controlled nanoribbon, similar lasing spectra to those from the nanowire as previously shown in Fig. 10.10 appeared. Figure 10.13b plots the lasing threshold as a function of the length of nanoribbon. Upon increasing the length, the lasing threshold is decreased. It can be explained from the relationship that the threshold gain is inversely proportional to the cavity length in (10.7) namely the decrease of distance between the two optical mirrors requires the higher optical excitation for population inversion. However, the lasing did not appear in some nanoribbons with the length below 6 µm or did not follow above the relationship of (10.7). The reason was believed to have resulted from the essential optical loss in the nanoribbon in spite of high pumping or from the deformation of the end facet during the FIB etching.

The modulated optical cavity was studied from the GaN nanowires by cutting and reforming to the ring shape using micromanipulator [48]. Two type of optical cavity modes for the Fabry–Perot and the ring cavity were observed from the GaN ring as shown in Fig. 10.14a. The mode spacing values for two cavities are obtained as 0.69 nm at 375 nm (wire) and 1.61 nm at 377 nm (ring). The threshold value of ring cavity is about two times higher than that of the wire (near 75 µJ cm^{-2}). But the precise comparison of the lasing threshold for two cavities was difficult because of the significant in-plane lasing at the ring cavity. The cavity quality Q factor in (10.8) was evaluated to be higher than 10^3 from 0.08% single shot acquired spectrum. In contrast, the Q factor of the nanowire by decoupling of the ring was decreased by about 40% due to a reduction of the photon-confinement lifetime and diffraction limited reflectivities as shown in Fig. 10.14b. Through the examination for ring diameter dependent laser emission characteristics, it was found out that the lasing modes were shifted to longer wavelengths and the

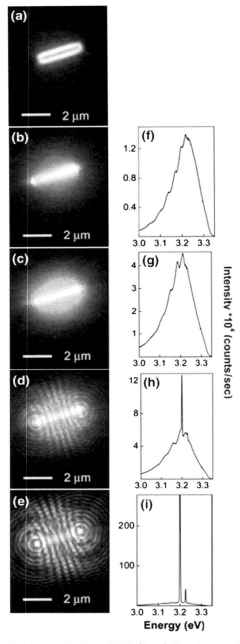

Fig. 10.12. (a) Optical image in dark field, (b–e) Panchromatic PL images of ZnO nanowire with increasing excitation intensity of (b) 24 W cm^{-2}, (c) 93 W cm^{-2}, (d) 139 W cm^{-2}, (e) 268 W cm^{-2}. (f–i) Corresponding PL spectra of images of (b–e). Reprinted with permission from [46]. Copyright (2006), American Chemical Society

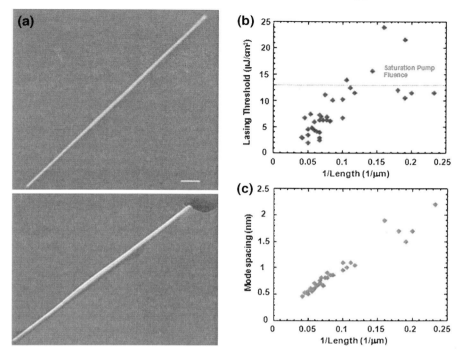

Fig. 10.13. (a) The length controlled single ZnO nanoribbon using FIB etching. Scale bar is 2 μm. The variation of (b) lasing threshold and (c) the spectral mode spacing in terms of the nanoribbon length. Reprinted with permission from [47]. Copyright (2003), Wiley

mode spacing was increased by decreasing the ring diameters. The red-shift of peak emission nearly by 10 nm was considered as the result of the improved coupling efficiency between tangentially overlapping nanowire ends.

10.3.2 Nanostructure-Based LED

ZnO and GaN nanostructures have been utilized as components for the realization of nano-LED or for an efficiency improvement of conventional LED because of their superior optical properties and morphological effects. The research on application to the LED can be divided largely into three schemes: (1) the formation of nanostructures on n- or p-type template. (2) the contact of nanostructure to nanostructure by utilizing the assembly methods, and (3) the fabrication of axial or radial heterostructures.

A simple LED structure with white luminescence was accomplished by the ZnO nanowires on fluorine-doped SnO films at room temperature [49]. The nanowires were grown by electrodeposition in aqueous solution. For insulating of the SnO_2 layer and homogeneous filling between the nanowires, polystyrene

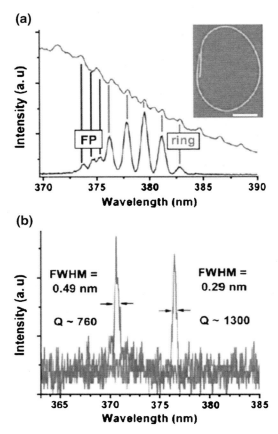

Fig. 10.14. (**a**) Photoluminescence (*upper*) and lasing mode (*lower*) of GaN ring resonator. Inset shows the SEM image of GaN ring. (**b**) A single-shot acquired spectrum for a single nanowire (FWHM = 0.49 nm) and ring type (FWHM = 0.29 nm) cavity. Reprinted with permission from [48]. Copyright (2006), American Physical Society

with high-molecular-weight were employed. The film thickness of polystyrene was controlled by the condition of spin coating, concentration and molecular weight. After a polymerization of the film by UV curing, the exposure of the top of the nanowires was possible by soaking in toluene and rinsing in toluene. A poly(3,4-ethylene-dioxythiophene)(PEDOT)/poly(styrenesulfonate) was coated on the nanowire as a p-type material. Current was transported by bias voltage between a deposited Au film on PEDOT/PSS and SnO_2 layer. Visible emission of white light appeared at a current density of $\sim$100 mA cm^{-2} at the nanowire tip. The PL and EL spectra of the LED revealed the broad defect-related emission centered at 620 nm. The stable light emission from the LED remained for 1 h at ambient conditions.

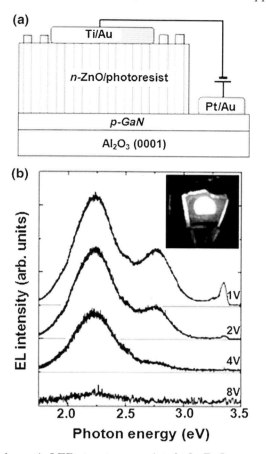

Fig. 10.15. (a) Schematic LED structure consisted of a ZnO nanorod arrays on GaN layer, (b) EL spectra of a p-GaN/n-ZnO nanorod device by reverse-bias voltage. The inset is a light emission image from the LED at a bias voltage of 5 V. Reprinted with permission from [50]. Copyright (2004), Wiley

The heterojunction with assembled nanorods on the template was attempted to enhance the carrier injection through nanosized junctions [50]. Figure 10.15a shows the schematic heterojunction structure of n-type ZnO nanorod on p-type GaN. The ZnO nanorod arrays were grown on p-GaN layer with a carrier concentration of $2 \times 10^{17}\,\text{cm}^{-3}$ and a mobility of $10\,\text{cm}^2\text{V}^{-1}\text{s}^{-1}$ using metal-organic vapor phase epitaxy (MOVPE) without the catalyst. ZnO nanorods grew vertically on the GaN layer due to a low lattice misfit of about 1.8%. Metal (Pt/Au and Ti/Au) layers for Ohmic contacts were deposited on the GaN and the tip nanorods. At the same time, the area around the nanorods were covered with a photoresist and then the tips of the nanorods exposed by surface etching using oxygen plasma. Electroluminescence (EL) spectra from the ZnO nanorod array on p-GaN layer were measured at various reverse-bias voltages as shown in Fig. 10.15b. On the reverse-bias voltage of 3 V, the yellow emission centered at 2.2 eV appeared. Upon increasing the reverse-bias voltage, the intensity of the yellow emission increased and blue emission at 2.8 eV and UV emission peak at 3.35 eV were also observed. The origin of EL emission peaks of the yellow and blue emission was suggested as a defect-related radiative transition and a radiative recombination related to Mg acceptor in Mg-doped GaN, respectively, by comparing it to the PL results of ZnO nanorod array on p-GaN layer.

The single ZnO nanowire-based LED was also demonstrated by employing the p-type template [51]. The ZnO nanowire was dispersed on a heavily doped p-type silicon (p-Si) substrate. Poly(methylmethacrylate)(PMMA) was covered on the ZnO nanowire and the substrate. The PMMA on the ZnO nanowire was removed in an optimized developed condition including PMMA thickness, e-beam dose, and development time. The existing PMMA layer on the side of ZnO nanowire acted as an insulating layer. The metal (Ti/Au) layers were deposited on the revealed surface of ZnO nanowire. Figure 10.16a shows the SEM images of the fabricated device through this process. The light emission was observed when the electrons and holes were injected through the Si substrate and the nanowire upon metal contact as shown in Fig. 10.16b. The p–n junction characteristics were confirmed by I–V plots for three devices with the same types (Fig. 10.16c). The different electrical behaviors between the three samples were regarded as the results of different diameters and length of the nanowires and the contact areas. The EL and PL spectra for the individual ZnO nanowires are shown in Fig. 10.16d. The major emission with a broad range from 400 to 800 nm and a weak emission at 380 nm due to excitonic recombination are observed in the EL spectrum of the three devices. In PL measurements, the broad peak centered at 556 nm with FWHM of 170 nm detected. It was suggested that the origin of broad emission might have resulted from the defects and surface states.

In the case of GaN nanowire, a similar LED structure was reported [52]. The n-type GaN nanowire and the p-type GaN film were utilized for the fabrication of nanoscale homojunction LED. The nanowires were aligned by dielectrophoretic method on the surface of a p-type GaN film. Besides parts that were in contact with the metal was passivated by SiO_2. The devices showed a stable UV light emission with 365 nm-wavelength and operation time more than 2 h.

The concept of nano-LED using junction nanowire was demonstrated on the dielectric material coated-substrate or the polymer by utilizing assembly methods as previously described in the Sect. 10.2.2. Figure 10.17 shows the GaN nanowire-based nano-LED [53]. The n-type GaN nanowire was formed by unintentionally doping and the p-type GaN nanowire with maximum hole mobilities of about $12\,cm^2\,V\cdot s^{-1}$ was prepared by Mg doping via CVD. Electrical characterization of the crossed GaN nanowire p–n junctions showed the characteristic of p–n diodes as shown in Fig. 10.17b, which is different from the

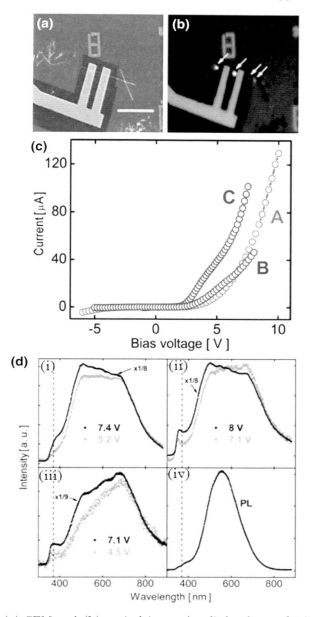

Fig. 10.16. (a) SEM and (b) optical image (applied voltage of +7 V) of single ZnO nanowire LED. Scale bar is 1 μm. (c) I–V characteristics about three prepared devices. (d) EL (i)–(iii) and PL spectra (iv) of the single ZnO nanowire LED device. Reprinted with permission from [51]. Copyright (2006), American Chemical Society

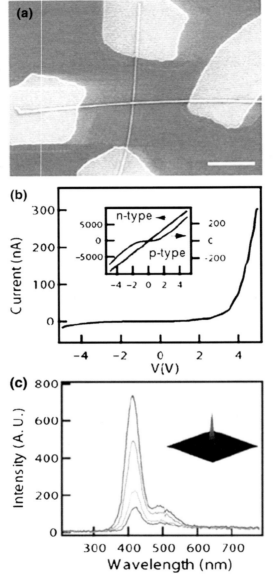

Fig. 10.17. (a) SEM image of a crossed n- and p-GaN type nanowire. (b) I–V plot recorded for the p–n junction and (inset) for the individual p- and n-type nanowire. (c) EL spectra from a p–n junction of the crossed GaN nanowire at the injection current of 61, 132, 224, and 344 nA, respectively. Inset shows the emission image from the crossed part. Reprinted with permission from [53]. Copyright (2003), American Chemical Society

symmetric behavior observed from the individual nanowire–metal contacts as shown in inset. The turn-on current at 3.5 eV is consistent with the bandgap of GaN. Electroluminescence (EL) spectra under the forward biased p–n junction of nanowires showed a dominant emission peak centered at 415 nm due to radiative recombination from the conduction band to deep levels associated with the Mg-doped p-type GaN. The emission at the 493 nm originated from a recombination with another impurity that appeared in (Fig. 10.17c). The near band-edge emission was increased by higher injection current.

Another concept on LED using 1D nanostructure is a radial structure with a high contact area [54]. Figure 10.18 shows the schematic procedures for the fabrication of a core/shell/shell (CSS) shaped nanowire using MOCVD via a catalytic growth. Each material of CSS structure is composed of (Si-doped)/n-GaN InGaN/(Mg-doped) p-GaN heterostructure, respectively. The electric contacts to the n-type core and p-type outer shell of individual CSS nanowires were achieved by selective etching using a FIB (inset of Fig. 10.18b). The I–V plots obtained from the CSS nanowires with electric contact shows typical characteristics of a p–n diode with a shape turn-on at 4 V in forward bias (Fig. 10.18b) EL spectrum of the CCS nanowire under forward bias exhibits strong emission at 456 nm with a FWHM of 50 nm (Fig. 10.18c). The blue emission was observed at the two positions of the nanowire as shown in inset of Fig. 10.18c. Comparing the optical image to the upper image of the inset of Fig. 10.18c, a strong emission between the metal contact and another weak emission occurred at the end of CCS nanowires. The emission originated from the recombination of injected electrons and holes in an InGaN active layer. Weak emission at the end of CCS nanowire is probably attributed to the waveguide effect.

10.4 ZnO and GaN Nanostructures for Electronic and Sensing Device Applications

10.4.1 Field Effect Transistors (FET)

The active devices based on large bandgap semiconductor materials such as ZnO and GaN are attractive candidates for high-power/high-temperature electronics. A nanowire based FET is generally fabricated by a dispersion of nanowire on an insulator-coated Si substrate. The metal is deposited on two sides of the nanowire as a drain and source. The Si substrate functioned as a back gate. The characteristics of fabricated FET were evaluated by the gate capacitance, transconductance, carrier mobilities, and subthreshold gradient from measured electrical data.

The gate capacitance (C_g) is given by

$$C_\mathrm{g} = 2\pi \frac{\varepsilon \varepsilon_0 L}{\ln(2h/r)} \tag{10.9}$$

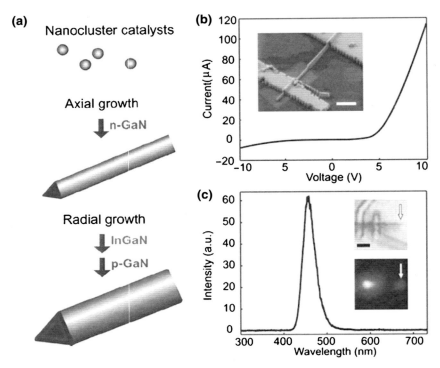

Fig. 10.18. (a) The radial heterostructured GaN based nanowire. (b) I–V plot of n-GaN/InGaN/p-GaN(CSS) nanowire. Inset shows the SEM image of CSS structure with metal contact at to n-type GaN and p-GaN. Scale bar is 1 μm. (c) EL spectrum of CSS junction under forward bias at 7 V. Inset images show bright field (*upper*) and EL images (*lower*, 4 K with a forward bias of 12 V). Reprinted with permission from [54]. Copyright (2004), American Chemical Society

where ε_0 is the permittivity of free space, L is the nanowire length, r is the nanowire radius, and h is the thickness of gate insulator with dielectric constant of ε.

The subthreshold gradient (S) is given by

$$S = \left(\frac{\mathrm{d}\log I_{\mathrm{ds}}}{\mathrm{d}V_{\mathrm{g}}}\right)^{-1}, \qquad (10.10)$$

where I_{ds} is the drain current and V_{g} is the gate voltage.

The mobilities of the electron and hole (μ_{e} and μ_{h}) can be estimated from transconductance (g_{m}) of the FET,

$$\mathrm{d}I_{\mathrm{ds}} = \mu \left(\frac{V_{\mathrm{ds}}}{L}\right)\left(\frac{C_{\mathrm{g}}\mathrm{d}V_{\mathrm{g}}}{L}\right), \qquad (10.11)$$

where V_{ds} is the drain voltage.

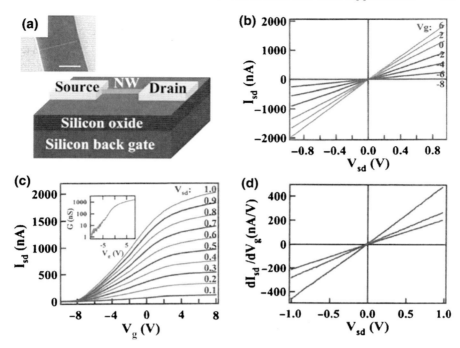

Fig. 10.19. (a) Schematic illustration of typical nanowire FET. The inset shows SEM image of GaN nanowire FET. (b) The gate voltage dependent $I_{sd} - V_{sd}$ curve for FET using GaN nanowire with a diameter of 17.6 nm. (c) $I_{sd} - V_g$ curve for different V_{sd}. The inset shows a conductance (G) of GaN nanowire vs gate voltage. (d) The plot dI/dV_g vs V_{sd} for three GaN nanowire FETs. Reprinted with permission from [55]. Copyright (2002), American Chemical Society

Figure 10.19a shows the FET based on individual GaN nanowire [55]. The GaN nanowires with diameters and lengths in the orders of 10 nm and 10 μm were synthesized using a laser-assisted catalytic growth. Source and drain electrodes were fabricated at the both ends of nanowire on oxidized Si substrate (resistivity of 1–10 Ω cm, SiO$_2$ thickness of 600 nm) by the deposition of Ti/Au (50/70 nm) via electron beam lithography. Typical $I_{sd}-V_{sd}$ curve obtained from the GaN nanowire FET at room temperature is shown in Fig. 10.19b. All curves show linear characteristics due to Ohmic contacts to the GaN nanowire independent of V_g. Figure 10.19c shows the $I_{sd}-V_g$ curves at various source-drain voltages. The conductance of GaN nanowire is increased by increasing the gate voltage. This behavior indicates that the GaN nanowire has n-type characteristic. The electron carrier density in the range of 10^{18}–10^{19} cm^{-3} was estimated from the total charge ($Q = CV_{th}$, where C is the capacitance of nanowire, V_{th} is the threshold voltage). And the mobility of the carriers in the GaN nanowire can be estimated by (10.10). The calculated mobility value using the plot dI_{sd}/dV_g vs V_{sd} in Fig. 10.19d was ranged from 150 to 650 cm^2 Vs^{-1}. These mobilities are comparable to or larger than the reported values for thin film GaN materials with similar carrier concentrations.

Vertical type FET was demonstrated using a direct, bottom-up integration of ZnO nanowire [56]. The growth of ZnO nanowire was selectively positioned by the catalyst pattern using EBL shown in Fig. 10.5a. The small lattice misfit of about 4% with SiC substrate led to the vertical growth of ZnO nanowire with a growth direction of c-axis. Figure 10.20a, b display the 3D and cross-sectional schematic illustration of the vertically surrounded-gate (VSG) FET, respectively. The typical fabrication process of VSG–FET is as follows: The individual ZnO nanowire was encapsulated with a SiO$_2$ thin layer of ~20 nm for the gate oxide of the VSG–FET by CVD. The Cr metal (~40 nm) layer was deposited using a conformal ion-beam on the surface or SiO$_2$. Further deposition of SiO$_2$ on Cr/SiO$_2$/ZnO and a chemical mechanical polishing made the flattened surface and exposed the top of the nanowire, respectively. The height control of the Cr gate metal was performed by selective wet chemical etching to avoid the direct contact between Cr gate and top drain electrodes. After the filling by the SiO$_2$, Cr was deposited on the top of the ZnO nanowires

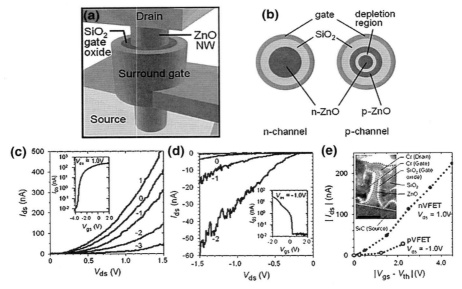

Fig. 10.20. (a) 3D and (b) cross-sectional schematic illustration of VSG–FET using ZnO nanowires. The curve of I_{ds} vs V_{ds} (c) n-VSG–FET and (d) p-VSG–FET for different V_g. The insets shows the plot of I_{ds} vs V_g at $V_{ds} = 1$ V and $V_{ds} = -1$ V, respectively. (e) The plot $|I_{ds}|$ vs $|V_g - V_{th}|$ for n- and p-VSG–FETs. The inset shows cross-sectional SEM image of a VSG–FET. Reprinted with permission from [56]. Copyright (2004), American Chemical Society

for the drain electrode. Figure 10.20c and d show the I_{ds} in terms of V_{ds} recorded from n-VSG–FETs and p-VGS–FETs, respectively. The inset shows I_{ds} vs V_g at constant V_{ds} of 1.0 V. The turn-on and off voltage (threshold voltage: V_{th}) of n-VSG–FET was $V_g = -3.5$ V, while the V_{th} of p-VSG–FET was close to zero (0.25 V). The g_m of nanowire was calculated to be 50 and 35 nS at $|V_{ds}| = 1$ V for the n- and p-channels, and the on-to-off current ratio (I_{ON}/I_{OFF}) for the n- and p-VSG–FET above 10^4 and 10^3 was obtained. The relationship between I_{ds} and $|V_g - V_{th}|$ at $|V_{ds}| = 1$ V is shown in Fig. 10.20e. The n-VSG–FET shows the linear characteristics, while p-VSG–FET shows a nonlinearity. The difference was explained as followings: in the n-channel, the variation of gate-induced charge dQ involves mainly electrons and I_{ds} is proportional to the electron density. On the other hand, in the p-channel, dQ involves holes as well as ionized impurities in the depletion region. The estimated values of μ_e and μ_h using (10.10) are 0.53 and 0.23 cm^2 Vs^{-1}, respectively. The subthreshold gradient (S) in (10.12) is 170 and 130 mV per decade for the n- and p-VSG–FET. From these results, it was suggested that the VGS–FET performance using the ZnO nanowire was comparable to the previous result of CNT–FET but could not outperform the Si based the state-of-the-art MOSFETs yet.

Further improvement of the FET performance could be expected by the control of nanowire properties, reduction of the gate-oxide thickness, and lowering of the contact resistance at source and drain. Recently the study on control of gate gap or shape between nanowires was performed. The plane gate electrode with a nanogap was fabricated using poor coverage and adhesion of Cr metal on ZnO nanowire. Figure 10.21a shows the ZnO nanowire FET with a nanosized gap between nanowire channel and gate [57]. The channel length and the width, the air gap between nanowire and gate, and the gate width in the ZnO nanowire-based FET structure was 968, 26, 60, and 360 nm, respectively. Niobium (Nb) with work function of 4.30 eV was used for the source and the drain electrodes because of its well matching with the electron affinity to the ZnO (4.35 eV). Figure 10.21b, c show the electric characteristics recorded from the ZnO nanowire FET. From these results and (10.10), an on/off current ratio by the gate, S, and a V_{th} were calculated to be 10^6, 129 mV per decade, and 0.4 V, respectively. And a g_m and normalized g_m by channel width were 3.06 µS and 51.2 µS µm^{-1}. The value of C was estimated to be 3.86×10^{-17} F from the 3D finite element method simulation. And the μ_e was estimated to be 928 cm^2 Vs^{-1}. This was the improved results compared with other ZnO nanowire based FET [56, 58].

Figure 10.22a and b show schematic and SEM image of the omega-shaped-gate ZnO nanowire FETs [58]. The structure was fabricated by typical photolithography process. Ti (30 nm) and Au (100 nm) layers were deposited on the ZnO nanowire as electrode metals. A gate insulator of Al_2O_3 layer with a thickness of 17 nm was deposited on surroundings of nanowire by atomic layer deposition. Typical I_{ds}–V_{ds} plots as a function of gate bias are shown

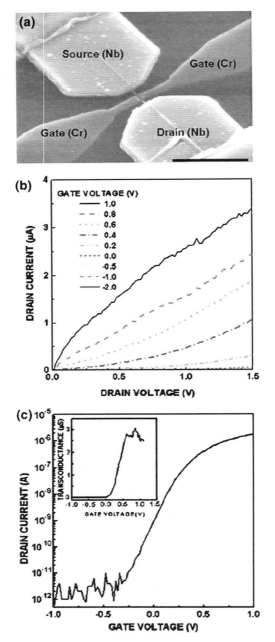

Fig. 10.21. (a) SEM image of ZnO nanowire FET with naonsize air gaps. Scale bar is 2.5 μm. (b) $I_{ds} - V_{ds}$ and (c) $I_{ds} - V_{g}$ characteristics of ZnO nanowire FET. Inset shows the g_m as a function of gate voltage. Reprinted with permission from [57]. Copyright (2006), American Institute of Physics

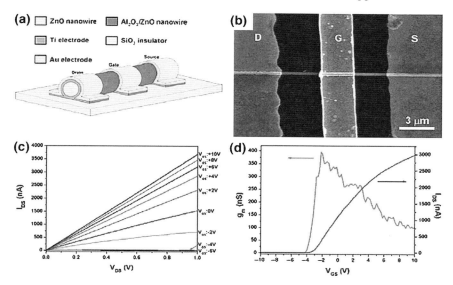

Fig. 10.22. (a) Schematics and (b) SEM image of the ZnO nanowire based OSG FET. (c) $I_{ds}-V_{ds}$ and (d) g_m-V_{gs} and $I_{ds}-V_{gs}$ characteristics of the OSG FET. Reprinted with permission from [58]. Copyright (2006), American Chemical Society

in Fig. 10.22b. The linearly increased current characteristics in the range of positive gate bias indicate that the nanowire is n-type and has Ohmic contact with the electrodes. From the $I_{ds}-V_g$ curve in Fig 10.22c, an I_{on}/I_{off} ratio was evaluated to be 10^7 and a V_{th} of 130 mV per decade was obtained at $V_{ds} = 1$ V. The transconductance curve reveals that a g_m is 0.4 µS at $V_g = -2.2$ V. The C_g of omega-shaped gate was estimated to be 5.2 fF from the (10.9) with the FET structure of $L = 3$ µm, $r = 55$ nm, and $h = 17$ nm. The concentration of charge carrier density is calculated to be 4.9×10^{18} cm^{-3}. The mobility is estimated to be 30.2 cm^2 Vs^{-1} from the (10.11) at $V_{ds} = 1$ V. The mobility, peak transconductance and I_{on}/I_{off} ratio for the omega-shaped-gate ZnO nanowire FET were respectively evaluated as 3.5, 32, and 10^7 times better than those of the back-gate FET using nanowire with the same growth conditions.

The effect of GaN nanowire diameter on conductivity in FETs was investigated [59]. Figure 10.23 shows the normalized conductivity by gate bias for the different nanowire FETs. The obtained conductivity at $V_{sd} = 1$ V was normalized to the maximum value. The conductivity of all FETs showed a decrease at negative gate bias. It indicates that the GaN nanowires have n-type properties. The large conductivity modulation of ∼75% by gate bias was obtained from the small diameter nanowire FET, while the modulation was reduced by increasing the diameter of the nanowire. The narrow nanowires have a small amount of carrier density within the nanowire, which leads to sensitive variation by external conditions such as surface electric field and functionalization.

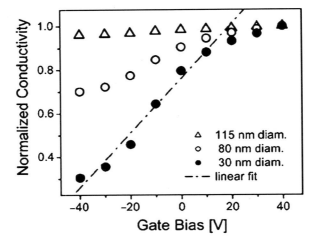

Fig. 10.23. Plot of normalized conductivity as a function of gate bias for three GaN nanowire FETs with diameters of 115, 80, and 30 nm. Reprinted with permission from [59]. Copyright (2007), American Institute of Physics

10.4.2 Light Sensor

Many studies on photoconduction of GaN and ZnO nanowires have been performed using different growth, assembly methods, and measurement conditions. Through the research on nanowire-based photodetector, the two main factors that can lead to this gain are: at first the large surface-to-volume ratio and the presence of deep level surface trap sites in nanowires that greatly prolong the photocurrent lifetime. Second, the reduced dimensionality of the active area in nanowire devices that shortens the carrier transit time. These lead to the substantial photoconductive gain. In general, the mechanism of the photocurrent has been explained by a trapping mechanism of oxygen at the surface of the thin film [60]. In the dark, the adsorbed oxygen molecules on the ZnO surface capture free electrons existed as negative charged ion in the n-type ZnO, and a low-conductivity depletion layer is formed near the surface. Upon illumination of UV-light with larger photon energy than the ZnO bandgap, surface electron-hole recombinations are generated by the holes migration to the surface and discharge the adsorbed oxygen ions. At the same time, the oxygen is adsorbed from the surface and the unpaired electrons significantly increase the conductivity of the nanowire.

Figure 10.24a shows the I–V plot under dark and UV illumination with wavelengths below 380 nm (UV lamp, $0.3\,\mathrm{mW\,cm^{-2}}$) on the single ZnO nanowire with a 60 nm diameter [61]. The resistivity of the nanowire was decreased by typically with 4–6 orders of magnitude. The power dependent photoresponse was investigated using the third harmonic of a Nd:YAG laser with a wavelength of 355 nm as a laser source. The relationship between incident

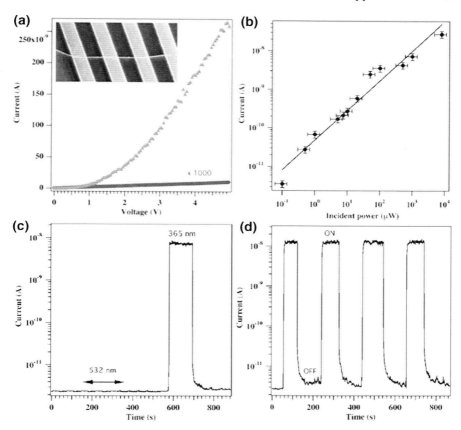

Fig. 10.24. (a) I–V curve of a single ZnO nanowire under dark and UV exposure. Inset shows the SEM image of ZnO nanowire with 60 nm diameter on four Au electrodes. (b) Incident power dependent photocurrent for a ZnO nanowire. (c) The photocurrent variation as a function of wavelength of 365 and 532 nm. (d) Switching behavior of ZnO nanowire by on and off of UV-lamp. Reprinted with permission from [61]. Copyright (2002), Wiley

power (P) and photocurrent (I) was expressed by the power law equation of $I_{pc} \propto P^{0.8}$ (Fig. 10.24b). Figure 10.24c shows the photocurrent variation when the light of 532 and 365 nm-wavelength was illuminated on the nanowire. It could be concluded that the light with larger photon energy than the bandgap of ZnO could not affect the photoconductions of nanowire. Figure 10.24d shows the optical switching behavior in terms of on and off time of the UV lamp. The current is sharply increased to the constant value and then decreased with a decay time of about 1 s.

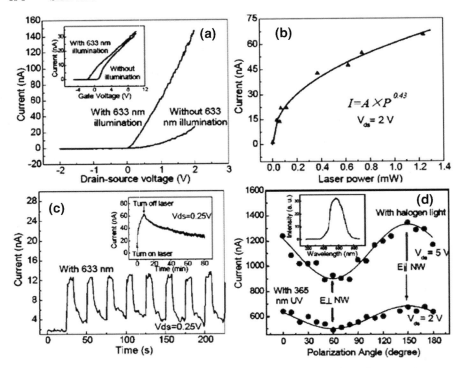

Fig. 10.25. (a) I–V curve of a single ZnO nanowire under dark and 633 nm laser exposure. (b) Incident power dependent photocurrent for a ZnO nanowire. (c) Switching behaviors in air and vacuum (inset) by the turn-on and off of 633 nm laser. (d) The photocurrent as a function of polarization angle of light for halogen light and 355 nm UV. Inset shows the light intensity for wavelength of halogen light. Reprinted with permission from [63]. Copyright (2004), American Institute of Physics

Contrary to these results, some researchers have observed the photoconduction variation of ZnO nanowire under the below-bandgap light illumination [62, 63].

Figure 10.25 shows the photoconduction characteristics for ZnO nanowire FET [63]. The variation of photoconductivity in Fig. 10.25a was obtained under 633 nm-He–Ne laser illumination of $\sim$0.2 W cm^{-2}. The conductance of nanowire under the laser illumination increased from 13.1 nS in the dark to 73.4 nS at $V_{ds} = 2$ V. At the same time, the transconductance was decreased from 2.9×10^{-9} to 2.6×10^{-9} A/V, therefore electron mobility also dropped from 23.4 to 21.0 cm^2 Vs^{-1}. These results were thought as the result of the enhancement of electron–electron scattering by increasing carrier concentration under laser illumination. The power dependent photocurrent behavior in Fig. 10.25b also obeys the power law dependence of $I_{pc} \propto P^{0.43}$. The photocurrent was repeatedly changed by switching illumination of 633 nm laser.

The photocurrent decay time of photocurrent was estimated to be 8s in air, while the decay time is much increased in vacuum as shown in Fig. 10.25c. It is probably due to the difference of oxygen amount between air and vacuum. Considering the oxygen trapping mechanism, small amounts of oxygen in the vacuum result in longer recovery time above 1 h. Figure 10.25d shows the photocurrent as a function of the polarization of the incident light. The photocurrent is changed by the angle (θ) between the polarization and a longitudinal direction of the nanowire with the relation of $\cos^2 \theta$.

The nanowire diameter affects the electronic properties of nanowire as introduced in Fig. 10.23. The diameter dependent photocurrent characteristics of nanowire were investigated by using the MBE grown GaN nanowires with diameter in the range from 20 to 500 nm as shown in Fig. 10.26a [64]. Above the 100 nm diameter, the photocurrent is gradually increased by increasing the diameter. In contrast, the value is sharply dropped by decreasing

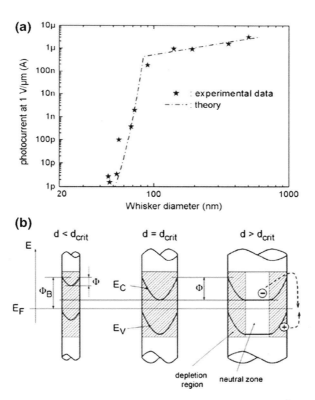

Fig. 10.26. (a) Photocurrent under UV illumination of 15 W cm^{-2} at a bias voltage of 1 V versus the diameter of GaN nanowire. (b) Schematic illustration of dependence of depletion region in GaN nanowires with different diameter. Reprinted with permission from [64]. Copyright (2005), American Chemical Society

the diameter below approximately 80 nm. The existence of a persistent photocurrent in the thick nanowire above 100 nm was also confirmed through the comparison between dark current at initial and after UV illumination. In the case of thin nanowire with a diameter of 70 nm, the photocurrent decay time is fast and the persistent photocurrent could not be observed in the dark. The diameter dependent photocurrent characteristics were interpreted by considering the difference of recombination barrier, which arises from the interplay between nanowire diameter and space charge layer extension at the surface. Figure 10.26b shows the schematic illustration of the depletion region in n-type nanowires with different diameters (d) and the band structure consisting of Fermi-level (E_c), conduction band (E_c) and valence band (E_v). Under the light illumination, holes migrate to the surface along the potential line due to band bending and combine with electrons at the surface. According to the model in Fig. 10.26b, the photocurrent remains in the nanowire while exceeding the critical diameter (d_{crit}) of 80–100 nm. Below the critical diameter, however, surface barriers (Φ) are shrunken and photocurrent drops due to the enhanced surface recombination.

High photocurrent gains in the ZnO and the GaN nanowires have recently been reported [65,66]. In the photoconductive properties of CVD grown ZnO nanowires with diameter of 150–300 nm and length of 10–15 μm for visible-blind UV photodetectors, high photoconductive gain of $\sim 2 \times 10^8$ was achieved. In the case of GaN nanowire, the gain values of 5×10^4–1.9×10^5 for the CVD grown m-axial GaN nanowires with diameters of 40 $\sim$ 135 nm were reported. These values are approximately three times the values of 5×10^1–1.6×10^2 for the estimated values from the thin film [67,68].

10.4.3 Gas and Solution Sensor

The surrounding ambient of ZnO and GaN nanowires changes the electronic properties due to the variation of depletion region in nanowires. This characteristic is applicable to fabrication of the chemical sensors for gas, solution, and bio material detection.

The sensing effects of assembled or single ZnO nanowire have been investigated for the realization of high-sensitive and minimized sensor. Figure 10.27 shows the ZnO nanowires-based gas sensor and its sensing characteristics [69]. The structures of sensor were fabricated by employing a microelectromechanical system including the silicon-based membrane embedded with Pt interdigitating electrodes and heater (Fig. 10.27a). ZnO nanowires shown in inset of Fig. 10.27a were obtained by the thermal evaporation of zinc metal pellets at 900°C. Figure 10.27b shows the switching characteristics induced by an exposure to ethanol gas with concentration in the range of 1 $\sim$ 200 ppm at 300°C. At the low concentration of 1 ppm, the response and recovery is fast. By increasing the concentration of ethanol gas, the difference of resistance also increased. The sensitivity (S_g) of ZnO nanowire-based gas sensor was estimated to be 47 for 200 ppm and 1.9 for 1 ppm with the definition of

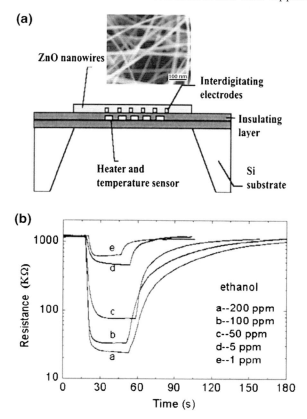

Fig. 10.27. (a) Schematic of ZnO nanowire based sensor. Inset shows SEM image of as-grown ZnO nanowires. (b) Resistance variation of ZnO nanowires by an exposure to ethanol gas with concentration of 1–200 ppm at 300°C Reprinted with permission from [69]. Copyright (2004), American Institute of Physics

$S_g = R_a/R_g$, where R_a and R_g are the electrical resistance of the sensor in air and in ethanol–air mixture ambient, respectively. The sensitivity of ZnO nanowire is certainly higher compared to the ZnO ceramic gas sensor showing the value below 2 for 200 ppm [70].

Hydrogen was selectively detected using the assembled ZnO nanowires [71]. Moreover, the sensitivity was enhanced by surface modification with Pd clusters. The resistance of the ZnO nanowires changed by the concentration of exposed hydrogen in the range from 10 to 500 ppm. When the Pd clusters were deposited on the ZnO nanowires, the hydrogen concentration below 10 ppm was detectable due to the catalytic effect of Pd metal. The initial resistance is recovered within 20 s by the removal of the hydrogen and exposing it to the oxygen.

Strong chemically stability of GaN is negatively effected on application of chemical sensor. To overcome the problem, the gold particle was deposited on the surface of GaN nanowire as shown in Fig. 10.28a [72]. The gold particles with an average particle size of 5 nm played as the catalyst on reaction with gas. The curve 1 in Fig. 10.28b shows the typical I–V curve for the as-grown GaN nanowires device. The I–V characteristic with Ohmic behavior was independent of the chemical environment. However, the GaN functionalized with the gold particles exhibited the nonlinear I–V curve in vacuum (curve 2). The I–V curves slightly changed by argon (curve 3) and by nitrogen (curve 4) atmosphere. It was considered as the result of the contamination by unintentional components in gas. When the gold nanoparticles-decorated GaN nanowires were exposed to the methane gas, the relative current variation was clearly observed as shown in curve 5. The gas sensing mechanism for the gold particles attached GaN nanowire was considered as followings (Fig. 10.28c): at first, the variation of I–V curve by the deposition of gold particles on GaN nanowires could be understood by the formation of a depletion layer of depth W_0 by nanoparticles (II). In contrast, the conducting region is decreased by increase of the depletion depth, which produces the conductivity drop of the functionalized nanowires. Secondly, the gas physisorption onto the surface of the gold particles induces the potential (V_g) at the surface of nanowire that changes the depletion depth from W_0 to W_a (III and IV). Consequently, the conductivity of nanowire is more reduced by the gas physisorption.

The absorption of polar molecules in the liquid leads to the change of surface potential at the interface between the semiconductor and liquid. By this principle, pH measurement using single ZnO nanowires was successfully demonstrated [73]. Figure 10.29a shows the single ZnO nanowire device with two electrodes (Al/Pt/Au) at the both ends of the nanowire and microfluidic channels. The pH solutions were prepared by the titration method using NHO_3, NaOH, and distilled water. The change of conductance by pH was evaluated in the dark and under the UV illumination as shown in Fig. 10.29b. The conductance of the ZnO nanowire was clearly decreased upon increasing the pH value of polar liquids. The difference of conductance according to the pH is enhanced by the UV illumination but the relative variation in conductance is independent to the UV illumination.

10.4.4 Biosensor

GaN and ZnO nanostructures have received much attention in the field of biosensors although many studies have not been performed yet. Especially, the ZnO nanostructures have the nontoxicity, biosafe, biocompatible properties as well as large specific surface area. The optical and electrical interactions between the ZnO and the biomaterial including conducting and piezoelectric properties show the communication features in biodevices.

ZnO with a high isoelectric point (IEP) of about 9.5 is suitable for adsorption of low IEP proteins or enzyme [74]. Uricase and glucose oxidase (GO_x)

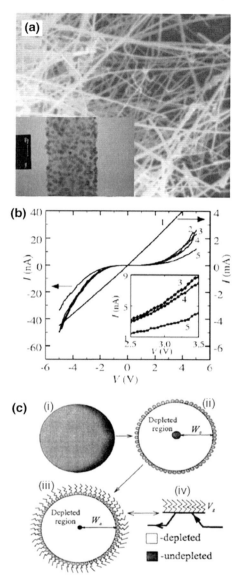

Fig. 10.28. (a) SEM image of as-grown GaN nanowires. Inset shows a TEM image of the GaN nanowire functionalized with Au particles. (b) I–V characteristics of as-grown GaN nanowire (1), Au particles attached GaN nanowire in vacuum (2), argon (3), nitrogen (4), and methane (5), respectively. Inset shows enlarge view of I–V curves. (c) The schematic illustration of variation of conducting and depleted region in bare GaN nanowire (i), the gold particles attached GaN nanowire (ii), gas physisorbed on the surface of the particles (iii), and a circuit diagram of the operation of the nanowire as a gas-sensitive FET (iv). Reprinted with permission from [72]. Copyright (2006), American Institute of Physics

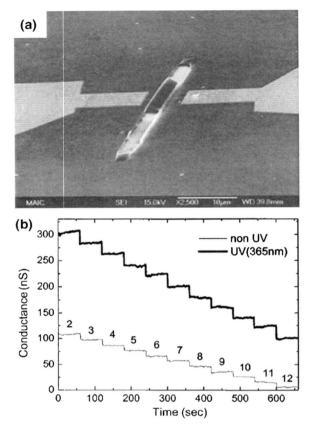

Fig. 10.29. (a) SEM image of the ZnO nanowire for pH sensor with integrated microchannel. (b) The conductance variation by pH in the range from 2 to 12 at $V = 0.5$ V. Reprinted with permission from [73]. Copyright (2005), American Institute of Physics

were immobilized on the ZnO nanowires matrix for biosensing [75, 76]. At the time, ZnO nanostructures not only provide a friendly microenvironment for immobilization of negatively charged biomaterial but also promote the electron transfer between the immobilized materials and the electrode [75]. The cyclic voltammograms of GO_x/ZnO nanowires/Au electrode by addition of glucose in a 0.01M phosphate buffer solution (PBS, pH 7.4) were changed as shown in Fig. 10.30a. Figure 10.30b displays amperometric response by additions of the glucoses with 0.01, 0.03, 0.05, 0.1, 0.5, and 1 mM per step in PBS buffer under an applied bias of 0.8 V. The current is fast and clearly increased by the addition of glucose. The glucose sensor using the ZnO nanowires achieved the 95% of steady-state current within 5 s. It indicates that the electron exchange between GO_x and ZnO is very fast. The key parameters in the

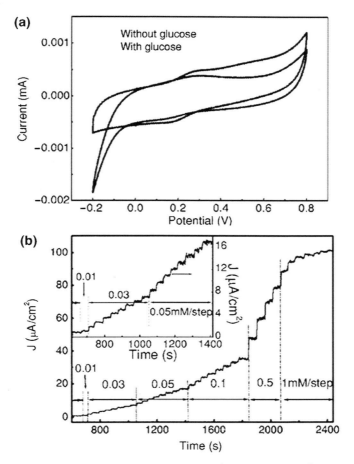

Fig. 10.30. (a) Cyclic voltammograms of GO_x/ZnO nanowires/Au with or without 5 mM glucose in 0.01M PBS buffer. (b) Current density response of GO_x/ZnO nanowires/Au electrodes by additions of glucose at applied potential of +0.8 V. The inset shows the enlarged plot for low concentration range. Reprinted with permission from [76]. Copyright (2006), American Institute of Physics

previously reported glucose biosensor are summarized in Table 10.1. Comparing each data in Table 10.1, it can be recognized that the ZnO nanowire-based biosensors show comparable or higher sensitivity and small Michaelis–Menten constant than those of previously reported glucose sensors.

It was reported that the ZnO nanowires provided the enhanced fluorescence detection of protein interactions using various fluorescein-conjugated proteins [77]. Figure 10.31 shows the interaction between the protein G (PG) and fluorescein isothiocyanate conjugated antibovine IgG (FITC-antiIgG) on the ZnO nanowire platforms. Both proteins have a concentration of

Table 10.1. Summary of previous results for glucose electrochemical biosensor fabricated by using various GO_x-modified nanomaterials as the working electrodes. Reprinted with permission from [76]. Copyright (2006), American Institute of Physics

Electrode material	Applied potential (V)	K_M^{app} (mM)	Sensitivity ($\mu A\,cm^{-2}\,mM^{-1}$)	LDD (μM)	Response time (s)
Titania sol–gel membrane	0.3	6.43	7.2	70	<6
Carbon nanotubes	0.55	...	30.14	0.5	<3
Au nanoparticles	0.3	4.3	8.8	8.2	<8
TiO$_2$ nanoporous film	0.7	6.08	4.58	...	<30
ZrO$_2$/chitosan film	0.5	3.14	0.028	10	<10
ZnO nanocombs	0.8	2.19	15.33	20	<10
ZnO nanowires	0.8	2.9	23.1	10	<5

* K_m^{app} Michaelis–Menten constant by Linewaver–Burk equation
LOD The limit of detection

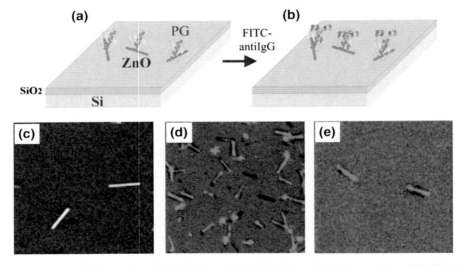

Fig. 10.31. Schematics of (a) PG and (b) FITC–antiIgG interaction. (c) SEM image of ZnO nanowire on Si substrate and (d, e) enhanced fluorescence emission through the interaction between PG and FITC–antiIgG on ZnO nanowires. Reprinted with permission from [77]. Copyright (2006), Wiley

$200\,\mu g\,mL^{-1}$. No emission was detected from the spatially separated ZnO nanowire (Fig. 10.31c) on substrate and the PG-adsorbed ZnO nanowire in the visible wavelength range, while a strong green emission was observed after the reaction with FITC-antiIgG in confocal images in Fig. 10.31d, e. The dominantly fast radiative decay caused by the modification of decay rates of radiative decay rate and/or the reduction of self-quenching of the fluorophore was suggested as a role of the ZnO nanowires.

The morphological effect of ZnO nanostructures with high specific surface area in biosensor technology was also demonstrated by DNA immobilization. The detection of immobilized DNAs on ZnO nanotips was performed by surface acoustic wave (SAW) response [78]. Here, the ZnO nanotips provided the enhanced immobilization of DNA by a factor of 200 compared to the ZnO film with smooth surface.

References

1. J.B. Baxter, E.S. Aydil, J. Cryst. Growth **274**, 407 (2005)
2. T. Kuykendall, P.J. Pauzauskie, Y. Zhang, J. Goldberger, D. Sirbuly, J. Denlinger, P. Yang, Nat. Mater. **3**, 524 (2004)
3. X.Y. Kong, Y. Ding, R. Yang, Z.L. Wang, Science **303**, 1348 (2004)
4. H.Q. Le, S.J. Chua, Y.W. Koh, K.P. Loh, Z. Chen, C.V. Thompson, E.A. Fitzgerald, Appl. Phys. Lett. **87**, 101908 (2005)
5. M. Law, L.E. Greene, J.C. Johnson, R. Saykally, P. Yang, Nat. Mater. **4**, 455 (2005)
6. L. Vayssieres, Adv. Mater. **15**, 464 (2003)
7. J. Yahiro, Y. Oaki, H. Imai, Small **2**, 1183 (2006)
8. H.T. Ng, T. Yamada, P. Nguyen, Y.P. Chen, M. Meyyapan, Nano Lett. **4**, 1247 (2004)
9. C. Wang, J. Summers, Z.L. Wang, Nano Lett. **4**, 423 (2004)
10. H.J. Fan, B. Fuhrmann, R. Scholz, F. Syrowatka, A. Dadgar, A. Krost, M. Zacharias, J. Cryst. Growth **287**, 34 (2006)
11. D.F. Liu, Y.J. Xiang, X.C. Wu, Z.X. Zhang, L.F. Liu, L. Song, X.W. Zhao, S.D. Luo, W.J. Ma, J. Shen, W.Y. Zhou, G. Wang, C.Y. Wang, S.S. Xie, Nano Lett. **6**, 2375 (2006)
12. L.H. Greene, M. Law, D.H. Tan, M. Montano, J. Goldberger, G. Somorjai, P. Yang, Nano. Lett. **5**, 1231 (2005)
13. S.H. Lee, H.J. Lee, D.C. Oh, S.W. Lee, H. Goto, R. Buckmaster, T. Yasukawa, T. Matsue, S.-K. Hong, H.C. Ko, M.-W. Cho, T. Yao, J. Phys. Chem. B **110**, 3856 (2006)
14. S.D. Hersee, X. Sun, X. Wang, Nano Lett. **6**, 1808 (2006)
15. D.S. Kim, R. Ji, H.J. Fan, F. Bertram, R. Scholz, A. Dadgar, K. Nielsch, A. Krost, J. Christen, U. Gosele, M. Zacharias, Small **3**, 76 (2007)
16. S.H. Lee, T. Minegishi, J.S. Park, S.W. Park, J.-S. Ha, H.-J. Lee, H.J. Lee, S.M. Ahn, J.H. Kim, H.S. Jeon, T. Yao, Nano Lett. **8**, 2419 (2008)
17. Y. Haung, X. Duan, Q. Wei, C.M. Lieber, Science **291**, 630 (2001)
18. D. Whang, S. Jin, Y. Wu, C.M. Lieber, Nano Lett. **3**, 1255 (2003)
19. A. Motayed, M. He, A.V. Davydov, J. Melngailis, S.N. Mohammad, J. Appl. Phys. **100**, 114310 (2006)
20. S.-W. Lee, M.-C. Jeong, J.-M. Myoung, G.-S. Chae, I.-J. Chung, Appl. Phys. Lett. **90**, 133115 (2007)
21. R. Agarwal, K. Ladavac, Y. Roichman, G. Yu, C.M. Lieber, D.G. Grier, Opt. Express **13**, 8906 (2005)
22. P.J. Pauzauskie, A. Radenovic, E. Trepagnier, H. Shroff, P. Yang, J. Liphardt, Nature Mater. **5**, 97 (2006)

23. G. Yu, A. Cao, C.M. Lieber, Nat. Nanotech. **2**, 372 (2007)
24. M. Huang, S. Mao, H. Feick, H. Yan, Y. Wu, H. Kind, E. Weber, R. Russo, P. Yang, Science **292**, 1897 (2001)
25. S. Chang, N.B. Rex, R.K. Chang, G. Chong, L.J. Guido, Appl. Phys. Lett. **75**, 166 (1999)
26. B.E.A. Saleh, M.C. Teich, *Fundamentals of Photonics*, 2nd edn. (Wiley, New York, 1991), pp.337–338
27. A.V. Maslov, C.Z. Ning, Appl. Phys. Lett. **83**, 1237 (2003)
28. J.C. Johnson, H. Yan, P. Yang, R.J. Saykally, J. Phys. Chem. **107**, 8819 (2003)
29. A.W. Synder, D. Love, *Optical Waveguide Theory*, (Kluwer, Boston, 1983)
30. J.I. Pankove, *Optical Processes in Semiconductors*, (Dover, New York, 1971)
31. H. Yoshikawa, S. Adachi, Jpn. J. Appl. Phys. **36**, 6237 (1997)
32. J. Wiersig, Phys. Rev. A **67**, 023807 (2003)
33. T. Nobis, E.M. Kaidashev, A. Rahm, M. Lorenz, M. Grundmann, Phys. Rev. Lett. **93**, 103903 (2004)
34. D. Wang, H.W. Seo, C.-C. Tin, M.J. Bozack, J.R. Williams, M. Park, Y. Tzeng, J. Appl. Phys. **99**, 093112 (2006)
35. P. Bhattacharya, *Semiconductor Optoelectronic Devices*, (Prentice-Hall, USA, 1994)
36. H. Yan, J. Johnson, M. Law, R. He, K. Knutsen, J.R. Mckinney, J. Pham, R. Saykally, P. Yang, Adv. Mater. **15**, 1907 (2003)
37. A.V. Maslov, M.I. Bakunov, C.Z. Ning, J. Appl. Phys. **99**, 024314 (2006)
38. L. Chen, E. Towe, Appl. Phys. Lett. **89**, 053125 (2006)
39. L.R. Hauschild, H. Kalt, Appl. Phys. Lett. **89** 123107 (2006)
40. D.M. Bagnall, Y.F. Chen, Z. Zhu, T. Yao, S. Koyama, M.Y. Shen, T. Goto, Appl. Phys. Lett. **70**, 2230 (1997)
41. M.A. Khan, D.T. Olson, J.M. Van Hove, J.N. Kuznia, Appl. Phys. Lett. **58**, 1515 (1991)
42. M.H. Huang, S. Mao, H. Feick, H. Yan, Y. Wu, H. Kind, E. Weber, R. Russo, P. Yang, Science **292**, 1897 (2001)
43. D.J. Sirbuly, M. Law, H. Yan, P. Yang, J. Phys. Chem. B **109**, 15190 (2005)
44. P. Zu, Z.K. Tang, G.K.L. Wong, M. Kawasaki, A. Ohtomo, H. Koinuma, Y. Segawa, Sol. Stat. Commun. **103**, 459 (1997)
45. A.V. Maslov, C.Z. Ning, Opt. Lett. **29**, 572 (2004)
46. L.K.V. Vugt, S. Ruhle, D. Vanmaekelbergh, Nano Lett. **6**, 2707 (2006)
47. H. Yan, J. Johnson, M. Law, R. He, K. Knutsen, J.R. McKinnery, J. Pham, R. Saykally, P. Yang, Adv. Mater. **15**, 1907(2003)
48. P.J. Pauzauskie, D.J. Sirbuly, P. Yang, Phys. Rev. Lett. **96**, 143903 (2006)
49. R. Konenkamp, R.C. Word, C. Schlegel, Appl. Phys. Lett. **85**, 6004 (2004)
50. W.I. Park, G.-C. Yi, Adv. Mater. **16**, 87 (2004)
51. J. Bao, M.A. Zimmler, F. Capasso, Nano Lett. **6**, 1719 (2006)
52. A. Motayed, A.V. Davydov, M. He, S.N. Mohammad, J. Melngailis, Appl. Phys. Lett. **90**, 183120 (2007)
53. J. Zhong, F. Qian, D. Wang, C.M. Lieber, Nano Lett. **3**, 343 (2003)
54. F. Qian, Y. Li, S. Gradecak, D. Wang, C.J. Barrelet, C.M. Lieber, Nano Lett. **4**, 1975 (2004)
55. Y. Huang, X. Duan, Y. Cui, C.M. Lieber, Nano Lett. **2**, 101 (2002)
56. H.T. Ng, J. Han, T. Yamada, P. Nguyen, Y.P. Chen, M. Meyyappan, Nano Lett. **4**, 1247 (2004)

57. S.N. Cha, J.E. Jang, Y. Choi, G.A.J. Amaratunga, G.W. Ho, M.E. Welland, D.G. Hasko, D.-J. Kang, J.M. Kim, Appl. Phys. Lett. **89**, 263102 (2006)
58. K.-H. Keem, D.-Y. Jeong, S.-S. Kim, M.-S. Lee, I.-S. Yeo, U.-I. Chung, J.-T. Moon, Nano Lett. **6**, 1454 (2006)
59. B.S. Simpkins, P.E. Pehrsson, M.L. Taheri, R.M. Stroud, J. Appl. Phys. **101**, 094305 (2007)
60. Y. Takahashi, M. Kanamori, A. Kondoh, H. Minoura, Y. Ohya, J. Appl. Phys. **33**, 6611(1994)
61. H. Kind, H. Yan, B. Messer, M. Law, P. Yang, Adv. Mater. **14**, 158 (2002)
62. K. Keem, H. Kim, G.-T. Kim, J. S. Lee, B. Kim, K. Cho, M.-Y. Sung, S. Kim, Appl. Phys. Lett. **84**, 4376 (2004)
63. Z. Fan, P.-C. Chang, J.G. Lu, E.C. Walter, R.M. Penner, C.-H. Lin, H.P. Lee, Appl. Phys. Lett. **85**, 6128 (2004)
64. R. Calarco, M. Marso, T. Richter, A.I. Aykanat, R. Meijers, A.v.d. Hart, T. Stoica, H. Luth, Nano Lett. **5**, 981 (2005)
65. R.-S. Chen, H.-Y. Chen, C.-Y. Lu, K.-H. Chen, C.-P. Chen, L.-C. Chen, Y.-J. Yang, Appl. Phys. Lett. **91**, 223106 (2007)
66. C. Soci, A. Zhang, B. Xiang, S.A. Dayeh, D.P.R. Aplin, J. Park, X.Y. Bao, Y.H. Lo, and D. Wang, Nano Lett. **7**, 1003 (2007)
67. B. Shen, K. Yang, L. Zang, Z.Z. Chen, Y.G. Zhou, P. Chen, R. Zhang, Z.C. Huang, H.S. Zhou, Y.D. Zheng, Jpn. J. Appl. Phys. Part 1 **38**, 767 (1999)
68. E. Munoz, E. Monroy, J.A. Garrido, I. Izpura, F.J. Sanchez, M.A. Sanchez-Garcia, B. Beaumont, P. Gibart, Appl. Phys. Lett. **71**, 870 (1997)
69. Q. Wan, Q.H. Li, Y.J. Chen, T.H. Wang, X.L. He, J.P. Li, C.L. Lin, Appl. Phys. Lett. **84**, 3654 (2004)
70. B.B. Rao, Mater. Chem. Phys. **64**, 62 (2000)
71. H.T. Wang, B.S. Kang, F. Ren, L.C. Tien, P.W. Sadik, D.P. Norton, S.J. Pearton, J. Lin, Appl. Phys. Lett. **86**, 243503 (2005)
72. V. Dobrokhotov, D.N. McIlroy, M.G. Norton, A. Abuzir, W.J. Yeh, I. Stevenson, R. Pouy, J. Bochenek, M. Cartwright, L. Wang, J. Dawson, M. Beaux, C. Berven, J. Appl. Phys. **99**, 104302 (2006)
73. B.S. Kang, F. Ren, Y.W. Heo, L.C. Tien, D.P. Norton, S.J. Pearton, Appl. Phys. Lett. **86**, 112105 (2005)
74. E. Topoglidis, A.E.G. Cass, B. O'Regan, J.R. Durrant, J. Electroanal. Chem. **517**, 20 (2001)
75. F. Zhang, X. Wang, S. Ai, Z. Sun, Q. Wan, Z. Zhu, Y. Xian, L. Jin, K. Yamamoto, Anal. Chim. Acta **519**, 155 (2004)
76. A. Wei, X.W. Sun, J.X. Wang, Y. Lei, X.P. Cai, C.M. Li, Z.L. Dong, W. Huang, Appl. Phys. Lett. **89**, 123902 (2006)
77. A. Dorfman, N. Kumar, J.-I. Hahm, Adv. Mater. **18**, 2685 (2006)
78. Z. Zhang, N.W. Emanetoglu, G. Saraf, Y. Chen, P. Wu, J. Zhong, Y. Lu, J. Chen, O. Mirochnitchenko, M. Inouye, IEEE Trans. Ultrason. Ferroelectr. Freq. control **53**, 786 (2006)

Index

A

AAS. *See* Atomic absorption spectroscopy
Absorption, 37
Acceptor bound exciton (A^0X), 320
Acidic, 48
Acidic ammonothermal, 48, 51, 58, 59
Acidic mineralizers, 24, 48–51, 58
  NH_4X (X = Cl, Br, I), 50
Activation energy, 110
Adducts, 140
Admittance spectroscopy (AS), 398
AES. *See* Auger electron spectroscopy (AES)
AFM. *See* Atomic force microscope; Atomic force microscopy
AlCl, 174
AlCl$_3$, 174
AlGaN, 136, 174
Alkaline mineralizers, 24
AlN. *See* Aluminum nitride
AlN, 131, 174
Al$_2$O$_3$, 21, 71
Aluminum gallium nitride ($Al_xGa_{1-x}N$), 174
Aluminum nitride (AlN), 174, 210
Ammonia (NH$_3$), 24, 49–51, 58, 132
Ammonothermal, 23, 47, 49, 51–55, 58–62
  growth, 22–26, 48

Amplitude of electron wave
  from imperfect crystal, 268
  from perfect crystal, 267
Anion-polar, 187
Anisotropic compressive strain, 342
Anisotropic growth, 463
Anisotropy, 234
Annealing, 38, 358, 363
a-plane, 274, 280
  GaN film, 233, 235, 239
  ZnO film, 247
Aqueous solution, 35
Arrhenius plot, 110
Atmospheric pressure
  growth, 132
  MOCVD, 133, 155
  reactor, 149
Atomic absorption spectroscopy (AAS), 37
Atomic force microscope (AFM), 76
Atomic force microscopy (AFM), 31, 35
Auger electron spectroscopy (AES), 75, 187, 198
Autoprotolysis, 52
Azimuth angle, 197

B

Background electron concentration, 373, 375
Backscattering geometry, 344
Band edge emissions, 73, 109, 315–320, 332–334
Band-edge energies, 315

Bandgap, 10, 105, 106, 111, 112, 125–129
  engineering, 68, 101, 102, 124
Basal growth sector, 36
Basal plane SFs, 242, 245
Basal SFs, 245, 248
Basic, 48
  ammonothermal, 48, 51
  mineralizers, 28, 48, 49, 51, 59
Beam, 40
  flux, 70
Bending of dislocations, 300, 302
BeO, 111
BeZnO, 111
Binding energy, 17
  of excitons, 333
Biosensors, 498–503
Bir-pikus hamiltonian, 12
Biscyclopenta-dienyl-magnesium (Cp_2Mg), 134
Blue emission, 322
Bohr radius, 17, 317
Bound excitons, 317–320, 333–334
  emission, 73
  peak, 82
Bowing parameter, 10
Bridgman type autoclaves, 23
Brillouin zone, 10
Buffer layer, 85, 206, 208, 236, 304
Built-in electric field, 328
Bulk property, 187
Burgers vector, 263–266, 270
Burying of crack, 164

C
CAICISS. See Coaxial impact collision ion scattering spectroscopy
Capacitance–voltage (C–V) measurements, 368
Carrier concentration, 44, 45, 481
Carrier localization, 324
Carrier mobility, 44
Catalytic growth, 460
Cathodoluminescence (CL), 36, 82, 327
Cation-polar, 187
CBED. See Convergent beam electron diffraction
Centrifugal flow planetary rotation, 148
  reactor, 152
Charge compensation, 38
Chemical lift-off (CLO), 171, 442
Chemical-mechanical polishing (CMP), 33–37, 54
Chemical treatments, 361
Cleavage planes, 165
CMP. See Chemical-mechanical polishing
Coalescence, 132
  of island, 165
Coaxial impact collision ion scattering spectroscopy (CAICISS), 96, 195, 196, 207
CO_2 laser, 114
Compressive strain, 264
Conduction band, 10
Conductivity, 44
  control, 382
Contact resistance, 356
Control of film polarity, 95
Convergent beam electron diffraction (CBED), 96, 192
Cooling shower head reactor, 147
Core/shell/shell (CSS) shaped nanowire, 485
Cp_2Mg. See Biscyclopenta-dienyl-magnesium
Crack, 163–165
  burying, 165
  formation, 165
Crack-free thick GaN, 165
Cracking, 164, 165
  under biaxial tensile stress, 164
  of GaN films, 164
Cr-compound intermediate layers, 210
Critical thickness, 264, 277
CrN nanoislands, 305
Cross-sectional view, 270, 274
Crucible, 70
Crystal-field interaction, 314
Crystal-field splitting, 11, 315
Crystal habit, 39
Crystallographic tilt, 293, 296
Cubic AlN layer, 209
Cubic perovskite structure, 121
Current blocking layers, 370
Current leakage, 369
  mechanisms, 369
Cylindrical gate capacitance, 485

D

DAP. *See* Donor-acceptor-pair transition
Deep centers, 323, 338
Deep levels, 323, 397
  emission, 73, 78, 82, 92, 323–324, 338
Deep-level transient spectroscopy (DLTS), 397
Defect reduction, 240
Deformation potentials, 11, 13
2DEG. *See* Two dimensional electron gas
Degree of polarization, 340
Determination of burgers vectors, 275
Determination of polarity, 191
2D growth, 80, 83, 93
3D growth, 82, 83, 93
Dielectric constant, 17, 49
Dielectrophoretic (DEP), 467
Diffraction patterns, 270
Dipole approximation, 313
Direct heteroepitaxial lateral overgrowth (DHELO), 302
2D island, 80
Dislocations, 34, 44, 253, 262, 270, 274, 299, 305, 306
  density, 29, 62, 262
  line direction, 266
  scattering, 383, 385, 393
  in wurtzite structure, 265
Displacement vector, 268
2D layer-by-layer growth of ZnO film, 85
2D nucleation, 73
Donor-acceptor-pair transition (DAP), 321, 334
Donor binding energy, 320
Donor bound exciton, 92, 120
  transition, 119
Doping concentration, 358, 359, 364
Double elog, 294
Double fault, 267
Double filament cell, 70
Double-heterostructure, 416, 421
D^0X_A, 334
Driving force, 159
Droplet formation, 113
2D ZnO growth, 81

E

$e-A^0$, 335
E-beam lithography (EBL), 434–436, 465
Edge dislocation, 263, 264, 273
Effective mass, 11, 13
Efficiency droop, 424, 425
Effusion cell, 70
E_2-high mode, 345
Elastic constants, 191
Elastic energy, 7
Elastic stiffness constants, 6
Electrical resistivity, 44–46
Electric field, 339
Electroluminescence (EL), 424
Electron blocking layer (EBL), 422
Electron diffraction pattern, 270, 275
Electron holography, 201
Electron paramagnetic resonance (EPR), 37, 38
Electron-polaron radius, 18
Electron reservoir layer (ERL), 424
Electron transport mechanism, 382
Electrostatic force microscopy (EFM), 201
ELOG. *See* Epitaxial lateral overgrowth
E_2-mode, 345
End on dislocations, 274
Envelope function approximation, 328
EPD. *See* Etch pit density
Epitaxial lateral overgrowth (ELOG), 132, 226, 262, 288, 294
Epitaxial relationship, 123, 169, 247, 248, 254, 255
Epitaxy, 264
Epitaxy-ready ZnO wafer, 46
EPR. *See* Electron paramagnetic resonance
Equilibrium partial pressures, 157, 159
  of GaCl, 159
Etched surface morphology, 201
Etching, 35
  rate, 201, 203
Etch pit density (EPD), 35, 36
Etch rate, 35
Ethanol gas, 496
Evaporable buffer layer (EBL), 169
Excimer lasers, 112
Exciton, 17

Exciton binding energy, 127, 315, 333
External quantum efficiency (EQE), 339, 425, 429, 431, 433, 435
Extraction efficiency, 430, 433, 434
Extra half plane, 278
Extraordinary, 40
Extrinsic, 267

F
Fabry-Perot cavity, 472
FACELO. *See* Facet-controlled epitaxial lateral overgrowth
Facet, 132
Facet-controlled epitaxial lateral overgrowth (FACELO), 297–299
Facet formation, 152
Factional mode power, 471
Fast-pulsed laser deposition (FPLD), 116
Femtosecond laser, 114
Fermi's golden rule, 313
FET. *See* Field effect transistor
Field effect transistor (FET), 214, 485–492
Flip-chip, 437–440
  package, 437
Flow rates, 52, 53
Fluorescence detection, 501
Fluorescence microscopy, 36
Forced flow vortex, 150
Fourier-filtered image, 280
Frank partial dislocation, 252, 285
Frank type, 266
Free excitons, 315, 332
  emission, 73
  peak, 82
Free-standing (FS), 21, 47
  GaN substrate, 156, 166–170, 172
  GaN film, 167
  GaN wafer, 21, 168, 169
Free-to-bound transitions (e-A^0), 322, 335–337
Fresnel loss, 432

G
GaCl, 157–159, 161
GaCl$_3$, 157–159
Ga-face, 240, 245

GaN, 131
  buffer layer, 131
  growth zone, 160
Ga$_2$O$_3$ interfacial layer, 207, 210
Ga source, 160
  zone, 160
g · b criteria, 269, 270
γ-LiAlO$_2$, 255
  substrate, 229
Glucose oxidase, 498
Grain size, 115
Green luminescence band (GL), 338
Growth
  mechanism, 29, 36
  from melt, 53
  mode, 83
  rates, 23–26, 29–31, 38, 51–53, 60, 61, 159, 203
  sector, 36
  from solution, 21
  temperature, 157, 375

H
H$_2$, 157, 159
Hall, 44
HCl, 157–159
Hcp structure, 267
Heteroepitaxy, 85, 130, 131, 264
Heterointerface, 146
Heterojunction, 421, 445–448
Heterostructure, 102, 107–109
Heterostructure field effect transistor (HFET), 214
Hexagonal lattice, 4
High isoelectric point (IEP), 498
High-pressure-solution-growth, 163
High-resolution TEM (HRTEM), 278
High speed flow horizontal reactor, 151
High speed rotation disc, 147
H$_2$O, dielectric constant, 23
Hole-polaron radius, 18
Hollow pipe, 281
Holographic lithography (HGL), 467
Homoepitaxial growth of m-plane ZnO films, 256
Homoepitaxy, 83, 93, 264
  ZnO films, 30, 31, 93
Homojunction, 421, 450
Horizontal dislocations, 292

Horizontal flow reactor, 146, 148, 150
Horizontal system, 156
Hot lip cell, 70
HRTEM. See High-resolution TEM
HVPE. See Hydride vapor phase epitaxy
Hybrid beam deposition (HBD), 452
Hydride vapor phase epitaxy (HVPE), 21, 47, 50, 51, 55, 56, 59, 60, 156, 446
  of AlN, 174
Hydrogen, 364, 395, 497
Hydrostatic pressure, 7
Hydrothermal, 22, 23, 27–29, 32, 48, 52, 53, 60, 61
  ammonothermal growth, 22
  growth, 22–24, 29, 30, 32, 41, 49
  ZnO, 39, 41
Hysteresis piezoresponse curves, 200

I

Ideal-diode equation, 418
Ideality factor, 419
II/VI ratio, 85, 90, 93
InCl, 174, 175
InCl$_3$, 174, 175
Incorporation rate, 85
Indium nitride (InN), 174, 210
Inductively coupled plasma mass spectrometry (ICP-MS), 37
Inert gas, 157
Infrared absorption spectroscopy, 37
InGaN, 138, 139
InGaN/GaN quantum well, 324
InN. See Indium nitride
InN layer, 175
In-plane orientation relationship (epitaxial relationship), 247
Input partial pressure, 157
Input V/III ratio, 157, 159
Inserting layer, 304
In-situ lift-off process, 172
In-situ monitoring of the growth process, 70
Interface, 55
Interfacial layer, 186, 206
Interlayer, 305
Intermolecular collision, 141
Internal parameter, 3

Internal quantum efficiency (IQE), 139, 324, 329, 416, 419, 421, 422, 424–426, 429, 431
Intrinsic, 267
Intrinsic limit of growth rate of GaN, 154
Inversion domain boundary (IDB), 135, 261, 285
  Faceted IDB, 286
Inversion symmetry, 96, 186, 206, 207, 210
Ionicity, 332
IQE. See Internal quantum efficiency
Island coalescence, 164
  mechanism, 164
I-V measurements, 367

J

Junction temperature, 439, 440

K

K-cell. See Kundsen cell
Kelvin force microscopy (KFM), 201
KrF excimer laser, 121, 124, 168
KrF laser, 114
Kundsen cell (K-cell), 69

L

Laminar flow, 146
  gas injection, 151
  gas injection reactor, 151
Langmuir-Blodgett compression, 467
Laser diode (LD), 136, 261
Laser emission, 474
Laser irradiation, 361
Laser lift-off (LLO), 166, 441, 443
  LLO process, 167, 168
Laser molecular beam epitaxy, 68
Lasing, 67
  operations, 262
Lateral epitaxial overgrowth (LEOG), 239
Lateral growth, 132, 300

Lateral overgrowth rate, 240
Lateral polarity heterostructure, 198, 199, 201
Lattice and thermally matched substrate, 21
Lattice constraint, 139
Lattice-matched, 22
  substrate, 41, 117
Lattice misfit, 71, 117, 254, 255, 264, 277, 278, 280
Lattice mismatch, 75, 77, 117
Layer-by-layer growth, 82
LEOG. See Lateral epitaxial overgrowth
LEO m-plane GaN, 245
(100) $LiAlO_2$, 255
Light emitting diode (LED), 131, 136, 261, 262
$LiNbO_3$, 216
Line defect, 263
Line direction, 264
Liquid phase epitaxy (LPE), 23, 30, 32, 41–44, 47, 62
$LiTaO_3$, 216
Localization energy, 318, 320
Local mode, 347, 348
Longitudinal exciton, 332
Longitudinal optical (LO) phonon, 321
LO-phonon-plasmon coupled mode, 346
Low dislocation density, 262, 288, 301
Low pressure MOCVD, 135
Low pressure solution growth (LPSG), 47
Low temperature grown GaN nucleation layer, 130
Low-temperature hydrothermal technique, 30
LPE. See Liquid phase epitaxy
LT buffer, 385
LT GaN capping layer, 371
LT ZnO buffer, 80
Luttinger-like parameters, 13
Lyddane-sachs-teller relation, 17

M
Macrocrack, 164
Majority carrier, 417, 419, 421
Mask, 288
  region, 295, 302
Maskless LEO, 242

Maskless PE, 296, 299
Mass transfer coefficient, 159
Matthiessen's rule, 382
Melt growth, 45, 47, 60
Metal organic chemical vapor deposition (MOCVD), 130, 188, 429, 446, 448
Metastable phase, 60, 125
$MgAl_2O_4$, 71
Mg_3N_2, 206
Mg–N bonding, 347
Mg_3N_2 interfacial layer, 206
MgO buffer, 68, 85, 89, 172, 304
  layer, 208
$Mg_xZn_{1-x}O$, 127
MgZnO, 338
Microcrack, 164
Microcrystals, 30–32, 41
Microfluidic channels, 466, 467
Micropipe, 281
Mineralizers, 22–24, 26, 29, 46, 48, 50–54, 58, 62
Minority carrier, 418, 419
Misfit, 264, 277
  dislocation, 248, 250, 264, 277, 278, 280
  strain, 280
Mixed dislocation, 263, 264, 273
Mobility, 45, 481
MOCVD. See Metal organic chemical vapor deposition
Mode, 473
Mode cut-off diameter, 470
Modulation-doped superlattice, 262
Molecular beam epitaxy (MBE), 68, 70, 188, 445
Mole fraction of hydrogen relative to inert gas atom, 158, 159
Monoclinic Ga_2O_3, 96
Morey, 23
Morphology, 31
  of film, 31
Mosaicity, 73, 77
m-plane, 274
  GaN, 245
  GaN film, 229
  ZnO, 254
  ZnO film, 247, 254, 255
MQW. See Multi quantum well
Mud-cracking pattern, 164

Multi-beam diffraction, 276
Multiple tapered nozzle, 147
Multiple zone furnace, 156, 160
Multi quantum well (MQW), 138, 422
Multi-wafer susceptor, 147

N
N_2, 159
Nanoisland, 305
Nano particle, 150–151
　generation, 140, 150
Nanopipe, 261, 281
Nanosized junction, 481
Nanosphere lithography (NSL), 465
Nanostructure, 460
Nanostructure-based LED, 479–485
NBE emission, 92
$NdGaO_3$ substrate, 168, 169
Nd:YAG laser, 113
Nd:Yttrium Aluminium Garent, 112
Near band edge (NBE), 92
　emission, 78
Near-field scanning optical microscopy (NSOM), 327
Neutral acceptor, 320
Neutral donor (D^0X), 317
N-face, 240, 245
NH_3, 50–53, 136, 157–159, 161
NH_4Cl, 52, 161, 169–170
NH_4X, 48
N incorporation mechanism, 378
Nitridation, 206
　temperature, 206, 209
Nitrogen vacancy, 388
Nonlinear optical device, 216
Nonpolar, 56, 139, 221
　film, 339, 340
　ZnO, 41
Nonradiative recombination, 110, 261, 262
Nonradiative recombination center, 288
n-type conductivity, 388
Nuclei, 131, 132

O
Offset angle, 78
Ohmic contact, 213, 356
Oligomer, 140, 150
Oligomer formation, 140
Omega-shaped-gate, 489
O-polar, 92
Optical cavity, 472
Optical-phonon, 17
Optical transmittance, 39
Ordinary, 40
O-rich flux condition, 93
Orientation relationship, 73, 75, 77, 78, 85
Orientation relationship (epitaxial relationship), 254, 255
Orifice, 70
Overlap integral, 328
Oxygen, 389
Oxygen molecule, 70
Oxygen radical, 70
Oxygen vacancy (V_O), 37, 364, 394

P
Package, 437, 439, 441
PAMBE. *See* Plasma-assisted molecular beam epitaxy
Parasitic nucleation, 26
Parasitic reaction, 134, 136, 138, 140, 141, 149, 151, 155
Partial dislocation, 263, 264, 266, 282
Particulate, 150, 154
PE. *See* Pendeo-epitaxy
Pendellösung fringe, 85, 116
Pendeo-epitaxial (PE) growth, 288
Pendeo-epitaxy (PE), 294
Perfect dislocation, 263, 264, 270, 276
Periodically polarity inverted (PPI), 200, 216, 467
PFM image, 201
Phase factor, 268
Phase-separation, 324
pH measurement, 498
Phonon scattering, 45
Phosphorous, 364
Photocurrent, 492
Photoelectron emission microscopy (PEEM), 201

Photoluminescence (PL), 35, 37, 40, 41, 43, 50, 59, 60, 73, 317, 332, 425
Photonic crystal, 433, 434
Piezoelectric and spontaneous polarization effects, 214
Piezoelectric coefficient, 189, 191
Piezoelectric effect, 188
Piezoelectric fields, 324, 329
Piezoelectric moduli, 190
Piezoelectric polarization, 188, 189, 191, 328
Piezo polarization, 7
Piezoresponse, 198
Piezoresponse force microscopy (PFM), 198
PL. See Photoluminescence
Planar-mirror resonator, 472
Planetary rotation, 148
Plan-view, 274
Plasma-assisted molecular beam epitaxy (PAMBE), 68–70, 72, 73, 80, 93
Plasma source, 70
Plasmons, 346
PL lifetime, 330
p–n junction, 416–418, 422, 439, 449
Point defect, 39
Polar angle, 197
Polar axis, 23, 186
Polarity, 68, 187, 188, 192, 193, 195, 197, 198, 201, 203, 204, 225, 328, 358, 369, 373, 375, 378, 467
  controlled ZnO film, 96
  dependence, 371
  inversion, 206
  inverted heterostructure, 210
Polarity control, 205, 208, 210
Polarization, 188
  anisotropy, 340
  field, 339
  induced 2DEG density, 214
  induced electric field, 214, 225
  induced sheet charge, 189
  induced surface charge, 188
  selectivity, 342
Polaron, 18
Porous TiN template, 305
Positron annihilation spectroscopy, 37
Potassium dihydrogen phosphate (KDP), 216

Potential fluctuation, 324
PPI GaN, 216
PPI ZnO, 217
Precursor, 24–26, 29, 46, 50, 51, 58
Pre-exposure, 73
Pregrowth, 80
Pre-reaction, 161
Pressure-melt-grown, 33, 37, 44
Pressurized melt growth, 29
Prismatic growth sector, 36
Propagation of dislocation, 305
Protein, 501
Psedomorphic, 264
P-type, 47
Pulsed laser deposition (PLD), 68, 445
Pyramidal growth sector, 36
Pyramidal shape, 289

Q

Q-factor, 474
Quantum chemical calculation, 142, 150
Quantum-confined Stark effect (QCSE), 228, 328
Quartz, 160

R

Radio Frequency (RF) plasma source, 69
Raman-active, 344
Raman scattering, 344
Raman selection rule, 344
Reactant, 159
Reaction product, 159
Reactive ion etching (RIE), 166
Reactor, 160
Reciprocal lattice, 6
Recombination, 417, 419, 420, 424, 425, 445, 446, 451
(3×3) Reconstruction, 87, 88
Reconstruction pattern, 87
Rectangular shape, 290
Reduced mass, 17, 333
Reflection high-energy electron diffraction (RHEED), 70, 73, 75, 76, 78, 80–82, 85, 93, 102, 104, 106, 122, 123, 193, 194
Refractive indices, 40
Regrowth, 294
Regular, 48

Regular type solubility, 51
Relaxed film, 264
Residual strain, 345
Retrograde type, 48
Retrograde type solubility, 51
RHEED. See Reflection high-energy electron diffraction
RHEED intensity oscillation, 80, 81, 85, 104, 105
RHEED oscillation, 102
RHEED pattern, 247
Rhombohedral Cr_2O_3, 210
Rocking curve (RC), 73
Rocksalt, 4
Rocksalt CrN, 210
Rocksalt MgO, 208, 209
Room temperature (RT), 73
  optical pumped lasing, 67
Root mean square (RMS) roughness, 35, 76, 93, 119
Rotated domain, 73, 74, 76, 78, 80, 85

S

SBH. See Schottky barrier height
SC. See Supercritical
Scalability, 22
$ScAlMgO_4$, 71
(0001) $ScAlMgO_4$ (SCAM) substrate, 68, 117
$ScAlMgO_4$ substrate, 68
(0001) SCAM substrate, 118
Schottky barrier height (SBH), 212
Schottky barrier, 356, 367
Schottky contact(s), 213, 214, 366
Schottky diode, 213
Scintillator, 44, 47
SC NH_3. See Supercritical (SC) ammonia (NH_3)
ScN layer, 305
Screening, of piezoelectric fields, 327
Screw dislocation, 55, 263, 272, 275
SCVT. See Seeded chemical vapor transport
Secondary ion mass spectrometry (SIMS), 36
Second harmonic generation (SHG), 216
Seeded chemical vapor transport (SCVT), 29, 33, 37, 41, 45
Selection rule, 312, 317

Self-separation, 170
Semipolar, 226
Semipolar GaN film, 239
Semipolar plane, 238
SFs. See Stacking fault
Sheet charge, 189
SHG. See Second harmonic generation
Shockley partial dislocation, 266, 285
Shockley type, 266
SiC, 21, 71
SiC substrate, 166
Sidewall lateral epitaxial overgrowth (SLEO), 239
Single fault, 267
SiN_x interlayer, 305
SiN_x nanomask, 234
$\alpha-SiO_2$, 22
Slate surface morphology, 233, 248
SLEO. See Sidewall lateral epitaxial overgrowth
Slip plane, 263
Snell's law, 433
Solid solution, 22, 31
Solubility, 22, 24, 29, 31, 48, 51, 52, 62
Solute, 22, 52
Solution, 22, 28, 30, 32
Solvent, 22, 23, 26, 30, 31
Solvothermal growth, 22
Spin-orbit splitting, 11, 315
Spontaneous and piezoelectric polarization, 225
Spontaneous (P_{SP}) and piezoelectric (P_{PE}) polarization, 188
Spontaneous nucleation, 24
Spontaneous polarization, 4, 7, 188, 191, 328
Spontaneous polarization coefficient, 188
Spotty, 81
$SrTiO_3$(STO), 121
Stacking fault energy, 267
Stacking fault (SF), 233, 250, 261, 266, 281
Stacking rule, 267
Stacking sequence, 285
Stagnation point flow, 147, 148
Stagnation point flow reactor, 146, 147
Step-flow growth, 82
Sticking, of Zn atoms, 203

Stoichiometric flux condition, 93
Stokes shift, 127
Strain, 6, 341, 345
Strained film, 264
Strained layer superlattice (SLS), 133
Stranski-Krastanov growth mode, 123
Streaky, 81
Stress, 6
Subcritical to near-supercritical hydrothermal, 32
Substrate cracking, 165
Subthreshold gradient, 486
Supercritical (SC), 21–23, 49
Supercritical (SC) ammonia (NH_3), 49, 52
Supercritical water (SCW), 24, 53
Supersaturation, 22, 28, 52
Surface conductivity, 44
Surface energies, of polar and nonpolar GaN and ZnO surfaces, 227
Surface morphology, 81, 248
Surface polarity, 359
Surface property, 187
Surface roughening, 433, 434
Surface roughness, 115
Surface texturing, 433
Surface treatment, 205, 206, 358, 359, 364, 373
Susceptor, 140, 150, 160

T
Tapered shape reactor, 147
TD density, 242, 245
TDs. See Threading dislocations
Technique, 44
Temperature dependence of PL, 326
Tensile cracking, 164
Tensile strain, 264
Tensile stress in GaN film, 165
TES. See Two-electron satellite
Thermal annealing, 35
Thermal expansion coefficient, 4
Thermodynamic analysis, 157
Thermodynamic equilibrium, 22
Thermodynamics, 157
Thermodynamic solubility limit of MgO in ZnO, 125
Thick GaN film, 163, 165
Thin SiN_x layer, 304

Threading dislocation density, 287
Threading dislocations (TDs), 55, 60, 131, 168, 233, 242, 248, 250, 261, 264, 269, 270
Three-dimensional (3D) growth, 73
Three-layer gas injection, 151
Three-layer laminar flow gas injection, 136, 138
Threshold gain, 473
Time-lapse red-shift, 327
Time-resolved spectroscopy, 42
TiN film, 168
TiN island, 168
TiN layer, 305
TiN nano-net, 168
Ti:sapphire laser, 112
TMA. See Trimethyl aluminium
TMG. See Trimethyl gallium
TMI. See Trimethyl indium
Total polarization, 188
Transconductance, 486
Transfer length method (TLM), 356
Transition probability, 329
Transmission electron microscopy (TEM), 55, 192
Transmittance, 39, 40, 125
Transparent conductive oxide (TCO), 433
Transverse exciton, 332
Trapezoidal shape, 291
Triethyl gallium (TEG), 139
Trimethyl aluminium (TMA), 134, 136, 140
Trimethyl gallium (TMG), 132, 140
Trimethyl indium (TMI), 139
Triple fault, 267
TRPL, 60
TRS. See Time-resolved spectroscopy
Tuttle type autoclaves, 23
Two-beam condition, 269, 270, 272, 274
Two dimensional electron gas (2DEG), 189, 213–215
Two-dimensional (2D) layer-by-layer growth, 68
Two-electron satellite (TES), 319, 334
Two-flow reactor, 147
Two-layer model, 383
Two-step growth, 235
Two step growth technology, 130

Two-step overgrowth, 297, 302
Type III stacking fault, 267, 285
Type II stacking fault, 251, 267, 285
Type I stacking fault, 267, 282, 285

U
Uricase, 498
UV LED, 171
UV lithography, 47

V
Valence band ordering, 331
Vapor growth, 47
Vapor-liquid-solid (VLS), 460
Vapor species, 157
Varshni formula, 10
VAS. *See* Void-assisted separation
Vertical flow reactor, 146, 147
Vertical high speed rotation disc reactor, 138
Vertical LED, 172, 441, 443
Vertical surrounded-gate (VSG) FET, 488
Vertical system, 156
V/III ratio, 385
Virtual crystal approximation (VCA), 128
Visible and invisible criteria, 252 of dislocation, 252
V_{max}, 52, 53
Void, 168, 170
Void-assisted separation (VAS), 166
Void formation, 305
Voltage-piezoresponse distance (V–Z) curve, 200, 201
VUV applications, transparent electronics (TCO), 47

W
Wafer, 21
Waveguide effect, 470
Wet etching rate, 203
Whispering gallery mode (WGM), 473
Window, 289
Window region, 289, 291, 294, 301
Wing region, 295
Wurtzite (WZ), 1
Wurtzite MgO, 208, 209

X
XPS. *See* X-ray photo electron spectroscopy
X-Ray Diffraction (XRD), 60, 73, 195
X-ray photo electron spectroscopy (XPS), 198
X-ray reciprocal space map, 76
X-ray rocking curve (XRC), 31–33, 36, 55, 56, 116
X-ray topography, 34, 35
XRC. *See* X-ray rocking curve
XRD. *See* X-ray diffraction

Y
Yellow luminescence (YL), 323

Z
Zigzag contrast, 274
Zincblende, 1
Zinc vacancy (V_{Zn}), 37, 394
ZnBeO, 111
Zn_3N_2, 207
ZnO, 71
ZnO buffer, 119
Zn-polar ZnO substrate, 83
$Zn_XCd_{1-x}O$, 68
$Zn_XMg_{1-x}O$, 68
Zone axis, 270